The Birds of
LONDON

The Birds of LONDON

ANDREW SELF

BLOOMSBURY
LONDON • NEW DELHI • NEW YORK • SYDNEY

First published in 2014

Copyright © 2014 text by Andrew Self
Copyright © 2014 illustrations by Fraser Simpson and photographs as individually credited

Publication of this book has been sponsored by
the London Natural History Society

The right of Andrew Self to be identified as the author of this work has been asserted
by him in accordance with the Copyright, Designs and Patents Act 1988.

All rights reserved. No part of this publication may be reproduced or used in any form or
by any means – photographic, electronic or mechanical, including photocopying, recording,
taping or information storage or retrieval systems – without permission of the publishers.

Bloomsbury Publishing Plc, 50 Bedford Square, London WC1B 3DP

www.bloomsbury.com

Bloomsbury Publishing, London, New Delhi, New York and Sydney

A CIP catalogue record for this book is available from the British Library

Commissioning editor: Jim Martin
Project editor: Jasmine Parker
Design and typesetting by Susan McIntyre

ISBN (print) 978-1-4081-9404-1
ISBN (epub) 978-1-4081-9405-8
ISBN (epdf) 978-1-4729-0514-7

Printed in China by South China Printing Company

This book is produced using paper that is made from wood grown in managed sustainable
forests. It is natural, renewable and recyclable. The logging and manufacturing processes
conform to the environmental regulation of the country of origin.

10 9 8 7 6 5 4 3 2 1

CONTENTS

Foreword	6
Acknowledgements	7
INTRODUCTION	8
DISCOVERIES AND DATES	10
Chronological highlights	10
Calendar of notable birds	15
HABITATS AND GEOGRAPHY	17
IMPORTANT LONDON NATURALISTS AND ORNITHOLOGISTS	21
GREEN SPACES IN LONDON	25
MAP OF LONDON	33
SPECIES ACCOUNTS	34
Wildfowl	35
Gamebirds	85
Divers	90
Seabirds	92
Cormorants	98
Herons	101
Storks and ibises	111
Grebes	115
Raptors	122
Rails and crakes	142
Crane	149
Bustard	150
Waders	150
Skuas	199
Gulls	204
Terns	222
Auks	234
Sandgrouse	237
Pigeons	238
Parrot	245
Cuckoos	246
Owls	247
Nightjars	254
Swifts	256
Kingfisher	258
Bee-eater	260
Roller	260
Hoopoe	261
Woodpeckers	262
Oriole	266
Shrikes	267
Crows	271
Crests	282
Penduline Tit	285
Tits	285
Bearded Tit	292
Larks	293
Swallows and martins	298
Warblers	303
Waxwing	323
Nuthatch	324
Treecreepers	325
Wren	326
Starlings	327
Dipper	331
Thrushes	332
Chats and flycatchers	341
Accentors	354
Sparrows	355
Wagtails	360
Pipits	365
Finches	372
Buntings	391
APPENDIX 1 – ESCAPES	399
APPENDIX 2 – SPECIES NOT ACCEPTED	411
Bibliography	414
Gazetteer	418
Index	424

FOREWORD

This is a book about the birds of one of the greatest cities in the world, in which the birdlife has probably been studied more intensively and over a larger time period than in any other city on the planet.

Having studied birds in the London Area, and particularly those of the Brent Reservoir, for over half a century, I was pleased to be invited to write the foreword to this book. I have worked with Andrew over a number of years, gathering information about the birds of the Brent Reservoir and through the work of the Welsh Harp Conservation Group helping to manage the site for birds.

A series of books has been published on London's birds, but until now none of them has given us the whole picture. Some, such as the London Natural History Society's (LNHS) *Birds of the London Area* (1957, revised in 1964), have covered the whole area but for only a limited period of time, in this case from 1900, and are now considerably out of date. Others have covered only part of the area, such as Glegg's *A History of the Birds of Middlesex* (1935), which goes back to when records began, but which is again now very much out of date.

With the publication of this book at last we have a comprehensive and detailed account from when records began, which covers the whole of the LNHS recording area, defined as being within a 32-kilometre (20-mile) radius of St Paul's Cathedral and covering some 3,250 square kilometres.

This is truly a monumental work on the changing birdlife of the area over centuries, which covers all known records going as far back as the 2nd century with the first record of Red Kite in Roman London. It is as up to date as is possible, with some records of rarities found in 2012.

Andrew Self has made a major contribution to increasing our knowledge of the birds of the London Area with its amazing variety of habitats. The book is comprehensively researched and clearly written, and is of the standard of the best of the country's avifaunas to date. In order to maintain a standard, the author has included only records accepted by the LNHS Rarities Committee. For completeness, however, mention of unacceptable records is also included where appropriate.

The scene is set in the introduction, which outlines the history of London together with a historical account of bird recording in the area. The table of chronological highlights provides a fascinating timeline of dates of first sightings in the London Area of each species (all 369 of them), along with relevant references.

For those seeking information on extreme rarities, a calendar of notable birds, that is species recorded three times or less, shows that in each month of the year since 1858 at least two rarities have been recorded in the area. There are also chapters on some of the capital's best sites, escaped birds and the changes brought about by urbanisation. Famous past London ornithologists and naturalists are also profiled, together with their major publications.

The main body of the book is devoted to a very detailed systematic list, where each species account has been thoroughly researched. It covers the changing status of each species, including its current breeding status, and its distribution and population size in the study area, as well as special mention of the situation in Inner London, where appropriate. Details are also included of the largest flocks seen and, in the case of migrants, the earliest and latest dates on which they have been seen in the London Area.

The accounts enable the reader to clearly understand the sometimes major changes in the population of each species. For example, we can see the welcome increase in some of our birds of prey such as Buzzard, Red Kite and Peregrine, as well as many waterbird species such as Little Egret and several ducks. Conversely, the catastrophic declines in some of our songbirds, such as Wood Warbler, Redstart, Willow Tit, House Sparrow, Red-backed Shrike, and other species such as Turtle Dove, are also documented in detail. What this book clearly illustrates is that the London Area is still a good place for birds.

Andrew Self is well qualified to write this book as not only has he travelled widely in search of birds, but he has also been birdwatching in the London Area for nearly 30 years. He held the positions of editor of the London Bird Report from 1998 to 2006 and Chairman of the LNHS Rarities Committee.

This is a book that everyone who watches birds in the London Area or who is interested in London birds should have, and I can thoroughly recommend it. Be warned, though, that once you start reading it is very difficult to stop.

Leo Batten B.Sc Ph.D C.Env MIEEM
August 2012

ACKNOWLEDGEMENTS

I would like to thank the following individuals who have helped me with this book: Leo Batten and Roy Beddard for loaning copies of papers; Barry Reed for verifying the position of some Amwell records; Bob Watts and the London Bird Club Rarities Committee for reviewing records and potential additions to the London List; Andrew Moon for additional research on the historical status of Chough; David Lindo for initial support for this project; Hein Van Grouw at the Natural History Museum in Tring; and all the London birders who have expressed their interest in the book and given me the encouragement to complete it. I am grateful to the late Mark Hardwick for getting me involved with recording birds in London in the first place. I am especially grateful to the countless naturalists in the LNHS whose records provide the factual basis of much of the text.

Also thanks to Fraser Simpson for his excellent illustrations – more of his work can be seen at www.fssbirding.org.uk – and to all of the photographers whose work can be found on the colour plates.

I am grateful to the London Natural History Society for their sponsorship of this project and the LNHS editorial team, David Darrell Lambert, Jan Hewlett, David Howdon and Andrew Moon for their advice and constructive review of the text. Many thanks are also due to everyone involved in the production of this book especially Jim Martin, Jasmine Parker and Tim Harris.

Andrew Self, October 2013

INTRODUCTION

London is one of the most famous cities in the world. The very name conjures up images of iconic landmarks – the Houses of Parliament, Tower Bridge, the Tower of London and Buckingham Palace – but ask people about its birdlife and most will be hard pushed to associate London with much more than pigeons, and the only green spaces that most tourists are familiar with are the central London parks. But for the resident or visiting birdwatcher London has much more to offer. Even most well-travelled and knowledgeable birdwatchers would struggle to name a county that has breeding Nightingales and Black Redstarts, and has been visited by extremely rare birds such as Lesser Kestrel, Western Sandpiper, and both Bridled and Sooty Terns, and hosted Britain's first Iberian Chiffchaff. Yes, the London Area has seen all these and many more besides.

This book attempts to document the changes to the London Area's birdlife over the centuries as well as providing a definitive list of all accepted occurrences of rare birds, so if you want to know when Corncrakes last bred, what the population of Tree Sparrows is, if Spoonbills have ever bred or how many American thrushes have been seen in the London Area, then this book has all the answers.

LONDON – BOTH CITY AND COUNTY

London has been in existence for around 2,000 years. Although there is some evidence of earlier settlements in the area, the first major settlement was founded by the Romans in AD 43 and was centred in the area now known as the City of London. At its height during the 2nd century, Roman London had a population of around 60,000. Further settlements were established next to the Roman city, and although they were initially separate from London as they were outside the City walls, over a period of time this conglomeration of districts became known as London. However, the City of London retained its own identity, as it still does today.

By the early 11th century, London had become the largest and most prosperous city in England, and the district of Westminster was chosen to be the seat of the royal court and government. London's population continued to expand, reaching 100,000 by the year 1300, although almost a third of its inhabitants succumbed to the Black Death in the mid-14th century.

Continued expansion and overseas trade saw London becoming the principal North Sea port; this helped to increase the population to about 225,000 by 1605. Two major disasters hit London in the 17th century, starting with the Great Plague in 1665, which claimed the lives of about 100,000 people. This was followed by the Great Fire of London in 1666, which destroyed the homes of 70,000 inhabitants, mostly within the City of London. Rebuilding took place over the following ten years and more bridges across the Thames saw London expand southwards and the Port of London spread to the east.

Over the next few centuries the population continued to increase and London was the world's largest city from about 1831 to 1925. London grew outwards as the population eventually reached an estimated peak of 8.6 million in 1939 although this was to drop significantly during, and following, the Second World War. Further outward expansion was prevented by the establishment in law of the Green Belt. At the last census in 2011, the total population in Greater London was 8.1 million.

In 1889 the Administrative County of London was formed from the City of London and parts of Middlesex, Kent and Surrey. In 1963 the Administrative County was replaced by Greater London, which also took in most of the rest of Middlesex and parts of Essex and Hertfordshire, as well as some county boroughs. Greater London is the official name of the county; it encompasses all of the London boroughs and the City of London, and is the area governed by the Greater London Authority. It covers an area of 1,570 square kilometres (607 square miles).

BIRD RECORDING IN LONDON

The London Natural History Society (LNHS) uniquely does not use the county area as its recording area; instead it covers a circle with a 20-mile (32km) radius centred on St Paul's Cathedral, giving it a total size of 3255 square kilometres (1,256 square miles). This has given the LNHS a fixed recording area that has not been subject to change, unlike many other counties whose boundaries have altered over time. This is often referred to as the 'London Area' and is the area used for this book.

The LNHS also uses the Watsonian vice-counties for recording so that, for example, there are separate bird recorders for each 'county' within the LNHS area, representing Buckinghamshire, Essex, Hertfordshire, Kent, Middlesex and Surrey. Because of many changes to the boundaries of administrative counties in the intervening years, many Watsonian county boundaries no longer match the modern county boundaries. This is particularly true of the London Area, where parts of counties such as Kent and Surrey have been swallowed up by Greater London, while the Watsonian county of Middlesex no longer exists at all as an administrative entity, and the area of Buckinghamshire within the LNHS Area now comes under Berkshire. This does introduce the odd anomaly. For instance, records from Staines are recorded in Middlesex even though that particular district is now in Surrey, and it is even more complex in the Lea Valley where the current border between Essex and Hertfordshire is no longer the same as the vice-county border. For recording purposes there is also an Inner London sector – a rectangle in central London, 8 miles by 5 miles, centred on Charing Cross – which was created in 1928. Additionally, for mapping projects such as the London Bird Atlases where data is collected in tetrads (2 x 2-kilometre squares), a polygon is used based on Ordnance Survey 10-kilometre grid squares and tetrads containing 856 tetrads approximating the LNHS area.

Little was written about the birdlife of the London Area until Victorian times, when Bond compiled a list of the waterbirds of Kingsbury Reservoir (now known as Brent Reservoir) in 1843. The reservoir had not long been completed and Bond shot many rarities there as did Harting, who went on to write the *Birds of Middlesex* in 1866. Swann's book *The Birds of London*, published in 1893, could be considered to be the first account of London's avifauna; it described the status of all the birds recorded within 12 miles of London. Although it drew heavily on Harting's records it does contain some information from east and south London.

In 1898 Hudson wrote *Birds in London*, which was mainly a personal study of birds in central London. A second edition of Harting's book was prepared but it was never published although the manuscript was purchased by Glegg, who used it to write a paper 'The Birds of Middlesex since 1866', which was published in 1930. Glegg later wrote *A History of the Birds of Middlesex*, which was published in 1935; he had previously written *A History of the Birds of Essex* in 1929, and county avifaunas have also been published for all the other counties that overlap with the London Area.

From 1921, annual reports on the birds of the London Area were published in the LNHS publication *The London Naturalist* and later ones were reprinted as a separate booklet. In 1936 the first London Bird Report was published and this has continued to be produced annually. In 1949, Richard Fitter wrote a book in the New Naturalist series called *London's Birds*. It was not a comprehensive listing of all species, but more a study of the regular birds seen in London; Fitter also chose to create his own study area, which was broadly the Greater London area.

In 1957 the LNHS published *Birds of the London Area*, which documented the status of every species recorded in the London Area since 1900. A revised edition was published in 1964 containing a new chapter detailing changes for the years 1955–61. Since then the only other books on the London Area's birds have been the Atlases – breeding ones covering 1972–74 (published in 1977) and 1988–94 (published in 2002). There is also one in preparation for the years 2007–12, which will cover both breeding and wintering birds.

This book, therefore, is the first to include a historical account of all the birds ever recorded in the entire London Area, as well as bringing the current status of species up to date.

DISCOVERIES AND DATES

The following table lists the year that each species was first recorded in London, along with the relevant reference.

CHRONOLOGICAL HIGHLIGHTS

Year	Species	Reference
2nd century	Red Kite	Cocker & Mabey
1003–1066	Nightingale	Morant (1768)
12th century	Teal	Fitter
1246	Mute Swan	Fitter
1317	Raven	Fitter
1385	Rock Dove	Fitter
1517	Grey Partridge, Pheasant	Harting, 1866
1523	Grey Heron, Spoonbill	Glegg, 1935
1604	Jackdaw	House of Commons Journals
1629	Cormorant	Parker
1638	Magpie	Harting, 1866
1666	Rook	Fitter
1666	Reed Bunting	Bucknill
1666–67	Hoopoe	Glegg, 1935
c. 1673	Red-legged Partridge	Wheatley
17th century	Hawfinch	Mountford
1703	Blackbird	Pigott
1720s	Snipe, Woodcock	Fitter
1722	Purple Heron	Morris
1731	Canada Goose, Nightjar	Albin
1732	Chaffinch	Albin
Pre-1738	Goldcrest	Wood
1738	Barn Owl, Long-eared Owl, Hooded Crow	Albin
Pre-1743	Bearded Tit	Edwards, 1743–51
Pre-1751	Mandarin Duck, Black-necked Grebe, Rose-coloured Starling	Edwards, 1743–51
1756–57	Common Crossbill	Harting, 1866
1758	Little Owl	Edwards, 1758–64
1760	Water Pipit	Edwards, 1758–64
1766	Red-breasted Goose	Latham, 1781–1802
1766	Grey Plover	Glegg, 1935
1767	Swallow, House Martin	White
1773	Dartford Warbler	Pennant
1774	Swift, Sand Martin	White
1776	Ortolan Bunting	Brown
1778	Tawny Owl, Green Woodpecker	White
1779	House Sparrow	White
1781	Waxwing	Ticehurst
1782	Little Bittern	Montagu, 1802
1782	Night Heron	Pennant
1782	Dunnock	White
1783	Reed Warbler	Glegg, 1935
1784	Sandwich Tern	Ticehurst
1785	Ruff	Ticehurst
1785	Cuckoo, Common Redstart	White

Chronological highlights

Year	Species	Reference
1786	Spotted Flycatcher	White
1787	Red-backed Shrike	White
1788	Twite	Latham, 1781–1802
1790	Great Skua	Latham, 1790
1794	Wood Warbler	Glegg, 1935
1798	Gannet	Ticehurst
18th Century	Storm Petrel	Christy
18th Century	White Stork	Ruegg
18th century	Peregrine	Glegg, 1935
1801	Bittern, Curlew Sandpiper	Latham, 1781–1802
1805	Little Grebe	Glegg, 1935
1808–16	Osprey, Great Grey Shrike	Hayes
c. 1809	Golden Eagle	Graves
1812	Marsh Harrier, Montagu's Harrier, Common Tern, Puffin, Pied Flycatcher	Graves
1812	Rough-legged Buzzard	Pennant
1812	Siskin	Ticehurst
1812	Little Crake	Harting, 1866
1812	Richard's Pipit	Harting, 1866
1813	Little Gull	Montagu, 1813
1813	Spotted Crake	Graves
1813/14	Great Northern Diver	Ticehurst
1816	Nutcracker	Ticehurst
1817	Dotterel	Graves
1817	Alpine Accentor	Christy
1819/20	Smew	Christy
c. 1820	Hen Harrier	Graves
1821	White-tailed Eagle	Latham, 1821
1821	Slavonian Grebe	Graves
1823	Leach's Petrel	Yarrell, 1826
1823	Stone-curlew	Taylor, et al
1823	Grasshopper Warbler	Sweet
1824	Grey Phalarope	Harting, 1866
1824	Arctic Skua	Harting, 1866
Pre-1824	Great Crested Grebe	Ticehurst
1825	Robin	Jennings
1826	Wren, Song Thrush	Jennings
1827	Kingfisher	Jennings
Pre-1828	Starling	Jennings
Pre-1828	Pomarine Skua	Christy
1828	Woodpigeon	Jennings
1828	Wryneck	Wood
1828	Tree Sparrow	Barrett-Hamilton
1828	Lapland Bunting	Bond, 1843 (B)
1828–32	Short-eared Owl	Barrett-Hamilton
1829	Hobby	Christy
1829	Black Redstart	Harting, 1866
1829/30	Whooper Swan, Long-tailed Duck	Ticehurst
1830s	Quail, Sparrowhawk, Lesser Spotted Woodpecker, Carrion Crow	Blyth
1830s	Kestrel, Garden Warbler	Bucknill

Year	Species	Reference
1830	Glossy Ibis	Ticehurst
1830	Cirl Bunting	Montagu, 1831
1831	Common Gull, Razorbill	Jesse
1831	Woodlark	Barrett-Hamilton
1831	Whinchat, Stonechat, Ring Ouzel	Doubleday
1831	Lesser Whitethroat, Blue Tit, Siskin, Lesser Redpoll	Montagu, 1831
1831	Parrot Crossbill	Bucknill
1832	Green Sandpiper	Christy
1832	Great Spotted Woodpecker	Doubleday
1832	Long-tailed Tit	Jesse
1832	Roller	Harrison, J.M.
1833	Scops Owl	Harting, 1866
1833	Common Redpoll	Blyth
1833/34	Chiffchaff	Wheatley
1834	Alpine Swift	Bucknill
1834/35	Common Buzzard	Bucknill
1835	Corncrake	Jesse
1836	Merlin, Grey Wagtail	Blyth
1837	Kittiwake	Wheatley
1838	Greylag Goose, Great Snipe, Snow Bunting	Torre
1838	Red-throated Diver	Meyer
1838	Honey Buzzard	Ticehurst
1838	Stock Dove	Jesse
1838	Goldfinch, Corn Bunting	Barrett-Hamilton
1839	Temminck's Stint	Yarrell, 1843
1839	Crested Tit	Wheatley
Pre-1840	Curlew	Meyer
1840	Squacco Heron	Harting, 1866
1840	Wood Sandpiper	Meyer
1840	Brambling	Christy
1841	Black-throated Diver	Sharpe
1841	Spotted Redshank	Harting, 1866
1841	Great Black-backed Gull	Harting, 1866
1842	Wigeon, Ringed Plover, Greenshank, Yellow Wagtail, Dipper	Meyer
1842	Red-breasted Merganser	Yarrell, 1843
1842/43	Gadwall	Harting, 1866
1843	Brent Goose, Pintail, Scaup, Goldeneye, Goosander, Oystercatcher, Little Stint, Black-tailed Godwit, Bar-tailed Godwit, Lesser Black-backed Gull, Little Tern, Arctic Tern	Bond, 1843 (A)
1843	Rock Pipit	Bond, 1844
1843	Nuthatch, Pine Grosbeak	Yarrell, 1843
1844	Knot, Sanderling	Bond, 1844
1846	Golden Plover	Barrett-Hamilton
1846	Pectoral Sandpiper	Glegg, 1935
c. 1846	Two-barred Crossbill	Wood
1847	Bean Goose, Guillemot	Ruegg
1847	White-fronted Goose	Meyer
Pre-1847	King Eider	Harrison, J.M.
1849	Black Tern	Wheatley
1850s	Tree Pipit	Hamilton, 1879

Year	Species	Reference
1850s	Moorhen	Fitter
1850	Shag	Clark Kennedy
1850	Whimbrel	Harting, 1866
1850	Red-necked Phalarope	Christy
1850	Golden Oriole	Wood
Pre-1852	Spotted Sandpiper	Harting, 1866
1853	Woodchat Shrike	Bucknill
1854	Avocet	Harting, 1866
1854/55	Fieldfare	Harting, 1866
1855	Jay	Fitter
1856	White-rumped Sandpiper	Harting, 1866
1857	Skylark	Glegg, 1935
1858	Cream-coloured Courser	Hudson
1859	Serin	Glegg, 1935
1859	Firecrest	Ticehurst
1861	Marsh Warbler	Harting, 1866
1862	Sabine's Gull	Harting
1862	Bluethroat	Bucknill
1862	Common Whitethroat	Glegg, 1935
1862–73	Red-crested Pochard	Gould
1863	Egyptian Goose	Clark Kennedy
1863	Ferruginous Duck, Redshank	Harting, 1866
1863	Pallas's Sandgrouse	Wood
1864	Little Ringed Plover, Dunlin	Harting, 1866
1865	Turnstone	Harting, 1866
1865	Shore Lark	Harting, 1866
1865	Coal Tit	Hibberd
1866	Mallard, Shoveler, Pochard, Tufted Duck, Common Scoter, Water Rail, Coot, Lapwing, Jack Snipe, Common Sandpiper, Black-headed Gull, Herring Gull, Roseate Tern, Turtle Dove, Bee-eater, Meadow Pipit, Pied Wagtail, Wheatear, Redwing, Mistle Thrush, Sedge Warbler, Blackcap, Willow Warbler, Great Tit, Marsh Tit, Treecreeper, Greenfinch, Linnet, Bullfinch, Yellowhammer	Harting, 1866
1866	Manx Shearwater	Christy
1866	Mediterranean Gull	Harting, 1872
1870	Great Bustard	Glegg, 1935
1870	Pacific Golden Plover	British Birds (2:150)
1870	Common Rosefinch	Gould
1871	Purple Sandpiper	Wheatley
1872	Crane	Palin
1874	Baillon's Crake	Wheatley
1875	Shelduck	Glegg, 1935
1877	Tengmalm's Owl	Christy
1877	Red-necked Grebe	Swann
1878	Kentish Plover	Bucknill
1879	Glaucous Gull	Hamilton, 1879
1882	Rustic Bunting	Lilford
1883	White-winged Black Tern	Read, R.H.
1883	Little Auk	Warren
1884	Short-toed Lark	Swann

Year	Species	Reference
1892	Red-footed Falcon	Bucknill
1893	Black Stork	Meinertzhagen
1897	Willow Tit	Hartert
1897	Lesser Grey Shrike	Ticehurst
1899	Yellow-browed Warbler	Ticehurst
19th century	Goshawk	Bucknill
Pre-1900	Black Guillemot	Bucknill
1902	Barnacle Goose	Davis
1905	Pink-footed Goose	Ticehurst
1909	Velvet Scoter, Black Grouse	Ticehurst
1912	Macaronesian Shearwater	*British Birds* (6:314)
1917	Eider	Glegg, 1929
1918	Black-winged Stilt	Nichols
1924	Aquatic Warbler	Glegg, 1935
1927	Garganey	LNHS, 1964
1931	Bewick's Swan	*British Birds* (24:339)
1939	Red-breasted Flycatcher	LBR
1942	Iceland Gull	LBR
1947	Crested Lark	*British Birds* (41:345)
1950	Baird's Sandpiper	LBR
1953	Buff-breasted Sandpiper, Lesser Yellowlegs	LBR
1954	Fulmar, Tawny Pipit	LBR
1956	Little Egret	LBR
1956	Little Bunting	LBR
1957	Collared Dove	LBR
1958	Ruddy Duck	Wheatley
1958	Broad-billed Sandpiper	LBR
1961	Green-winged Teal, Caspian Tern, Whiskered Tern, Melodious Warbler	LBR
1963	Marsh Sandpiper, Snowy Owl	LBR
1964	Red-rumped Swallow, Red-throated Pipit	LBR
1965	Icterine Warbler	LBR
1966	Long-tailed Skua	LBR
1967	Solitary Sandpiper	LBR
1969	Ring-necked Parakeet	Wheatley
1971	Gyr Falcon, Sooty Tern	LBR
1972	Barred Warbler, Iberian Chiffchaff	LBR
1973	Western Sandpiper	LBR
1974	Ring-necked Duck, Lesser Kestrel	LBR
1975	Cetti's Warbler, Short-toed Treecreeper	LBR
1976	Sharp-tailed Sandpiper, Yellow-legged Gull, Subalpine Warbler, Arctic Redpoll	LBR
1977	Long-billed Dowitcher	LBR
1979	Black Kite, Ring-billed Gull, Savi's Warbler	LBR
1980	Snow Goose, Gull-billed Tern	LBR
1981	Blue-winged Teal	LBR
1982	Penduline Tit	LBR
1983	Collared Pratincole, Wilson's Phalarope	LBR
1984	Balearic Shearwater, Killdeer, Common Nighthawk	LBR
1985	Sociable Plover, Pallas's Warbler	LBR
1986	Black-headed Bunting	LBR

Year	Species	Reference
1989	Desert Wheatear	LBR
1990	Naumann's Thrush	LBR
1991	Bridled Tern, Yellow-billed Cuckoo	LBR
1992	Cattle Egret, Olive-backed Pipit, Sardinian Warbler, Pine Bunting	LBR
1993	Citrine Wagtail	LBR
1994	Isabelline Shrike	LBR
1995	American Golden Plover	LBR
1996	Lesser Scaup	LBR
1997	Pied-billed Grebe, Great White Egret	LBR
1998	Caspian Gull	*British Birds* (97:584)
1999	Paddyfield Warbler	LBR
2000	Franklin's Gull, Radde's Warbler	LBR
2001	Blyth's Reed Warbler	LBR
2004	Hume's Warbler	LBR
2005	Grey-cheeked Thrush	LBR
2006	Laughing Gull, American Robin	LBR
2007	Glaucous-winged Gull	LBR
2009	Brown Shrike	*British Birds* (103:605)
2010	White-tailed Plover	*British Birds* (104:578)
2010	Dusky Warbler	LBR
2011	Eastern Crowned Warbler	*British Birds* (105:595)
2012	Bonaparte's Gull, Buff-bellied Pipit	

The full references of each publication are listed in the Bibliography. LBR refers to the London Bird Report for that year.

CALENDAR OF NOTABLE BIRDS

This section lists the first dates when extremely rare birds have been discovered in the London Area, i.e. those species which have been recorded no more than three times.

As might be expected, most rarities have been found in October but, over the years, every month has produced at least two rare birds. Perhaps the most surprising result is that March, which is usually considered to be a quiet month, has recorded more extreme rare birds than April, which is one of the main arrival months for spring migrants. It is also interesting that the first and last months of winter (December and February) have only recorded six rarities combined while the middle month, January, has been more productive with seven.

January

- 4th: Lesser Scaup, Stocker's Lake (2003)
- 6th: Naumann's Thrush, South Woodford (1997)
- 11th: Hume's Warbler, Fairlop Waters (2004)
- 19th: Naumann's Thrush, Woodford Green (1990)
- 26th: Pied-billed Grebe, South Norwood Lake (1997)
- 31st: Killdeer, Beddington SF (1984)

February

- 9th: Rustic Bunting, Beddington SF (1993)
- 12th: Pine Bunting, Dagenham Chase (1992)

March

- 2nd: Snowy Owl, Sewardstone (1963)
- 6th: Gyrfalcon, Queen Mary Reservoir (1971)
- 8th: Sociable Plover, Dartford Marsh (1985)
- 8th: Crested Lark, between Chiswick and Hammersmith (1947)
- 14th: Gyrfalcon, Chessington (1972)
- 16th: Pied-billed Grebe, Walton Reservoirs (1999)
- 21st: Isabelline Shrike, Richmond (1994)

April

- 7th: Lesser Scaup, Tyttenhanger GP (1996)
- 13th: Franklin's Gull, Crossness (2000)

13th: Desert Wheatear, Barn Elms Res (1989)
18th: Glaucous-winged Gull, Beddington SF (2007)
28th: Marsh Sandpiper, Broxbourne GP (1984)

May

1st: Hume's Warbler, Brent Reservoir (2004)
2nd: Pacific Golden Plover, Beddington SF/London Wetland Centre (2010)
19th: Bonaparte's Gull, Crossness/Barking Bay (2012)
21st: Lesser Grey Shrike, Banstead Downs (1956)
23rd: Roller, Oxshott (1959)
26th: Short-toed Treecreeper, Epping Forest (1975)
26th: Bonaparte's Gull, Crossness (2012)
31st: Lesser Kestrel, Hampstead Heath (1992)

June

2nd: Bridled Tern, West Thurrock (1991)
2nd: Sardinian Warbler, Hogsmill SF (1992)
3rd: Iberian Chiffchaff, Brent Reservoir (1972)
10th: Icterine Warbler, Sear's Park (1983)
16th: Blyth's Reed Warbler, Fishers Green (2003)
19th: Collared Pratincole, Staines Reservoir (1983)

July

2nd: Collared Pratincole, Rainham Marshes (2005)
7th: White-tailed Plover, Rainham Marshes (2010)
9th: Caspian Tern, Walthamstow Reservoirs (1961)
21st: Western Sandpiper, Rainham Marshes (1973)
26th: Gull-billed Tern, Barn Elms Reservoirs (1980)
27th: Broad-billed Sandpiper, Rye Meads (1958)
29th: Caspian Tern, Staines Reservoir (1979)
30th: White-rumped Sandpiper, Perry Oaks SF (1984)
31st: Lesser Kestrel, Rainham Marshes (1974)

August

6th: Sharp-tailed Sandpiper, Staines Reservoir (1976)
9th: Icterine Warbler, Staines Reservoir (1965)
11th: Caspian Tern, Staines Reservoir (1982)
17th: Balearic Shearwater, Island Barn Reservoir (1984)
18th: Sooty Tern, Staines Reservoir (1971)
20th: Macaronesian Shearwater, Welling (1912)
22nd: Citrine Wagtail, King George V Reservoir (1994)
24th: Citrine Wagtail, Beddington SF (1993)
24th: Black-headed Bunting, Bromley-by-Bow (1986)

26th: Marsh Sandpiper, Swanscombe Marsh (1963)
26th: Marsh Sandpiper, Chigwell Sewage Works (1966)
26th: Roller, Osterley Park (1968)

September

2nd: American Golden Plover, King George VI Reservoir (1995)
3rd: Baird's Sandpiper, Rainham Marshes (1977)
5th: Wilson's Phalarope, Staines Reservoir (1997)
14th: Wilson's Phalarope, Staines Reservoir (1983)
17th: Baird's Sandpiper, Perry Oaks SF (1950)
19th: Laughing Gull, Hilfield Park Reservoir (2006)
26th: Roller, Staines Reservoir (1959)

October

1st: Long-billed Dowitcher, Staines Reservoir (1977)
1st: Radde's Warbler, Fairlop Waters (2000)
4th: Long-billed Dowitcher, Staines area (1987)
6th: Lesser Grey Shrike, Perry Oaks (1957)
6th: Blyth's Reed Warbler, Canary Wharf (2001)
11th: Brown Shrike, Staines Moor (2009)
14th: Baird's Sandpiper, Staines Reservoir (1982)
17th: Yellow-billed Cuckoo, Oxted (1991)
19th: Cream-coloured Courser, Stratford (1858)
23rd: Common Nighthawk, Barnes Common (1984)
23rd: Olive-backed Pipit, Chingford (1992)
26th: Paddyfield Warbler, Seventy Acres Lake (1999)
29th: Pallas's Warbler, Wandsworth Common (1985)
30th: Eastern Crowned Warbler, Hilfield Park Reservoir (2011)

November

12th: Pacific Golden Plover, Epsom (1870)
13th: Grey-cheeked Thrush, Northaw Great Wood (2005)
18th: Tengmalm's Owl, near Dartford (1881)
26th: Laughing Gull, Amwell GP (2006)

December

4th: Sociable Plover, Rainham Marshes (2005)
12th Buff-bellied Pipit, Queen Mother Reservoir (2012)
26th Buff-bellied Pipit, Queen Mother Reservoir (2012)
28th: Great Bustard, Feltham (1870)

HABITATS AND GEOGRAPHY

London's best-known geographical feature is the River Thames, a tidal river that crosses the city from the south-west to the east. The Thames Valley is a floodplain surrounded by gently rolling hills, although this is hard to visualise with the amount of development along the river. The Thames has been extensively embanked for much of its length in London since the Victorian era. It was once a much broader, shallower river with extensive marshlands; at high tide, its shores reached five times their present width. Many of the tributaries that flow into the Thames in London now flow underground.

Sadly, the extensive marshes that bordered the Thames have either totally disappeared or are a mere fraction of what they used to be. We know that there were marshes at Fulham, which included breeding Spoonbills in the 16th century. Today, there are only remnants of marshes left alongside the Thames, in the east of the area at Barking and Rainham on the north side of the river, and Crossness, Crayford, Dartford and Swanscombe on the southern side. Even these have been subject to further destruction, such as at Rainham with its silt lagoons, bypass and enormous rubbish dump, while much of Erith Marshes disappeared under the extensive housing development of Thamesmead, leaving a few fragments of what is now known as Crossness.

The water quality of the river has, however, improved markedly since the 1960s due to improvements in sewage treatment, and this should improve further if the proposal for a Thames Tideway Tunnel, which will relieve pressure on overloaded combined sewers at times of heavy rainfall, goes ahead. Fifty years ago, the river was essentially fish-less in central London, but more than 118 fish species have been recorded in recent years, including an important nursery ground for Dover Sole. Improved fish stocks have had a direct effect on fishing birds, with Grey Heron nesting along or nearby the river and Cormorants fishing regularly even in central London.

As well as the many rivers in the London Area, there are many man-made bodies of water, which are an important habitat for the capital's flora and fauna. Some of these are over two centuries old and their birdlife has been intensively studied, making an important contribution to the knowledge of the London Area's avifauna. One of the best documented of these is Brent Reservoir in north-west London, which was one of several reservoirs built to supply water to the Grand Union Canal. Brent Reservoir attracted the attentions of some of London's earliest naturalists such as Frederick Bond and James Harting, who both lived nearby.

Later, reservoirs were built to supply drinking water, e.g. Staines Reservoir, which was constructed in 1901. Some of these concrete-banked reservoirs are more important for wintering ducks than as breeding areas, although in years when the water levels are low they are often used by breeding and passage waders. A few reservoirs have natural banks, like at Brent where there are marshy areas, reed beds and willow carr; these increase the biodiversity. The complex of reservoirs at Walthamstow also has several tree-clad islands, which are home to breeding Cormorants, Grey Herons and Little Egrets.

The reservoirs provide a major attraction for birds on passage, particularly those that are clustered together such as in the Lea Valley and in south-west and west London. These large expanses of water regularly attract wildfowl, waders, skuas, gulls and terns, as well as many migrant passerines. The reservoirs and associated wetlands in the Colne Valley and the Lea Valley have been designated as a Special Protection Area for Birds under European legislation, and both Brent Reservoir and the London Wetland Centre (former Barn Elms Reservoirs) are SSSI, in recognition of their national importance for wetland birds.

Another important man-made water habitat is gravel pits. Many of these were created in the 1930s to provide building materials for the ever-expanding suburbs and even more were dug during the war to construct aerodromes; by 1948 there were almost 200 gravel pits in the London Area. Even during their working lives many of these provided Little Ringed Plovers with places to breed. After the gravel extraction had finished, the pits filled with water and were landscaped. The London Area is fortunate to be home to several large complexes of gravel pits, notably in the Colne and Lea Valleys, and around Sevenoaks. Whilst some of the pits are managed as nature reserves, others are used for water-sports and are less hospitable for birds.

In central London the most obvious open habitat is parks. The five Royal Parks that are within the Inner London recording zone – Green Park, Hyde Park, Kensington Gardens, Regent's Park and St James's Park – make up a total of 461 hectares and were all acquired by Henry VIII. Green Park and St James's Park were originally marshy swamps; the former was a burial ground and was added to a private estate before being made into a Royal Park by Charles II. St James's Park was drained by James I and he used it to house exotic animals;

there were aviaries along the south side, hence the modern name, Birdcage Walk. Later changes were made, including landscaping of the lake, and it was opened to the public by Charles II. Hyde Park was enclosed as a private hunting ground before being opened to the general public in 1637; Kensington Gardens was created from within the park in the eighteenth century. Regent's Park was also set aside as a hunting park and then let out in smallholdings for hay and dairy produce until the leases expired in 1811; the park was first opened to the general public in 1845.

The Royal Parks constitute a major green space in central London and are important refuges for birds. Apart from the rather sterile Green Park, they all have a waterbody and shrubbery for nesting and foraging. A surprisingly wide variety of breeding birds can be found in the parks, from waterbirds, woodpeckers and owls, to many of the common passerines. During spring and autumn these parks offer birds an oasis in an otherwise unfriendly urban environment and migrants can also be found there, mainly in the early morning before they get too disturbed. There are wildfowl collections in Regent's Park and St James's Park, as well as a small one in Buckingham Palace Gardens. During the early to mid-20th century many ducks and geese bred in these parks and their offspring remained free-flying. This led to populations of nationally scarce species such as Pochard and Gadwall colonising the London Area.

Away from central London there are three other Royal Parks: Bushy Park, Greenwich Park and Richmond Park. Bushy Park is the second largest Royal Park in the London Area, a status it was given in 1529. Henry VIII enclosed the area with a brick wall and turned it into a deer park. In later years water features and gardens were added, and this has helped provide plenty of habitat for a number of birds, including scarce London Area breeding birds such as Lesser Spotted Woodpecker, Swallow and Stonechat, and it also has a healthy population of Little Owls.

The oldest Royal Park is Greenwich Park. When it was enclosed as a deer park in 1433, it mostly comprised common land with scrub oak, thorns, birch, gorse, broom and heath, but it was landscaped in the 17th century. It is the least watched Royal Park and was one of the last suburban sites to have breeding Spotted Flycatchers.

Richmond Park, the largest of the Royal Parks, was first given its charter by Charles 1st in 1625. It retains much of the original deer park landscape, with magnificent veteran trees set amongst open grassland. Although it is perhaps most famous for its deer, the park gained SSSI and National Nature Reserve status principally on account of the rare beetle assemblages found in its veteran oak trees. The mature trees and fine open grassland are also valuable for birds, with more than 60 breeding species, including all three woodpeckers, Skylarks and Stonechats; Dartford Warblers are often present in winter.

The parks were well watched during the 20th century and for each one there was, until recently, an official position of Park Recorder. Various studies have taken place in the parks, including autumn counts in Hyde Park/Kensington Gardens, which first started in 1925. Regular migration watches in Regent's Park were first conducted in the 1950s.

There are many other parks scattered across the London Area although their usefulness to birds varies greatly, depending on how they are managed. Most of them fall under the remit of the relevant London Borough, but some are managed by the City of London Corporation, which also looks after Epping Forest and Hampstead Heath. In the most built-up areas, small parks and garden squares offer nesting opportunities for a fair range of common birds. However, much depends on the landscape structure; those with a well-developed under-storey of shrubbery support more bird species than those which have just trees and grass. Many parks and other urban green spaces are underwatched but they are always capable of providing a surprise, such as the London Area's only Common Nighthawk on Barnes Common or a Pallas's Warbler on Wandsworth Common.

Gardens provide the largest habitat in the suburbs. Here, the density of species such as Blackbirds and Robins is significantly higher than it is in woodlands (Batten, 1972). Although there are dangers associated with gardens, especially from domestic cats, there are also significant advantages, not least the food that owners put out for birds, particularly in winter, and an assortment of nesting places, from nest-boxes and sheds to shrubbery and trees. Suburban gardens are the main habitat of London's House Sparrows although they have disappeared from many areas.

Probably the most overlooked habitat in the London Area is buildings. While many are inhospitable for birds, others provide places to breed for some of the city's rarest species, such as Peregrines and Black Redstarts, even in the heart of the city. During most of the 20th century huge numbers of Starlings roosted on buildings in central London but they can no longer be found there.

Today there is very little natural woodland left in the London Area. In the 12th century most of Middlesex was covered by the great Forest of Middlesex, which stretched for 32 kilometres (20 miles) north of the City of London. Fitzstephen described it in the late 12th century as a 'vast forest, its copses dense with foliage concealing wild animals – stags, does, boars, and wild bulls'. Deforestation started in the thirteenth century, however, some important fragments still remain including Ruislip Woods (now a National Nature Reserve), Scratch Wood, Highgate Wood and Queens Wood.

Perhaps the most important example of ancient forest that is still with us is Epping Forest. Situated in east London, it lies on a ridge between the valleys of the Lea and Roding and is now a National Nature Reserve and also a Special Area for Conservation under European legislation, designated on account of its importance for Stag Beetle and other insects associated with its fine veteran trees. The specialist woodland species have declined in Epping Forest as they did throughout the London Area during the 20th century, so birds such as Wryneck, Common Redstart and Red-backed Shrike no longer breed. By the beginning of the 21st century, Nightingales and Hawfinches had also ceased to breed in Epping Forest. A few pairs of Tree Pipits still hang on and Nightjars are sometimes seen in summer and may still breed. The ponds are one of the strongholds of the introduced Mandarin Duck.

As in so much of southern England, there is very little lowland heath left in the London Area. It is a landscape characterised by plants like common ling and gorse, and trees such as Scots pine and birch. There are a few heaths in the far south of the London Area, such as at Headley Heath on the North Downs, and Epsom, Esher and Oxshott Commons in the Thames Basin. Allied to this specific habitat are commons, of which there are more in the London Area. Examples of old commons are Bookham, Wandsworth, Wanstead Flats and Wimbledon. Many of the birds associated with these habitats, such as Common Redstart, Dartford Warbler, Nightjar, Tree Pipit, Wood Warbler and Woodlark, have either become extinct in the London Area or have been reduced to just a handful of pairs. Away from south-west London there are only a few other small patches of heath, including Black Park and Hounslow Heath. Hampstead Heath retains very little of the requisite vegetation to be deemed a heath and much of Blackheath now comprises closely mown grass, suitable for nothing except large numbers of foraging Carrion Crows.

Despite being dominated by the metropolis, there is still a significant amount of farmland in the area. This habitat surrounds much of suburbia, particularly to the south, north and east; the built-up area extends so far west and with huge areas taken up by Heathrow Airport, that there is very little room left for farmland. A notable exception is, however, the surviving farmland surrounding Osterley Park which supports typical birds of an agricultural landscape. Across most of the southern area and the north-west the traditional type of farming has been pasture, whereas in the north and east it is arable. This is still largely true today and helps to explain why the majority of rookeries are located in the north-east. Farmland birds have declined across Britain more than any other types of bird and it is unlikely that Corncrake and Cirl Bunting will ever return to the London Area. Grey Partridges, Turtle Doves, Tree Sparrows, Yellowhammers and Corn Buntings are perilously close to extinction in the London Area and may soon also be lost forever. Fortunately it isn't all bad news as Common Buzzards have colonised the London Area and may be followed by Red Kites and Ravens; the last two species now breed in Hertfordshire and are increasingly seen in London itself.

For much of the 20th century sewage farms were an important habitat in the suburbs. The settling tanks attracted insects, and pipits and wagtails fed on them; some areas were much wetter and provided refuges for ducks and other waterbirds. Emergent vegetation gave Reed Warblers somewhere to nest, and the scrub that grew up around the edges provided breeding areas for many other birds. Waders were often seen on migration and in winter, and the larger sewage farms such as Beddington and Perry Oaks soon became renowned for their ability to attract rarities. In the period between 1936 and 1962 about 20 small sewage farms were closed in north-west London; by the 1970s due to modernisation of the sewerage system many of the traditional sewage farms were no longer needed. Some of the sites were developed into public amenity areas like at Elmers End, which became South Norwood Country Park, while others were transformed into nature reserves like at Rye Meads. Beddington is also used as a landfill tip, but there are long-term plans to gradually convert the area to a mixture of open space and nature reserve. Access to the much-beloved Perry Oaks SF was restricted in the 1980s because of tighter security at Heathrow Airport and the site was eventually demolished in 2002. Now, instead of wading birds refuelling before continuing their journeys, the area throngs with air passengers waiting to fly off to foreign destinations from Terminal 5. One of the few remaining sewage farms is Hogsmill, part of which is managed as a nature reserve by Thames Water.

Probably the most valuable sites in the London Area are nature reserves. This is because they have a level of protection so they cannot be developed; in many cases they preserve much-declined habitats. There are three large reserves owned by national bodies: the RSPB reserves at Rainham and Rye Meads, and the London Wetland Centre owned by the Wildfowl and Wetlands Trust.

Rainham Marshes is one of the few remaining grazing marshes in the Thames Estuary. It consists of wet grassland and ditches and, along with rank grassland and scrub, supports many breeding and wintering birds. Rye Meads, in the Lea Valley, was formed when Rye House RSPB reserve acquired the adjacent Rye Meads SF and the two were amalgamated. It comprises mainly wetland habitats, including lagoons, pools, reed beds and scrapes. The London Wetland Centre was built on the site of the Barn Elms Reservoirs after they were decommissioned by Thames Water. It is made up of a number of large lakes, small pools, reed beds, seasonally-flooded wetlands, fen meadows and wet woodland, and is nationally important for wintering Gadwall and Shoveler. It is also the closest site to the centre of London with breeding Redshank.

The London Area also has many smaller nature reserves within the built up area. Perivale Wood in Ealing was established as a nature reserve by the Selborne Society in 1902 and is one of the best studied woodlands for its bird life. London Wildlife Trust manages more than 40 nature reserves, many of them, such as Camley Street Natural Park near Kings Cross and Gunnersbury Triangle in Chiswick, are quite small sites, but offer birds a nesting opportunity in areas which would otherwise be unsuitable as well as providing urban communities a chance to experience nature at close hand. There are also some small areas owned by the Woodland Trust and many locally managed reserves, some of which have already been mentioned in this chapter. Others include East India Dock Basin, at the confluence of the River Lea and the Thames which is the only site in the inner suburbs with a salt marsh.

IMPORTANT LONDON NATURALISTS AND ORNITHOLOGISTS

This chapter includes a short summary of important naturalists and birdwatchers who are no longer with us but have made a significant contribution to our knowledge of London's avifauna, either from writing books and major articles or being involved with the study and recording of its birds.

Frederick Bond (1811–1889)

An ornithologist and entomologist who lived in Kingsbury, Bond was one of the founders of *The Zoologist* magazine. He was elected a member of the Entomological Society in 1841 and a member of the Zoological Society in 1854. Bond wrote a number of papers for *The Zoologist*, including a list of waterbirds seen at Kingsbury Reservoir in 1843. He was a collector of British Lepidoptera and owned a large collection of stuffed birds, much of which is now in the Natural History Museum in Tring.

Stanley Cramp (1913–1987)

Although he was best known as the first Chief Editor of the encyclopaedic nine-volume handbook *The Birds of the Western Palearctic*, Cramp also wrote several papers about the birds of Inner London, including 'The Birds of Kensington Gardens and Regent's Park', which was published in the 1948 *LBR*. He held various roles in the LNHS, RSPB and BTO, and was also Editor of *British Birds*.

Mike K. Dennis (1949–2006)

Mike was a well-known birder in east London who had a passion for Rainham Marshes. He wrote a series of papers about the birdlife of Rainham for the *LBR* and was deeply involved with the conservation of the area during its transition to an RSPB reserve. For many years he chaired the Ornithology Research Committee in the LNHS and was also the Essex sector Recorder. Mike also found London's only Short-toed Treecreeper.

Charles Dixon (1858–1926)

Dixon collected specimens for Seebohm and discovered the St Kilda Wren. He wrote *The Bird Life of London*, which was published in 1909.

Richard S. R. Fitter (1913–2005)

An all-round naturalist, Fitter was the author of many books on all aspects of natural history, including two about London. From a birdwatching point of view, his most famous books were field guides. His *Pocket Guide to British Birds* was one of the first modern field guides, and his later guide *Birds of Britain and Europe with North Africa and the Middle East* is still in print today. He held a number of posts in the LNHS, including Editor of the *London Bird Report*, Recorder/Joint Recorder for London from 1939 to 1943 and Chairman of the Records Committee from 1944 to 1946. He was later President of the LNHS.

His published works include:
London's Natural History, 1945
London's Birds, 1949
The Pocket Guide to British Birds, 1952
Birds of Town and Village, 1953
Britain's Wildlife: Rarities and Introductions, 1966
Guide to Bird Watching, 1970

William E. Glegg (1879–1952)

Glegg was born in Scotland and moved to London in 1903, living near Epping Forest. He was the London Bird Recorder from 1913 to 1917 and President of the LNHS. One of his presidential addresses was on 'The

Birds of Middlesex since 1866', which he based on Harting's unpublished manuscript. He also wrote a paper for the *London Naturalist* in 1928 'The Thames as a Bird-migration Route'. He is best known for his two county avifauna books.

His published works include:
A History of the Birds of Essex, 1929
The Birds of the Lea Valley reservoirs, 1933–34
A History of the Birds of Middlesex, 1935

Peter J. Grant (1943–1990)

Grant is best known for his book *Gulls: a Guide to Identification*, which he wrote and illustrated. Grant started birding in London and later co-wrote *The Thames Transformed* in 1976, about the birdlife that had returned to the river. He held a number of posts within the LNHS, including Editor of the *London Bird Report*. Grant also served on the British Birds Rarities Committee, later being its chairman for ten years.

Edward Hamilton (1815–1903)

Hamilton was a Vice-President of the Zoological Society of London for 20 years and wrote a number of articles about birds in London. These included 'Rooks and Rookeries in London, Past and Present' and 'The Birds of London, Past and Present, Residents and Casuals', both of which were published in *The Zoological Journal*.

Mark Hardwick

Hardwick moved to London after university to seek work. He adopted Hampstead Heath as his local patch and among his many finds there was London's second Lesser Kestrel. He was editor of the *London Bird Report* from 1990 to 1994 and was also Inner London Recorder. He later emigrated to Australia.

James E. Harting (1841–1928)

Harting was an ornithologist who lived in Kingsbury for most of his life before moving to Weybridge, Surrey. He was a regular visitor to Kingsbury [Brent] Reservoir, where he shot many birds and built up a large collection of specimens. He worked as a solicitor from 1868 to 1878, but gave this up and wrote several books on natural history. He was Editor of *The Field* from 1871–1928 and *The Zoologist* from 1877 to 1896, and was considered to be the best living authority on British birds at that time. He was both Assistant Secretary and Librarian to the Linnean Society of London. In a London context, his most important book was *The Birds of Middlesex*, the first avifauna for the county. He wrote a second edition, which was never published although the completed manuscript was bought by Glegg.

His published works include:
The Birds of Middlesex, 1866
The Ornithology of Shakespeare, 1871
A Handbook of British Birds, showing the distribution of the resident and migratory species in the British Islands, 1872
Our Summer Migrants: an account of the migratory birds which pass the summer in the British islands, 1877

Rupert Hastings (1955–1993)

A well-known London-born birder who lived in the city for most of his life, Hastings started work at Kew Gardens in 1978. He spent much of his time birding in west London, especially at Barn Elms and Staines Reservoirs. He became the first person to see 200 species in London in a single year (in 1985) and found the capital's first Ring-billed Gull as well as identifying London's only Balearic Shearwater. Between 1981 and 1988 he was the bird recorder for the Surrey section of London and was also on the *London Bird Report* editorial team.

Arthur Holte Macpherson (1867–1953)

Author of 'A List of the Birds of Inner London' (*British Birds*, vol. 22, 1929), Macpherson also wrote annual supplements on the birds of Inner London from 1929 to 1940 and a number of other reports on London's birds. He was official Bird Observer for Hyde Park and Kensington Gardens from 1928 to 1939.

Richard C. Homes (1913–1978)

Dick, as he was known to everyone, moved to London from Somerset to take up work. After joining the LNHS he was quickly appointed as a bird recorder and compiled the first separate issue of the London Bird Report in 1936. He later held a number of the other posts in the LNHS including President. He oversaw the committee that produced the LNHS publication in the *New Naturalist* series *The Birds of the London Area since 1900* that was published in 1957 and the revised version in 1964, the predecessor to this book.

William H. Hudson (1841–1922)

Born in Argentina after his parents settled there from the USA, Hudson spent his youth studying the local wildlife before settling in England in 1869. He wrote a number of books based on his ornithological studies, notably *Birds in London*, which was based on his observations of a select number of birds that he studied when he stayed in London in the 1890s. He was a founding member of the Royal Society for the Protection of Birds.

His published works include:
British Birds (1895)
Birds in London (1898)
Birds in Town and Village (1919)

John R. Jefferies (1848–87)

Raised in Wiltshire, Jefferies moved to Tolworth in 1877 and wrote a number of books and essays on natural history, as well as novels. From a London context his most important work was *Nature Near London* (1883), which he wrote from the studies he made while exploring the local countryside.

Peter Naylor (1950–2012)

Naylor lived virtually his entire life in London and was a familiar figure on the west London birding scene, especially at Staines Reservoir where he found several rare birds including London's first Collared Pratincole. He held the post of Middlesex Recorder longer than anyone else, for 18 years between 1975 and 1992. He also served on the Rarities Committee and wrote the 2002–03 double issue *London Bird Report* single-handedly.

Edward Max Nicholson (1904–2003)

Born in Ireland, Nicholson moved to England in 1910, settling in Staines. Through his interest in natural history and later birdwatching, he developed into a keen conservationist. He went on to make a career in conservation, eventually serving as Director General of the Nature Conservancy from 1952 to 1966. He was Chairman of the British Trust for Ornithology from 1947 to 1949, Senior Editor of *British Birds* from 1951 to 1960 and President of the Royal Society for the Protection of Birds from 1980 to 1985.

He was also instrumental in setting up Britain's first urban ecology park, the William Curtis Ecological Park, which was created on the site of a derelict lorry park near Tower Bridge. He began the bird counts in Kensington Gardens in 1925–26.

His published works include:
Birds in England, 1926
The Art of Bird-Watching, 1931
The Birds of the Western Palearctic, 1977–1994 (Chief Editor)
Birdwatching in London – a Historical Perspective, 1995

Ken Osborne (1930–2003)

Osborne was a keen birder and active member of the LNHS where he held various posts, including seven years as Editor of the *London Bird Report*, and Inner London recorder. He developed a unique way of creating illustrations, and his maps and drawings were heavily featured in many bird reports.

Sir Thomas Digby Pigott, C.B. (1840–1927)

Pigott's main published work was *London Birds and Other Sketches* (1884), but he also wrote many letters to *The Times* about birds.

Frederick D. Power

Power pioneered the study of birds in a local area and was one of the first to record birds passing over his garden. He published these studies in 1910 in his book *Ornithological Notes from a South London Suburb 1874–1909*.

Eric Simms (1921–2009)

An ornithologist, naturalist, broadcaster and conservationist, Simms was best known as the author of many books about birds, particularly in the New Naturalist series. He lived in Dollis Hill, north-west London, for most of his life, where his study of local birds led to his book *Birds of Town and Suburb*. He was featured birdwatching in Gladstone Park in Sir John Betjeman's 1973 TV documentary *Metro-land* about the Metropolitan railway line. He later retired to Lincolnshire.

His published works include:
Bird Migrants. Some Aspects and Observations, 1952
Woodland Birds, 1971
Wild Life in the Royal Parks, 1974
Birds of Town and Suburb, 1975
The Public Life of the Street Pigeon, 1976
British Thrushes, 1978
Birds of Town and Village, 1978
A Natural History of Britain and Ireland, 1979
Natural History of Birds, 1982
A Natural History of British Birds, 1983
British Warblers, 1985
British Larks, Pipits and Wagtails, 1992

Harry K. Swann (1871–1926)

The author of *The Birds of London* (1893), Swann was also an authority on the bibliography of British ornithology.

William Yarrell (1784–1856)

Yarrell lived in London and was an all-round naturalist, also being known for his proficiency as an angler and shooter. He was elected a Fellow of the Linnean Society and was one of the original members of the Zoological Society of London. In 1833 he was a founder of what became the Royal Entomological Society of London. He made many important contributions to ornithology, notably publishing a two-volume book in 1843, *A History of British Birds*, which he later revised several times. In 1830 he was also the first person to describe Bewick's Swan, distinguishing it from the Whooper Swan. The British subspecies of White Wagtail, Pied Wagtail *Motacilla alba yarrellii*, was named after him.

His published works include:
'Notice of the Occurrence of Some Rare British Birds', vol. 2, *The Zoological Journal*. 1825.
A History of British Birds, 1843.

GREEN SPACES IN LONDON

This section details some of the main sites in London that are referred to in this book. In most cases it gives the historical background to the site, including main changes to the habitats and/or birds, and gives a flavour of what important birds in a London context may be found there. A list of other sites that have previously been known by other names is also included. Note that some of these sites have limited or no access; for further information about the sites and other sites in London refer to *Where to Watch Birds in the London Area* (Mitchell, 1997) or the London Bird Club wiki. Published books and papers about each site are also referenced here.

Aldenham Reservoir

This was originally called Elstree Reservoir and is London's oldest reservoir, having been completed in 1797 to top up the canal system. It used to cover an area of 26 hectares but was reduced to its current size of 22 hectares. It is part of Aldenham Country Park, a 70-hectare parkland and woodland which was created in 1971. The reservoir's heyday was mainly in the Victorian era although it later hosted London's first Bluethroat in 1942. The reservoir is used for sailing and is mainly surrounded by trees so is less attractive to wildfowl and waders which tend to prefer open conditions with clear flight lines.

Alexandra Park

The park dates back to the 1860s when it was built on farmland as a rival to Crystal Palace; it was opened in 1863. Although it has a mainly open, typical urban parkland landscape, its location and height mean that it is a particularly good site for attracting migrants. It contains some small areas of scrub and woodland, and a small wetland area consisting of Wood Green Reservoir and the adjacent filter beds. In the 1970s up to four Woodlarks wintered there.

The Birds of Alexandra Park and Wood Green Reservoir, Bowman, N.B. LBR No. 51 (1986)

Barn Elms Reservoirs

The four concrete-banked reservoirs at Barn Elms were completed in about 1896 and covered an area of 40 hectares. With a favourable position adjacent to the Thames, the reservoirs were a haven for birds and birders alike. During the 1950s flocks of up to almost 100 Smew used to overwinter. Many rarities have been recorded there, including London's only Desert Wheatear, as well as rare waders such as Spotted Sandpiper and Buff-breasted Sandpiper. Divers, grebes, wildfowl, gulls, terns, passage waders and the occasional seabird have all occurred at Barn Elms.

The large numbers of waterfowl using the site gave it SSSI status, but by the beginning of the 1990s the numbers had declined. Around the same time Thames Water completed construction of the London Ring Main and the reservoirs had become redundant. A partnership was formed with Thames Water, the Wildfowl and Wetlands Trust and a property company to redevelop the site, and in 1995 the site was demolished. It reopened five years later as the London Wetland Centre (see separate listing).

Battersea Park

This park is situated on the south bank of the River Thames opposite Chelsea and was formerly a marsh. It was opened in 1858 and covers an area of 83 hectares; it is owned by Wandsworth Council. It has a heronry with around 30 nests.

Beddington Sewage Farm

This suburban site is also known as Beddington Farmlands. The whole site covers about 162 hectares of mixed habitats including a small lake, reed beds, scrubland, sludge beds, shallow pools and meadowland. In the early days the area was a mixture of farmland and meadows as well as the traditional sewage farm. In the middle of the twentieth century, most of the area was used for sewage treatment but there were still extensive areas of wet meadows which had breeding Snipe, Redshank, Lapwing, Grey Partridge, Barn Owl, Yellow Wagtail, Skylark and Meadow Pipit.

From the mid-1950s onwards the area was intensively watched. It was soon realised that being completely surrounded by a suburban sprawl made it a magnet for unusual birds, and it attracted Little Bittern, Black-winged Stilt and Bee-eater amongst others. In 1962 over 5,000 Snipe were counted on the meadows, and between 1963 and 1965 Spotted Crakes probably bred. During the late 1960s and 70s due to changes to the way sewage was treated and other management work the wet meadows were lost and the site was much less hospitable to breeding birds; consequently the area received fewer visits from birders.

By the 1980s the new sludge beds were proving to be attractive to passage waders and the rarities returned, including London's only Killdeer. The number of breeding birds increased as management work became sensitive to the needs of wildlife. In 1992 the site was fenced off and access was only permitted to a limited number of birders. Six years later the site was transformed when gravel extraction and landfill began, and the following year saw the start of large-scale tree-planting. In the 21st century Beddington claimed another London first with a Glaucous-winged Gull, and it remains one of only two sites in London where Tree Sparrows breed. Further changes will continue until 2023, when the site is scheduled to no longer be in commercial use and it is expected to become a country park; it is planned to include a flooded grassland area in the future.

Brent Reservoir

This reservoir was built to supply water to Regent's Canal and construction was completed on 24 November, 1835; two years later a small extension was added. It was originally known as Kingsbury Reservoir. In subsequent years it was locally known as the Welsh Harp after the inn of the same name which was responsible for the leisure activities during its Victorian heyday; it still retains this local name today.

With its natural grass banks it has a very different look to it compared to most other London reservoirs. In its early days it was a haven for shooters such as Bond and Harting, who both lived nearby and often visited the reservoir in the hope of adding new birds to their collections. A total of 250 species has been recorded at Brent, including a number of rarities from the Victorian era such as Little Bittern, Squacco Heron, Spotted and White-rumped Sandpipers and Pallas's Sandgrouse. There was less coverage in the first half of the 20th century but since the 1960s the reservoir has been intensely watched and more rare and scarce birds have been seen, including the UK's first Iberian Chiffchaff and London's first Great White Egret. Notable second records for London include Hume's Warbler and Penduline Tit.

'The Past and Present Bird Life of Brent Reservoir and Its Vicinity', Batten, L.A. (*The London Naturalist*). 1972
Birds of Brent Reservoir, Batten, L.A., Beddard, R., Colmans, J. and Self, A., eds (2002)

Bushy Park

This Royal Park is situated in south-west London and dates back to 1514, when the land was enclosed as a deer park for Hampton Court Palace; it was opened to the public in 1838. About 150 species have been seen in Bushy Park and the adjoining Hampton Court Park. The park contains a number of wooded plantations, a river and ponds. Up to six pairs of Lesser Spotted Woodpeckers once bred but only a single pair now remains. Dartford Warblers are seen in most years and scarce migrants such as Hoopoe, Wryneck and Woodchat Shrike have been recorded.

Birds of Bushy and Hampton Court Parks, Betton, K.F. LBR 58 (1993).

Crossness Nature Reserve

The marshes at Thamesmead were first reclaimed in 1178 but continued to flood until a better sea wall was constructed in the 19th century, when the area was then turned into a grazing marsh. Crossness is the current name for all of the marsh areas in the Thamesmead region that were previously called Plumstead Marshes and Erith Marshes. Much of these marshes, along with an extensive reedbed, disappeared when the area was redeveloped for a massive housing programme in the 20th century, beginning with Abbey Wood in the early 1900s and Thamesmead in the 1960s. Prior to that there was encroachment due to the expansion of Woolwich Arsenal, and in 1860 construction of Crossness Sewage Works began.

The nature reserve was established in 1996 and is a 20 hectare site with a range of habitats including grazing marsh, reed-bed, a network of ditches, open water, scrub and rough grassland. Across the road from the reserve, a larger fragment of Erith Marshes, known locally as Southern Marsh, extends the available habitat. With its location by the Thames it is a good area for migrant and wintering wildfowl, waders and gulls, notably Caspian Gull. In recent years rarities such as Squacco Heron, Purple Heron and Bonaparte's Gull have also been recorded.

Summer's Evening Rambles Around Woolwich, Ruegg, R. (1847)
The Birds of Thamesmead, Wheatley, N. LBR No. 55 (1990)

Dartford Marsh

Although this area was subject to much development work during the 20th century due to its position on the southern side of the Thames it still attracts many birds. Littlebrook Power Station was built on the marsh in 1937. The area contains a wide range of habitats, including salt and fresh marshes, lakes, woodland and farmland. Over the years it has recorded several rarities, including Sociable Plover and Woodchat Shrike.
  The Birds of the Dartford District, Davis. W.J. 1904
  The Birds of Crayford, Dartford and Stone Marshes, Morris, A.J. & Wright, B.E. LBR No. 57 (1992)

Epping Forest

This is the largest area of woodland in London and covers an area of some 2,400 hectares. The forest stretches for 11 miles from Wanstead north-eastwards to Epping. It is mainly mature deciduous woodland with a mixture of beech, oak, hornbeam and birch with a number of ponds and open areas. Although the entire forest is protected, the bird population has been in decline since the middle of the 20th century. Wryneck, Common Redstart, Whinchat, Wood Warbler, Willow Tit, Spotted Flycatcher, Red-backed Shrike and Hawfinch have all been lost as breeding species, and there are very few Lesser Spotted Woodpeckers, Tree Pipits, Nightingales and Marsh Tits left. Nightjars used to breed in good numbers and may also be extinct in the area, although occasional ones are sometimes still reported. The Chingford area is one of the most birded parts of Epping Forest and has produced three first records for London: Olive-backed Pipit, Naumann's Thrush and Short-toed Treecreeper.
  A Profile of the Winter Bird Community in Epping Forest, Wallace, D.I.M. LBR No. 36 (1971)

Greenwich Park

This is the oldest enclosed Royal Park and was originally created as a deer park in 1433 when it was mostly heathland. It covers 73 hectares and is one of the largest single green spaces in south-east London. Some of the trees date back to the early 1600s when the park was laid out in the French style. It is notable as one of the few sites in the inner suburbs of SE London where Jackdaws are seen regularly, and has been for many years a breeding site for Spotted Flycatcher.
  The Birds of Greenwich Park and Blackheath, Grant, P.J. LBR No. 31 (1966)

Hampstead Heath

This is a large, ancient park in north London covering 320 hectares and is managed by the City of London Corporation. The Heath dates back over a thousand years and there have been many changes to the site, the last of which was in 1928 when Kenwood House and grounds were added. The highest point is Parliament Hill, which is 134 metres above sea-level. The Heath is a mixture of open grassland, woodland and ponds; those on the east side of the heath are collectively known as Highgate Ponds. One of the most wooded areas is Kenwood, which was originally called Caen Wood.

The Heath gets very crowded, especially at weekends, and receives less attention from birders because of the disturbance. However it is a good area for watching visible migration and has an impressive list of migrants and rarities including Little Bittern and London's second Lesser Kestrel. Migrant passerines such as Pied Flycatcher and Wood Warbler used to be annual, but the latter have become rare in London now. Other scarce migrants recorded there recently include Richard's Pipit, Golden Oriole and Red-backed Shrike.

Lesser Spotted Woodpeckers used to breed annually on the Heath and may still do so; otherwise the breeding birds are typical parkland species but with good populations of hole-nesting birds such as Stock Dove, Ring-necked Parakeet, Nuthatch and Jackdaw.
  Bird Life. Goodchild, H. Hampstead Scientific Society's Hampstead Heath. 1913

Hilfield Park Reservoir

This reservoir was completed in 1955 and covers 46 hectares. It is owned by Three Valleys Water and managed by the Herts & Middlesex Wildlife Trust; it became a Local Nature Reserve in 1969. It is essentially a man-made reservoir with large marshy areas and woodland. The area is of national importance for its summer populations of moulting ducks and breeding Common Terns. It also has a large number of wintering ducks and is the main roost site for gulls in north-west London.
  The Birds of Hilfield Park Reservoir, Elton, G.S. LBR 57 (1992)

Hogsmill Sewage Farm

This was originally called Berrylands SF and is managed by Thames Water; it includes a sewage farm and nature reserve.

Hyde Park and Kensington Gardens

Hyde Park was acquired by Henry VIII from the monks of Westminster in 1536 and preserved as a royal hunting ground. The park was opened to the public in 1637 and covers over 140 hectares. In 1728, Queen Caroline formed Kensington Gardens from Hyde Park and separated the two parks with a long ditch. She also established a new landscape fashion and created a lake in Hyde Park called the Serpentine by damming the Westbourne Stream. Further changes were made in the 1820s, when a new road called West Carriage Drive was installed which formally split Kensington Gardens from Hyde Park.

Kensington Gardens covers 111 hectares and includes the Round Pond, created in 1728, and Long Water which is the top part of the Serpentine, the border being only a bridge that crosses it. The parks are important in sustaining a fair variety of birds in the city centre, and some surprising species can turn up during migration.

The Changing Status of Birds in Kensington Gardens, Sanderson, R. LBR 32 (1967)

Island Barn Reservoir

This reservoir was opened in 1911 and covers an area of about 50 hectares. It is managed by Thames Water and is one of a number of reservoirs in south-west London. Due to disturbance from sailing activities some birds move around between these reservoirs. Island Barn hosted London's only Balearic Shearwater in 1984 and regularly attracts divers and grebes in addition to a large number of wildfowl.

Kempton Park Nature Reserve

Originally known as Hampton Reservoir, the two main reservoirs were completed in 1907 but were decommissioned in the late 20th century. Kempton Park East Reservoir was turned into a nature reserve owned by Thames Water; it was the site where Avocets first bred in London.

King George V Reservoir

This is situated in the Lea Valley and lies just to the north of Girling Reservoir; collectively they are known as the Chingford reservoirs. Construction was started in 1908 and completed in 1912; the two concrete basins cover an area of 170 hectares. Often abbreviated to KGV Res, it is owned and managed by Thames Water and they issue birdwatching permits. Sailing takes place on the southern basin and consequently most birds are seen on the northern basin.

The reservoir is a Site of Special Scientific Interest (SSSI) and is a major wintering ground for wildfowl, including nationally important numbers of some species. The water also forms a moult refuge for a large population of ducks during the late summer. In winter, and sometimes on passage, scarce grebes and the occasional diver may be recorded and Common Scoter is almost annual. Rarities such as Spotted Sandpiper and Citrine Wagtail have been recorded there along with two separate Long-tailed Skuas which were both present for several days.

King George VI Reservoir

Situated just to the west of Staines Reservoir, KGVI Res covers 141 hectares and was opened in November 1947. It is owned by Thames Water and there is currently no access to the reservoir. The reservoir forms part of the Staines Moor SSSI and hosts nationally important wintering populations of Tufted Ducks, Pochard, Goosander and Goldeneye. It has attracted rarities such as London's only American Golden Plover as well as Long-billed Dowitcher.

Lea Valley

The Lea Valley stretches 26 miles along the River Lea from Ware in Hertfordshire (just beyond the London Area), through to the Thames at East India Dock Basin. The Lea Valley comprises land owned or managed by the Lee Valley Regional Park (LVRP) and Thames Water; most of the sites managed by the LVRP are open access. It is often split into three sections: Lower Lea Valley, Upper Lea Valley and North Sites. The names Lea and Lee are interchangeable.

The Lower Lea Valley includes the large reservoirs of King George V, Walthamstow and William Girling (which are listed separately in this section), along with Bow Creek Ecology Park, East India Dock Basin, Middlesex Filter Beds, Three Mills Island, Tottenham Marshes, Walthamstow Marsh and Waterworks NR (previously known as Essex Filter Beds). The Upper Lea Valley consists of Cornmill Meadows, Gunpowder Park, Rammey Marsh and Sewardstone Marsh.

The North Sites include Amwell Nature Reserve (informally known as Amwell Gravel Pits), the complex of gravel pits and lakes at Cheshunt/Fishers Green, and Rye Meads RPSB Reserve amongst others. This is potentially a confusing area as many of the sites are contiguous and birds frequently move about them. It is also very difficult from a recording point of view as the county border between Essex and Herts runs through the Lea Valley and has changed since the Watsonian Vice-counties system was introduced; additionally the London border dissects Amwell NR.

The whole Lea Valley is an important area for birds in London; parts are also nationally and internationally important - a Special Protection Area for Birds under European legislation and a Ramsar Site. It has good numbers of wintering wildfowl, grebes and Bitterns; is a major roost site for gulls; has many breeding birds including Little Ringed Plover and Nightingale; and is a main migration route.

London Wetland Centre

This was built on the site of Barn Elms Reservoirs and covers 42 hectares. It is managed by the Wildfowl and Wetlands Trust and opened in May 2000. The site, which is adjacent to the Thames, includes an extensive network of large lakes, small pools, reed beds, seasonally flooded wetlands, fen meadows and wet woodland. It has regular wintering Bitterns and breeding Little Ringed Plover, Redshank and Lapwing. Marsh Warbler has bred once and Cetti's Warbler bred for the first time in 2010. It is also nationally important for wintering Shoveler. An artificial nesting bank for Sand Martins has around 90 breeding pairs, making it the largest colony in London. Since opening it has attracted a number of rare birds, including London's second Pacific Golden Plover.

Perry Oaks Sewage Farm

Perry Oaks was constructed in 1934 and demolished in 2002 to make way for Heathrow Terminal 5. The drying beds originally covered 72 acres and were a magnet for waders, particularly during the spring and autumn passage. Apart from the number of common and scarce waders which sometimes gathered there, e.g. Ringed Plover, Ruff and Wood Sandpiper, Perry Oaks was arguably the number one site in London for rare waders, and between 1950 and 1990 attracted Baird's Sandpiper, White-rumped Sandpiper, Long-billed Dowitcher, Solitary Sandpiper, Buff-breasted Sandpiper and Lesser Yellowlegs. For many years the site hosted London's only breeding colony of Black-headed Gulls.

Queen Elizabeth II Reservoir

This is one of a number of reservoirs in south-west London that are owned by Thames Water. When it was under construction it was known as Walton South Reservoir. It was completed in 1962 and covers 128 hectares. It has a major Cormorant roost in winter of up to 480 birds.

Queen Mary Reservoir

Previously known as Littleton Reservoir, it covers 283 hectares and is the largest reservoir in London. It is owned by Thames Water and was completed in 1925; at that time it was the largest reservoir of its kind in the world and a breakwater was constructed to restrain the water in high winds. Because of its size and depth it rarely freezes over and consequently holds a large number of wildfowl in severe winters. However, the effects of continually deepening the reservoir for gravel extraction and sailing activities have resulted in fewer water-birds using the reservoir now.

The large numbers of wintering Goosander and Goldeneye and the vast flocks of Great Crested Grebes are all a thing of the past. Even passage birds are no longer found here in the same numbers although that might be due to decreased coverage, for example 13 White-winged Black Terns occurred here between 1961 and 1978 but none have been recorded since.

Queen Mother Reservoir

This is one of the west London reservoirs and it covers 192 hectares or about 1 kilometre in diameter. It was built in 1976 and originally called Datchet Reservoir until 1981 when it was renamed; it is managed by Thames Water. The London boundary cuts through the reservoir near the far west corner. The reservoir lies within the Colne Valley regional park.

Rainham Marshes

Rainham Marshes is the collective name for an extensive area covering over 600 hectares which includes a RSPB reserve, a working landfill site and salt-marsh along the Thames between Mar Dyke and the River Ingrebourne. The Thames riverside path runs alongside the entire site.

Despite its SSSI status, for many years the future of Rainham hung in the balance, once its former use as a military firing range ceased, with threats of a theme park among others. Conservation organisations lobbied hard to have the site protected and eventually in 2000, the RSPB were able to purchase Rainham, Wennington and Aveley Marshes with the help of lottery funding, opening a 350 hectare nature reserve in 2006. There are long-term plans to landscape the landfill area once operational activities cease.

This is arguably London's premier birding site. With its Thames-side location and mixture of habitats, including damp grazing marshes, old silt lagoons, grassland and scrub, Rainham Marshes attracts more species of bird every year than any other site in London; a record 209 species were seen in 2010. It is important for breeding, wintering and passage wildfowl and waders, and the RSPB reserve has attracted a number of major rarities, including London's only White-tailed Plover plus Collared Pratincole and Sociable Plover amongst others; it has also had regular over-wintering Penduline Tits.

It was largely ignored by ornithologists until 1938. In 1959 mud was dredged from the Thames and dumped onto Rainham, transforming a part of the wet pastureland into muddy lagoons which attracted many waders and wildfowl during the winter and passage months. In the 1970s the area also attracted Lesser Kestrel and Western Sandpiper.

Birds of Rainham Marsh, Noble, K. LBR No. 35 (1970)
Birds of Rainham Marsh – Ten Years On, Dennis, M.K. LBR No. 46 (1981)
Birds of Rainham Marsh Part Three, Dennis, M.K. LBR No. 55 (1990)
Birds of Rainham Marshes – Part 4, Dennis, M.K. LBR No. 65 (2000)

Regent's Park

Regent's Park (along with Primrose Hill) covers an area of 166 hectares. It was acquired by Henry VIII in 1538 as a hunting park, but was not landscaped as a park until the early 19th century. The bird life was first intensively studied in the 1950s when it was discovered that many migrants were seen in the park or flying over. The park's mature trees and lake support a wide variety of birds, including breeding Tawny Owl and Little Owl, a heronry with around 20 pairs, and, thanks to the recent planting of reed-beds around the lake and wetlands, several pairs of Reed Warbler. *The Regent's Park and Primrose Hill*, Webster, A.D. 1911.

Richmond Park

This is the largest green space in Greater London and covers 1,000 hectares. It is 12 miles from St Paul's Cathedral and includes open grassland on low hills, woodland gardens and ancient trees. The area has royal connections dating back some 700 years. It was turned into a deer park by Charles I and the land was enclosed in 1637. The Park has changed little over the centuries but there has been some more recent landscaping, such as the Isabella Plantation, which was created after the Second World War from existing woodland. Richmond Park has been designated as a SSSI and a National Nature Reserve. Over 60 species breed in the park including a healthy population of Skylark.

Ruislip Lido and Woods

Originally called Ruislip Reservoir, the Lido was constructed in 1811 as a top-up for the Grand Union Canal. As one of London's oldest reservoirs it attracted good numbers of wildfowl but after it was turned into a lido in 1936, increased recreational activities resulted in it becoming too disturbed for many birds.

The adjacent Ruislip Woods became London's first National Nature Reserve in 1997 and comprises 305 hectares of ancient semi-natural woodland. The area includes Bayhurst Wood, Copse Wood, Mad Bess Wood and Park Wood as well as other surrounding land.

St James's Park

This is the oldest Royal Park in central London. The area was originally a marsh, fed by the Tyburn River, before being acquired by Henry VIII in 1532 and stocked with deer. The waterfowl collection was started by Charles II and he could often be found in the park feeding the ducks. Early in the 19th century the canal was transformed into a more natural-looking lake covering an area of 5 hectares. In 1837 the Ornithological Society of London presented some birds to the park and erected a cottage for a bird-keeper.

The park is best known for its wildfowl but also supports a fair range of common breeding birds. Gadwall, Red-crested Pochard, Pochard and Tufted Duck all breed regularly, and are derived from birds introduced into the collection. Little and Great Crested Grebe breed most years. The presence of so many wildfowl often attracts other visiting waterbirds such as Long-tailed Duck and Slavonian Grebe. Up until the 1980s the park had a large population of several hundred House Sparrows which were fed by the vast number of visitors but these are now extinct in the area.

St James's Park – A Water Bird Refuge, Oliver, P.J. LBR No. 51 (1986)

Sevenoaks Wildfowl Reserve

This reserve was established by the Wildfowlers' Association of Great Britain and Ireland in 1956 when it took over the management of a 40-hectare site of gravel workings. It introduced Greylag Geese and a population soon became established there. Landscaping of the area produced many habitats and by 1972 there was a total of 107 pairs of wildfowl there.

The Sevenoaks Gravel Pit Reserve, Harrison, J. LBR No. 37 (1972)

South Norwood Country Park

This was created from Elmers End SF after it became disused in 1966–67; it now covers an area of 50 hectares. Conversion to a country park began in 1988 when the area was extensively landscaped. Over 150 species were recorded between 1935 and 1988, including an overwintering Great Grey Shrike.

The Birds of South Norwood Country Park, Birkett, J. 1991 (LBR 56)

Staines Reservoir

The reservoir was completed in 1902; its surface area of 170 hectares is divided in two by a causeway approximately 1 kilometre in length. The south basin has a maximum depth of 9 metres and the north basin 12 metres. The fame of Staines Reservoir began with a series of papers in *The Zoologist* in 1906 and 1908, written by G.W. Kerr. There were few observations made in Edwardian times and regular observations were resumed in the early 1920s, by W.E. Glegg and A. Holte Macpherson, who established Staines's reputation as a major site for wintering waterfowl.

Since then it has become one of London's most-watched sites and consequently a significant number of scarce and rare birds has been found there. Black-necked Grebes are regular between July and April, and divers are often present in winter. Depending on water levels waders may be present, and when one of the basins is partially drained the site acts as a magnet for waders. It was during one of these times that Britain's first wintering Baird's Sandpiper was present; the reservoir has also attracted a number of other American waders including Buff-breasted Sandpiper, Long-billed Dowitcher, Wilson's Phalarope and Lesser Yellowlegs, as well as Sharp-tailed Sandpiper and Collared Pratincole. Migrant gulls and terns are regular while seabirds and large raptors are recorded, most years; rarities include Sooty, Caspian, Whiskered and White-winged Black Terns, as well as Roller and Bee-eater.

Stocker's Lake

The Stocker's Lake complex lies in the Colne Valley near Rickmansworth and consists of three disused gravel pits, Stocker's Lake, Springwell Lake and Stocker's West. These date back to the early 1920s and gravel continued to be extracted until the Second World War. Two lakes situated to the east, Bury and Batchworth, are often grouped within this complex.

Bird Life of the Stocker's Lake Gravel Pits, Chesterman, D.K. LBR No. 42 (1977)

Stoke Newington Reservoirs

These were built in 1833 and cover 17 hectares of open water. Adjacent to the East Reservoir is a community nature reserve run by London Wildlife Trust and opened in 2008.

Thamesmead – see Crossness

Surrey Docks

After these docks at Bermondsey were closed to shipping in 1970 the area was left undeveloped for several decades and became the most important bird area in Inner London. During the first winter of shipping inactivity, over

2,500 ducks used the abandoned docks; most of these consisted of Tufted Duck and Pochard, but rare birds such as Ferruginous Duck were also present. Passerines soon made use of the surrounding 'wasteland' and up to a thousand fed on rough ground during the cold winter in January 1971. Much of the area has been redeveloped but there are still suitable areas for birds such as Canada Water where Common Terns breed.

Walthamstow Reservoirs

This site in the lower Lea Valley consists of 11 reservoirs, the seven on the south side which were constructed at various times between 1853 and 1904, and four on the north side which were built later. Banbury Reservoir is often included in this complex although it has separate access arrangements. The reservoirs are owned by Thames Water and are also managed as a fishery. The reservoirs are designated a SSSI and are internationally important for their wildfowl; they are also a Special Protection Area (SPA) and a RAMSAR site. Plans were announced in 2012 to improve public access and create more wildlife habitat.

Together with the neighbouring Chingford Reservoirs to the north, Walthamstow Reservoirs are a major post-breeding refuge for Tufted Duck during their post-breeding moult. The wooded islands on some of the smaller reservoirs host one of the five largest heronries in the country, as well as large numbers of breeding Cormorants and a small colony of Little Egrets; Lesser Black-backed Gulls breed on one of the islands. The reservoirs have a small wintering population of Common Sandpipers. In 2010 a Dusky Warbler frequented bushes at the north end of the site and Purple Heron has been seen on more than one occasion; rare ducks such as Green-winged Teal and Blue-winged Teal have also been recorded.

Wanstead Flats

At the southern apex of Epping Forest, this 135-hectare mosaic of rough grassland, hawthorn and broom scrub, deciduous copses and football pitches, just north of Forest Gate, proves surprisingly attractive to migrants in spring and autumn. Northern Wheatear, Whinchat, Common Redstart, Ring Ouzel, and Spotted and Pied Flycatchers are all regular in autumn, and Stone Curlew, Wryneck and Dartford Warbler have occurred in recent years. Lesser Spotted Woodpeckers sometimes breed in the copses, and there is a healthy breeding population of Skylarks and Meadow Pipits.

Walton Reservoirs

Originally a complex of concrete-banked reservoirs situated in southwest London consisting of the Chelsea and Lambeth Reservoirs (originally called the East Molesey Reservoirs) along with Bessborough and Knight Reservoirs (originally known as West Molesey Reservoirs). Chelsea and Lambeth have since become redundant and have been landscaped to become a 60 hectare wetland reserve.

Bessborough and Knight Reservoirs are situated close to Queen Elizabeth II and Island Barn Reservoirs and there is a considerable movement of birds between all of these water bodies, particularly when sailing is taking place on the latter. The reservoirs attract a large number of wildfowl as well as passage waders and seabirds.

William Girling Reservoir

Although construction work began in 1936, the reservoir (named after the Chairman of the Metropolitan Water Board) was not completed until 1951. This reservoir, usually just called Girling Reservoir, and the adjacent King George V Reservoir, are known collectively as the Chingford reservoirs. It covers 135 hectares and is a major wintering ground for waterbirds, including nationally important numbers of some species. A flock of Black-necked Grebes, numbering in the 20s, is present during autumn and winter on the reservoir and divers are occasionally seen. There is no access to the reservoir but there is limited viewing from vantage points such as Mansfield Park. There is a vast gull roost here with up to 50,000 birds present during winter.

Wimbledon Common

This site also includes Putney Heath and Lower Putney Common, and in total covers 460 hectares. The Common is the largest expanse of heathland in Greater London; the area also includes a bog and mature mixed woodland. Beverley Brook runs along the western edge of Wimbledon Common and there are a number of ponds on the common.

The Birds of Wimbledon. Olive, C.D. Wimbledon and Merton Annual. 1904

The Birds of Wimbledon Common and Putney Heath 1974–96. Wills, D.L. and Kettle, R.H. LBR No. 61 (1996)

River Thames, Westminster (*Andrew Self*). In central London the Thames has artificial banks; shingle banks exposed at low tide are regularly used by gulls.

Aveley Bay, Rainham (*Dominic Mitchell/www.birdingetc.com*). Intertidal mudflats can be found on both sides of the Lower Thames and support wintering and passage waders.

Crayford Marshes (*Dominic Mitchell/www.birdingetc.com*). Although the banks along the Lower Thames are heavily industrialised there are marshes on both sides of the river.

Rainham Marshes RSPB (*Jill Aldred*). More species are recorded here annually than at any other site in London, a testament to its prime riverside location and range of habitats.

Dagenham Riverside (*Dave Morrison*). In winter, large flocks of waders regularly use artificial roosting sites like this jetty at Dagenham, as well as the stone barges at Rainham.

East India Dock Basin Nature Reserve (*John Archer*). This small reserve is located where the River Lea enters the Thames; it has attracted migrating Roseate Terns.

King George VI and Staines Reservoirs (*Dominic Mitchell/www.birdingetc.com*). The concentration of reservoirs in west London is an important wetland habitat for wildfowl, especially in winter.

Brent Reservoir (*Andrew Self*). Its natural banks provide large marshy areas and reed beds, in contrast to the concrete banks of most other London reservoirs. Consequently, it has more breeding species.

Walthamstow Reservoirs (*Andrew Self*). The wooded islands at Walthamstow hold more breeding pairs of Grey Heron, Little Egret and Cormorant than any other site in London.

Lea Valley (*Jill Aldred*). Much of this area consists of a large complex of disused gravel pits that have been landscaped for wildlife. Additional reed beds have been planted in the hope that the wintering Bitterns will stay and breed.

London Wetland Centre (*Dominic Mitchell/www.birdingetc.com*). This WWT reserve hosts the closest breeding Little Ringed Plover, Lapwing and Redshank to central London and is totally unrecognisable from the four square, concrete-banked reservoirs that were originally there.

Canada Water (*Richard Bonser*). This is part of the old Surrey Docks complex and is the only site in Inner London where Common Terns breed.

St James's Park (*Andrew Self*). The lake in this central London Royal Park attracts wintering ducks and Black-headed Gulls from northern and eastern Europe.

City of London, from Tower 42 (*Andrew Self*). Despite the seemingly hostile terrain, this area has breeding Peregrines and Black Redstarts.

Enfield (*John Dobson*). Farmland in London is largely confined to the outer edges. In north and east London it is mostly arable, while south of the Thames it is mainly pasture, especially along the North Downs.

Wanstead Flats (*Jonathan Lethbridge*). This extensive area of rough grassland holds the largest breeding population of Skylarks in Greater London.

Epping Forest (*Andrew Self*). This is the largest area of ancient woodland in London, but species such as Common Redstart, Wood Warbler and Hawfinch no longer breed there.

Richmond Park (*Jill Aldred*). London's largest Royal Park supports a wide range of breeding and wintering birds.

Stave Hill (*Richard Bonser*). There are hundreds of small parks and urban nature reserves dotted around London, such as this one in Rotherhithe.

Wormwood Scrubs (*David Lindo*). Just 10 kilometres from the centre of London, Wormwood Scrubs has breeding Meadow Pipits and a significant population of Linnets.

Beddington Farmlands (*Peter Alfrey*). This area is undergoing a major landscaping project and is destined to become a country park and nature reserve. Being such a large area surrounded by suburbia it acts as a magnet for birds on passage.

Staines Moor (*Andrew Moon*). One of the largest areas of unimproved neutral grassland in England, this is important for both breeding and wintering birds although it became famous for hosting the UK's longest-staying Brown Shrike.

Bewick's Swans, Rainham (*Paul Hawkins*). A few are seen on passage or in hard weather in most years.

Egyptian Goose family, St James's Park (*Andrew Self*). This species is now well established, even in central London where it competes for handouts from the public.

Tundra Bean Geese, Rainham Marshes (*Paul Hawkins*). Although still rare in London, they have become more frequent since 2000.

Lesser Scaup, Queen Mother Reservoir (*Andrew Moon*). The abundance of waterbodies in London helps to attract unusual wildfowl, including three Lesser Scaup to date.

Great Crested Grebe, Lea Valley (*Andrew Self*). This species is a widespread breeder in wetlands throughout London, even on some of the park lakes in the centre.

Grey Heron, London Wetland Centre (*Czech Conroy*). Some 450–500 pairs breed in London and they are regular at most wetland sites, even in the central London parks.

Little Bittern, Stocker's Lake (*Andrew Moon*). Rare herons are seen in London in most years but this Little Bittern in 2012 was the first for 15 years.

Above: **Red Kite** (*Andrew Moon*). Following their reintroduction into the UK, they are now breeding on the edge of London and are often seen throughout the capital.

Left: **Peregrine** (*Dave Morrison*). This has now become the most frequently seen falcon in some parts of London.

Right: **Common Buzzard** (*Andrew Moon*). Buzzards are breeding in increasing numbers and are regularly seen on passage over London.

Below: **Waders, Rainham Marshes** (*Dominic Mitchell/ www.birdingetc.com*). Large flocks of waders, such as this mixed group of Ringed Plover, Dunlin and Sanderling, use the Lower Thames Marshes in winter and on passage.

Little Ringed Plover, Beddington (*Peter Alfrey*). Sewage farms and gravel pits have provided good habitat for a healthy breeding population.

Turnstone and Dunlin (*Andrew Moon*). Migrant waders are regularly seen at most reservoirs despite the lack of mud.

Gulls, Beddington Tip (*Dominic Mitchell/www.birdingetc.com*). The rubbish tips at Beddington and Rainham attract thousands of gulls, but both are scheduled for closure.

Caspian Gull, Rainham Marshes (*Dominic Mitchell/www.birdingetc.com*). The Lower Thames Marshes and Beddington are among the most reliable sites in the UK for Caspian Gull. This colour-ringed Polish bird appeared in two successive winters.

Bonaparte's Gull, Crossness (*Andrew Moon*). This is one of the two Bonaparte's Gulls that were seen on this stretch of the Thames in 2012.

Gannet, Walthamstow Reservoirs (*Andrew Self*). Seabirds such as this are seen in most years, typically after strong winds.

White-winged Black Tern (*Richard Bonser*). This juvenile spent several days in Hyde Park and Kensington Gardens in 2010, showing that rare birds can even be found in the centre of London.

Shore Lark, Queen Elizabeth II Reservoir (*Andrew Moon*). Reservoirs often attract scarce migrants, such as this Shore Lark.

Ring-necked Parakeets at roost (*Peter Alfrey*). The roost in Hersham held almost 7,000 birds but dispersed after the trees were cut down.

Lesser Spotted Woodpecker, Wanstead (*Jonathan Lethbridge*). There has been a big decline in the breeding population but they can still be found at traditional sites.

Nightjar (*Andrew Moon*). This individual held territory in Teddington in 2006. Only one or two pairs now breed in London, typically in the outer fringes.

Hoopoe, Farthing Down (*Paul Hawkins*). A scarce migrant, one or two are seen in most years.

Above: **Brown Shrike, Staines Moor** (*Andrew Moon*). This individual spent over two months in west London and was only the seventh Brown Shrike to be seen in the UK

Left: **Red-backed Shrike, Greenwich** (*Richard Bonser*). This former breeder is still seen on migration in most years.

Top right: **Black Redstart, Rainham Marshes** (*Dominic Mitchell/www.birdingetc.com*). Several pairs breed in central London and elsewhere in industrial areas.

Right: **Tree Pipit** (*Dominic Mitchell/www.birdingetc.com*). The breeding population now consists of only a few pairs.

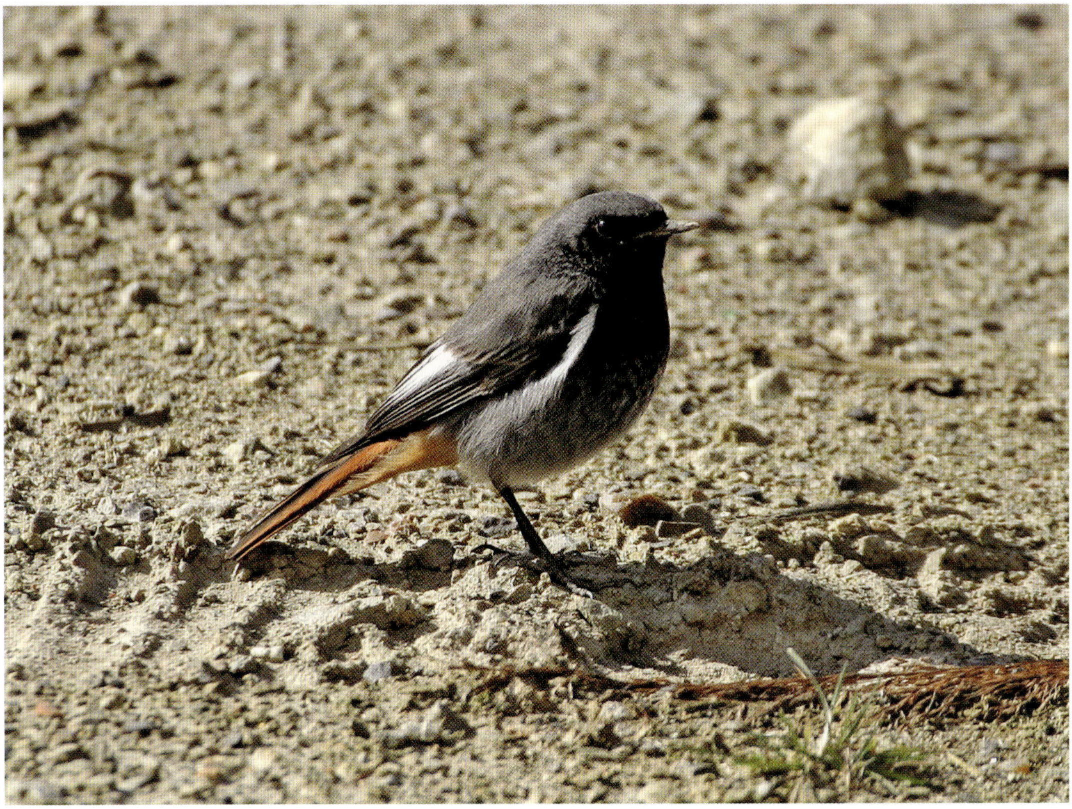

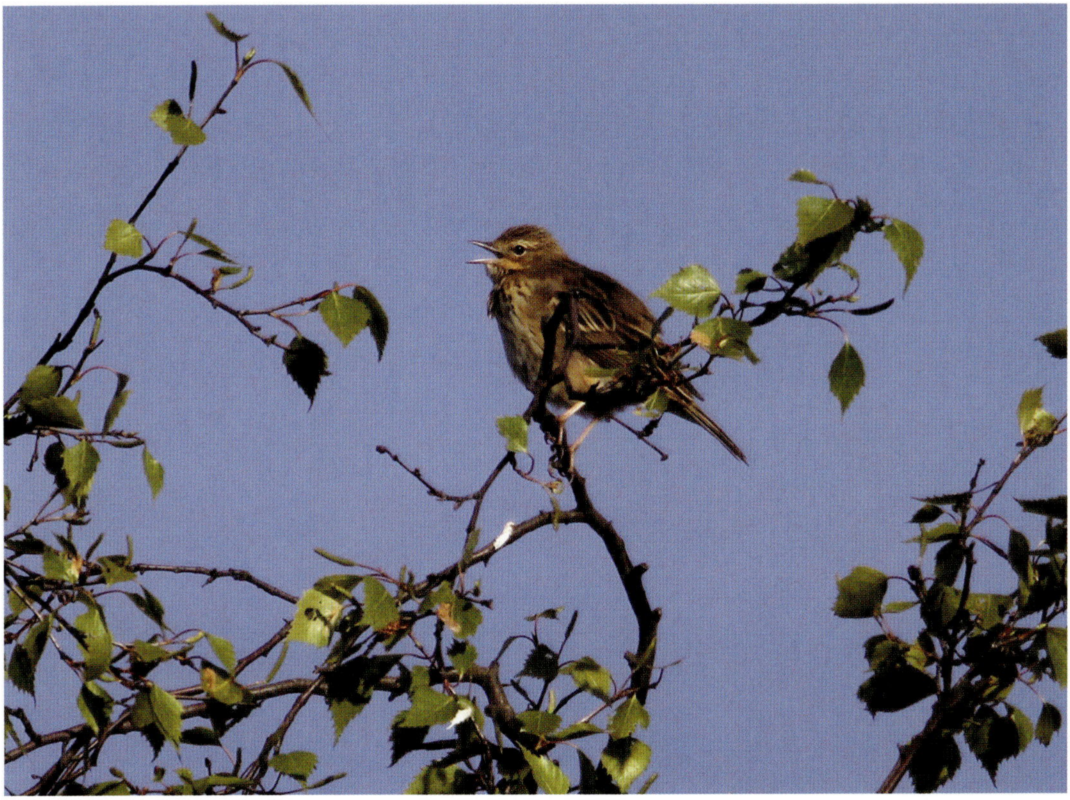

Above: **Bearded Tit, Rainham Marshes** (*Paul Hawkins*). This is mainly a wintering species in London but it has bred on a few occasions.

Left: **Penduline Tit, Rainham** (*Jonathan Lethbridge*). These are seen at Rainham in most years.

Top right: **Eastern Crowned Warbler, Hilfield Park Reservoir** (*Jack Fearnside*). The rarest bird to be pulled out of a mist net in London, this was just the second British record.

Right: **Willow Warbler** (*Denis Tuck*). There has been a dramatic decrease in breeding numbers in southern England recently and it is now largely a passage migrant.

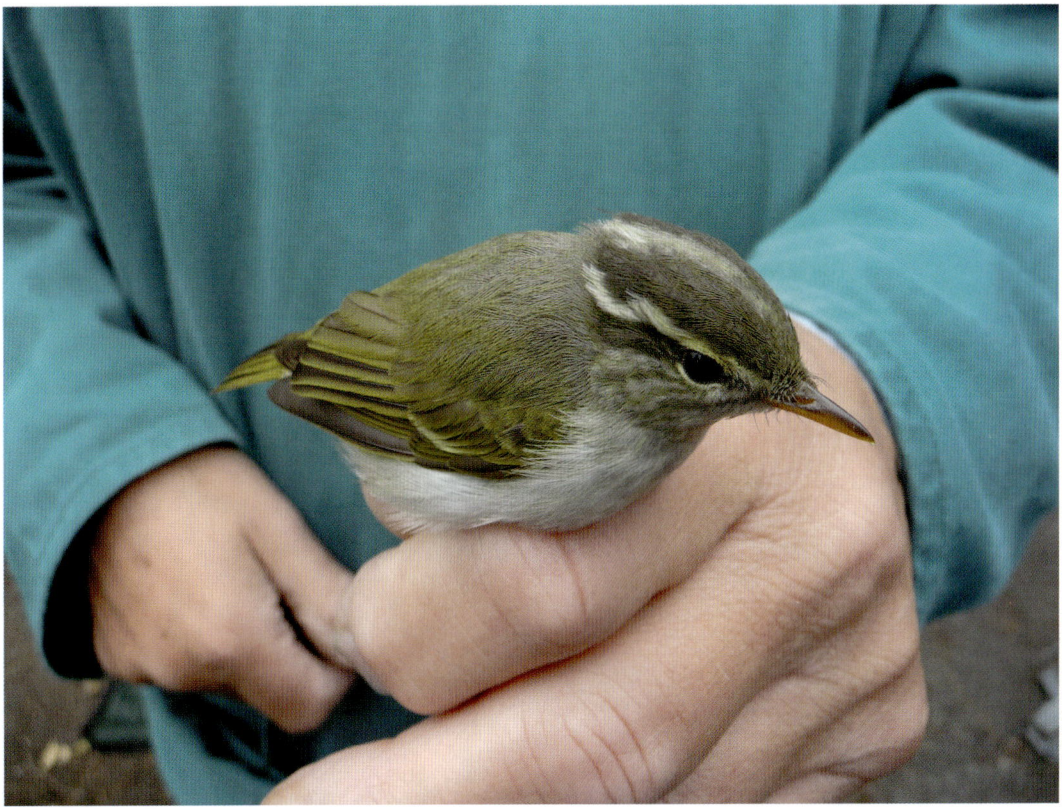

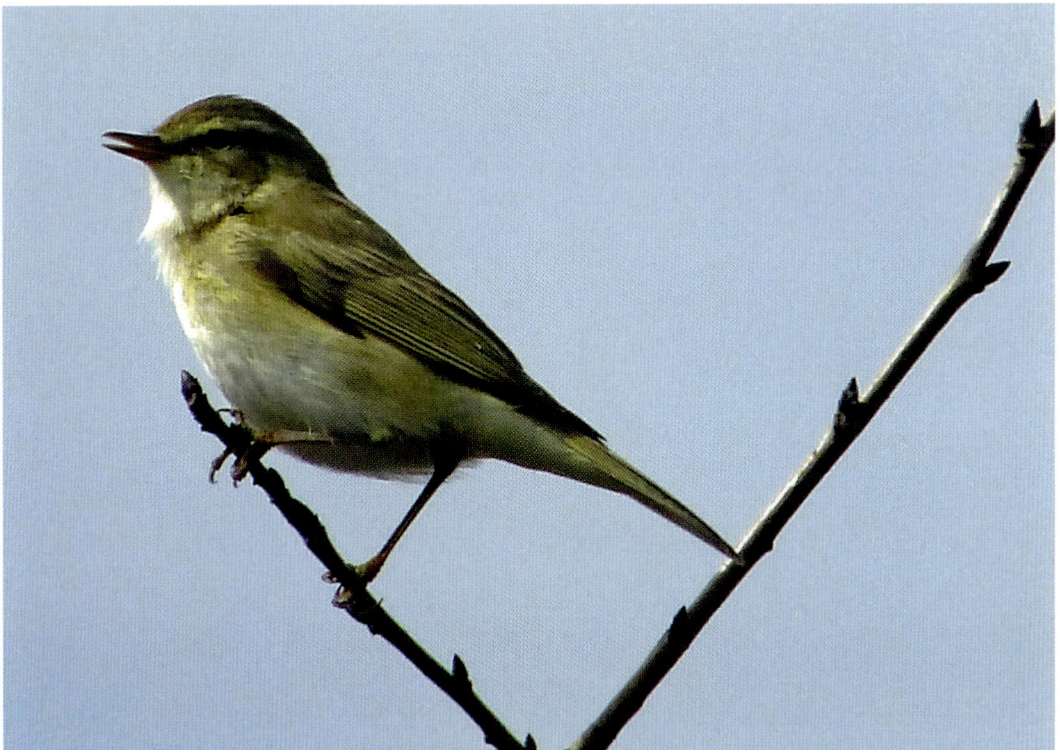

Melodious Warbler, Leyton (*Martin Blow*). This scarce migrant spent two days in a tiny, scrubby patch of urban east London in 2012.

Spotted Flycatcher, Wanstead (*Jonathan Lethbridge*). Although widespread on passage, the breeding population has dwindled to just a few pairs.

Linnet, Rainham Marshes (*Jonathan Lethbridge*). Like many other passerines, the breeding population in London has seriously declined.

Tree Sparrow, Beddington (*Peter Alfrey*). Beddington is one of only two breeding sites in London.

Reed Bunting, Rainham Marshes (*Paul Hawkins*). A widespread but declining breeding species.

Baillon's Crake, Rainham Marshes (*Dominic Mitchell/www.birdingetc.com*). This juvenile in autumn 2012 was the first in London for over 120 years.

Little Owl, Brent Reservoir (*Andrew Self*). A pair bred at this site in 2009 and is part of a recent colonisation of the more urban areas of London.

Buff-bellied Pipit, Queen Mother Reservoir (*Dominic Mitchell/www.birdingetc.com*). This is the most recent addition to the London list, but perhaps not unexpected following the large influx in autumn 2012.

Northern Wheatears (*Dominic Mitchell/www.birdingetc.com*). These birds are regularly seen on migration throughout London and there are occasional breeding attempts.

House Sparrows (*Dominic Mitchell/www.birdingetc.com*). After a widespread population crash they have since recolonised most of the suburbs but are still largely absent from central London.

Map of London

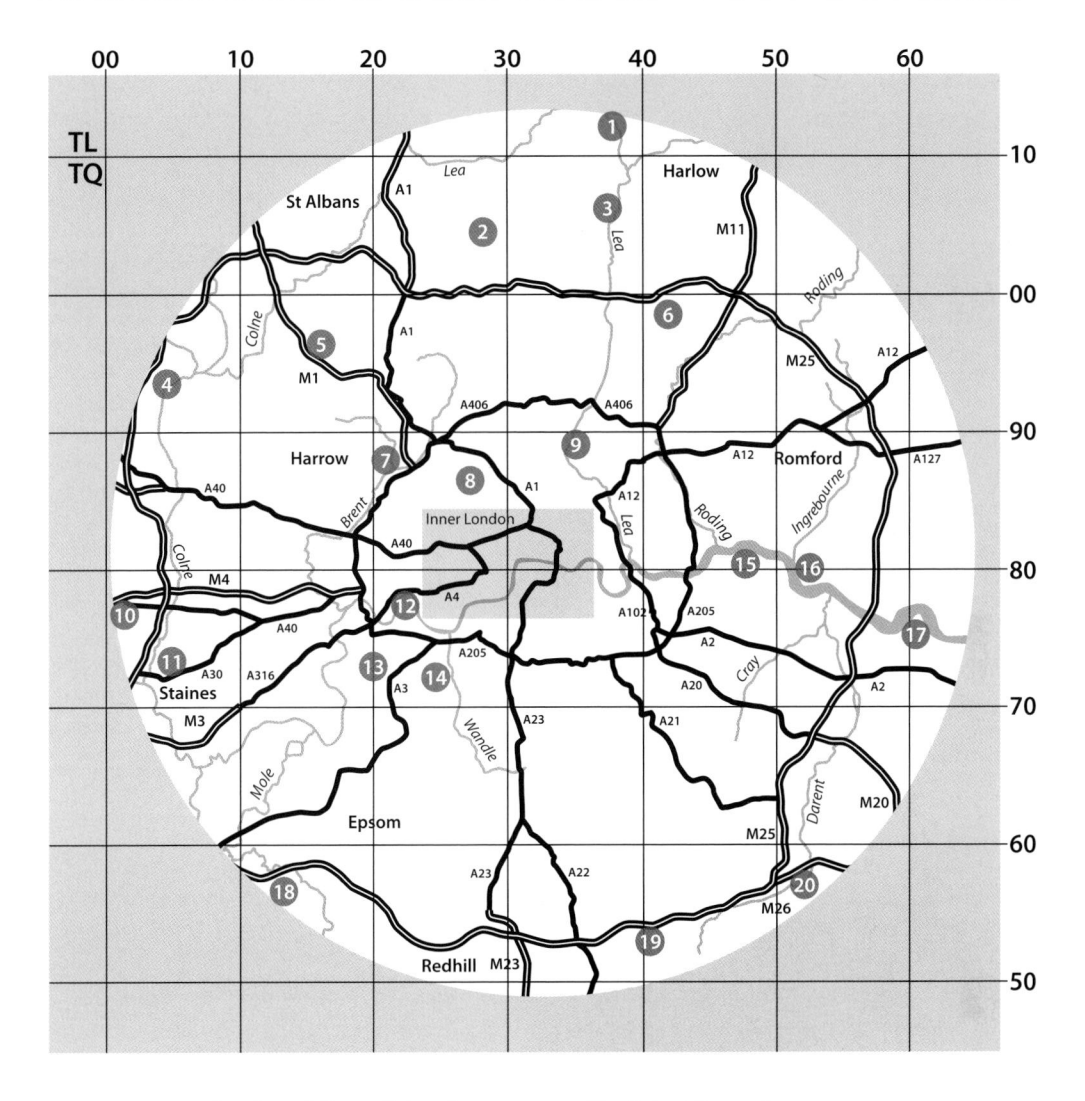

1. Amwell Quarry Nature Reserve
2. Northaw Great Wood
3. Cheshunt Quarry Nature Reserve
4. Stocker's Lake
5. Hilfield Park Reservoir
6. Epping Forest
7. Brent Reservoir
8. Hampstead Heath
9. Walthamstow Reservoirs
10. Queen Mother Reservoir
11. Staines Reservoir
12. London Wetland Centre
13. Richmond Park
14. Wimbledon Common
15. Crossness
16. Rainham Marshes
17. Swanscombe Marsh
18. Bookham Common
19. Limpsfield Chart
20. Sevenoaks Wildlife Reserve

SPECIES ACCOUNTS

INTRODUCTION

The area covered by this book follows the recording area used by the London Natural History Society (LNHS), i.e. a 20-mile radius from St Paul's Cathedral; this is known as the 'London Area'. Watsonian vice-counties are also used by the LNHS but these are not used in this book to make it easier for the reader. References to counties, therefore, e.g. Essex, refer to all of that county and not just the portion that lies within the 20-mile limit. Additionally, reference is made to Inner London, which is a designated LNHS recording area in central London, being a rectangular area measuring 8 miles east to west and 5 miles north to south and centred on Charing Cross.

There are 369 species on the London List, all of which are described in this section. This is followed by a summarised list of escaped birds and details of other birds that have not been accepted onto the London List. Research for this book has uncovered three new additions to the London List: Snow Goose, Black Grouse and Black Guillemot, while the records of American Wigeon were reviewed and rejected. Most records of rare birds since the middle of the 20th century have been considered by the British Birds Rarities Committee or at a local level by the LNHS and/or the relevant county bird club. These decisions have generally been followed in this book and any exceptions are detailed. Earlier records, especially prior to 1900, have usually been subject to critical review by the authors of the respective county avifaunas. In some cases these are not always in agreement; contentious records are discussed within the text. There are also a number of records that have been listed in earlier London Bird Reports or in the LNHS book *Birds of London* that do not fall within the 20-mile limit; these have been omitted from this book. All additions/deletions to the London List, along with other significant records, have been agreed with the Rarities Committee of the London Bird Club, which is part of the LNHS.

The taxonomy generally follows the BOU British List, although some familiar English names have been used as an alternative. Monthly charts have been used for some species and show the estimated total number of individual birds seen each month, summing the monthly totals for ten years (or in the case of exceptional rarities for all years in which they have been recorded).

CURRENT STATUS AND BREEDING DEFINITIONS

Each account includes a brief statement about the abundance of the species in the London Area, its presence throughout the year and, where applicable, a breeding status; the latter definition covers the years 2000–2009.

Abundance

Very rare – occurred less than five times in total
Rare – occurred 5–25 times in total
Scarce – occurred more than 25 times in total, not seen annually
Uncommon – annual, seen at some sites
Common – annual, seen at many sites
Very common – annual, seen at almost all sites

Resident – present all year round
Winter visitor – present in winter months and on passage
Summer visitor – present in breeding season and on passage
Passage migrant – mainly seen on passage
Partial migrant – some resident, some migratory
Visitor – occasional visitor
Vagrant – accidental visitor

For rare and some scarce birds, a total is also given. This figure is the total number of individuals known to have occurred in the London Area. In most cases it is the all-time figure but where the species was formerly more common in the London Area, the total applies from a specific date. In some cases the total cannot be exactly calculated due to unspecific numbers being recorded, so the minimum figure is given.

BREEDING STATUS IN THE LONDON AREA

Historic – only bred prior to 1900
Former – bred in the 20th century but not since 2000
Rare – only a few records since 2000
Scarce – less than 10 pairs annually
Localised – 10–100 pairs annually or less than 50 colonies/sites
Widespread – over 100 pairs annually at many sites

To give each species some context, a short definition of the range is provided.

ABBREVIATIONS

The following abbreviations have been used in this book:

BBRC	British Birds Rarities Committee
BTO	British Trust for Ornithology
GP	Gravel Pit(s)
RSPB	Royal Society for the Protection of Birds
SF	Sewage Farm
SP	Sand Pits
WeBS	Wetland Birds Survey

Mute Swan *Cygnus olor*

A common resident and widespread breeder

Range: resident across northern and central Europe.

Historical

Mute Swans appear to have been given royal status in the 12th century, after which, if a privately owned swan escaped it became the property of the crown. The office of 'Keeper of the King's Swans' has been in existence since at least 1378.

The earliest reference in the London Area was in 1246 when Henry III issued a mandate referring to swans belonging to the King and to the Hospital of Hampton. They were mentioned again in the reign of Edward III who made a grant 'to the Warden and College of the King's Free Chapel of Wyndesore of all swans flying, not marked, within the water of Thames between Oxford and London Bridge, as fully as these should pertain to the King by reason of his right and prerogative' on 20 June 1356. In the winter of 1496–97, the secretary of the Venetian Ambassador wrote about how beautiful it was to see one or two thousand swans on the Thames.

In the early 16th century Dunbar wrote about them in his poem 'In Honour of the City of London'. Swan-keeping had largely fallen out of favour by 1650 and by 1793 general ownership had ceased; the last three owners became the sole keepers of all the swans on the Thames. Towards the end of the 19th century Mute Swans were still taken from the Thames to be fattened up for the royal larder at Windsor.

Yarrell described the swanneries in Middlesex and gave a full account of the Mute Swan in his book on British birds. In the 19th century all Mute Swans came under the control of swanneries, which were owned by the Crown, public companies and private individuals. Swans on other waterways were owned by different companies. For example, during the 19th century those on the River Lea at Hackney were owned by the East London Waterworks Company. Swan-upping was an annual event in which unmarked swans on the Thames between Battersea and Staines were rounded up, caught, marked and then released; the highest total of newly marked birds was 167 in 1894. The population of swans on the Thames remained fairly constant between the mid-19th century and the mid-20th century at 400–450 apart from in the 1880s when it dropped to 350 and in the 1920s when it peaked at 732 in 1927.

20th century

During the first two decades the numbers that were rounded up varied from 97 to 222. After this there was a steady increase and by 1938, a total of 418, including 60 cygnets, were rounded up. During and after the war there was a marked decrease in numbers and they continued to drop until 1951 when 112 were counted between Putney and Staines. However, on the stretch of river between Putney and Waterloo, 157 were present that year; it is thought that they were attracted to the food provided by visitors to the Festival of Britain celebrations. Over the next few years, a large increase was noted and the population between Putney and Staines reached 479 in 1954.

Elsewhere in the London Area they were fairly widespread but not common. In the 1930s Glegg stated that it was a common resident wherever there was suitable water and was most numerous on the Thames, but also bred on the River Colne, on reservoirs, gravel pits and many ponds. He added that all of these birds were descended from captive individuals but sometimes birds arrived in the London Area that he considered to be wild. There had been an increase in breeding sites since the previous decade, some of which was due to the creation of flooded gravel pits; other swans began colonising ponds in built-up districts.

By the middle of the century, large numbers could be found away from the Thames in Kensington Gardens and Richmond Park where they exceeded 50. In 1956/57 the peak count between Putney and Richmond was 426. In 1956 a census located 188 nests in London, a much greater breeding density than the national average. However, disaster struck that year when oil barges sank in the Thames at Battersea on two separate occasions. After the first incident in January, 200 swans were contaminated with oil and a number died. During the second incident in December, more than 800 were collected for treatment and at least 243 died. Numbers had recovered to some extent by the winter of 1959/60, but began to decline in the 1960s. The vast majority of Mute Swans on the Thames were resident in west London so a group of 43 off Millbank on 28 August 1963 was most unusual. By the mid-1960s the number on the Thames had decreased to little more than 100. However, these figures did not include the Lower Thames, where numbers increased markedly in the late 1960s–early 70s. For example, in late 1969 unprecedented large numbers were seen by a sewer outfall on the Lower Thames at Woolwich; by November there were 200.

In 1970 there were signs of a continuing decline as none bred in Richmond Park for the first time since the war; the largest concentration was on the River Lea at Bow Creek where about 300 were present on 22 November. The dangers of overhead wires became very apparent when nine died within two weeks from collisions in the Broxbourne area in 1971. On the Thames at least 208 were counted at Woolwich in October 1971 but the large numbers wintering on the western section had declined to a mere handful. In 1973 the maximum count on the Thames was 467 between Bow Creek and Woolwich Ferry. A census in 1978 discovered a total of 69 breeding pairs.

By the start of the 1980s the large numbers that had been seen on the Thames had substantially dwindled. The main cause of the decline was attributed to lead poisoning as a result of swans ingesting anglers' discarded lead weights. Lead weights were banned from 1987 and by the late 1990s the breeding numbers reported in the LBR had recovered to 119 pairs. There was very little change in the distribution between the two breeding atlases; the bulk of the population was along the Thames and in the Colne and Lea valleys.

21st century

By 2003 around 166 breeding pairs were located along with more than 600 non-breeding birds. Since then there appears to have been a reduction as just over 100 pairs were reported in 2010, although it is unclear how far this figure reflects incomplete coverage.

The largest concentration was 182 at Harrow Lodge Park in February 2009, whilst on the Thames there were 93 between Eel Pie Island and Teddington Lock in 2003. In central London there is a large flock in Kensington Gardens which peaked at 84 in 2010.

Bewick's Swan *Cygnus columbianus*

A scarce winter visitor

Range: breeds northern Russia, mostly winters western Europe.

Historical

In the 19th century this species was often confused with Whooper Swan and there were no documented sightings of Bewick's Swan in London.

20th century

The first record was in 1931 at Gravel Pit Pond, Wanstead, from 5 February; the bird remained there until 14 April, although it was occasionally absent. Four years later, an immature was seen on Staines Reservoir on 27 October. In the 1940s there was a sudden increase in records, beginning in 1944 when a family party of six stayed at Ruislip Lido from 17 to 28 February. They were seen annually from 1946 to 1949, including a flock of 16 which spent two days on Barn Elms Reservoirs in 1946 and one on the Serpentine, Hyde Park, in Inner London on 6 March 1948.

After a gap of six years with no records, there were four sightings in early 1954, all on large reservoirs in the western half of London. There was a large influx in early 1956 during very cold weather when their normal wintering area in Holland froze over. A flock of 49 was seen on both of the Chingford Reservoirs on 18 February; 25 were present at Cheshunt GP on 26 February; and there were 13 at Stoke Newington Reservoirs on 11 February. During the northward passage that year a flock of 33 was seen at Hilfield Park Reservoir on 1 April, four of which remained until 20 April, and 22 were at Walthamstow Reservoirs on 1 April. There were another three records in the late 1950s, following which sightings increased.

In the early 1960s most records came from the Lea Valley but the largest flock was 19 at Brent Reservoir on 5 January 1962. There was another influx in the very cold winter of 1963, particularly in the Barnes area where 17 arrived on 12 January, most of which remained for two weeks, occasionally being seen on the Thames; elsewhere, a group of 24 was seen at Queen Elizabeth II Reservoir on 20 January and small numbers were recorded at another six sites. In 1967 a flock of 17 birds was tracked flying over Dartford, Queen Mary Reservoir and Staines Reservoir on 8 January.

By the start of the 1970s Bewick's Swans had become virtually annual visitors, typically in small flocks. In 1979 there were nine records, including flocks of 80 and 40 flying over Hyde Park/Kensington Gardens on 23 February, 19 at Nazeing on the same day and 13 over Holmethorpe Sand Pits on 1 January; additionally, a flock of 20 wild swans flew over St James's Park on 3 January.

In the 1980s they were scarce but regular winter visitors; typically flocks were seen flying over or just staying for a day but occasionally they would linger. In 1980 a flock of 38 was seen at Staines Reservoirs on 10 February and up to 13 were seen at seven other sites, including seven that remained in the Lea Valley between 29 November and 7 December. In 1984 a flock of 35 flew east over Bushy Park on 4 March and there was an influx in early 1985 with sightings at 12 localities, including flocks of 19 and 15; several birds remained for up to a week. In 1987 there was an influx of about 38 in January due to very cold weather. In 1989 a flock of 29 flew over East Molesey on 11 March and there was a good arrival in the last three months of the year with seven sightings.

There was a surge of records in 1991, including a small influx on 27 October, with 18 over Fairlop Waters and several records in the Lea Valley. In 1992 a flock of 34 flew over Cobham on 16 March and there were 21 at Hilfield Park Reservoir on 16 February. In 1993 the peak count was 30 on Great Stew Pond, Epsom, on 9 March, one of the largest flocks actually grounded in London. In 1994 a colour-ringed bird was seen at Maple Cross in December, which had been ringed on the Pechora River, Russia, in 1992. There was a large influx of 101 at the beginning of 1997, including flocks of 40 over Queen Elizabeth II Reservoir and 32 over Amwell on 1 January and a flock of 19 in the Lea Valley between 3 January and 2 February.

21st century

Between 1995 and 2005 there was a decrease of 27 per cent in the numbers wintering in north-west Europe, possibly linked to climate change, and this was reflected by fewer sightings in London. Typically, most flocks are either seen flying over or are short-staying. For example, in 2001 there were 38 in the last two months of the year, none of which remained for more than a day; the largest flock was 27 at Netherhall Gravel Pit on 14 December. There were two large flocks in 2005 when 22 were at Rainham Marshes on 28 December and 20 flew north over Elstree on 7 March. An influx involving four flocks and totalling 66 birds occurred on 3 December 2010, including 27 past Crayford and Rainham.

Most are seen on wetland sites in the river valleys, especially along the Lower Thames, Colne and Lea, but they are occasionally seen in more urban settings like the flock of eight that landed on the lake in Battersea Park on 5 November 2005.

Facts and Figures

The earliest returning group was of five birds at Rye Meads on 4 October 1962.
The largest flock was of 80 birds over Hyde Park/Kensington Gardens on 23 February 1979.

Whooper Swan *Cygnus cygnus*

A scarce winter visitor

Range: breeds Iceland and northern Europe, winters Europe.

Historical

Fossil evidence of this species dating back some 300,000 years has been found at Grays and Ilford. In the 19th century Harting described this as 'formerly an occasional winter visitant to the Thames and our reservoirs, but none have been seen for many years', adding that it was only during severe weather that they came so far inland. However, some early records of 'Wild Swans' could have related to either Whooper Swans or Bewick's Swans as they were frequently confused.

The earliest record was of one shot at Dartford Marshes in the winter of 1829/30. In 1843 Bond included it on a list of birds that had occurred at Brent Reservoir; at the same location there was a flock of five in 1846, one of which was captured, and three on 28 January 1871. In 1847 Ruegg stated that it was regular on the marshes at Woolwich. In 1862 two were killed at Wraysbury and in 1892 a flock of about 30 flew over Hertford in late December.

20th century

There were no further records until 1941 when four were seen on Connaught Water on 26 January. The only other record in the first half of the century was three years later in 1944 when there was one on a gravel pit at Riverhead on the relatively late date of 9 April.

From the mid-1950s onwards there was a large increase in the numbers occurring in London. In March 1955, eight were seen in the Lea Valley, and the following year there was a large national influx associated with very cold weather in February; they were seen at around 20 locations in London, mainly in small groups but there were also 22 at Fishers Green on 22 February; 11 at Cheshunt GP on 26 February; 11 at Panshanger on 11 March; and flocks of 29 and 19 flew over King George V Reservoir on 1 April.

During the 1960s the typical pattern was for fairly small groups or single birds to be seen briefly with occasional sightings of larger numbers flying over. In 1961 a flock of 22 was seen at King George V Reservoir during a cold spell on 28 December. During the extremely cold winter in early 1963 they were seen at 16 locations between January and March; most of these were fairly small groups or singletons but on 22 March about 50 flew over Chessington; these were probably migrants returning to their breeding grounds.

The only large flocks in the 1970s were of 13 birds at the Chingford Reservoirs on 1 December 1973 and 14 at Staines Reservoir on 1 January 1979. During the 1980s most sightings were of very small numbers in the Lea Valley. They remained scarce but more widespread in the 1990s with the highest count being a flock of 13 flying past Dartford Marsh on 23 November 1990. About 19 were seen in 1997, including two flocks on Brent Reservoir in December.

21st century

By the beginning of the new millennium they had become much rarer in the London Area and none were seen until 2002 when five were on Island Barn Reservoir on 18 October. One or two are seen in most years but there were 16 in 2005, including 10 at Walthamstow Reservoirs on 21 January. Most occur on reservoirs and gravel pits beyond the suburbs, particularly in the Lea Valley. They are rarely seen in Inner London but two were seen flying over Paddington Green on 18 December 2005.

Escapes

In 1968 three were seen at Hampstead Heath on 12 December but were later discovered to be captive birds. Two were present during the summer at Higher Denham in 1984. In 1993 there were three reported escapes, including one at Stanborough Lakes from July to at least December 1994. Since then one or two escapes are seen in most years and are often long-staying birds. For example, one was resident at the Wetland Centre between at least 1997 and 2006.

Facts and Figures

The largest flock was of about 50 birds over Chessington on 22 March 1963.

Bean Goose *Anser fabalis*

A rare winter visitor

Range: The tundra form breeds in the Siberian tundra and small numbers occasionally winter in the UK. The taiga form breeds in the taiga forests of northern Europe and Siberia; there are two regular wintering flocks in the UK, in Scotland and the Norfolk Broads.

Historical

Harting thought that it was more than probable that Bean Geese, which were more frequently seen in winter than Greylag Geese, had 'many times been killed in Middlesex', but had no direct proof. Bond, writing in 1843, knew of several records of grey geese being killed at Brent Reservoir but was unable to examine the specimens to assign them to species. The first definitive record came in 1847 when Ruegg stated that they were regular in the Woolwich area. A single Bean Goose was said to have been shot on the lake at Munden Park during the winter of 1890/91 but no further details about this bird are known and it was not accepted by Sage as a confirmed record. In 1892 one was shot in winter at Cannon Farm, Fetcham.

20th century

There were no more sightings until 1940 when a flock of eight was seen on the iced-over Staines Reservoir on 10 February; the number had increased to 11 by 15 February and they remained there for a further week. In 1958 there were two at Beddington on 4–6 January; they were described as having noticeably long necks. During the severe winter in 1963 a flock of seven remained on Queen Elizabeth II Reservoir from 21 to 26 January and a single bird was at Walton Reservoirs on 2 March. In 1978 one flew over Wraysbury Gravel Pits on 31 December; it was considered to be wild due to the onset of severe weather, which subsequently resulted in an influx of Bean Geese into England in early 1979; two were seen at Maple Cross on 16 February.

The only records in the 1980s were in successive years in the middle of the decade, all in west London. A flock of 11 was on Staines Moor for a day during a period of severe weather on the Continent in 1984. The following year one flew east over Wraysbury Reservoir and two were present on New Butts Green Farm, Horton; these records again coincided with a large movement of birds escaping the very cold weather on the Continent.

The first confirmed records of Tundra Bean Goose were in the 1990s. In 1993 a group of five was present at Holmethorpe on the morning of 28 November for about two hours before flying off north-west; there was an influx of Bean Geese in the country at the time and a flock of White-fronted Geese was also present at the same site. In 1997 there was one at Beddington on 1 February.

21st century

Since 2000 sightings have increased and the species has become near-annual; most of these have been Tundra Bean Geese. There were two sightings in 2003 including a flock of 17 that flew north-east over Queen Mother Reservoir, the largest group to be seen in London. In December 2004 there was a small influx of Tundra Bean Geese on the East Coast; two birds at Rye Meads and one at Staines were probably all of this form although only the latter was confirmed. There were flocks at both ends of the year in 2007 at Rainham; the flock of 12 in December arrived on the same day as two flocks of Barnacle Geese.

'Tundra Bean Goose' *A. (serrirostris) rossicus*

(27)

1993: Holmethorpe Sand Pits, five on 28 November
1997: Beddington Farmlands, on 1 February
2003: Island Barn Reservoir, on 1–2 and 20 March
2004: Staines Reservoir, on 8–9 December
2007: Rainham Marshes, five on 27 January
2007: Rainham Marshes, 12 on 21–22 December
2010: Rainham Marshes, two on 27 January

Unassigned to race:

Pre-1847: Woolwich
1892: Fetcham, shot
1940: Staines Reservoir, eight on 10–11 February and 11 on 15–22 February
1958: Beddington SF, two on 4–6 January
1963: Queen Elizabeth II Reservoir, seven on 21–26 January
1963: Walton Reservoirs, one on 2 March
1978: Wraysbury Gravel Pits, one on 31 December
1979: Maple Cross, two on 16 February
1982: Staines Moor, 11 on 16 January
1985: Wraysbury Reservoir, on 13 January
1985: Horton, two on 20–21 January
2001: King George V Reservoir, six on 20 January
2003: Queen Mother Reservoir, 17 on 24 February
2004: Rye Meads, two on 2 December
2010: Beddington SF, 14 on 1 December

There have been no confirmed records of 'Taiga Bean Goose' *fabalis*.

Escapes

There have been a number of Bean Geese that were presumed to be escapes; most were long-staying single birds, although it is possible that some may have been wild. Most records are from the Colne and Lea Valleys and Sevenoaks Wildfowl Reserve.

Pink-footed Goose *Anser brachyrhynchus*

A scarce winter visitor

Range: breeds Greenland, Iceland and Svalbard, winters western Europe.

20th century

This was a rare winter visitor in the first half of the 20th century and was recorded on just seven occasions, the first being a group of five in the Darenth Valley in October 1905, one of which was shot at South Darenth. There were no further sightings until 1929 when four were seen on Staines Reservoir on 18 January and were still present the following day; later that year six flew over between Runnymede and Stanwell on 3 December. At the end of 1938, there were seven at Beddington on 22 December, and probably a different flock of 10 on 23 December, which left the following morning; there were also three at Barn Elms from 27 to 30 December, which were described as very tame although they were considered to be wild birds due to the very cold weather at the time. In 1941, two skeins totalling about 100 birds flew over Great Parndon on 15 March.

In the 1950s they started to be seen a little more frequently. In 1950 there was one at Staines Moor from 24 to 27 December, often accompanying a flock of White-fronted Geese. A flock of at least 30 flew over Uxbridge on 30 January 1954 and four flew over St Albans on 11 November in the same year. During the cold winter in 1956 a skein of 75–80 flew south over Hainault on 28 January and a similar-sized flock flew north over Coulsdon on 26 February.

No more were seen until 1962 when a flock of 54 flew south-east over Brent Reservoir on 5 January. The following winter was extremely severe and on 1 January a flock of at least 70 flew north over St Pancras Mortuary and singles were seen at Beddington on 6 January and Girling Reservoir on 23 February; in spring four were seen flying north over Rye Meads on 11 May. Between 1965 and 1980 there were a further 10 records, none of which exceeded five birds. The only sightings in the 1980s and 90s were three that flew over Stanstead Abbotts on 19 January 1982, and skeins of 15 and 45 that flew over Charlton during stormy weather on 2 January 1998.

21st century

Since the start of the century they have remained a scarce visitor. In 2001 three were seen on Staines Moor with White-fronted Geese on 16 February. There were three small groups in 2003, including two on the lake on Wimbledon Common on 26 February. There were no further records until 2009 when nine flew over Chafford Hundred on 16 September. In early 2010 there were several flocks on the Lower Thames Marshes including 71 at both Belvedere and Rainham on January 21st.

Escapes

Single Pink-footed Geese seen during the summer or remaining for long periods are likely to be escaped birds. These have been recorded in the London Area in most years since 1969. In 1990–91, for example, there were probably four different escapes present for long periods, including in midsummer.

White-fronted Goose *Anser albifrons*

A scarce winter visitor

Range: breeds Greenland and northern Europe, winters Europe, Asia and North America.

Historical

Harting quoted an experienced wildfowler who said that White-fronted Geese did not venture as far inland as other grey geese, preferring wetland areas near the coast where they were regular winter visitors. The earliest record was of one shot on the Thames at Chertsey in February 1847. There were only two other 19th-century records: one of a bird in St James's Park in April 1859; and one of a flock of 50 in a large field in Acton on 25 December 1860, about 20 of which remained for several days.

20th century

In the first half of the century there were about 16 records, the first of which was on 2 March 1909 when a flock was heard flying over Belgravia. In the 1920s and 30s there were five records, none of which exceeded a dozen birds. There was a gradual increase in sightings beginning in the 1940s when they were seen in six different years, including a flock of 53 over Sundridge Park on 2 January 1940 and a flock of 15 over Lord's cricket ground on 7 February 1945.

From the 1950s they started to be seen more regularly. On 20 December 1950 around 60 were present on Staines Moor with some remaining there for a week. In 1954 a flock of 64 flew over Sutton and Beddington on 1 February, and three skeins totalling 80–100 birds flew east over Ruislip on 20 March. The latter record hints at a likely source of some of the White-fronted Geese seen over-flying the London Area as they probably wintered at Slimbridge and were returning to their breeding grounds in northern Russia. The same year, a first-winter bird remained on Hampstead Ponds from 3 January until 7 May; despite the unusual location and length of stay it was considered to be a wild bird as there were no breeding ones known in captivity anywhere in Europe at the time. In 1956 severe weather brought an influx of wildfowl into the London Area and White-fronted Geese were recorded flying over at five sites.

In 1960 a group of 11 landed in Regent's Park on 27 January and joined the captive geese before flying off again. In 1961 two skeins totalling 67 birds flew north-east over Epsom on 6 February. In the severe winter of early 1963 there was an unprecedented influx of geese into London, including hundreds of White-fronts; the largest skein was 200 over Uxbridge on 16 January. The next influx was in January 1966 when about 100 flew over Sevenoaks on the 14th; a large flock flew over Brent Reservoir after dark three days later and 60 flew north over Romford on the 23rd. On 4 March 1968 about 180 flew north-east over Ewell. In 1969 there was a widespread movement of White-fronts in January with five flocks of up to 50 birds, including about 30 over Regent's Park on the 13th, and in February 65 flew over Stanwell Moor.

During the 1970s only small numbers were seen until 1979. Cold weather brought a large influx into the London Area beginning on 27 January when 26 flew south over Girling Reservoir; the next day nine flocks totalling about 350 flew over various sites. On 23 February about 200 flew over Hyde Park and up to 41 arrived at Rainham Marshes where they stayed until 4 March, when 200 were seen over Dartford Marsh.

There were no further large flocks until 1984 when about 70 flew over Kew Gardens on 9 January. There was an influx in January 1985 due to severe weather on the Continent; the peak count was on the 12th when a flock of about 210 flew over Dartford Marsh, Queen Mary Reservoir and Stanwell Moor. Another large influx occurred in early 1987 when there were freezing conditions on the Continent; on 1 January there were 65 at Rainham Marshes and on the 12th a huge flock of about 700 flew south-west over Regent's Park. At the end of the year 75 flew east over Wraysbury GP on 8 December.

In the 1990s there were typically just a few sightings each year apart from in 1993 when there was an influx between late January and March with records at 11 sites, including a few small groups that stayed for several days in the Colne and Lea Valleys; and in 1996 about 15 flocks were seen. The maximum count of the decade was a flock of 137 over Dartford Marsh on 6 January 1998.

21st century

Although they are recorded virtually annually, the numbers vary considerably with occasional influxes during cold weather. Most are only seen flying over but small groups sometimes remain for several weeks. In early 2000 there was a large influx into the London Area on 16 January; a group of 73 flew north over Dartford Marsh and a few minutes later 354 flew west over Rainham Marshes where they split up: 244 were seen shortly afterwards over Barking Bay and the other 110 flew over Dagenham Chase and half an hour later were seen flying north over Brent Reservoir. Also that morning, probably the same flock was seen over Southgate after leaving Brent and about 400 grey geese were seen over Regent's Park. Very few were noted during the next five years apart from a flock of over 100 at Beddington on 6 February 2005. There were records from 16 localities in 2006, and some of the birds were present for several days.

Facts and Figures

The largest flock recorded was of 700 birds over Regent's Park on 12 January 1987.
The earliest returning flock was on 19 October 1958 at Queen Mary Reservoir.

Greylag Goose *Anser anser*

A common resident and localised breeder; naturalised population

Range: breeds northern Europe and Asia, winters Europe; naturalised population in England.

Historical

Fossil remains of this species from the Pleistocene epoch were found at Grays and Ilford. Wild Greylag Geese were formerly not uncommon in England in winter but by the mid-19th century they had become rare. Harting noted that occasionally a small skein of grey geese would drop down onto the London reservoirs in hard weather but they rarely stayed for more than a day due to disturbance or being shot.

The earliest documented record was in 1838 when Torre included it on his list of Middlesex birds. Glegg chose not to accept any records of Greylag Goose due to the risk of confusion with other species of grey geese. However, a record of two that were shot from a group of 30 in a field by the River Brent in Hendon in the winter of 1860/61 is acceptable since Harting stated that he had received an accurate description of them. The only other dated record was of two shot in the Lea Valley in the severe winter of 1895.

20th century

In the early 20th century there were two records, of three birds in 1918 at South Weald Park from 23 March to 20 April, and one on Eagle Pond, Snaresbrook, on 7–8 February 1919, during very cold weather. These may have been the last wild birds to be seen in the London Area as all subsequent records are considered to be of feral birds; possibly apart from a few isolated sightings up to the end of the 1960s.

By the time of the next record in 1947, when two were on Banbury Reservoir on 12 February, wild Greylags had ceased wintering in southern England. There were also full-winged flocks in East Anglia and escapes had been seen on the lake in Regent's Park.

In the early 1960s semi-domesticated Greylags were being kept at a number of places south of the Thames and were responsible for the occasional sighting elsewhere. During the cold winter of 1962/63 a family party arrived at Barn Elms Reservoirs on 5 January and the birds were described as being 'very tired'; they were possibly of wild origin. Later that year Greylag Geese were introduced to Sevenoaks Wildfowl Reserve and within a few years they had begun to breed regularly. In 1964 three at Hilfield Park Reservoir were reported to be 'very wild'. By 1965 the population at Sevenoaks had reached 26 birds and as they were also widely kept in collections it was no longer possible to determine if any birds seen elsewhere were wild or not; the only likely group of wild birds seen that year were six that flew over Old Parkbury on 7 February. In 1968 two of the eastern race were seen at Sevenoaks, whereas the resident population was of the western race; this resident group raised 11 young in 1968.

During the 1970s the population at Sevenoaks continued to increase and reached 360 in October 1977. In 1974 a pair bred in the Colne Valley and by 1977 a free-flying population had become established in St James's Park. By 1983 the Sevenoaks population had increased to 659 and across the London Area the population in autumn was about 840. The other main population in St James's Park increased from 28 in 1981/82 to 66 in 1985/86; they also started to breed in Regent's Park in 1983. The Sevenoaks population continued to increase and by 1988 had reached 895.

During the 1990s the population expanded across London. The main groups were at Sevenoaks and in central London where they move around between St James's Park, Buckingham Palace Gardens, Hyde Park and Kensington Gardens (Sanderson, 1997). A survey of Greylag Geese during the late summer moult in 1990 located 261 birds, including 44 juveniles, in the 15 10km squares which make up the BTO's London Recording Area; this includes Walthamstow and much of the Colne Valley but excludes much of the outer part of the LNHS recording area including Sevenoaks (Baker 1991).

Elsewhere, eight pairs bred in the Colne Valley, seven pairs bred in the Lea Valley and breeding occurred at a further three sites. There was a 1,000 per cent increase in the breeding distribution between the breeding Atlas surveys in 1968–72 and 1988–94. The population at Sevenoaks had spread up the Darenth Valley and across the Thames. The other main area was in the Lea Valley and this may have been colonised from the population in the central London parks.

21st century

Greylags are controlled at some sites, particularly in central London; the largest breeding population is at Walthamstow Reservoirs, where up to 29 pairs breed. The winter flock in the central Royal Parks disperses in spring as pairs look elsewhere for nesting sites and regroups later in the year. The highest counts since 2000 were 445 birds at Sevenoaks and 305 in Hyde Park/Kensington Gardens, both in 2007.

Facts and Figures

The largest count was 895 birds at Sevenoaks Wildfowl Reserve on 17 September 1988.

Grey goose sp. *Anser* sp.

There are too many records of unidentified groups of grey geese to list. However, as there are very few records before 1900, these are all listed. Also listed are high counts from Inner London and exceptionally large numbers elsewhere.

In January 1864 Harting observed some small skeins of grey geese flying over very high, and in the first week of March 1865, 24 flew over Kingsbury. In February 1897 a flock flew low down the Thames past the Tower of London.

In 1952 more than 100 grey geese over Kensington on 9 March were probably White-fronted. In 1955 about 475 grey geese in four skeins flew north over King George V Reservoir on 14 March. In the extremely severe winter of 1962/63 large numbers of geese were seen throughout London. At the beginning of the very cold weather a flock of about 100 flew west over Leatherhead shortly after midnight on 30 December, and later in the day about 250 flew west over Swanscombe. On 1 January some 100–120 flew over St Paul's Cathedral. On 19 January 1969, 120 flew over Beddington and at the end of the year 80–100 flew over Weybridge; both flocks were thought to be White-fronted Geese.

In January 1970 at least five skeins of geese were seen, with the largest flock – 52 over Staines Reservoir – thought to be Pink-feet. In 1979, 70 flew over Regent's Park on 28 January and 150 passed over the following day, while 80–100 flew over Hyde Park on 23 February; these were probably all White-fronts as there was a large influx at the time. In 1982 about 200 flew over Hillingdon on 1 February. In 1987 a flock of about 1,000 passed north-east over North Finchley on 15 March at 10:00 hrs; these were probably White-fronts as the main wintering flock at Slimbridge left earlier that morning on its way back to the Continent.

Snow Goose *Anser caerulescens*

A very rare vagrant (17)

Range: breeds Greenland, Canada and eastern Siberia, most winter North America, vagrant in western Europe.

The only record of presumed wild birds occurred in 1980 when a flock of 17 flew east past Swanscombe Marsh on 1 March. These birds were all white morphs and consisted of 13 adults and four immatures; they were also seen at Cliffe, Kent, later that day. Subsequently, a flock of 13 adult and four first-winter white-morph birds and one blue-morph goose were present in Holland from 18 to 26 April and were presumed to be the same flock. One of the white-morph adults was colour-ringed, indicating that it was a male 'Lesser Snow Goose' ringed as a gosling in Manitoba, Canada, in 1977. It has since become established that a few wild Snow Geese regularly winter in Holland.

Although there were a few escapes present in the London Area and Kent at the time there were no flocks numbering anywhere near this size, and the entire flock of 17 was presumed to be of Canadian origin. This record was published in the 1980 London Bird Report but it was never added to the London List, an oversight that is hereby corrected.

1980: Swanscombe, 17 on 1 March

Escapes

There have been various escapes, seen in most years since 1963 when one began a year-long residency at Sevenoaks. Most records are of one or two birds, some of which have remained for long periods. Up to six were seen in the Lea Valley in 1980–81; up to four were seen at Tyttenhanger Gravel Pits in 1993; and in 1996 four flew over Girling Reservoir on 22 October and were seen at Dungeness later the same day. A flock of 40 that flew over Chelsham on 17 March 2008 was considered to be feral, although its origin was never traced.

Canada Goose *Branta canadensis*

A common resident and widespread breeder; population derived from escapes

Range: North America; naturalised population in north-west Europe.

Historical

This North American species was first introduced into Britain in the late 17th century when King Charles II had it imported for his collection in St James's Park. This was probably the source of one that was shot on the Thames near Brentford prior to 1731. By the end of the 18th century Canada Geese were breeding on many estates in the country. They were still very rare in the 19th century and only two records are known: five at Brent Reservoir during very severe winter weather prior to 1866, and two in Syon Park in about 1882.

20th century

In the early years of the century there were a number of established populations in the country although it remained a rare visitor to London. A pair bred at Godstone in 1905; there was one at Westerham in 1906; a pair was seen at Walthamstow Reservoirs in about 1907–08; and four were seen in Gatton Park in 1908 or 1909. No more were documented until the 1920s although it is assumed that most went unrecorded.

During the 1930s they became established in Gatton Park; by 1933 the population there had increased to over 100, and by 1936 to over 130. Many of these had bred elsewhere in the district, such as at Limpsfield, where they were resident throughout 1934. Between 1936 and 1938, an effort was made to control the numbers and nests were destroyed on the estate, but this appeared to have little impact as by 1938 the numbers had reached 200. Almost all of these birds left in March of that year and as a result of military occupation during the war did not return to breed; instead, they spread into adjacent areas. Nesting occurred at Cobham, Godstone, Oxted and Reigate Priory. Elsewhere in London, two pairs were first reported breeding at Walthamstow Reservoirs in 1936.

In the early 1950s Canada Goose was admitted to the British List and the UK population was between 2,200 and 4,000 birds. They bred in the collection at Regent's Park and their offspring were left full-winged.

An introduced pair bred at Valentines Park in Ilford in 1953. In 1955, 17 full-winged birds were introduced in Hyde Park. They were still rare visitors to most parts of the London Area, e.g. in the Lea Valley the only record was of one at Cheshunt Lock on 14 August 1956. The same year 12 were introduced to Sevenoaks Wildfowl Reserve and quickly established a breeding population there. In the late 1950s they bred also at Cobham, Kelsey Park and Walthamstow Reservoirs.

They slowly spread across the London Area during the 1960s, helped by the fact that the young of breeding captive pairs were left full-winged and by introductions such as at Chipstead where 59 birds were present in 1962. By 1965 the central London population had expanded to 150 birds. At least 36 pairs bred in 1967 and there were three separate flocks exceeding 100 birds. One bird seen in Hyde Park and Kensington Gardens in 1968 was part of the original introduction there in 1955. At the end of the decade the British population had increased to over 10,000. There were then at least 23 breeding pairs across London, and populations of 200 or more in the Royal Parks and at Sevenoaks.

A national survey in 1975–76 put the population at 19,400, almost double the number seven years earlier. The first count of 500 was made in 1977 on Staines Reservoir where they had roosted overnight. At the beginning of the 1980s the population was still expanding and had exceeded 2,500 in London; flocks of over 100 were becoming commonplace. In 1983 about 1,250 goslings were recorded during a breeding census and the total population had reached about 3,400. In September 1986 the first counts of over a 1,000 were made: 1,139 on King George VI Reservoir and 1,087 at Amwell.

By the early 1990s the London Area population had reached over 5,000 birds, and concern was raised at the possible effects on other breeding water birds, as well as the impact on park management e.g. fouling and bank erosion around lakes. Over 2,000 eggs were pricked in an effort to curb the population. At the time of the first Atlas in 1968–72 the breeding population was mainly confined to the southern half of London; during the second Atlas in 1988–94 they had spread throughout the London Area and were breeding in one-third of all tetrads. At Walthamstow Reservoirs 49 pairs bred successfully in 1991, contributing to a peak count of 1,157 birds. Further north in the Lea Valley, 70 pairs nested at Fishers Green. The same year there was an attempt made to curb the growing population in the Royal Parks when 186 were relocated from St James's Park to the North Kent Marshes in July; the following year many nesting birds had their eggs pricked.

21st century

By 2000 Canada Geese had become one of the most recognised birds in London, probably second to the Feral Pigeon, thanks to their presence in most parks where people feed birds. Efforts have continued to prevent the population from increasing at some sites, and this has probably had some effect locally, but its impact on the overall London Area population must surely be limited while the birds are apparently left to breed uncontrolled at other major nesting sites. The largest population is in the Lea Valley, particularly at Walthamstow Reservoirs where 57 pairs bred in 2009 and the highest count was 916 in 2006.

Facts and Figures

The largest ever count was 1,400 birds on Staines Reservoir on 27 August 1996.

Barnacle Goose *Branta leucopsis*

A rare winter visitor; small breeding population of escapes (6 records/133 birds)

Range: breeds Arctic Circle, winters western Europe.

20th century

None were seen until 1902 when one was shot at Chelsfield; escapes were unknown at the time. In 1959 four flew over Beddington on 5 April. In 1979 a flock of 13 remained at Rainham Marshes with White-fronted Geese between 23 February and 4 March; these birds are likely to have been wild since there had been an exceptionally large arrival of geese in England from the Continent due to severe weather.

21st century

Since 2000 there has been an increase in the number of sightings in London. It is not known if they are genuinely wild birds from the wintering population on the near Continent or if they are from the burgeoning feral flocks on the East Coast but as the latter are a self-sustaining population they are counted here as acceptable records. There is also another population in the Midlands, some of which have been seen in north London.

In 2002 a flock of 16 flew in from high to the east and landed on the Thames at Rainham on 22 December; after a short while they flew back east. In 2005 a flock of 11 was present on Verulamium Lake, St Albans, on 9 September; one of these had a colour ring on it which was traced to a self-sustaining population in Bedfordshire. In December 2007 there was a large arrival of Barnacle Geese into Europe. This coincided with an influx at Rainham where, on the 21st, two flocks numbering 32 and 50 flew in from the east and landed; by the 24th they had increased to 88. Apart from a short visit to Dartford Moor on the 26th they remained there into the New Year, although the numbers had decreased to 37 by 30 December.

1902: Chelsfield, shot
1959: Beddington, four on 5 April
1979: Rainham Marshes, 13 on 23 February–4 March
2002: Rainham Marshes, 16 on 22 December
2005: Verulamium Lake, 11 on 9 September
2007: Rainham Marshes, 82 on 21–23 December, 88 on 24–29 December and 37 on
30 December–16 January 2008

Escapes

The first known escapes were seen in 1941 at Aldenham Reservoir. In the winter of 1949/50, 10 escaped from a collection at Wormley Bury, near Broxbourne, and one of these was subsequently seen at Hamper Mill. No more were seen until 10 years later when one wintered at Brent Reservoir. During the 1960s the number of escapes began to increase, particularly in the Sevenoaks area. In the mid-1970s several took up residence in the Colne and Lea Valleys. By 1983–84 up to eight were at Holmethorpe and Gatton Park, and in 1985 there were up to eight in the Lea Valley and 11 at Wraysbury GP.

Breeding occurred for the first time at Fishers Green in 1986–87. In 1989 a flock of 14 tame birds was present in Osterley Park on 23 November. By 1990 escaped Barnacle Geese were just beginning to establish a resident population in the London Area and it had become virtually impossible to distinguish these from potentially wild birds. A pair bred again at Fishers Green from 1991 to 1994. Three flocks were seen in 1993 that may have originated from feral populations in southern England: 11 at Dagenham Chase on 16 February, 12 at Thorpe Water Park on 21 March and 20 at Waltham Abbey on 19 September.

Since 2000 the species has bred in most years but there is currently no sign that a self-sustaining population is developing as the breeding pairs are scattered across the London Area and the total numbers remain fairly low.

Brent Goose *Branta bernicla*

An uncommon winter visitor

Range: breeds Arctic Circle, winters western Europe and North America.
All records relate to 'Dark-bellied Brent Goose' *bernicla* unless otherwise stated.

Historical

Although 'immense flocks' wintered on the North Kent Marshes, they were very rare in the London Area in the 19th century, with only two known records, of a bird at Brent Reservoir on an unknown date prior to 1843 and of one shot near Greenford in November 1892. Additionally, two were shot on an unknown date in Lullingstone Park.

20th century

There were no were further records until the 1930s when eight were seen on Island Barn Reservoir on 17 December 1938 during a period of very cold weather. Two earlier records in 1935 and 1937 were both thought to be of escapes. After this the number of sightings gradually began to increase.

There were three records in the 1940s, including of two birds on the Thames at Chiswick on 14–16 February 1940 during severe weather, the first time that the species had remained anywhere in the capital for more than a single day; the same year a flock of 40–50 geese, which were probably Brent, flew over Regent's Park on 16 December. In the 1950s they started to appear on the Lower Thames Marshes but most sightings were of birds just passing through. The largest group, of 21, flew south over Ruxley GP on 13 December 1959.

During the 1960s larger flocks started to appear, including 38 at Staines Reservoir in 1962 and 60 over Walthamstow Reservoirs in 1968. The following year there was a small influx in March, starting with a flock of 50 grazing in a flooded field at Hainault on the 8th, followed by 22 on Staines Reservoir on the 11th. By the 1970s they had become an annual, but irregular visitor, especially along the Thames. By the 1980s there was a passage through the London Area in October and November in most years. In January 1987 severe weather brought about 125 into London, including a flock of 75 up the Thames past Northfleet on the 11th.

The peak decade was the 1990s. In 1991 several very large flocks were seen, including at least 550 birds at Rainham Marshes on 9 February and 200 over both Beddington and Carshalton on 10 October. In 1994 there was an obvious passage across the London Area on 5 November, with 89 birds in four flocks over Queen Mother Reservoir, 61 at Staines Reservoir, 42 over Brent Reservoir and several smaller flocks elsewhere; also, 85 flew over King George V Reservoir on 11 November. In 1995 there was a widespread influx during the last two months of the year when 130 were seen, including a flock of 15 in Inner London; these flew over the Thames at Southbank. In 1997 about 450 were seen, including one flock of around 250 that flew south-west over Beddington on 20 October; these were part of a movement of at least 327 through the London Area that day.

21st century

Fewer birds have been seen since 2000 compared to the 1990s, reflecting a decrease in numbers nationally. Most years only a few are recorded, but there are occasional influxes, typically during the late autumn passage, e.g. there were just 12 in 2001 but 211 in 2005. Although sightings are widely spread across London, they are most often noted on the Lower Thames Marshes, particularly at Rainham.

Escapes

Occasional escaped birds have been seen; for example, one summered around Barnes and Richmond in 1984.

Facts and Figures

The earliest migrants were on 21 September 1996 at Hilfield Park Reservoir.

'Pale-bellied Brent Goose' *B. b. hrota*

There have only been two records in London:

1998: Beddington, flew over on 7 February
2003: Staines Reservoir, on 29 December

'Black Brant' *B. b. nigricans*

There have been three sightings of the North American race. The first was on Staines Reservoir between 27 October and 23 November 1984; although it associated with Canada Geese during its stay its arrival fits the pattern for naturally occurring wild birds. In 1990 one was on King George V Reservoir with eight other Brent Geese on 18 March; and in 1999 one was seen at both Dartford and Rainham Marshes on 13 November.

Red-breasted Goose *Branta ruficollis*

A very rare vagrant (1)

Range: breeds Siberia, winters eastern Europe; vagrant to UK.

A rare winter visitor to the UK, often with flocks of wild geese; an attractive and colourful species, this diminutive goose is often included in wildfowl collections hence escapes are occasionally seen. The only record was of one shot 'near London' in the severe winter of 1766. The specimen is in the Hancock Museum at Newcastle-upon-Tyne and was used by Thomas Bewick for his engraving in his book *British Birds*. This was the first sighting of the species in the UK, although there is another record from Yorkshire around the same time.

Unfortunately, the exact location of where the bird was shot in unknown. Harting included this in his *Birds of Middlesex* as did Glegg, although there is no evidence to corroborate this. It was originally mentioned by Latham, who lived at Dartford, so perhaps it is more likely to have been shot on the marshes along the southern side of the Thames. At the time London did not extend eastwards beyond Stepney, and most of the Thames was surrounded by marshes on both sides of the river. There are also fossil remains of this species from Grays, and one was shot in Essex in 1871.

1766: 'near London', one shot early in the year

Escapes

There have been a few escapes seen since the 1970s, although most have been more recent. There were singles at Gatton in February 1970 and Walton-on-the-Hill in early 1972. In 2000 one was at Dagenham Chase in April and then at Fairlop between May and September. In 2005 five were seen on the River Frays at Uxbridge on 21 December. A pair was present at the London Wetland Centre from March to May 2008. Singles were present at Sevenoaks Reserve in November 2010 and at Rainham and Crossness in March 2011.

Egyptian Goose *Alopochen aegyptiaca*

An uncommon resident and localised breeder; naturalised population

Range: Sub-Saharan Africa; naturalised population in western Europe.

Historical

The English population originated from pinioned birds in Holkham Park, Norfolk, which were first introduced in the 17th century. Escapes from other collections may also have subsequently bred and contributed to the spread of this species. There are three dated records in the London Area in the 19th century: one shot at Colnbrook in the winter of 1863; one shot at Brent Reservoir in January 1888; and seven in Clissold Park in November 1893. Several were also seen in London parks in the 1890s, having escaped from collections.

20th century

In the first half of the 20th century one was shot at Chingford prior to 1905; a pair frequented Kensington Gardens in 1924; and in 1937 a pair was resident on Connaught Water.

Egyptian Goose was formally added to the British List in 1971 when the population was largely restricted to the Norfolk Broads. During the 1970s there was a slow increase in the number of sightings in London. In 1972 there was one at Sevenoaks on 1 June; a pair at Walthamstow Reservoirs in July and August; and probably the same pair at Wanstead Flats in late November and December. In 1977 up to four were present at Kempton Park and one or two were at two other sites. In 1979 up to five were present in the Lea Valley from mid-March onwards and a pair attempted to breed at Fishers Green.

At the start of the 1980s three were present in the Fishers Green area but one was later shot; however, the first breeding occurred there the following year. In 1982 three adults and two immatures were at Shepperton GP in September and in 1986 a pair bred again at Fishers Green.

At the beginning of the 1990s Egyptian Geese were fairly scarce visitors to the London Area but that began to change after six birds were released at Broxbourne. They were most often seen in the upper Lea Valley and by 1993 there were pairs breeding at Amwell and Nazeing GP. Elsewhere, there were various escaped birds, including a pair at South Norwood Country Park and three at Walton Reservoirs in 1995; the latter may have escaped from a nearby collection at Sunbury-on-Thames where many wildfowl were unpinioned. They were introduced on an estate at Oxted and a pair bred there in 1996. Another pair bred at Bushy Park in 1999. Breeding success during the 1990s was quite poor, particularly in the Lea Valley, and the population remained relatively low.

21st century

At the start of the 21st century Egyptian Geese had just established themselves in the London Area; the breeding population was eight pairs and small numbers were seen widely across the London Area. The focus of the population had moved from the upper Lea Valley to south-west London where the highest count was 15 at Hampton Court Park in autumn 2000. The population has continued to increase annually since then; 10 pairs bred in 2002, all but one in the west and south-west parts of London.

Breeding attempts have been more successful than in the past. For example, in 2004 eight pairs reared at least 49 young, and this has enabled the population to expand across the London Area. They have become more frequent in the Royal Parks in central London and by 2008 had started to breed there. In 2007–08 the total breeding population was at least 15 pairs.

Facts and Figures

The highest count was 63 birds in Hyde Park/Kensington Gardens on 12 January 2013.

Shelduck *Tadorna tadorna*

An uncommon partial migrant and localised breeder

Range: Europe, Asia and North Africa; partially migratory.

Historical

Remains of Shelduck found at Southwark were thought to be post-Roman. They were rare in the 19th century and the earliest record was in 1875 when two were seen at Brent Reservoir on 27 September. The only other records were from Aldenham Reservoir, where single birds were seen in December 1883 and December 1896; at Ashtead where one was shot in about 1894; and at Fetcham where several were on the River Mole prior to 1900.

20th century

The next record was in 1902 when Cornish wrote that one had stayed for several days on the Thames at Chiswick Eyot in June, although it is unknown which year he was referring to. A pair was shot in May 1905 in Darenth Wood where the birds were presumably prospecting for a nest site. Elsewhere, there were six on Staines Reservoir in the winter of 1906/07 when one also occurred in Clapton and another was on the Lea Valley reservoirs. One was picked up dead at Godstone in 1914 and a pair was seen there on 19 December 1920.

From the mid-1920s onwards they became occasional visitors to the west London reservoirs between October and April, particularly at Staines where there were records in eight years up to 1935. One individual remained at Staines Reservoir from 16 October 1932 to at least December 1934.

Most of these sightings were of single birds or small groups, but the occasional flock was seen, such as 14 at Staines on 1 September 1946 and 12 at Barn Elms on 26 August 1929. An apparently wild Shelduck was seen on the Serpentine on 19 March 1943.

There is a surprising lack of records from the Thames-side marshes in the first half of the 20th century and they appear to have been genuinely absent. From 1949 onwards parties of up to 10 birds were seen at Stone and Swanscombe Marshes, and up to eight were present during the breeding season in the Rainham area. The first breeding record was in 1954 when a pair was seen with two downy young at Rainham Marshes and in the same year there were two pairs in the breeding season at Stone Marshes; two immatures were present there in September.

During the late 1950s and into the 60s they were seen more frequently on the Thames marshes and on 30 December 1961 there were 71 at Swanscombe during hard weather, the largest count in the London Area up to that time. They also bred at the latter site in 1959 and 1961–62. During the severe winter in early 1963 unprecedented numbers were seen along the Thames, culminating in a count of over 500 at Swanscombe on 3 February. In the summer of 1963 a pair bred on Aveley Marsh and another bred again at Swanscombe. Breeding became almost annual from then on. In 1965 the numbers on the Thames hit a new peak when about 600 were counted at Stone on 21 February. In 1966 there was an influx to many inland sites, including 16 at Barn Elms Reservoirs on 22 January. In 1969 a pair in the collection at Regent's Park fledged seven young, which were left full-winged and were responsible for sightings in the surrounding area. On the Thames there were 1,000 at Belvedere and Swanscombe on 4 January and into February with another 550 at Rainham Marshes.

At the start of the 1970s at least 3,000 Shelduck were wintering on the Lower Thames Marshes between Barking and Swanscombe but there were fewer in subsequent winters. In 1974 two pairs bred at Rainham and two pairs attempted to breed on Erith Marshes; additionally, two pairs bred at Queen Mary Reservoir and one pair bred at Wraysbury GP, the first breeding to take place away from the Lower Thames Marshes. By 1979 the breeding population had increased to 10 pairs; additionally, captive birds bred in St James's Park and their young were often left full-winged, which helped to account for the increase in sightings elsewhere in the London Area.

During the 1980s the breeding population increased dramatically and by 1989 26 pairs bred successfully, including nine in west London. However, the numbers wintering on the Lower Thames Marshes decreased to about 300. By the start of the 1990s they had also colonised the Ingrebourne and Lea valleys and in 1992 the breeding population had reached 50 pairs. In the second breeding survey of 1988–94 breeding birds were located in 60 tetrads compared to just four in 1968–72. The population had spread from the Lower Thames up the Lea Valley, and into west London, particularly in the Colne Valley, at the large reservoirs and nearby gravel pits. There were also isolated breeding pairs away from these core areas. By the end of the decade there was a population of 11 full-winged birds in St James's Park.

21st century

At the beginning of the 21st century the breeding population was at least 55 pairs, including 41 along the Lower Thames Marshes. However, the breeding population declined during this first decade and by 2009 had dropped to 25 pairs.

They are widely seen across the London Area outside the breeding season, particularly on spring migration. The highest counts are on the Lower Thames, particularly at Crossness, where the post-breeding flocks can exceed 300 and about 200 overwinter in the area.

Facts and Figures

The largest concentration was of about 1,600 birds on the Thames in the Barking and Woolwich/Crossness area in January 1970.
The largest flock seen away from the Thames marshes was of 30 birds at Walthamstow Reservoirs on 22 April 2009.

Mandarin Duck *Aix galericulata*

An uncommon resident and localised breeder; naturalised population

Range: East Asia; naturalised population in UK and Europe.

Historical

Mandarins were first introduced into Britain in the mid-18th century and one was seen on Richmond Green at some point between 1743 and 1751. In 1834 they first bred successfully in captivity at London Zoo.

20th century

A small colony of free-flying birds was formed at Woburn, Bedfordshire, from 1900. Another population became established around Virginia Water in Windsor Great Park from birds that had escaped from a collection in Cobham. From these sites, which are about 3 kilometres beyond the western boundary of the London recording area, Mandarins very slowly spread into the capital. The first London Area record was on 18 November 1928 at Staines Reservoir, an atypical habitat for a species that usually prefers small, tree-lined waterbodies.

In 1930 there was an attempt to establish Mandarin Ducks in London when 99 young birds imported from China were released in the Royal Parks, Hampton Court and the grounds of Buckingham Palace. However, instead of staying to breed most of them departed. Another 15 were released into Regent's Park in 1931 but they also flew off with only one or two seen there the following year. No more were reported until the 1940s when one was seen on the Round Pond in Kensington Gardens on 9 January 1940. In 1946 a nest with eight or nine eggs was discovered in an oak tree near Thorpe, the first breeding record in the London Area. A drake wintered on Connaught Water between November 1947 and February 1948. From 1953 to 1955 at least six were present on ponds in Epping Forest and breeding was thought to have occurred.

During the 1960s sightings became more frequent. In 1961 breeding occurred at Addlestone and on the River Mole at Burhill; there were also records from eight other locations, all in the south of the London Area. In 1962 family parties were seen at Silvermere and on the River Mole near Leatherhead; in the autumn a flock of 24 was seen feeding on grain stubble near Cobham on 23 September. In 1963 there was a large increase in the number of records, although most of the birds were still around Cobham/Leatherhead. In 1968 there were seven pairs at Stoke d'Abernon and six pairs at Great Bookham. A detailed search in 1969 found 55–60 pairs along or near the River Mole between Mickleham and Esher; the largest flock seen was 42 feeding on acorns at Stoke D'Abernon on 29 October.

Mandarin Duck was officially admitted to the British List in 1971. The following year a pair bred north of the Thames for the first time, at Wrotham Park. During the 1970s the population began to expand across the London Area and breeding was first recorded at Wraysbury GP in 1973, at Greenwich Park and Hampton Court Park in 1974, and on Darlands Lake, Totteridge, in 1975. Additionally, up to three were present for much of 1974 on Connaught Water, an area of Epping Forest that they would later colonise; the first breeding in Epping Forest occurred at High Beech in 1979.

The population continued to increase during the 1980s and in the autumn of 1983 there was a record count of 105 roosting at Buckland Sand Pit. Also that year five pairs bred in Epping Forest, a pair bred in a garden in Southgate and at least one pair bred in Trent Park. By 1986 the population in north and east London had overtaken that of the one south of the river; there were 31 in Grovelands Park and 16 in Epping Forest, compared to 20 on Epsom Common.

By the start of the 1990s the total population was around 200 and still increasing. In 1993 there were at least 24 breeding pairs. Peak counts in the mid-1990s were 133 on Epsom Common; 61 at Grovelands Park; 50 on Connaught Water; and 30 in Panshanger Park. In 1996 10 captive-bred juveniles at St James's Park were left full-winged and within three years up to 18 could be seen there. In November 1998 there were 101 on Connaught Water.

21st century

By the year 2000 Mandarin Ducks had become reasonably widespread across the London Area with a breeding population of at least 50 pairs. The largest breeding population is in Epping Forest where at least 21 pairs bred in 2009. They have also begun colonising Inner London; in 2007 a pair bred in Kensington Gardens and the following year five pairs bred in the Royal Parks.

Facts and Figures

The largest count was 180 birds at Grovelands Park in 2003 and 2004.

Wigeon *Anas penelope*

An uncommon winter visitor

Range: breeds northern Europe and Asia, winters Europe, Asia and North Africa.

Historical

The earliest reference was in 1842 when Meyer stated that flocks occurred on flooded meadows near Chertsey. Also in the 19th century, Harting described Wigeon as 'a regular winter visitant, arriving towards the end of October, their numbers increasing as the season advances', adding that in most years they were the most numerous duck. Typically, they were found on reservoirs in small parties but 20–30 had been seen at Brent Reservoir. They were uncommon winter visitors in Richmond Park from 1883 onwards and were also recorded at several other sites in south London.

20th century

In the early years of the century small numbers wintered at Staines Reservoir and the first large flock occurred in 1909 when up to 200 were on Barn Elms Reservoirs. Elsewhere, flocks of over 100 were occasionally seen at Queen Mary Reservoir and around Molesey.

During the 1930s Glegg stated that it was 'a local winter resident' and in Inner London it was 'an uncommon visitor in hard weather'. They arrived from early September, increasing until the peak in January; numbers remained large until March and some occasionally stayed until early May. By this time up to about 200 usually wintered at Staines Reservoir. However, during a particularly cold spell, there was estimated to be about 700 on

19 December 1933. Typically, around 80 per cent of the Wigeon recorded on wildfowl censuses were at Staines, where they had access to a safe refuge and feeding areas on the grassy banks and nearby at Staines Moor; only small flocks were present at other sites.

There was little change during the 1940s and even during the severe winter of 1946/47 there were only about 300 at Staines Reservoir. By the early 1950s at least 100 were wintering on the reservoirs and gravel pits around Molesey and they had also taken to using the Thames, particularly around Chiswick, and in the east of the area at Swanscombe Marsh. In 1951 three pairs summered at Staines Reservoir. On several occasions wild birds were seen in the Royal Parks or flying over the Thames in Inner London. More than 600 were present in a very cold spell in February 1956, and birds moved around a lot as the weather conditions changed.

By the early 1960s the wintering numbers had increased to around 550, mostly in the Staines area. In the severe winter of 1962/63 there were a lot fewer present and the peak count from the reservoirs was just under 300 in February; however, more were seen on the Thames, including 180 at Rainham. At the beginning of the 1970s there were more Wigeon on the Lower Thames Marshes: up to 200 at Rainham and 160 at Swanscombe. There was an influx in early 1979 due to very cold weather, including 500 on Staines Moor on 6 February.

During the 1980s cold weather on the Continent produced several large influxes. For example, in January 1982 there were 420 at Rainham Marshes and 308 on King George VI Reservoir. In early 1985 some 3,300 were present in the London Area on 19–20 January, including 750 on King George VI Reservoir. Even higher numbers were present the following year with 823 on King George VI Reservoir and 460 on Kempton Park Racecourse in February. In 1987 at least 3,500 were seen, including 900 at Wraysbury GP, 759 at Staines Reservoir, 650 at Walton Reservoirs, 570 at Fairlop and 500 at both Dartford and Rainham Marshes.

In 1992 a pair was seen mating in the Lea Valley at Hall Marsh on 19 March and the male was still present in June but there were no further signs of breeding activity. There was a large passage across the London Area on 5 November 1994 when 180 arrived on Stocker's Lake, 160 flew over Queen Mother Reservoir and 101 passed over Beddington. In January 1997 about 3,000 were present across the London Area. Also in 1996 three captive-bred juveniles at St James's Park were left full-winged and by the end of the decade there were up to 24 full-winged birds present; these were sometimes seen on the Thames and in the Barnes area.

21st century

At the beginning of the 21st century up to 3,000 wintered in the London Area. Numbers have increased significantly at Rainham Marshes following habitat restoration and now more than 1,000 winter there. As well as September being the arrival month for some of the wintering flocks, it is the peak time for passage migrants and small groups of Wigeon may appear on almost any suitable wetland. The shallow lakes and flooded grasslands at Barnes Wetland Centre attract good numbers, with occasionally over 100 present, and small flocks can often be seen at Pen Ponds in Richmond Park.

Facts and Figures

The largest flock recorded is of about 1,068 birds at Rainham Marshes on 12 February 2006.

Gadwall *Anas strepera*

A common resident and localised breeder

Range: Europe, Asia, North America, Africa.

Historical

In the 19th century this was a rare winter visitor to England. The first record for the London Area was of a male shot at Brent Reservoir in the winter of 1842/43. The only other definite wild bird was shot on Bromley Common in 1845. In the years 1854–65 up to three were recorded on 22 occasions in the central London Royal Parks, all between October and May; the origins of these birds are unknown.

20th century

Gadwall continued to be a rare visitor in the first third of the century. There were no known sightings until 1920 when one was at Navestock on 7 November. The same year one wintered on the Round Pond in Kensington Gardens and then returned every year until 1933 and once, in the autumn of 1922, was joined by a second bird; towards the end of its stay it became a year-round resident. There were also records from Aldenham and Staines Reservoirs.

The status of Gadwall changed in the 1930s. A pinioned pair was introduced into St James's Park in 1931 and raised eight young. A further four pairs were introduced the following year and 15 young were raised and left full-winged; during the autumn the lake was drained and some of these moved elsewhere. By 1934 there were 32 free-flying birds in the park. These were the source of birds which later bred at Barn Elms in 1936 and Beddington from 1938 onwards. From then on it became increasingly difficult to know if birds seen at other locations originated from these escapes or were wild. Meanwhile, breeding continued in St James's Park and over the next three years another 46 more young were left full-winged. It seems likely that the colonisation of the London Area came from those birds that were raised in St James's Park. Up to three pairs bred at Barn Elms and the adjacent Lonsdale Road Reservoir on a regular basis. Elsewhere in the London Area, the number of records gradually increased from this time.

By the early 1960s the population in the Barnes area had risen to 25 and the birds were slowly spreading out. From the mid-1960s the local wildfowling association released many Gadwall annually at Sevenoaks; between 1965 and 1972 over 100 locally bred birds were ringed and while many remained in the area, others moved around Kent and there were six recoveries in France and one in Germany. In 1966 only small numbers were seen in the London Area away from the populations at Barnes, Sevenoaks and the Royal Parks. In 1969 the population at Barn Elms had risen to 49; a pair bred at Godstone; and breeding was recorded for the first time in the Lea Valley, at Cheshunt. At the start of the 1970s the population exceeded 100 birds for the first time. The species bred for the first time at Ruxley GP and Walthamstow Reservoirs in 1971. By 1973 the maximum count at Barn Elms had reached 130. No breeding was recorded in the years 1977–78 and the resident population appeared to have stabilised.

In 1980 the breeding population away from the Royal Parks was still fairly small at about seven pairs but by 1988 it had increased to about 18 pairs. The wintering population increased significantly during this period, with the wildfowl count showing a rise from 189 in 1980/81 to 516 in 1989/90. Cold weather in early 1987 brought an influx into the London Area and about 1,300 were seen. In 1989 the highest count was 453 at Cheshunt where the wintering population had trebled in three years.

At the start of the 1990s the population was still on the rise and there had been a 25 per cent increase on the wildfowl counts between 1986/87 and 1990/91. The largest numbers were in the Lea Valley where there were 569 between Amwell and Cheshunt in November. The only other site to record more than 200 was Richmond Park. The breeding population had risen to at least 24 pairs, including 10 at Rye Meads. In the winter of 1994/95 more than 2,300 were present in the London Area. In 1996 almost 200 young were reared at Rye Meads.

21st century

By 2000 the wintering population had increased to about 2,500 birds, a massive increase from 100 birds of 30 years earlier. The Lea Valley is one of six areas in the UK that is internationally important for this species, with an average of more than 600 in winter. The breeding population has also increased to more than 70 breeding pairs, most of which are in the Lea Valley, although odd pairs are scattered throughout the area. This species prefers still eutrophic waters and has undoubtedly benefitted from the large number of gravel pits resulting from construction work around the London Area.

Facts and Figures

The highest count was 453 birds at Cheshunt GP in December 1989.

Teal *Anas crecca*

A common winter visitor and rare breeder

Range: Europe and Asia, also winters Africa.

Historical

Remains of Teal have been found at a number of sites in south London dating back to Roman Britain. Also, remains have been found at Bermondsey Abbey dating to 1140–1680 and to Southwark dating to 1500–1700. As Teal were hunted for food these remains were not necessarily locally killed birds. The earliest reference comes from the 12th century when a falconer from the Bishop of London's household attempted to get his hawk to catch a Teal on the edge of the Thames; the bird ignored the Teal and captured a Pike instead.

In the 1860s Harting wrote that "of late years Teal have much decreased in numbers, and, though formerly regular winter visitants, are now very uncertain in their appearance". He stated that flocks of 20 to 30 were then remarkable. He also observed that the species was most plentiful in February and preferred rivers and quiet pools that were least disturbed. There are records of eight over Clapham Common on 12 July 1897 and three on Dulwich Park Lake on 6 October 1898.

20th century

Early in the 20th century wintering numbers began to increase, possibly due to the construction of large reservoirs, and by 1906 a few had started wintering on Staines Reservoir. In the 1930s Glegg stated that it was 'a winter resident in considerable numbers', mainly on the reservoirs where up to 400 had been counted on Queen Mary and about 300 at Staines; elsewhere it was uncommon. In 1930 Seth-Smith stated that some liberated birds were now residing in Regent's Park and sometimes visited the Zoo. During the period 1930–32, regular counts took place at Staines; Teal arrived during the first week of August and gradually increased to a peak by early November. They remained until early March when many left and had all departed by the middle of April. At some sites there was evidence of a migration in March; for example, at Island Barn Reservoir a group of around 300 Teal was noted on 27 March 1937 after most wintering birds had already left.

In the 1940s and 50s numbers continued to increase and the peak counts during this period were about 650 at Staines Reservoir on 22 January 1949 and about 450 at Island Barn Reservoir on 7 February 1954. The highest count in the 1950s was during the severe winter in 1956 when 1,252 were counted on 15 January. Various duck censuses showed an average wintering population of around 850, almost all on the reservoirs. The censuses did not include all the London Area's waterbodies, and small flocks could be found on some unsurveyed gravel pits, sewage farms, watercress beds and parts of the Thames. However, the total number of Teal on these unsurveyed waters was unlikely to have been many more than 100. There was some evidence that Teal spent the day on the large reservoirs and flew off elsewhere to feed as flocks of up to 100 were seen feeding on flooded fields at night in the Rainham area.

In the early 1960s wintering numbers were about 1,300. By 1962 they began to be seen in Inner London and that year there were 14 in Regent's Park and seven over Hyde Park. During the severe winter in early 1963 unusually large numbers were present on the Lower Thames, including 300 at Swanscombe and 250 at Barking; further upstream there were about 250 at Chiswick. In subsequent winters the numbers remained relatively high despite them being mild. So, in 1964/65 the maximum count on King George VI Reservoir was 1,255 on 24 December and there was a total of 1,465 on the wildfowl counts in December 1965. During a cold snap in 1968 two flocks totalling 70 flew over Regent's Park on 14 December.

At the beginning of the 1970s a new record total count was made in January of 1,750 and 1,000 were seen at Rainham at the end of the month. Cold weather at the end of 1970 brought an influx with 1,500 on the Thames between Woolwich and Crossness. After the winter of 1974/75 there was a large decline in wintering numbers along the Thames so by 1977 the peak counts at Barking and Rainham were just 200. Cold weather towards the end of 1978 brought a large influx of Teal into the area and there were 3,000–4,000 at Rainham Marshes in December though numbers declined to 1,000 in the New Year. At the start of the 1980s the wintering numbers on the Lower Thames Marshes had recovered, and up to 2,000 were at Rainham; elsewhere, flocks of 100 or more were seen at 11 other localities. In early 1986 about 5,000 wintered in the London Area, including 2,000 on Staines Reservoir, whereas the Rainham flock had reduced to a maximum of 500 after its favoured lagoons had dried up. The popularity of the site increased again, however, so that by the end of the decade up to 1,200 were present at Rainham.

By the start of the 1990s due to a run of mild winters overwintering numbers were at a low ebb. However, in early 1991 there was a big influx and two sites recorded large flocks: Rainham Marshes had 1,200 in February and Kempton Park had 1,195 in March. In December 1996 there was a site record count at Beddington of 1,250.

21st century

At the start of the 21st century there were just over 2,000 wintering in the London Area. This increased to 7,000 by 2002 and exceptional numbers were present at Rainham Marshes in the winter of 2002/03 when they peaked at 3,475 in January. Although there have been fewer wintering in subsequent years, the largest flocks are still found on the Lower Thames, in the Barking, Crossness and Rainham area. Elsewhere, they are widespread on water bodies throughout the London Area, typically in flocks of less than 300. The larger reservoirs are mainly used as roosting sites with most feeding taking place in nearby shallow, well-vegetated waters and damp grassland.

Breeding

In about 1880 a pair bred at Epsom, the first breeding record in the London Area. Hudson stated that a pair bred in Richmond Park, but there is some doubt about this record. Teal may have bred in the Keston district prior to 1909 and in 1930 a pair nested near Cobham. The following year a female with three small young was seen on the River Colne on 28 May. Glegg stated that it was fairly certain that Teal had bred at this spot in other years. At Ruislip a nest was found in 1933 and two pairs bred in 1934; in subsequent years up to five pairs bred regularly in the area to about 1941.

No further breeding occurred until 1951 when a pair bred at Epsom SF, where a nest containing 18 eggs was found (this may have been the combined clutches of two females) and 15 later hatched. One or two pairs bred in the Lea Valley in 1958–59. In 1965 a pair probably bred at Belvedere SF as a party of nine birds, including juveniles, was later seen in Greenwich Park.

During the 1970s pairs summered in most years and breeding occurred from 1972 to 1975 on Black Pond, Esher, and in 1977 on Staines Moor. In 1984 two pairs summered at Rainham Marshes and one brood was seen; a pair also bred at Shepperton GP but there was some doubt as to the origin of these birds. At least 13 pairs summered in 1988 but there was no confirmed breeding.

In the 1990s up to 15 pairs summered annually and breeding was confirmed at Brent Reservoir, Hornchurch, Rainham Marshes and Walton Reservoirs. At the beginning of the 21st century five pairs were present in the breeding season, one of which bred at Rainham Marshes. One pair bred again at Rainham in 2001–02.

Facts and Figures

The highest count was 3,475 birds at Rainham Marshes in January 2003.

Green-winged Teal *Anas carolinensis*

A rare vagrant (10)

Range: North America; vagrant to UK.

Green-winged Teal is the American version of Teal; it was elevated to full species level by the BOU in 2001 although some authorities have since relegated it to a race of Teal. All the confirmed sightings in the London Area refer to adult males.

The first record was in November 1961 when one was present at Hilfield Park Reservoir. Ten years later there were two more sightings, probably involving the same bird, at Barn Elms on 11 April and at King George VI Reservoir two days later; at Barn Elms it was accompanied by a female that may also have been of this species but it was never confirmed. In 1985 one was present on Walthamstow Reservoirs in February and March, and was also seen on Girling Reservoir.

In 1997/98 there was one at Beddington; the following year one was present at two sites in the Lea Valley on 11–16 February, the third consecutive year one had been seen in the London Area. Since the beginning of the 21st century there has been an increase in the number of sightings. Two of the four individuals were seen only on one day – at Staines Reservoir on 9 March 2002 and at the Wetland Centre on 5 December 2004 – while the other two were seen at various sites during longer stays.

1961: Hilfield Park Reservoir, on 11 and 15 November
1971: Barn Elms Reservoirs, on 11 April; then at King George VI Reservoir on 13 April
1985: Walthamstow Reservoirs, on 28 February–27 March; also on Girling Reservoir on 9 March
1990: Sevenoaks WR, on 6 February–4 March
1997: Beddington SF, on 14 December and again on 17 January–8 February 1998
1999: Hooks Marsh, on 11–12 February; then at Cornmill Meadows on 12–16 February
2002: Staines Reservoirs, on 9 March
2003: Walthamstow Reservoirs, intermittently on 16 February–10 March; also seen at Girling Reservoir on 24 February, Cornmill Meadows on 17 March, Thamesmead on 23 March and Belvedere on 16 April
2004: London Wetland Centre, on 5 December
2005: Staines Reservoir, on 23–26 April; then at Beddington SF on 1–2 May

Mallard *Anas platyrhynchos*

A common resident and widespread breeder

Range: Europe and Asia, also winters Africa; resident feral population.

Historical

Fossil remains have been found from the Pleistocene epoch at Grays and in the peat at Walthamstow.

Known in the 19th century simply as 'Wild Duck', they were numerous winter visitors with only a few pairs remaining throughout the summer to breed. They would arrive from October, gradually increasing in number to peak in February, and would be all gone by the end of March except for a few breeding pairs. At some point in the 1800s they were introduced to the Serpentine and soon flourished, being fed daily by the park-keepers and public alike. Every December a thinning exercise took place and any ducks showing plumage signs of domesticity were removed.

20th century

At the beginning of the century, Kerr wrote that prior to the construction of the reservoir, only small flocks visited the Staines district in severe winters, when they could be found on the Thames. By 1906, four years after Staines Reservoir was finished, flocks of several hundred were wintering there.

In the 1930s Glegg described Mallard as a numerous resident, whose numbers increased in winter. He estimated that thousands wintered in Middlesex and that it was one of the species that had increased, despite the effects of urbanisation. It was also a common resident in Inner London, breeding in parks with lakes, even in New Square, Lincoln's Inn. An influx in late autumn was particularly noticeable on the large reservoirs, with Queen Mary and Staines Reservoirs each attracting around 1,000 in 1931. In addition to the large reservoirs, many congregated on the Thames between Kew and Richmond, particularly in autumn; almost 600 were counted there on 23 October 1949.

There was a gradual increase in wintering numbers from the 1930s to the 1950s, the highest count in 1950 being around 1,100 at Staines Reservoir in November. The only exception to this was in 1947 when there was a particularly severe winter and the numbers recorded were unusually low. By the winter of 1956/57 the population had increased to just over 6,000. In 1959 there were at least 50 breeding pairs in Regent's Park and about 175 young were raised.

Mallards are rarely thought of as passage birds but in 1962 about 900 flew south over Rye Meads on 21 October. During the winter of 1962/63 the population increased to a record count of 7,063 in January; as the reservoirs froze many moved onto the Thames, even in Inner London where at least 60 were present between Blackfriars and

Southwark bridges. Further upriver there were 3,630 between Putney and Teddington on 13 January. The following winters were fairly mild and numbers returned to normal level.

By the beginning of the 1990s the number wintering in the London Area had significantly reduced, especially on the Thames; the peak count was 848 at Amwell in January while the highest count on the Thames was 457 at Richmond in December.

21st century

Mallard remains a widespread and familiar duck throughout the London Area, breeding at most wetland areas. The wintering population, which is supplemented by wild immigrants from the Continent, has significantly decreased since the 1990s. By 2009 no site held more than 500 birds apart from Panshanger Park where they are artificially reared for shooting.

Breeding

Breeding Mallards have thrived in the London Area due to their ability to tolerate people and make use of unnatural habitats. While some have remained faithful to their ancestral genes and nested on the ground or in trees, others have made use of roof gardens, balconies near parks, flower-boxes, the timbers on a temporary bridge at Waterloo, on an old raft near Lambeth Bridge, on barges and boats, and on the static water tanks used by the fire service in the Second World War. They have also exploited the areas around reservoirs (for example, 28 broods were counted around the Barnes district in 1950) and in the many parks, suburbs and rural districts.

Female Mallards have been seen escorting their ducklings along pavements and across busy roads to water since the 19th century, sometimes even receiving an escort from the police or members of the public.

In the first Atlas from 1968–72, breeding Mallards were located in 470 tetrads. During the second Atlas in 1988–94 they had increased their range to 553 tetrads, most likely as a result of increased wetland habitat. However, the Breeding Bird Survey has recorded a decline in the breeding population in London of 16 per cent between 1995 and 2010 which contrasts with a positive trend in the surrounding regions.

Observations

Although Mallard rarely excite the experienced birder, they are immensely important in a social context as feeding the ducks is often the first point of contact with nature for many Londoners. Bird feeding brings its own problems when done to excess, as uneaten food pollutes the water which can lead to mortalities in many waterfowl as well as encouraging rats.

Facts and Figures

The largest flock recorded is of about 1,500 birds on Queen Mary Reservoir in January 1930.
The earliest that eggs have been seen was 20 January 1962 in St James's Park.

Pintail *Anas acuta*

An uncommon winter visitor; small breeding population of escapes

Range: breeds northern Europe, Asia and North America, winters Europe and North Africa and North America.

Historical

Harting wrote that Pintail has 'occasionally been killed on our reservoirs and fresh-water pools in winter, but its appearance is very uncertain'. The earliest reference was in 1843 when Bond included it in his list of birds observed at Brent Reservoir. The only dated records from the 19th century are of one shot on the Thames near Walton in February 1875 and one at Strand-on-the-Green in about 1878.

20th century

In the early part of the 20th century, Glegg described Pintail as 'an unusual and irregular winter visitor'. At Richmond Park there was a pair on 23 March 1907; there were four records between 1910 and 1926 in the eastern part of the London Area, and at Barn Elms there were four on 22 February 1924. In 1926–30 one or two were seen annually in Richmond Park and in 1926–33 up to four were seen on Staines Reservoir every year. Elsewhere, there were just a few scattered sightings. In the winter of 1938/39 a few were seen on the Thames for the first time and in February 1942 one was seen on the Thames at Chelsea. During the severe winter of 1946/47 there were 26 on Walton Reservoirs on 8 March.

Despite an increase in the number wintering in south-east England, they remained scarce in the London Area until 1954 when there was a small influx, including a group of eight which flew over the Lea Valley on 31 January. In 1958 higher numbers than usual were recorded, culminating with a flock of 29 at Rye Meads on 1 March, the largest number recorded in the London Area at the time. By the early 1960s one or two short-staying birds were being recorded on many of the reservoirs during the non-breeding season. In the very cold winter of 1962/63 larger numbers were seen, including 23 at Rainham on 27 January and a group of 14 that flew over Barn Elms Reservoirs on 12 January. During the winter of 1967/68 large flocks arrived on the Lower Thames Marshes, beginning with an unprecedented flock of 60 at Rainham Marshes on 3 December; up to 130 were seen at Woolwich in January and 100 flew over Dartford Marshes on 16 February.

By the beginning of the 1970s even higher numbers were wintering on the Thames with 350 between Woolwich and Crossness. A count from a boat between Woolwich and Swanscombe produced a total of 367 in January 1971. In 1973 the peak count was 485 at Barking in January. However, land reclamation at Woolwich led to a large reduction in the numbers on the Lower Thames and in 1974 the peak count was just 80. There was a slight recovery the following year when 150 were present in January and the numbers varied in subsequent years from 100 to 170.

At the start of the 1980s up to 150 were still wintering at Rainham Marshes and in 1981 a flock of 54 flew west past Northfleet on 24 October. After two years with far fewer birds, cold weather on the Continent in early 1985 brought an influx of birds into the London Area with 110 at Rainham in January, more than 100 at Thamesmead and 65 on Staines Reservoir on 3 February. By 1986 the lagoons at Rainham, which were much favoured by this species, had dried up and the peak was just 19. In 1988 only three were seen at Rainham and the London Area's highest count was at least 60 at Beddington on 17 October.

By 1990 they were being recorded on many of the reservoirs annually, albeit in very small numbers and often for quite short periods; the main exceptions were on King George V Reservoir where they were seen virtually throughout autumn with a peak of 31 on 25 September, and at Staines Reservoir where they were present between January and April and again in autumn. They were seen irregularly on the Thames marshes, particularly at Rainham where the maximum count was 35. Large flocks were occasionally recorded elsewhere, e.g. 26 over Stanstead Abbotts GP on 17 October 1992; 37 past Dartford Marsh on 14 October 1993; and 35 over Fishers Green on 11 November 1999.

In the mid-1990s a free-flying population had been established in St James's Park as the result of captive-bred birds that had been left full-winged. Some of these moved to the Barnes area where a resident population of up to 19 became established by 1999.

21st century

Early in the 21st century only a few wild birds were seen on the reservoirs. However, by 2002 they had begun to winter at Rainham again following wetland habitat creation since the marshes have been developed as a nature reserve. Numbers continue to fluctuate there, occasionally peaking at over 100. Elsewhere, they are only an irregular visitor in small numbers and are most often seen on passage; the highest count was a flock of 28 at Fishers Green on 7 March 2005.

Summering/Breeding

In 1975 a pair was present at Rainham Marshes on 12 June and the following year a female summered at Girling Reservoir. In the summer of 1990 up to three were present at Brent Reservoir from 8 June to 10 July but there were no signs of any breeding activity; a male was present from June to September the following year and a pair was also present at Rainham Marshes in April and May, and the birds were seen displaying but were not thought to have bred. In 2001 a pair from the feral population at Barnes bred successfully at the London Wetland Centre. In 2002 two pairs summered at Rainham Marshes.

Facts and Figures

The largest-ever count was 485 birds on the Thames at Barking in January 1973.

Garganey   *Anas querquedula*

An uncommon summer visitor, occasionally in winter and a rare breeder

Range: breeds Europe and Asia, winters Africa.

20th century

Garganey was unknown in the London Area until 1927 when a male was seen in Richmond Park on 17 March and another was in Kelsey Park on 10 April. The following year a pair was seen at Barn Elms Reservoirs on 31 March. Just four years after the first sighting, a pair bred at Aldenham Reservoir in 1931; then in 1932 a female was at the same site from 25 March to 26 April. There were another five sightings up to 1945, including three pairs on the Thames opposite Syon House on 21 March 1943.

After the Second World War there was a large increase in the number of sightings in the London Area and they became a regular annual visitor. Between 1946 and 1952 one or two pairs was seen every year in the Rainham area and may well have bred, and three pairs were seen in the Lea Valley in March 1948. There was also a regular spring passage in west London, including a flock of 10 on Barn Elms Reservoirs on 18 March 1947, although there was no further evidence of breeding. During the 1950s there was an increase in the number of birds summering, and breeding occurred at Weybridge in 1952 and Broxbourne GP in 1959. The first record from Inner London came in 1952 when three were seen on the Serpentine on 12 March. There was a large passage in spring and autumn in 1959 with flocks of 20 at Swanscombe Marsh on 25 August and 11 at Brent Reservoir on 17 August.

The first winter record was in 1962 when two drakes were at Hilfield Park Reservoir on 8 December. In 1964 a pair probably bred in the Staines Moor area. In 1970 seven pairs were seen at Sevenoaks Wildfowl Reserve on 25 March but none remained during the breeding season. In 1974 a pair nested at Sevenoaks but did not rear any young. There were no further breeding attempts until 1983 when a pair summered at London Colney. In 1988 two were seen in January, a male at Wraysbury GP on the 1st and a female at Perry Oaks on the 30th.

In the summer of 1990 there were signs of breeding at four sites: there was a displaying pair at Amwell; two males and a female at Beddington for a month; and territorial males at Brent Reservoir and Regent's Park. A pair was also seen displaying at Rainham Marshes the following year. In 1993 a pair attempted to breed at Beddington. In 1994 one remained at Barn Elms Reservoirs from 11 November to 22 December, leaving only when cold weather arrived. In 1995/96 one overwintered at Brent Reservoir. In 1997 one was in the Lea Valley in January and February, and again in December. A pair bred at Rainham Marshes in 1998, the only successful breeding of the decade.

21st century

Since 2000 there have been sporadic breeding attempts in most years, typically of one or two pairs although three pairs summered and probably bred at Rainham Marshes in 2002. Small numbers are seen on passage every year, and occasionally larger flocks are recorded; for example, there were up to eight on Staines Reservoir in early September 2005. Overwintering birds were present at Island Barn Reservoir in 2001/02 and at the Wetland Centre in 2007/08.

Facts and Figures

The largest gathering was of 20 birds at Swanscombe Marsh on 25 August 1959. The earliest migrant was on 5 February 2000 at Amwell GP.
The latest migrant was on 22 December at Barn Elms Reservoirs.

Blue-winged Teal *Anas discors*

A rare vagrant (5)

Range: North America; vagrant to Europe.

The first record for the London Area was in February 1981 when a drake was present at Barn Elms Reservoirs for four days; it arrived with an influx of Teal. Three years later a female was present on Stocker's Lake in April; it associated with a small group of Shoveler, one of which mated with it on one occasion.

In 1995 a female was found at Brent Reservoir on 9 December and remained until the 12th; it reappeared on the 28th and was present again from 6 to 30 January 1996. It was likely to have been present the whole time as it associated with a group of Teal which retreated into the back of one of the marshes once the ice had melted. In 1998 a male in eclipse plumage was seen at Hilfield Park Reservoir and in 2000 a female or immature was present at Walthamstow Reservoirs in the late afternoon of 5 November.

1981: Barn Elms Reservoirs, drake on 16–19 February
1984: Stocker's Lake, female on 20–27 April
1995: Brent Reservoir, female on 9 December–30 January 1996
1998: Hilfield Park Reservoir, drake on 25 August
2000: Walthamstow Reservoirs, age/sex uncertain on 4 November

Escapes

A pair was present on Connaught Water from March 1994 onwards, with three birds on 12 December 1995; in 1996 a pair built a nest there and was still present in 1998 with the female remaining until 1999. In 2002 another pair was seen on Connaught Water and other ponds in Epping Forest for most of the year; the male remained into 2003. All these birds were believed to have escaped from a nearby waterfowl collection.

Shoveler *Anas clypeata*

A common winter visitor and scarce breeder

Range: Europe and Asia, also winters Africa.

Historical

In the 19th century Harting stated that this was 'an occasional winter visitant, never appearing in any numbers', adding that he seldom saw more than five or six together. There are dated records from Wraysbury where one was shot in September 1867, and in south London there were four on Dulwich Park Lake on 8 October 1898; they were also seen at Cobham.

20th century

By the early 1930s Shoveler had become annual visitors, albeit in small numbers; they were mostly seen on the large reservoirs between October and March although they were never resident throughout this period. During the winter of 1937/38, around 30 remained on Staines Reservoir; this was the first flock to overwinter in the London Area. Towards the end of this decade they were recorded in larger numbers with 88 on Island Barn Reservoir and 81 at Staines Reservoir, for example.

During the war, access to the reservoirs was curtailed and in the autumn of 1945 when access was restored, the number of Shoveler at Staines had risen to 269. Elsewhere, only one or two birds were seen at five other sites that year. Single Shovelers in St James's Park on 16 March 1940 and on the Long Water in May 1947 were considered to be unusual records in Inner London at the time. Numbers peaked at Staines in autumn and up to 200 wintered; towards the end of winter many of these had left and there was often a noticeable passage in March, with the rest departing during April. The only real change during the 1960s was a build-up in numbers during autumn at other sites, for example 110 at Island Barn Reservoir in September 1961.

At the start of the 1970s the wintering population continued to increase, e.g. 350 were present on Queen Mary Reservoir; in 1971 winter numbers increased to 454 there. Comparing the wildfowl counts made from the winters of 1947–48 to 1971–72 there was a 400 per cent increase in numbers. In 1978 there was a record count of 720 at Wraysbury Reservoir in November. By 1984 about 1,000 were wintering on the west London reservoirs. In 1990 there was regular build up in numbers during autumn, which peaked at 1,771 at the main 13 sites in October, with the highest numbers in winter being 300 at West Thurrock. They had also taken to wintering in the Royal Parks with peak counts of 66 in Regent's Park and 25 in Kensington Gardens in 1991.

21st century

Compared with other dabbling duck, which tend to feed mainly in shallow water, Shoveler are able to exploit the feeding resources of deep reservoirs, filtering the water with their huge lamellated bill. They can often be seen swimming in circles, each one benefiting from small organisms or particles stirred up by the one in front. Two good sites to see this are the Round Pond in Kensington Gardens and Lonsdale Road Reservoir. Three sites hold nationally important numbers of wintering Shoveler in the London Area: the Lea Valley, London Wetland Centre and Staines Reservoir. Elsewhere, they are a widespread wintering bird. The largest count since 2000 was 469 on Staines Reservoir in the winter of 2005/06.

Breeding

Breeding first occurred at Langley in 1930; then at Beddington in 1932, Queen Mary Reservoir in 1935 and 1939, and Staines Reservoir in 1939. From 1948 onwards up to three pairs bred in the Staines area and by 1952 this population had increased to 10–12 pairs. However, in 1956 only one pair bred there. Elsewhere, a pair bred at Berwick Ponds in 1951 and birds summered there, and in the Lea Valley, in subsequent years.

During the 1960s Shoveler bred several times at Rye Meads and Walthamstow Reservoirs. A pair bred at King George VI Reservoir in 1962 and one or two pairs summered at a few other sites. There was very little change in the 1970s with several pairs present during the breeding season in most years although breeding was not confirmed every year. The small breeding population was restricted to Walthamstow Reservoirs and the Staines and Wraysbury areas.

There was a small increase in the summering population in the 1980s; the most successful year was 1985 when four pairs bred on the partly drained southern basin at Staines Reservoir and another pair bred at Brent Reservoir. During the period of the second breeding Atlas in 1988–94 breeding occurred in 20 tetrads, all but one north of the Thames; about half of these were in the Lea Valley. At the start of the 1990s a relatively large population was present during the summer; in 1990 about 28 pairs were present, five of which bred – three at Hall Marsh and single pairs at Brent Reservoir and Rainham Marshes.

Breeding numbers have fluctuated since 2000 but on average about five pairs breed successfully each year with several more pairs present during the summer months. In 2000 seven pairs bred including five at Rainham Marshes, which has become the main site for summering Shoveler; in 2002, 13 pairs were present in the breeding season. In 2007 nine pairs bred in the London Area, including five at Rye Meads.

Facts and Figures

The highest count was 1,134 birds on King George VI Reservoir in November 1995.

Red-crested Pochard *Netta rufina*

An uncommon breeding resident; naturalised population

Range: Europe and Asia, also winters Africa; naturalised population in parts of UK.

Historical

The earliest reference was in Gould's *Birds of Great Britain* (1862–73), which listed a flock of 18 on the Thames at Erith, one of which was shot and illustrated by Gould. This was the only 19th-century record in Kent – they were exceptionally rare in the country at the time.

20th century

The next record was in 1924 when a male was seen on the Lea Valley reservoirs on 16 February at a time when others were seen on the East Coast. No birds were held in local collections at the time so this may be the last definite record of a genuinely wild bird. This species was kept in the collection in St James's Park and bred at least twice in the 1930s. The only other record in the 1930s was from Staines Reservoir where one was seen in 1934 on 26 April and 10 May; at the time it was thought to be an escape.

By the 1940s wild birds were breeding in Western Europe in Denmark, Germany and Holland. It was later established that some of these birds formed moulting flocks in Holland and that their migration to these areas coincided with appearances in south-east England in autumn, particularly at Abberton Reservoir in Essex. Therefore it is not inconceivable that the occasional wild bird may have found its way into the London Area. There were three in the 1940s: at Beddington on 23 November 1940; in Kensington Gardens on 14 February 1941; and at Barn Elms Reservoirs on 2 May 1946. By this time there were free-flying individuals in central London and only the one at Beddington was possibly a wild bird.

In 1950 several pairs were introduced into Regent's Park and St James's Park; they bred and the young were left full-winged. By the end of the year they had been seen in Kensington Gardens and on Barn Elms Reservoirs. Breeding occurred in subsequent years in the park collections and up to a dozen were seen in Kensington Gardens. As an example of how far these birds moved, one ringed in St James's Park in 1952 was recovered in Holland the following year. There was also a collection at Woburn, Bedfordshire.

Birds that arrived in autumn or were present in winter were usually treated as wild birds; however, this is the same time that young birds bred in captivity would disperse so for all records from the 1950s onwards it is impossible to determine the origins of these sightings. All records during the summer months are assumed to be escapes and it is likely that almost all subsequent records were also escapes.

By the early 1960s full-winged birds that had been raised in the collections in the Royal Parks had started breeding in at least two of the parks. In January 1963 there was an influx of ducks during very severe weather and three Red-crested Pochards were seen: in Valentine's Park on 8–9 January, at Barn Elms Reservoirs on 9–10 January, with another nearby on the Thames on 26 January; these may have been wild birds that had moved in from the Continent. By 1964 there was a population of up to 30 full-winged birds in St James's Park.

In the early 1970s small groups began to be seen. For example, up to four were seen at Barn Elms throughout the year, and up to four were seen in Kensington Gardens and at Rickmansworth in autumn, while on Stoke Newington Reservoirs at least six different birds were seen during January; all of these were assumed to be escapes. During the 1970s several pairs of captive birds continued to breed in St James's Park and several young often evaded the bird-keeper and ended up free-flying. Additionally, a full-winged pair bred at Kew Gardens, rearing two young in 1972. Chandler and Osborne analysed all published records of this species in the London Area between 1955 and 1974 and found that there was no conclusive evidence that genuine wild birds had occurred.

They continued to breed in the Royal Parks throughout the 1980s and 90s; as an example of their success rate there were at least 18 full-winged juveniles in St James's Park in 1990.

21st century

At the beginning of the 21st century there were virtually no breeding pairs other than those in Regent's Park and St James's Park. However, in 2000 a pair bred in Kew Gardens and a pair introduced from Regent's Park to Bushy

Park bred and raised two young. In subsequent years a small breeding population developed at the last site. A pair also bred at Maple Lodge NR in 2007.

Single birds or small groups are increasingly seen throughout the London Area and more than 15 regularly winter in the Colne Valley. They had become so well established in the London Area that they were considered to be a self-sustaining population in 2010.

Pochard *Aythya ferina*

A common resident, more numerous in winter, and localised breeder

Range: Europe and Asia, also winters Africa.

Historical

In the 19th century, this was described by Harting as 'a winter visitor, varying much in numbers in different years'. There are very few dated records and the largest flock appears to have been of 25 birds at Gatton Lake in March 1875.

20th century

During the first two decades of the century wintering numbers began to increase. In February 1904 there were 45 on the Serpentine in Hyde Park. The species had also become frequent in Richmond Park and the first large flock was counted there in February 1906, when 121 were present. The construction of more large reservoirs during this period created new habitat for many ducks such as Pochard, which they were quick to exploit. At Barn Elms Reservoirs, more than 100 were counted in November 1909 and at Stoke Newington Reservoirs, there were more than 200 in February 1914.

In Inner London Pochard remained a scarce winter visitor, appearing only occasionally in small flocks in cold weather, and then staying only for a few days. From 1924 they were seen regularly in winter in Kensington Gardens. Elsewhere in the London Area in the 1920s there were peak counts of at least 500 at Barn Elms Reservoirs on 3 December 1925; about 300 on the Lea Valley reservoirs in 1929; 200 at Brent Reservoir in 1924; and 150 on the Thames in 1929. By the 1930s they had become more common and had started to breed. Glegg stated that they were more or less regular on the reservoirs and were sometimes seen on small ponds, pits and the Thames. They could be seen in every month of the year but the highest numbers were in winter. By the end of 1939 more than 1,000 were wintering in the London Area.

The increase continued during the war years. In the severe winter of 1946/47, there were 1,160 on Barn Elms Reservoirs on 24 December. When these eventually froze up they moved to Walton Reservoirs and 1,500–2,000 were counted there in early March; later, these reservoirs also froze over and all the ducks moved over to the Thames. In the following winter there were around 2,700 on the London Area's main reservoirs, including a flock of up to 1,256 that spent several weeks at Staines Reservoir in January. There were also increases elsewhere, with peaks of 505 at Brent Reservoir and more than 500 at Walthamstow Reservoirs.

An analysis of the records around this time shows that Pochard were largely absent between April and June, although some returned as early as June in a few years. Post-breeding moulting flocks built up in July and August. Numbers then decreased in September and remained low until a build-up from late October or early November. By the middle of November, more than 1,000 were present and, depending on the prevailing weather conditions, there would be a further increase in late December. They would then typically remain until the middle of February when a large number would depart, and by early March only 200–300 remained.

In February 1956, during very cold weather, about 3,000 were seen in the London Area. Cold weather at the beginning of 1962 produced a sudden increase in numbers at Barn Elms and by 6 January about 2,000 were present there. During the severe winter of 1962/63 there was no dramatic increase in the number of Pochards as there was in other diving duck, and the largest total was about 3,000 on Queen Elizabeth II Reservoir on 9 February; in Inner London a flock of 72 flew east over Westminster Bridge on 28 February.

In the winter of 1969/70 up to 2,500 were on the Thames between Woolwich and Crossness; elsewhere the maximum counts on the reservoirs were 850 at Barn Elms, 650 at Stoke Newington and more than 600 at Staines; along with counts from elsewhere, the winter population totalled 5,200. In 1971 Pochard began wintering on the recently closed Surrey Docks and up to 1,700 were present there in January. In 1972 about 4,000 were noted on the Thames at Woolwich on 4 February and at the end of the year 2,000 were present at Walthamstow Reservoirs. Cold weather in early 1979 brought a total of 5,835 to the Lower Thames, most of which were at Crossness. A coordinated count in mid-August 1979 found 814 moulting Pochard.

There was a large influx in the winter of 1979/80 with up to 5,000 at Staines Reservoir compared to the typical wintering population of 1,100 on all the main water-bodies. From the mid-1980s onwards there was a reduction in wintering numbers due to milder winters as more ducks remain on unfrozen waters in continental Europe. There was another large build-up on Staines Reservoir in the winter of 1994/95, which peaked at 4,106 in January.

21st century

Between 2000 and 2009 the winters were relatively mild, and as a consequence the wintering population was typically below 2,000 birds. The downturn reflects the national trend, with a 46 per cent decline from 1998/99 to 2008/09 (State of UK birds 2012). The largest concentrations were on the reservoirs and larger gravel pit complexes in west London, particularly when the smaller water-bodies have frozen over. Hilfield Park Reservoir is an important post breeding moult site with up to 200 birds present in summer.

Breeding

The first breeding occurred in 1877 in Regent's Park. Then there were no more breeding records until 1923 when a pair bred in Kensington Gardens. Also around this time unpinioned females bred with captive drakes in St James's Park. Breeding began at Barn Elms in 1927 and during the 1930s there was a gradual colonisation of south-west London; it is likely these were all descended from birds bred in St James's Park, though six pairs bred just beyond the London Area boundary in Windsor Great Park in 1907 so some wild birds may have moved into west London. In 1945 a pair bred in the eastern part of the area for the first time, at Sewardstone, and two years later a pair bred towards the northern boundary of the London Area, at Old Parkbury, near Radlett.

Prior to the 1950s, in central London Pochards were mainly restricted to St James's Park, but in 1953 two pairs of unpinioned birds bred in Regent's Park. The following year there were at least four broods in Regent's Park and a pair bred in Battersea Park for the first time. In 1957 eight pairs bred in the Royal Parks and in 1958 a female with a small duckling was seen on the Thames off Victoria Tower Gardens. As well as the burgeoning population in central London there was also immigration into other parts of the London Area from the expanding population in southern and eastern England. For example, four pairs summered at Cheshunt in 1953.

By the start of the 1960s there were 14 breeding pairs in the Royal Parks. However, after the cold winter in early 1963, the breeding population was reduced to just two or three pairs in the parks and one pair at Nazeing. There was a gradual recovery and in 1967 there were 19 broods in the central parks, three broods at Walthamstow Reservoirs and single pairs bred at Finsbury Park, Hamper Mill and Thorpe GP. By the end of the decade the breeding population stood at about 30 pairs, including nine at Walthamstow.

During the 1970s there was a gradual increase in the population, with at least 13 pairs at Walthamstow and 12 pairs at Finsbury Park. By the mid-1980 the breeding population had reached 55 pairs, including 25 in Regent's Park. At the beginning of the 1990s there were about 70 breeding pairs in the London Area, a major proportion of Britain's total breeding population; the main concentration was still in the Royal Parks. The distribution of breeding Pochard doubled between the two breeding Atlas surveys.

Since the beginning of the 21st century the breeding population has decreased slightly to about 50–60 pairs annually. The highest breeding concentrations away from central London are in the Ingrebourne Valley, the Lea Valley and the London Wetland Centre.

Facts and Figures

The highest count was 5,800 birds at Crossness in January 1979.
One ringed in Kensington Gardens in December 1977 was recovered in Russia, east of the Urals, in May 1979.

Ring-necked Duck *Aythya collaris*

A rare vagrant (17)

Range: North America; vagrant to Europe.

The first record for the London Area was of a drake at Broxbourne GP for three days in late April 1974; it was presumed to be the same bird that had wintered at Marlow, Bucks, where it had been until 17 April. In 1977 a male was seen at Sevenoaks Wildfowl Reserve in 15 April and probably the same bird was seen there again in November. Although these sightings were treated as separate birds by BBRC at the time, many rare ducks do revisit previous sites and given this species' rarity in the London Area at the time these records should be treated as relating to the same bird. The following year a drake was seen in the Royal Parks in May and June; the time of year does cast some doubt on the origins of this bird but the locations should not as both park lakes have attracted rare ducks before and since.

In 1981 a drake was at Amwell in April and May. The first female seen in the London Area wintered in early 1989 at Yiewsley GP and Little Britain Lake. Sightings of the species increased significantly during the 1990s when seven more were seen, and Ring-necked Duck was removed from the BBRC list at the end of 1993.

In 2000 a drake was present on Walton Reservoirs on 7 April. In 2001 three were seen: a drake on Hilfield Park Reservoir in May; a first-winter female on Island Barn and Walton Reservoirs from November into 2002; and another first-winter female at Bourne Hall Park, Ewell, from December into 2002. The last two birds roamed widely and the Ewell bird returned in 2002/03. Also in 2002, a drake was present on Broadwater Lake between May and August, the first time one had summered in the London Area.

1974: Broxbourne GP, male on 28–30 April
1977: Sevenoaks WR, male on 15 April and 27 November
1978: St James's Park, male on 23 May, also at Regent's Park on 14 June
1981: Amwell GP, male on 14 April–14 May and again on 25–26 May
1989: Yiewsley GP/Little Britain Lake, female on 14 January–2 May
1992: Lonsdale Road Reservoir, male on 20 April
1993: Stocker's Lake, male on 14–17 May, also at Broadwater GP on 15 May
1995: King George V Reservoir, female on 28–29 September, 21 October, 3, 7 and 15 November; also at Fishers Green on 8 October; seen again at KGV Reservoir on seven dates in 1996 up to 17 April and at Bowyer's GP on 8 and 19 April 1996
1997: Surrey Docks, first-winter male on 11 January–13 February; also at Blackwall Basin on 26–27 January and 4 February
1997: Walthamstow Reservoirs, male on 24 February
1997: Gatton Park Lake, male on 23 March
1999: Valentines Park, female on 20 February–8 May
2000: Walton Reservoirs, male on 7 April
2001: Hilfield Park Reservoir, male on 20 May
2001: Bourne Hall Park/Court Lake, Ewell, first-winter female on 22 December–17 February 2002; and again on 5 December–26 February 2003; also seen at Island Barn Reservoir on 1 March 2003
2001: Island Barn/Walton Reservoirs, first-winter female on 25 November–4 May 2002; also seen at Field Common GP and River Ember at East Molesey on several dates in 2002, London Wetland Centre on 3 February 2002, Beddington SF on 24 and 26 February 2002, Shepperton GP on 4–14 April 2002 and Charlton GP on 6 April 2002
2002: Broadwater Lake, male on 6 May–24 August

Ferruginous Duck *Aythya nyroca*

A scarce winter visitor (59)

Range: Europe and Asia, most winter Africa.

Historical

This was a rare winter visitor in the 19th century with only one well-documented record of a female shot at Brent Reservoir in 1863. This species was also on a list of wildfowl recorded in winter in London between 1860 and 1864 by Hussey, writing in *The Zoologist*. He recorded one or two in St James's Park and on the Round Pond in Kensington Gardens annually from 1860 to 1863; however, Harting did not consider that escaped and semi-domesticated birds had been ruled out and these records were also dismissed by Glegg.

20th century

A pinioned pair was added to the collection in St James's Park in 1912 and three more were added in 1930; in 1932 a pair raised five young which were later seen on other waterbodies in the London Area. In 1938 they raised seven young and a pair seen at Barn Elms on 17 July was assumed to have come from St James's Park.

There were no more acceptable records of wild birds until September 1928 when a pair was seen on Staines Reservoir. Glegg did not accept this record but given the absence of escaped birds at the time, it should be treated as a valid record. The next record was in 1947 when an adult drake was seen at Barn Elms in January during a severe winter. Although it is not known whether there were any remaining in St James's Park at the time, the Duke of Bedford stated that 'the war has had a devastating effect on collections of waterfowl ... for the next few years, at any rate, I think it will be fairly safe to assume that any full-winged duck, even if rare and unusual, is not one that has escaped from captivity'.

In the 1950s this species was regularly seen in eastern England and it was believed that these birds arrived from the near Continent. A few made it into the London Area and most of these were considered to be of wild origin. However, up to three seen in 1952 at Osterley Park and Southall Aviary from 8 to 11 October were most likely to have been escapes. By 1960 full-winged birds were known to be present in the collection in Regent's Park, and birds from there seen in St James's Park in August and at Brent Reservoir in September. Despite the influx of other diving ducks in the severe winter in early 1963 there were no reports of any Ferruginous Ducks in the London Area.

During the remainder of the 1960s and into the 70s there was a succession of records, most of which were considered at the time to be escapes, even ones occurring away from central London such as in the Lea Valley and at Staines Reservoir. Three presumed wild birds were seen in 1969, including a female in Hyde Park and Kensington Gardens from January to April, which did not join the other birds being fed by the public. In 1971 an unprecedented six birds were seen, beginning with a pair at Surrey Docks, which arrived with an influx of diving ducks during cold weather on 7 January. One presumably returned to Surrey Docks in subsequent winters and additional birds were also seen there.

Between 1978/79 and 1983/84 a drake wintered annually on Little Britain Lake and surrounding waters. There was a national influx of this species in late autumn 1986, with four birds arriving in the London Area, one of which wintered in the Lea Valley. The following winter two more overwintered, a female at Stocker's Lake and an immature drake at Walthamstow Reservoirs. They were seen in most years up until 1992, then no more were seen until 1999 when two were seen in the Lea Valley.

21st century

Since the start of the new millennium there has been an increase in sightings, with 10 accepted records. Of these, seven were in the eastern half of the London Area suggesting genuine arrivals from the Continent.

1863: Brent Reservoir, shot on 24 December
1928: Staines Reservoir, pair on 24 September
1947: Barn Elms Reservoirs, male on 12 and 29 January
1951: Dartford and Stone Marshes, male on 23 January–7 February and two males on 4 March

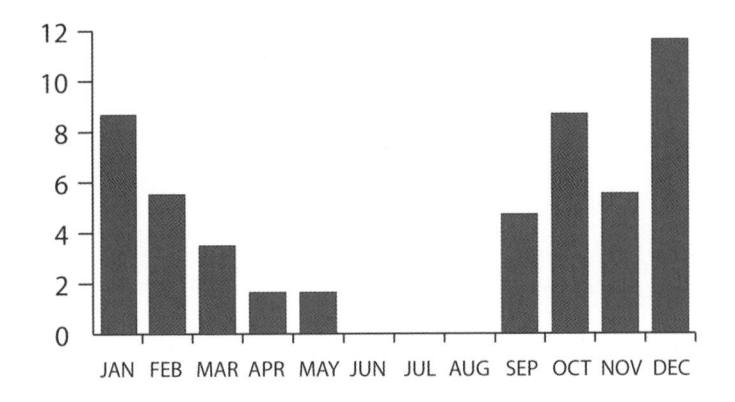

Seasonal occurrence of Ferruginous Duck: arrival dates (only birds considered to be wild).

1956: Barn Elms Reservoirs, on 19–20 February, then on Walton Reservoirs on 23–28 February
1958: Queen Mary Reservoir, on 8 December
1960: Barn Elms Reservoir, on 10–27 January
1961: Hilfield Park Reservoir, on 26 November
1963: Ruxley GP, on 27 October
1964: King George VI Reservoir, on 19–31 December; then Staines Reservoir on 2–3 January 1965
1965: Brent Reservoir, on 5 November
1969: Hyde Park and Kensington Gardens, on 31 January–2 April
1969: Barn Elms Reservoir, on 9–14 February
1969: Fishers Green, on 2 April; also 2 February 1970 and 29 October 1970
1970: Swanscombe, on 21 March
1971: Surrey Docks, pair on 7 January 7th, one until 10th; also from 1971/72 to 1974/75
1971: Barn Elms Reservoirs, drake on 24–27 January
1971: Staines Reservoir, drake on 29 October
1971: Cheshunt GP, on 14 November
1972: Walton and Queen Elizabeth II Reservoirs, on 15 October
1972: Wraysbury GP, intermittently on 29 December–17 February 1973
1973: Walthamstow Reservoirs, on 18–25 February and 15 December
1973: Barn Elms Reservoirs, on 9 and 23–24 December
1973: Brent Reservoir, on 26 December
1974: Surrey Docks, second bird on 11 January (see 1971 Surrey Docks record)
1975: Surrey Docks, female on 3–6 February
1975: Little Britain Lake, on 9 February
1975: Wraysbury GP, on 21 December
1978: Little Britain Lake, drake on 2–18 November and in subsequent winters until 25 March 1984; also on nearby waters.
1982: Wraysbury Reservoir, drake on 17 January
1982: Stocker's Lake, female on 30 October
1983: Cheshunt GP, drake on 17 September–23 October
1983: Walthamstow Reservoirs, female on 31 December–7 March 1984
1985: Weald CP, on 16 February
1986: Wraysbury Reservoir, on 20 September
1986: Staines Reservoir, on 26 October–26 November, with two on 2 and 15 November
1986: Lea Valley, on 15 November–2 January, then Walthamstow Reservoirs on 6 January–25 February
1987: Walthamstow Reservoirs, on 2 December–26 February
1987: Stocker's Lake, on 19 December–11 February 11th and again on 15–31 December 1989

1990: Beddington, on 11 December
1991: Amwell, on 9 September
1992: Amwell, from 8–10 November
1992: Stocker's Lake, on 23–29 December; also seen at Bury Lake
1999: Cheshunt, on 1 March
1999: Seventy Acres Lake, on 28 October–19 December
2000: Kempton Park NR, drake on 22 May
2001: Netherhall GP, female on 6–15 October; also at Amwell on 3–4 November
2002: Wraysbury GP, on 1 October; also at other sites to 30 November
2006: Rainham Marshes, female on 22 May
2007: Girling Reservoir, drake on 8–17 March
2007: Harrow Lodge Park, on 7–10 April; also Dagenham Chase on 7–8 April
2008: Amwell, drake on 11–31 October
2010: Burgess Park, on 28 November
2010: Queen Mary Reservoir, on 19 December
2012: Thamesmead, on 15 January

Escapes

During the 1950s to the 70s there were various escapes, mostly from local collections. In 1981 a male was seen in the Barn Elms Reservoirs area on various dates between 11 February and 9 November; due to the length of time it was present it should be treated as a probable escape. In 1993 a presumed escape adult drake was present at Cheshunt from 30 November to January 1994 and again from 26 December into 1995. In 2000 up to two were seen in Bushy Park. In 2007 one seen at Staines and King George VI Reservoirs between 14 May and 8 June was thought to have been an escape.

Tufted Duck *Aythya fuligula*

A common resident, more numerous in winter, and localised breeder

Range: Europe and Asia, also winters Africa.

Historical

Fossil remains dating back to the Pleistocene epoch have been found at Walthamstow. In the 19th century this was described by Harting as 'a regular winter visitant, but varying much in numbers in different years'. He added that it was usual to see 10 or a dozen but up to 30 had been seen at Brent Reservoir during hard weather. In south London they were rare and only three records are known, at Gatton Lake, at Poynter's near Cobham and on the Thames at East Molesey.

20th century

Early in the century they started to become more widespread in the London Area and slowly increased in number. For example, in 1902 there were more than 50 at Barn Elms Reservoirs and a flock of 90 was seen in Battersea Park in 1905. The first large flocks occurred in 1929 when almost 500 were on Walthamstow Reservoirs on 10 March and about 410 at Molesey in September.

    The wintering population continued to increase during the 1930s. Large flocks built up on the reservoirs during late summer before decreasing at the end of autumn; the onset of winter saw a new increase and these birds remained until early March when they started to leave, most having departed by the end of April. The largest flocks could still be found at Walthamstow, where they had increased to about 1,000 by 1932, and 600–700 birds wintered at Staines Reservoir. There were also counts of several hundred on some of the other reservoirs and occasionally on the park lakes in central London. A coordinated count in December 1937 produced a total of almost 2,500. The largest single count was at Barn Elms Reservoirs in December 1938 when 1,400–1,500 birds moved in following a period of very cold weather. By the end of the decade more than 3,000 were present on the main reservoirs in winter.

During the war years large numbers were recorded on the Thames in the very cold winters. Previously, small flocks were seen on the river, the maximum being about 80 at Chelsea in 1928, but in February 1942, 700–800 were counted at Chelsea and in 1945 there were about 2,500 upstream from Chiswick on 28 January. The winter of 1946/47 was exceptionally cold and by February many of the smaller waterbodies were completely frozen; more than 3,500 were then concentrated on Walton Reservoirs. The following four winters were mild by comparison and the average total population on the main reservoirs in winter was a more normal 2,500–3,000. In addition, another 300–400 wintered at St James's Park.

Between the winters of 1947/48 and 1955/56, the population increased by about 70 per cent. In February 1956 at least 6,400 birds were seen during very cold weather. In the severe winter of 1962/63 there was a dramatic influx in January with a total count of 6,632, most of these remaining well into February despite many reservoirs freezing over. In St James's Park around 2,000 were present in January and February. Most of the ducks concentrated onto the few remaining larger reservoirs with open water where they peaked at 3,350 on Walton Reservoirs on 4 March. At Staines Reservoir a passage of birds was noted during March. For example, in 1965 the numbers increased from 200 to 1,500 in two days, and dropped to 600 four days later. During the next few winters the typical overwintering population was about 5,000, with most birds on just two reservoirs, Barn Elms and Staines.

At the start of the 1970s large numbers were seen on the Lower Thames, with up to 800 between Woolwich and Crossness. By 1980 there were no longer large flocks wintering on the Thames but numbers on the reservoirs had increased, with up to 4,500 at Staines at the beginning of the year and 2,000 at King George V in November. A coordinated count to locate moulting Tufted Ducks on 18–19 August 1979 found just over 6,600, the highest figure in the country and the third highest in western Europe. The August moulting flocks peaked at 7,739 in 1986 including 2,914 on Staines Reservoir; later years were typically around 5,000.

21st century

The wintering population remained stable for much of the first decade of the 21st century but there was a significant increase in December 2010 when around 9,000 birds were present; this was the coldest December for a century. Three sites were listed as holding nationally important numbers (1 per cent of the UK population) in 2010/11: the Lea Valley gravel pits, Staines Reservoir and Walthamstow Reservoirs; the last two sites host post-breeding moulting flocks of more than 1,000 birds. Smaller numbers can be seen on many gravel pits and park lakes, where they appear quite tame, often joining other waterfowl in taking advantage of food offered by the public.

Breeding

The breeding population in the London Area can be traced back to 1831 when the species was first exhibited at London Zoo. It bred regularly on the lake in Regent's Park up to 1848. At St James's Park a pair raised four young in 1838; this family often flew over to Hyde Park and three broods were raised there, on the Serpentine, the following year.

Although captive birds bred in the central parks, the population of wild birds was spreading across the country at the time so today's breeders could be descended from either source. That said, the gradual spread of breeding birds was from central London outwards. The first breeding of full-winged birds took place in 1901 on the lake in Dulwich Park. A pair bred at Walthamstow Reservoirs in 1905 and breeding commenced at Victoria Park in about 1912; at Connaught Water, Finsbury Park and Winchmore Hill in 1913; Snaresbrook in 1916; Hampton Wick in 1918; and Kelsey Park in Beckenham in 1920. By 1913 full-winged birds were nesting freely in the London parks. The population continued to expand during the 1920s with breeding reported from Godstone in 1922; Barn Elms in 1925; Richmond Park in 1927; and on gravel pits in the Beddington district in 1929. The last four sites provide a good example of the species' successful colonisation since within a few years up to 25 pairs were breeding there. In 1929 about 100 ducklings were counted on the lake in St James's Park. One of the American White Pelicans in the park habitually ate young Tufted Ducks in the 1930s and some of the other pelicans followed suit.

During the 1930s Tufted Ducks colonised other reservoirs, both in the suburbs and beyond. Towards the end of the decade there was a minimum of 53 pairs breeding at 17 locations; this excludes the central London parks, which were not monitored. About two-thirds of the population was in south London, including 12 pairs at Gatton Lake. They bred regularly in Hyde Park and Kensington Gardens from 1924 to 1938. However, the Second World War put paid to this population and there was no further breeding there until 1954. Occasional pairs also bred in Buckingham Palace Gardens, Southwark Park and Victoria Park, and a pair bred for the first time in Battersea Park in 1955.

By the beginning of the 1950s the highest breeding population outside the central London parks was Walthamstow Reservoirs where there were up to 12 broods annually. In 1953, 10 pairs bred in Kew Gardens and in 1957 four pairs bred at Regent's Park, the first breeding there since 1944. By 1959 the Lea Valley breeding population had reached 33 pairs. In 1961 the breeding population in the London Area was more than 100 pairs; by 1967 there were 60 nests in St James's Park and 15 broods in Regent's Park. Away from central London the most important breeding area was the Lea Valley. The population continued to expand along the Lea on the many reservoirs and gravel pits, and by the end of the 1960s about 50 pairs bred there.

The first breeding Atlas in 1968–72 estimated a population of about 300 pairs. In 1982 the breeding population was estimated to be in the region of 400–500 pairs. A breeding survey in 1984 suggested that the actual population was between 783–1,027 breeding pairs (Oliver, 1984). By 1985 the Lea Valley was the most important area for breeding Tufted Ducks with 110 broods counted.

The second Atlas in 1988–94 found breeding pairs in 193 tetrads, an increase of almost 80 per cent since the first Atlas. The majority of these were concentrated in the built-up areas of the London Area as well as the Colne and Lea valleys, and the Lower Thames Marshes. Good numbers of young continued to be reported through the 1990s, with 200 or more broods most years, and over 250 in some years, but there was a slight reduction to 151–213 from 2006–2010.

Ringing Recoveries

Many Tufted Ducks have been ringed or recovered on the south-west reservoirs and in St James's Park. The results show that birds wintering in the London Area have come from Iceland, Sweden, Finland and Russia. It has also been shown that some birds move on during the winter as birds ringed in the London Area have been recovered elsewhere in Oxfordshire, the Thames Estuary and Co. Antrim later in the same winter.

Scaup *Aythya marila*

An uncommon winter visitor

Range: breeds northern latitudes, winters on European and North American coasts.

Historical

In the 19th century this species was rare in the London Area with about 11 records. Brent Reservoir accounted for four of these: two were shot on separate occasions either in or before 1843, and there was a female on 2 November 1867 and one in 1892. Elsewhere, a pair was shot at Munden, probably in the 1840s; singles were shot on the Thames at West Molesey in February 1874 and 1877; six or seven were at Gatton on 8 January 1871; one was shot at Cobham; a male was at Ruislip Lido on 30 January 1886; and five were on Dulwich Park Lake on 8 October 1898. Ticehurst stated that they were also taken several times at Dartford.

20th century

During the first two decades of the century it was still a rare bird in southern England and there was only one sighting in the London Area, of a female shot on West Thurrock Marshes in October 1905. From the mid-1920s it became more regular and up to nine were seen annually until the 1950s. Most records were of single birds, but in 1933 eight were seen at Walthamstow Reservoirs on 25 February. Even larger numbers were seen in severe weather; for example, in 1939 there were 17 at Swanscombe on 8 January. An adult drake Scaup wintered in the London parks every year from 1939/40 to 1946/47. During the winter of 1946/47 small parties were present for several months on the Thames and the reservoirs, including 17 at Chiswick Eyot on 9 March, and there was a combined count of 26 on 23 February at Barn Elms Reservoirs and on the Thames at Putney.

There was an influx in early 1954: birds were present at Queen Mary Reservoir from 6 February to 7 April, with a peak of 20 on 6 March, and small groups were present at several other localities. There was an even larger influx during severe weather in February and March 1956 with at least 58 including 24 on Staines Reservoir. There was little change of status during the late 1950s and into the 60s. One or two were seen in Inner London every winter, either on the park lakes or on the Thames. During the exceptionally cold winter of 1962/63 there was virtually no difference to the number seen on the reservoirs but on the Thames there was an unprecedented flock of at least 100 at Stone on 3 February.

In 1970 up to 18 were on the Thames during January in the Rainham and Dartford area. The following year a pair of Scaup was present at Brent Reservoir from 12 April to 17 May 1971 and was seen defending one of the nesting rafts from other birds. In early 1985 there was a large influx, with 45 seen from 14 to 19 January. Severe weather in early 1987 brought an influx into the London Area for the third year running, with around 60 seen including a flock of 11 on Queen Mary Reservoir on 18 January.

During the 1990s fewer were seen apart from in two years when there were large influxes. 1991 was probably the best year ever with about 130, many of which arrived in February and early March in cold weather; they were widely seen but with no particularly large flocks; in Inner London there were six on the Thames at Chelsea Reach on 8 February. About 100 were seen in 1997, including eight at Blackwall Basin on 20 January.

21st century

Since 2000 the average number of Scaup seen each year has been about 24 with a peak of 38 in 2001. Although predominantly a maritime species in winter, small numbers can be seen on almost any water-body; they are most frequently seen on the large reservoirs in west London. One male was resident in Regent's Park from the late 1990s to at least 2010 and occasionally visited other park lakes. A drake summered on Staines Reservoir in 2009.

Facts and Figures

The only ringing recovery is of a juvenile that was ringed in Iceland in 1947 and recovered at Lonsdale Road Reservoir on 5 March 1950.

The largest flock was of at least 100 birds on the Thames at Stone on 3 February 1963.

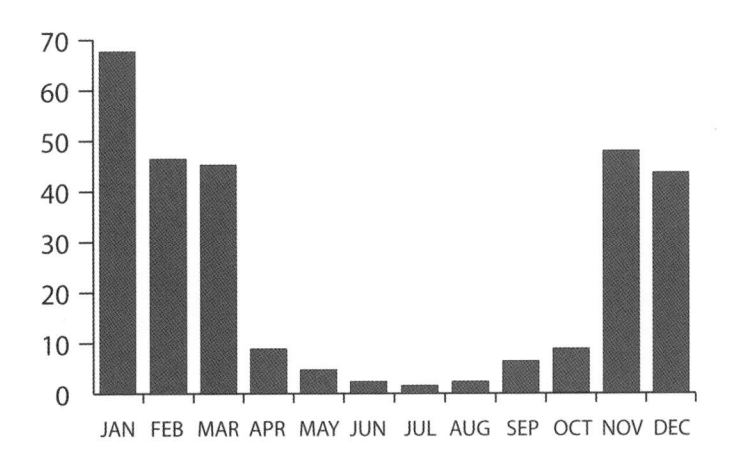

Seasonal occurrence of Scaup: 2000–2009.

Lesser Scaup *Aythya affinis*

A very rare vagrant (3)

Range: North America; vagrant to Europe.

There have been three records of this North American duck, all since the mid-1990s. The first was an adult drake at Tyttenhanger GP in April 1996. This was just nine years after the first ever British record, in Staffordshire in 1987. Despite the population decreasing in North America there has been a notable increase in sightings in Great Britain with over 150 accepted records up to 2010. Most arrivals nationally are between October and December.

In 2003 a first-winter drake was seen at Stocker's Lake on 4 January; it was not seen again until 14 February when it was discovered at Brent Reservoir, where it remained until the 23rd before disappearing again. On 3 March it was found in Regent's Park where it was seen intermittently until 8 April; during this period it spent three days in the Lea Valley, on Seventy Acres and nearby lakes from 23 to 25 March. It finally made one last visit to Brent Reservoir on 17 April. During its spell in Regent's Park it moulted into adult plumage. Five years later an adult drake was found at Wraysbury GP with a flock of Tufted Ducks in October; a few days later it was relocated on Queen Mother Reservoir, where it remained for eight days.

1996: Tyttenhanger GP, male on 7–18 April
2003: Stocker's Lake, first-winter male on 4 January; also at Brent Reservoir, Regents Park and Seventy Acres Lake on 14 February–17 April
2008: Wraysbury GP, male on 5 October, then on Queen Mother Reservoir on 8–15th

Eider *Somateria mollissima*

A scarce winter visitor

Range: breeds coastal northern Europe, Asia and North America; winters on coasts.

20th century

The first record was in 1917 during very cold weather when a flock of seven was seen on the River Roding near Ilford on 21 February; there were 10 present on 10 March. There were no further records until 1957 when six were on King George VI Reservoir from 12 November to 1 December; there was also one on Girling Reservoir between 30 November and 29 December.

In the 1960s the number of sightings began to increase. In 1961 a female was seen at Hilfield Park Reservoir on 17 December and on the same date two arrived at Brent Reservoir and remained until 27 December. In the severe winter of 1963 a flock of six was seen at Sevenoaks on 2 January and one was on Hilfield Park Reservoir on 5–6 January. No more were seen until 1973 when a drake was on Island Barn Reservoir on 16 December; it remained until 19 June 1976, a stay of 30 months. At least one was seen every year during the rest of the decade.

During the 1980s Eider were seen in every year except 1983; all were single birds apart from two flocks, four on Queen Mother Reservoir on 8 November 1980 and of 25 birds at Walthamstow Reservoirs during a snowstorm on 9 January 1982. Most were on reservoirs and lakes apart from a rather sick individual on a garden pond in Streatham and one on the Thames at Rainham.

In the 1990s there were influxes in December 1995, including a flock of 35 that flew upriver past Thamesmead on the 7th, and again in December 1997 when 23 arrived in the Lea Valley, the peak count being 18 at Cheshunt. The following year eight arrived at Queen Elizabeth II Reservoir on 21 November, five of these remaining until February 1999. Also in 1999, four immature drakes were on Queen Mary Reservoir between 29 October and 8 November.

21st century

The first sighting of the new millennium was the long-staying female on Queen Elizabeth II Reservoir from 1998 which remained until 13 March 2000. In 2002 a female flew downriver past Dartford Marsh on

14 December. The following year an immature male was seen on the Thames at Northfleet on 17 October. In 2007 two were seen on King George V Reservoir on 27 April.

Escapes

A pair of captive birds bred in 2004–06 in St James's Park; the young were left full-winged but remained in the park until most of the ducks died from botulism.

Facts and Figures

The largest flock was of 35 birds at Thamesmead on 7 December 1995.

King Eider *Somateria spectabilis*

A very rare vagrant (1)

Range: breeds Arctic Circle, winters northern Europe and North America; rare visitor to UK.

The only record was of an adult male shot on the Thames at Erith Reach. There had only been four other records in the UK at the time, three in Scotland and one in East Yorkshire; a number of earlier records from East Anglia have since been deemed unacceptable. The London specimen was bequeathed to the Oxford University Museum of Natural History in 1847.

Pre-1847: Erith Reach, male shot

Escapes

In 2002 a pair was present at Chalfont Park on 1 October.

Long-tailed Duck *Clangula hyemalis*

A scarce winter visitor

Range: breeds Arctic Circle, winters Europe and North America.

Historical

The first record was of one shot on Dartford Marshes in the winter of 1829/30.

20th century

There were no more records until almost a century later when one remained at Barn Elms Reservoirs from 13 November to 17 December 1928. From the early 1930s they suddenly became near-annual visitors. In the winter of 1932/33 there were up to four at Staines Reservoir and one at Walthamstow Reservoirs; presumably one of the same then wintered at Staines in 1933/34. Another four were seen at Staines on 18 October 1934, one of which remained until December. No birds were seen in 1935 but one wintered again every year from 1936/37 until at least 1939/40 when access to the reservoirs was denied during the war years.

The first post-war record was in the severe winter of 1946/47; up to two were seen from 25 February to 5 August at a number of locations, including the Thames at Hammersmith, Barn Elms and Walton Reservoirs, and St James's and Regent's Parks. Apart from in a couple of weeks in June, one of these remained in St James's Park from 23 March to 5 August; it swam around the lake with Mallards and ate bread offered by the general public, demonstrating that wild birds can become habituated very quickly.

From the early 1950s to the mid-1980s one or two were present almost annually on the reservoirs, occasionally staying all winter. In the first few months of 1968 six were present, including three on King George V Reservoir from 3 to 17 March, two of these remaining until 4 May. In 1972 one was on the Thames at Rainham and Erith from 22 to 29 October. In 1982 no birds were seen for the first time in 25 years.

The succession of mild winters at the end of the 1980s continued into the early 1990s resulting in fewer Long-tailed Duck sightings. In October 1991 one arrived in the Colne Valley and wintered there every year until 27 April 2000, usually on Broadwater Lake. In 1992 six were seen, all remaining for some time. In 1996 a summer-plumage male was on King George V Reservoir from 18 to 23 June.

21st century

After the demise of the regular Colne Valley bird, Long-tailed Ducks were infrequently seen in the London Area. In 2001 an immature drake was on Walton Reservoirs on 9 December; in 2002 a drake was at Brent Reservoir on 23 March; and there were two in 2003 on the south-west London reservoirs, one of which overwintered. In 2004/05 two wintered in east London, at Stubbers and Grange Waters. Most UK birds winter along the Scottish coast with only small numbers venturing further south than Norfolk. A combination of mild winters and a decline of 60 per cent in the Baltic Sea, which is their main European wintering area, between 1992/93 and 2007/09 contributed to a total absence of birds in the London Area from 2006–2009. Most unusually, an adult female over-summered in 2013 on various small urban water-bodies in southeast London.

Facts and Figures

The largest group was of five birds on Staines Reservoir from 4 to 13 April 1986.

Common Scoter *Melanitta nigra*

An uncommon passage migrant

Range: breeds northern Europe and Asia; winters at sea in Europe and North Africa.

Historical

In the 19th century Common Scoter were infrequently encountered in the London Area. In the 1860s Harting stated that he knew of two that were shot at Brent Reservoir in severe weather and was told that, many years earlier, a few were generally seen there every winter; subsequently they were recorded in 1890 and 1897. Elsewhere, the only other records were: one shot near Hoddesdon at Field's Weir in October 1868; one shot at East Molesey in 1878; a pair at Bushey Heath in February 1881; and one shot at Rickmansworth in November 1898. Ticehurst stated that they had been obtained at Dartford from time to time.

20th century

In the early part of the century there were records from 1909 and 1910, and one was seen on the Thames at Vauxhall in March 1912. Since the mid-1920s they have been seen every year. Glegg described them as 'an irregular visitor', adding that they had occurred in all months of the year excepting January and October. April accounted for the most records and the largest flocks – of 16 birds on 3 April 1932 and of 13 on 25 April 1933 – were both on Staines Reservoir.

Between 1930 and 1979 there was little change to the status of Common Scoter. Most records were of birds on the large reservoirs, particularly at Staines and on the adjacent King George VI Reservoir, where there were 18–20 on 7 June 1953. Occasional birds were seen on much small waterbodies, including on the Beverley Brook on Wimbledon Common, the River Wandle at Carshalton, a sludge bed at Beddington SF and the Long Water in Kensington Gardens (in July 1940). In 1956 about 37 were seen during severe weather in February and March, including 19 at Staines Reservoir. During the severe winter of 1962/63 only a dozen were seen in the London Area between January and March, six of which were found dead. In 1964 a flock of 27 flew in from the east and landed on Staines Reservoir on 27 July.

By the start of the 1980s small numbers were being seen annually. Exceptionally, on 6 November 1980 a flock of 79 females/immatures appeared on Walthamstow Reservoirs. An analysis of all records between 1940 and 1986 showed a large peak in April, which is the peak month for birds passing up the English Channel. In 1988

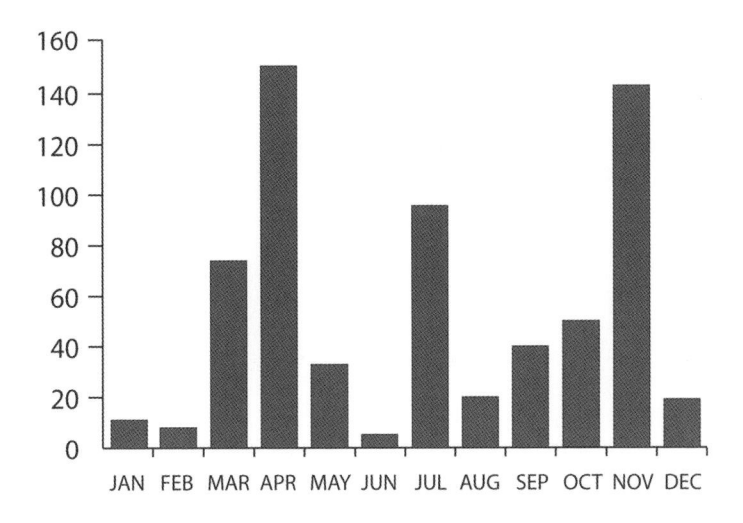

Seasonal occurrence of Common Scoter: 2000–2009.

about 80 were seen, including a flock of 33 on Girling Reservoir from 11 to 13 November; the flock gradually declined and two remained to 1 January 1989. About 195 were seen in 1996, with flocks of 65 on Girling Reservoir and about 50 over Barn Elms Reservoirs, both on 7 April.

21st century

Since 2000 they have become an irregular passage migrant, sometimes seen in large flocks, with occasional sightings in winter. They are seen in varying numbers each year, from about 40 to 150 with most occurring on the large reservoirs or on the Thames, particularly at Rainham.

Facts and Figures

The largest flock was of 90 birds on Queen Mother Reservoir on 8 November 2004.

Velvet Scoter *Melanitta fusca*

A scarce winter visitor

Range: breeds northern Europe, winters at sea in Europe.

20th century

The earliest reference was in 1909 when Ticehurst stated that 'they have even been seen up the Thames as far as Dartford'. In 1927 an immature was at Barn Elms Reservoirs on 3 December; this was probably the same bird that was later seen at Staines Reservoir on 13 December. In 1929 there were two in the Lea Valley from 16 February to 3 March and one at Staines Reservoir on 23 November. Only two were seen in the 1930s: an immature male at Walthamstow Reservoirs from 20 February to 9 April 1932 and a female at Staines Reservoir on 18 April 1937.

No more were seen until the severe winter of 1947 when one was seen at King George V Reservoir on 8 February; it was later found dead along with three others. In 1948 there was an unprecedented influx of 25–26 on 30 October, 20 at Staines, three or four at Brent and singles at Walton and Walthamstow Reservoirs. All had departed by the following day, although another was seen two weeks later, at Barn Elms on 14 November, and one was found dead at Brent Reservoir on 27 November. The last record of the decade was in 1949 when two drakes were seen at Staines on the unusual date of 7 July.

In the 1950s the number of sightings started to increase, and six were seen between 1950 and 1955, all in west London. Severe weather in early 1956 brought eight into the London Area, including four at Walton Reservoirs and one on the Pen Ponds in Richmond Park on 30 January. The winter of 1962/3 was even more severe and six were seen in the London Area between 12 January and 10 February, three of which were found dead. Towards the end of 1964 a flock of six was on Queen Mary Reservoir between 29 October and 4 November, two to three remaining until the 28th and one staying until the end of the year. In 1966 three were on King George V Reservoir on 19 February, and there were four at Brent Reservoir on 14 March.

In 1970 a flock of nine on Queen Mary Reservoir on 18 October gradually dwindled to one by 22 November. Another three or four were seen on the same reservoir the following year. In early 1979 six were seen, including three at two sites in the Colne Valley on 23 January. By the start of the 1980s Velvet Scoter had become an almost annual visitor, though mainly confined to the large reservoirs in west London and the Lea Valley. In 1984 a female was picked up in fog at Shooters Hill on 11 December and later released. An influx of about 17 occurred during early 1985 following freezing conditions on the Continent; the peak count was four at both Wraysbury and Walthamstow Reservoirs. In 1988 seven arrived on Girling Reservoir on 20 November and two remained from 22 November to 2 January 1989.

In the autumn of 1990 there was a small arrival of sea-duck, including Scaup and Long-tailed Duck. This also brought two Velvet Scoters to Girling Reservoir on 28 October, where they remained until 25 November. In 1994 three immatures were on Old Slade GP on 29 November.

21st century

This species is much scarcer than the Common Scoter in the London Area, generally less likely to be seen inland, its main UK haunts being the east coast of Scotland, with a few found on the east and south coasts of England. Since the beginning of the 21st century the species has become very rare and the only records between 2000 and 2007 were in the winter of 2001/02; an immature arrived at Walton Reservoir on 7 December, quickly followed by four immatures two days later on Island Barn Reservoir; all overwintered until April 2002 and occasionally joined up. There were two in 2008, both on Queen Mother Reservoir, and a flock of five on Staines Reservoir on 13 April 2009.

Facts and Figures

The largest flock was of 20 birds at Staines Reservoir on 30 October 1948.

Goldeneye *Bucephala clangula*

An uncommon winter visitor

Range: Europe, Asia and North America; in UK breeding restricted to Scotland, widespread in winter.

Historical

Harting described the Goldeneye as 'an uncertain winter visitant, never appearing in large flocks...but usually seen in little parties of ten or a dozen'. It was first recorded at Brent Reservoir by Bond, writing in 1843. Elsewhere, the only other records are of a pair shot at Wraysbury in about 1858; one in Hyde Park in January and February 1864; near Kew in 1898; and undated records from Fetcham Mill Pond and Poynter's, near Cobham.

20th century

In the early part of the 20th century they became a more regular winter visitor following the construction of several large reservoirs. Staines Reservoir was the favoured locality, a flock having first been recorded there in 1906. By the early 1920s wintering numbers there had increased to 25, then to an average of 40 by the early 1930s, with flocks of about 50 in 1931.

From 1925 onwards they became regular at Barn Elms although always in single figures, and by the 1930s there was a regular wintering flock at Walton Reservoirs. Elsewhere small numbers were regular on Kempton Park

Reservoirs and occasional birds were recorded from other sites. A Goldeneye on the Round Pond in Kensington Gardens from October 1934 to May 1935 was killed by Mallards. In the severe winter of 1946/47, more than 20 were seen on the Thames at Hammersmith.

By the 1950s they were still only regular winter visitors to the reservoirs at Barn Elms, Staines and Walton, with occasional flocks being seen in the Lea Valley. During very severe weather in early 1963 a total of 106 were counted on 13 January. By then many reservoirs had started to freeze over, most ducks had taken to the Thames and about 40 Goldeneye were seen in two flocks between Kew and Richmond. The wintering population gradually increased and by the beginning of the 1970s it was more than 100.

In the mid-1980s severe cold weather in Europe brought influxes into the London Area and in March 1986 there were at least 477, including 109 at Walton Reservoirs. There were two counts exceeding 100 in 1987: 147 on King George VI Reservoir in March and 120 on King George V Reservoir in April. There was little change during the 1990s with only occasional influxes during periods of very severe weather. The favoured location was Staines Reservoir where the peak count of 157 in March 1999 included birds on passage.

21st century

Since 2000 around 300–350 have overwintered on the large reservoirs and gravel-pit complexes. The peak counts usually occur in March when there is a notable passage at some reservoirs, especially on Staines.

Breeding

Goldeneye has never bred in the London Area but there have been some instances of birds summering. At Staines Reservoirs two immature drakes summered in 1929; three drakes remained from 18 June to early July 1938 at Walton Reservoirs; and one was present for a week in June the following year. In 1984–85 three summered in the Lea Valley. In 1989 one summered at Girling Reservoir. In 1994 a pair defended a territory at Maple Lodge NR for a few days in May and the female was seen sitting on a nest-box but they did not remain to breed.

Facts and Figures

The largest group was of 147 birds on King George VI Reservoir in March 1986.

Smew *Mergellus albellus*

An uncommon winter visitor

Range: breeds northern Europe and Asia, winters Europe.

Historical

The remains of one found in a ditch at Southwark have been dated to 1500–1700. In the 19th century Smew was a rare winter visitor to the London Area, usually only in severe weather. The earliest record was in the winter of 1819/20 when two males were taken alive in Bow Creek. At Brent Reservoir one was shot prior to 1843, three were shot in January 1849 and another was shot in the winter of 1860/61. In 1846 one was shot at Munden on 26 December and a female was shot on the Thames at Kingston on 31 January 1869.

20th century

In the early years of the century the only records were of a female shot at Kempton Park on 30 December 1908 and one on Ruislip Lido on 24 January 1909. From the early 1920s they became a regular winter visitor, particularly at the smaller reservoirs. By 1925 'large numbers' were noted at Walton Reservoirs at the beginning of the year and 48 were at Barn Elms on 6 December. They generally shunned the deep reservoirs and the Thames unless the smaller reservoirs froze over, as happened in 1929 when 14–15 were seen on the Thames on 16 February and 10 were at Staines Reservoir on 10 March.

Smew continued to increase during the 1930s; the favoured localities were then Barn Elms, the Lea Valley reservoirs, Lonsdale Road and Walton, where 51 were seen in the winter of 1933/34. There was a severe cold spell at the end of 1938 and at Walton the numbers increased to a new peak of 117 on 28 December. Elsewhere, they were reported from a number of other localities and were even seen in Inner London: there was one in Kensington Gardens on 20 December and three on the Serpentine on 24 December.

During the war years there was no access to some of the reservoirs so it was not possible to conduct counts. However, during a particularly cold spell in January 1940, 37 were seen on the Thames by Chiswick Eyot. In 1943 there were peaks of 70–80 at Brent Reservoir and 42 at Walthamstow Reservoirs; the following year there were about 60 in the Barn Elms area. In January 1946 the water level at Staines Reservoir was particularly low and attracted a flock of 48. In a typical mild winter an average of approximately 100 were present. In the severe winter of 1946/47, however, there were peaks of 125 at Walton, 89 at Walthamstow Reservoirs and 78 at Barn Elms. In 1946–47 and 1949, one or two red-heads were seen during the summer months and an adult male was seen at Poyle GP on 2 June 1951.

By the early 1950s they had largely deserted Walton Reservoirs and switched their allegiance to Brent and Barn Elms; on 21 January 1951 the peak count at Barn Elms was 93 and at Brent Reservoir there were 119 on 10 February. In 1956 severe weather in February and March brought record numbers into the London Area. On 28 February there were 144 at Brent Reservoir, and elsewhere there were maximum counts of 85 on King George V Reservoir, 65 at Girling Reservoir, 51 at Walthamstow Reservoirs and 44 at Barn Elms Reservoirs. Although the counts were made on different dates between 2 February and 11 March, so there may be some overlap, the possible maximum total was 472. After the record-breaking numbers of 1956, there were fewer Smew in subsequent winters and the highest counts were 50 at Barn Elms and just 25 at Brent Reservoir. However, they did become more widespread; for example, there were up to 13 on a small lake in Canon's Park, near Stanmore, in 1958.

At the start of the 1960s Smew had declined noticeably and the total count peaked at 70. There was a large influx during the severe winter of 1962/63, including 124 at Walthamstow Reservoirs on 4 March, and 55 at Walton Reservoirs and 52 at Stoke Newington Reservoirs on 27 January; elsewhere there were counts of over 20 at eight other locations, including 23 on the Thames at Kingston on 26 January. More than usual numbers were also seen in Inner London, with a peak of nine in Regent's Park. In subsequent winters numbers returned to the previous pattern and continued to decline; by 1968 the highest count had reduced to 29 at Staines Reservoir.

They continued to decline during the 1970s; in 1974 the only site to reach double figures was Brent Reservoir, where the maximum count was 16. There was an influx in early 1979 due to cold weather, with peaks of 36 at Wraysbury GP, 26 at Wraysbury Reservoir and 17 on Queen Mary Reservoir.

During the 1980s there were several cold winters and these brought influxes into the London Area such as in early 1985 and 1987 when there were about 200 counted. There was a similar pattern in the 1990s with influxes occurring only in the coldest winters. Most favoured the gravel-pit complexes in west London, in the Wraysbury and Horton area and at Thorpe Water Park, although they were sometimes seen on the larger reservoirs. They typically fed on the gravel pits during the day and roosted on Wraysbury Reservoir. A count of 58 at Wraysbury GP on 18 January 1998 was the highest count in the London Area since 1965.

21st century

The number of Smew wintering in the London Area has dramatically fallen since the beginning of the 21st century due to milder winters. They typically arrive in December and leave in early March with the peak numbers usually seen in the New Year. They are most reliably seen on the gravel pits and lakes in the Colne and Lea valleys. The Wraysbury GP complex has been identified as the most important site in the UK for this species during 1996–2005 (Parkin and Knox). The largest flock since 2000 was 29 at Wraysbury GP in the winter of 2000/01. Smew are exceptionally rare in summer but a moulting drake was present on the Chingford Reservoirs in August and September 2003.

Facts and Figures

The earliest migrant was on 30 August 2003 at King George V Reservoir.
The latest migrant was on 20 April 1962 at Staines Reservoir.
The flock of 144 at Brent Reservoir on 28 February 1956 was the largest ever recorded in the UK.

Red-breasted Merganser *Mergus serrator*

An uncommon winter visitor

Range: Europe, Asia and North America.

Historical

In the 19th century they were rare visitors and were only recorded eight or nine times. There were four records on the Thames in winter: of one shot near Chertsey in November 1842; one shot near Putney Bridge in 1843 or earlier; two shot at Hammersmith in January 1854; and one shot at Chiswick in 1855. Additionally, two were shot at Munden in the 1840s; one was shot at Wraysbury in the winter of 1854; two or three were shot at Brent Reservoir prior to 1866; and there was one at Merstham in April 1883.

20th century

No more were seen until 1916 when four were in Richmond Park on 20 February. The only other sightings in the 1920s were of a male in Hyde Park on 12 February and four on Staines Reservoir on 18 February 1922; and in the winter of 1929, there were five on the Thames at Hurlingham on 8 March and others at Barn Elms and Walton Reservoirs. Between the 1930s and 50s they were seen almost annually. There were two in Inner London, at St James's Park on 7 October 1933 and in Battersea Park on 29 March 1950. In very severe weather during February 1956 there was an unprecedented influx and more than 100 were seen, including 60–70 on the Lea Valley reservoirs, 24 at Walton, nine at Barn Elms and seven at Staines.

During the 1960s they averaged about 10 a year apart from in the severe winter of 1963 when about 22 were seen between January and March; these included at least six at Sevenoaks and four on the Thames between Putney and Barnes. In early 1979 there were two influxes due to severe weather, with peak counts of 17 at Broadness on 17 February, 11 at Wraysbury GP on 11 February and 10 at Staines Reservoir on 24–25 February; in the second half of February there was a total of 102 across the London Area.

Numbers were at their highest during the 1980s and 90s, especially in the colder winters, which often brought an influx into the London Area. As was the case with most other diving ducks there was a large influx in early 1985 due to the severe weather and about 90 were recorded, including flocks of 15 on the Thames at Barking and 11 at Queen Elizabeth II Reservoir. A smaller influx of about 52 occurred in 1987; they were widely seen in small numbers and the largest flock was of six birds on the Thames at Kew Bridge. There was a small movement on 5 November 1994 when flocks of 10 and four flew over Queen Mother Reservoir; the former flock was later found on Stocker's Lake. About 84 were seen in 1996, with a maximum count of nine on Girling Reservoir on 10 February. There was also an influx in early 1997, with 10 on Albert Basin on 12 January.

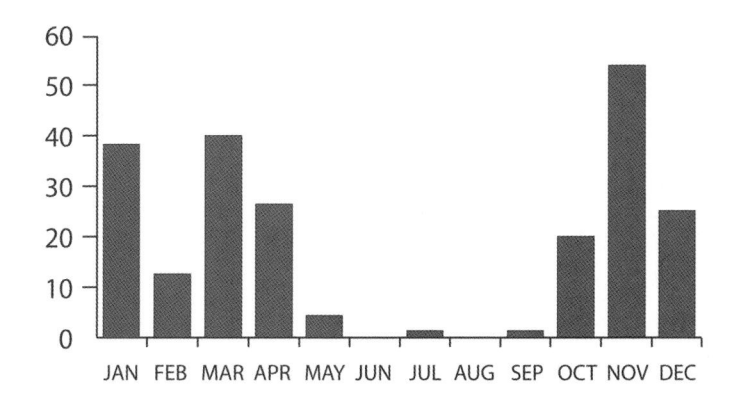

Seasonal occurrence of Red-breasted Merganser: 2000–2009.

21st century

In Europe this duck breeds in northern countries and winters on the coast; in the London Area; they are mostly seen in winter with the occasional bird seen on passage. Since 2000 there has been a reduction in the numbers recorded; almost 30 were seen in 2000 but since then the annual totals have been around 10–20. The only exception was in 2002 when in excess of 50 were seen, more than half of which arrived on 20 November, including a flock of 17 on King George V Reservoir.

Facts and Figures

The largest flock was of 36 birds on Walthamstow Reservoirs on 25 February 1956.

Goosander *Mergus merganser*

An uncommon winter visitor

Range: Europe, Asia and North America.

Historical

This was a rare winter visitor in the 19th century with about 13 records. The first was of a pair taken at Woolwich on 29 November 1811. This was followed by one shot at Erith in 1829/30 and one shot in Lullingstone Park in 1840.

A drake was shot at Brent Reservoir prior to 1843; Harting described how it was shot 'by a labouring man, who, thinking the bird unique, refused a crown for it, and carried it to London, expecting to sell it for a fabulous sum'. Two shot on the Thames at Hammersmith in January 1854 and one at Chiswick in 1855 were originally listed by Harting as Red-breasted Mergansers; they were subsequently reidentified. In the winter of 1856 one was obtained at Wraysbury. There was one at Walton Reservoirs on 10 January 1877, and in 1881 a female was at Gatton Lake for almost two months from 26 February. In the winter of 1884/85 there were 15 at Wanstead Park for five days and five were killed on Hackney Marsh in November 1892. Ticehurst also stated that several had been shot at Dartford.

20th century

In the first two decades the only record was of one on the Thames at Vauxhall Bridge in 1908. Despite the regular observations at the newly constructed reservoirs, there were no more sightings until 1922 when at least 60 were on Staines Reservoir. From then on it became a regular winter visitor to the reservoirs in west and south-west London, occasionally in very large numbers. Elsewhere, it became a regular visitor in small numbers; by 1929 about 100 were wintering across the London Area.

During the 1930s the species continued to increase; at Walton the peak counts rose from 52 in 1930 to 184 in 1938. A census at the main sites in February 1939 produced a total count of 263. They were sometimes seen in Inner London, with the first one being in Hyde Park on 14 December 1933. Occasionally they were seen during the summer months, including three immatures which over-summered on a small pond on Clapham Common in 1936. In 1942 they were regularly seen on the Thames between Hammersmith and Kew Gardens. After the war, when observations returned to normal, even larger numbers were recorded. In 1945, 320 were seen at Walton on 2 February during cold weather, and in the severe winter of 1946/47 about 600 were on Queen Mary Reservoir and at least 550 were at Walton in February, although these may have been part of the same flock.

Subsequent winters were less harsh and fewer birds overwintered. Even in the severe winter of 1956 the highest count was only 230 at Walton Reservoirs. During the 1950s the favoured sites were still the large concrete reservoirs in west London and in the Lea Valley. They typically fed elsewhere during the night and early morning and then returned to the reservoirs by late morning and spent the rest of the day there. Nearby feeding areas in west London included the Pen Ponds in Richmond Park, Hampton Court and Langley Park.

Between 1961 and 1966 an injured bird remained throughout on Walthamstow Reservoirs and in 1962 an early returning pair was seen on Hilfield Park Reservoir on 15 August. As the exceptionally cold winter of 1962/63 started to take effect there was an increase in the numbers seen in the London Area. On 28 December about 100 were on Walton Reservoirs and 65 flew south over Brent Reservoir on 30 December. The main influx came during the first week of January; by the end of the month there were about 1,000 in total, including at least 850 on Queen Elizabeth II Reservoir. Similar numbers remained throughout February. Large numbers were also seen on the Thames including one count of 139 between Walton and Chertsey on 12 February.

Modest numbers overwintered in subsequent years apart from during very severe weather when influxes occurred. However the peak counts did not exceed 130 until early 1985 which witnessed the largest influx since the winter of 1962/63. More than 670 were present in mid-January, the peak count being 440 on Queen Mary Reservoir. A smaller influx occurred in early 1987 and the maximum count was again on Queen Mary Reservoir – 218 in February. During the 1990s there were several small influxes in the coldest winters and the highest count was 218 on Queen Elizabeth II Reservoir in January 1997.

21st century

As in the case of the other sawbills, the wintering population has declined since the beginning of the 21st century, possibly due to a series of mild winters; this reflects the national trend of a decline of 31 per cent from 1998/89 to 2008/09. In January 2000 there were around 170–200 with the largest count being 57 on Queen Elizabeth II Reservoir, but by 2005 there were fewer than 100 wintering in the London Area. They are usually present from the beginning of November to early March with occasional records in the summer months; most are seen on the west London reservoirs and in the Lea Valley.

Facts and Figures

The largest count was at least 850 birds at Queen Elizabeth II Reservoir on 27 January 1963.
The earliest migrant was on 10 August 2003 at Staines Reservoir.
The latest migrant was on 20 May 2001 at Holyfield Hall Farm.
A number were ringed at Walton Reservoirs during the 1930s; three of these were recovered in Sweden, one in Russia and one in Finland.

Ruddy Duck *Oxyura jamaicensis*

An uncommon resident and scarce breeder; naturalised population

Range: North America; naturalised population in Europe and North Africa.

20th century

Ruddy Ducks were first introduced to wildfowl collections in Britain in the 1930s and at Slimbridge in 1948. They soon began to breed in the latter collection and free-flying birds began to escape in the 1950s; by 1963 some 74 young had escaped. They were first recorded in the London Area in February 1958 on Island Barn Reservoir. Escaped birds started breeding in Britain in 1960 and the same year a male summered at Walthamstow Reservoirs. In 1963 at least two were seen in the Lea Valley. Two years later a female was seen at Walton Reservoirs on 10 January, then on Island Barn Reservoir later in the month and on Staines Reservoir between 7 February and 6 March. Possibly the same bird wintered at Walton in 1965/66.

No more were seen in the London Area until 1971, when they were added to the British List, and one to two were seen at three localities outside the breeding season. In 1973 five were on Island Barn Reservoir on 16 June and over the next few years the number of sightings gradually increased. In 1976 the national population had reached about 375. In 1979 there was a large increase in the numbers seen in the London Area with up to 13 at Staines Reservoir and 10 at Wraysbury Reservoir; during the summer two drakes and a duck were present at one west London site but there was no proof of breeding.

In 1980 a pair bred for the first time in the London Area, in the Colne Valley, and the population continued to increase. In January 1982 up to 90 were seen with the peak being 20 at Walton Reservoirs; the first one seen in Inner London was in Regent's Park on 30 July. In 1983 a pair bred again in west London and breeding has occurred ever since. In 1985 two pairs bred in the Lea Valley and a captive pair bred in St James's Park; the peak count was 57 on Walton Reservoirs. In 1987 the first count of 100 was made on 31 January when 103 were on Walton Reservoirs and the following day there were at least 204 in the London Area. By the end of the 1980s the breeding population had reached at least eight pairs and their favoured site had become Brent Reservoir, where there were four broods in 1989.

The population continued its increase during the early 1990s, particularly in winter when flocks started to congregate on some of the larger reservoirs. By 1991 there were counts in excess of 100 at both Hilfield Park and Staines Reservoir, and in 1993 the wintering population was about 400. The breeding population at the start of the 1990s was still fairly small at just 10 pairs, including seven at Brent Reservoir. In 1992 a dozen pairs bred, including eight at Brent Reservoir where a late-nesting pair hatched a juvenile in November. By 1996 the breeding population had increased to 25 pairs and the peak winter count was 306 on Hilfield Park Reservoir.

21st century

At the beginning of the 21st century the breeding population was about 30 pairs with at least 720 overwintering; the peak count was 352 on Hilfield Park Reservoir. The national population had spread to other European countries and some had begun hybridising with native White-headed Ducks in Spain. As this was a threatened species the Government and leading conservation bodies agreed to reduce the British population of Ruddy Ducks. This began with a trial cull in some areas of the country but did not affect the numbers in the London Area, which continued to increase.

By 2003 the London Area wintering population had reached about 1,200, including 879 at Staines; this represented over a quarter of the national population of 4,400. A national cull began and its effects were soon noted in the London Area as the wintering numbers declined to fewer than 500 by December 2006. Large numbers were culled at the main wintering areas on Hilfield Park and Staines Reservoirs. This also led to a reduction in the breeding population and in 2007 only 15 pairs bred. In 2011 none bred in the London Area for the first time since 1980, and by the end of the year the population had been reduced to just 13 – including a flock of eight at Brent Reservoir. The national population had been reduced to around 100 by February 2012. This has been a controversial programme to control a species which, with its jaunty manner and turned-up tail was popular with the public. It is to be hoped that the Spanish population of White-headed Duck will now have a more secure future.

Black Grouse *Tetrao tetrix*

An historic breeder

Range: northern Europe and Asia.

Black Grouse were once widespread across much of England and used to breed in Kent, Surrey and Sussex up to at least the 18th century; they last bred in Surrey in 1905. Ticehurst, referring to Kent, stated in 1909 that 'the district where these birds lingered latest was the high ground known as Brasted Chart between Westerham and Sevenoaks', which is just within the south-eastern boundary of the London recording area.

Escapes

A female shot in Gullet Wood, near Watford, on 1 December 1906 was believed to have come from a nearby location where Black Grouse had been deliberately released.

Red-legged Partridge *Alectoris rufa*

An uncommon resident and localised breeder; population derived from releases.

Range: western Europe; naturalised population in UK.

Historical

Red-legged Partridge was first introduced into the UK during the 17th century and some were released into Richmond Park in about 1673. They were then introduced at Wimbledon between 1712 and 1729 but the colony was extinguished by a neighbour. Further introductions in Suffolk from 1770 onwards led to the colonisation of East Anglia and from there they spread south towards London. In the middle of the 19th century they were established in Hertfordshire and in 1865 a pair bred at Elstree and two were shot from a covey near Brockley Hill. They continued to spread further south and were recorded near Stanmore in 1867 and near Hampstead in 1871. By 1873 they had colonised Enfield Chase. Some were introduced in Essex prior to 1877. The next breeding record was in 1891 when a nest was found near Willesden. Also around this time stray birds were seen in a number of locations in the outer districts of London.

20th century

In 1906 Kerr stated that it was a rare visitor to the Staines district. However, on the other side of the London Area the numbers shot in Woolwich greatly exceeded those of Grey Partridge. Breeding also occurred at Chigwell in 1906 and 1907, at Enfield in 1907 and in Epping Forest in 1919. In the 1920s breeding was recorded from Curtismill Green and Dunton Green. The species could be found in the Colne Valley from Staines northwards; breeding was recorded on Staines Moor in 1928 and 1929, and Red-legged Partridges bred regularly at East Lodge and Ruislip. At one shoot in Stanwell 140 partridges were killed, of which 20–30 were Red-legged.

Although they had increased between 1925 and 1934, Glegg considered them to be unusual visitors and scarce compared to Grey Partridges. The only records from the inner suburbs in the 1930s and 40s were of one found dead at Barn Elms Reservoirs in 1930 and one in a garden in Southfields in 1936. By 1933 they had also started breeding in south London. In 1936 at least two pairs bred at Sewardstone. They were reported from seven other sites, including Staines Reservoirs where three or four birds flew over the water on 15 March, one of them alighting on the water and drowning. In 1937 up to three pairs bred in south-east London, at Hayes.

By the 1950s they had become a common resident across much of rural north London but were generally less numerous than Grey Partridge. However, in some areas they were much more common than the other species, notably in several eastern districts such as at Hainault, Grange Hill, in the Roding Valley, and in the upper Lea Valley. On the Thames-side marshes they had increased, e.g. at Stone they were thought to outnumber Grey Partridge. In west London they were increasing and 32 were seen at Perry Oaks in 1950. There were also some localised declines, such as on the North Downs where they were once quite common but were thought to have suffered from the effects of suburbanisation. Around this time the closest breeding birds to central London were at Osterley Park. In 1953 a pair bred in Kew Gardens.

They continued to increase slowly during the early 1960s although they had abandoned the suburbs. By the mid-1960s they were widespread in the rural areas of the London Area, particularly in the north and east. Up to 10 pairs were present around Fairlop and Hainault, for example. During the time of the first Atlas in 1968–72 they bred in 105 tetrads, mostly beyond the suburbs. In 1971 they were very quick to take advantage of the recently disused Surrey Docks and a pair bred, the first breeding record for Inner London. Up to two pairs bred there almost annually until 1977 and up to eight birds were still present in 1980. They disappeared when the area started to be redeveloped.

Little change was noted during the 1980s; in some areas, such as Ockendon and Sewardstone, they were released for shooting, the latter probably contributing to the population in the Lea Valley. The second Atlas in 1988–94 showed a range expansion of 70 per cent with breeding occurring in 179 tetrads. The vast majority of these were in the rural areas of north and east London. In 1996 at least 250 were released for shooting in the Tyttenhanger area.

21st century

There has been a rapid decline in the national population since the beginning of the 21st century. By 2010 the only breeding pairs in Greater London were in the Enfield and Rainham areas. They have also disappeared from much of the surrounding countryside. Releases for shooting occur in some rural areas, e.g. at Orsett Fen, where some 600 were seen on 4 September 2006. They occasionally appear in more urban settings, particularly in spring; it is not known where these birds come from.

Movements

Examples of birds in urban areas include: in Richmond Park on 11 April 1977, the first there for 12 years; at Staines Reservoir on 28 April 1980; on a fourth-floor window ledge of Buckingham Palace on 1 May 1986 and probably the same bird in the grounds a week later; another in Buckingham Palace Gardens for about two weeks in March 2003; in Alexandra Park, by Stoke Newington Reservoirs and Wormwood Scrubs in spring 2009. At Brent Reservoir there have been seven records since 1944, all between 28 March and 20 May.

Grey Partridge *Perdix perdix*

An uncommon resident and localised breeder

Range: Europe and Asia.

Historical

In the Middle Ages Grey Partridges were found in the districts surrounding the City of London. In 1517 Henry VIII issued a proclamation to preserve the partridges, pheasant and herons 'from his palace at Westminster to St Giles-in-the-Fields, and from thence to Islington, Hampstead, Highgate, and Hornsey Park'. Various remains have been found in archaeological excavations dating as far back as 1140–1350 at Bermondsey Abbey and other south London locations.

By the 19th century they were no longer found in the inner districts of Westminster and St Giles. There were still some in the outer districts, which by then were mainly grassland, but they were much more common in north London where there were more suitable arable fields. They were present in good numbers in Richmond Park where 165 were shot in 1839. In the spring of 1874 one was seen in Regent's Park. There were also a few records from Dulwich between 1874 and 1909.

20th century

In the early part of the century they were widely distributed across all but the built-up areas of central London. They were described as common in the Willesden district and not uncommon at Hampstead. On one estate in the east of London, up to 300 could be shot in a single day. In the 1930s they were still a common resident in suitable rural areas and were occasionally seen elsewhere, even in Inner London. They had started to decline in some areas as a result of suburbanisation, such as at Malden where they last bred in about 1936, and they had declined in areas where they once abounded. In Richmond Park, where there was unrestricted shooting, about five or six coveys were seen in 1936–37. A pair bred in a garden in Roehampton in 1936. During the war years a large area of the park was cultivated, making it more suitable for this species, and 113 were counted on the eastern side alone in 1950.

By the 1950s they had retreated further away from central London due to the loss of habitat as more areas were built upon. However, they could still be found in some suburbs where suitable habitat remained, such as at Brent Reservoir, Chiswick, Dulwich, Hither Green, Kew, Shooters Hill and Woolwich Common. In 1951 a pair summered at Wormwood Scrubs, only about 1 kilometre from Inner London. In the rural areas, they were still fairly common and could be found in a variety of habitats, including arable land, especially where corn was grown and there was grassland with light scrub. Towards the end of the 1960s there were widespread decreases. At North Weald, for example, there had been a population of 10–12 pairs but by 1967 they had all disappeared.

The largest breeding population in 1970 was in Osterley Park where there were 25 pairs, and the highest count was 60 at Dartford in November. During the first breeding Atlas in 1968–72 they were found breeding in a quarter of all London Area tetrads, mostly encircling the suburbs, although there were large areas in the east where they were unrecorded. By the mid-1970s a further decline had set in and they had become scarce in some areas. For example, the population in Osterley Park had decreased to 15 pairs by 1975.

At the start of the 1980s at least 42 breeding pairs were reported but the total was probably higher than this as during the second breeding Atlas from 1988-94 Grey Partridge was found to be breeding in 115 tetrads. This represented a range decrease of almost half since the first Atlas; the largest loss was in the southern half of the area. Most were located in the more rural areas although there were still some breeding in the outer London boroughs. In 1986, 50 captive-bred birds were released into Richmond Park; they bred there for several years but had all disappeared by the early 1990s.

21st century

At the start of the 21st century they were marginally commoner than Red-legged Partridge with at least 34 breeding pairs. The largest population was along the Lower Thames with 16 pairs at Barking Bay and six pairs at Rainham. However, by 2002 there had been another decline, particularly on the Lower Thames Marshes, and in 2006 there were just six breeding pairs reported. Pairs at three sites in west London in 2007 were thought to have been introduced.

Across the UK as a whole, there has been a 91 per cent decline since 1970 and the species is Red listed as a Bird of Conservation Concern. The decline has been attributed mainly to changes in farming practice, particularly the use of herbicides which reduce the abundance of weeds which supported the insects on which the young feed.

Movements

As with the previous species there is some evidence of birds moving through the built-up areas of London, particularly in spring; some of these may be released birds. In Inner London there were several sightings during the twentieth century and even more recently, one was found in a gutter not far from the Marylebone Flyover on 20 January 2010.

Quail *Coturnix coturnix*

A scarce summer visitor and former breeder

Range: breeds Europe, Asia and Africa, winters Africa.

Historical

Quail bred sporadically in the outer districts of London in the 19th century. They were recorded in the Tooting area in around the 1830s. Harting described them as being 'of rare occurrence'. They bred near Harrow in 1842 and 1848; breeding occurred at Romford in 1866; five pairs bred between Hampstead and Hendon in 1870; a nest with 16 eggs was found at Epsom in 1870; three nests were taken between Hampstead and Kentish Town in 1871; a pair bred in Hampstead in 1883; and pairs bred at Orsett in 1885, Epsom Downs in 1886 and Little Bookham in 1893. In south-east London they bred at Crayford, Dartford and Green Street Green, and were considered 'not uncommon' around Blackheath and Eltham until 1859.

Apart from these known breeding records calling birds were occasionally heard in a few other years, sometimes in good numbers when they were also likely to have bred, e.g. in 1877 a dozen were heard around Hampstead and there were five on Epsom Downs in 1893. Introductions were occasionally made, which slightly cloud the picture; for example, a dozen were released at Epsom in 1882. An unseasonal bird near Hertingfordbury in late December 1892 may have been released.

20th century

By the start of the century there had been a marked decline and the only confirmed breeding record was in about 1900 when a clutch was taken at Dartford Marshes. Territorial birds were seen or heard in only six years during the first two decades of the century: at Haileybury in 1901, Banstead in 1907, Old Oxted in 1909, near South Croydon in 1912, in Richmond Park in 1913 and at Warlingham in 1917.

Few were noted from the 1920s onwards, which was thought to be the result of suburbanisation and the consequent reduction in cornfields. There was little change reported until after the war when there was an increase in the amount of arable land in the London Area and a general upturn in the numbers visiting the UK. During the 1950s up to five were heard calling annually but most of these appeared to be migrants and there was no evidence of any breeding.

1964 was a good 'quail year' and 13 were noted calling in the spring and summer. Of these, at least three remained on territory for several weeks or more; at Perry Oaks there were three on 2 August. There was another good 'quail year' in 1965, which was especially evident in the Darent Valley where at least 10 were calling in June and July and probably bred. In 1967 a local farmer reported that up to four had called annually at Clement Street for the previous 10–20 years.

Quail were much scarcer during the 1970s and 80s, apart from in 1970 when up to four were heard calling in a cornfield at Thorpe, and in 1989 when there was a particularly large national influx and six were recorded in the London Area, two of which were on territory. The only confirmed breeding was at Epping in 1974. During the 1990s one or two calling birds were heard in most years. There were six in 1997, including one on territory at Hornchurch CP for at least a fortnight. There were three spring migrants in 1998, including two calling as they flew over at around midnight at Woodford and Rickmansworth.

21st century

In line with a national decline, Quail have become scarcer in the London Area since 2000 and are no longer annual visitors. Most calling birds move on after one day and there has been no sign of any breeding. However, in 2007 one summered at Rainham Marshes from 5 May to 2 July.

Inner London Records

Migrants have been seen on three occasions in Inner London: near the Royal Exchange in the City in July 1884; one caught at Mile End on 11 May 1915 was thought to be a genuine migrant as the importation of Quail had ceased during the war; and one picked up alive in Warwick Square, Pimlico, on 18 September 1947. Quail used to be imported live for food and have escaped from time to time; one found alive on the roof of the Royal Agricultural Hall, Islington, in 1906 or 1907 may have been an escape.

Facts and Figures

The earliest record was on 17 April 1912 at Haling Down, near South Croydon.
The latest record was at Ponders End SF on 15 November 1923.
One wintered at Romford SF in 1967 between 16 January and 26 February.

Pheasant *Phasianus colchicus*

An uncommon resident and localised breeder; naturalised population supplemented by releases

Range: Asia; naturalised population in Europe.

Historical

Pheasants were first thought to have been introduced into Britain by the Romans, so have probably been in the London Area for some 2,000 years. The earliest reference dates back to 1059 when Pheasant was on the menu for the guests of the Canon of Waltham Abbey. There are no further known references until 1517 when Henry VIII issued a proclamation in order to preserve the partridges, pheasant and herons 'from his palace at Westminster to St. Giles-in-the-Fields, and from thence to Islington, Hampstead, Highgate, and Hornsey Park'. The Pheasantry on Kings Road, Chelsea, was a rearing area for Pheasants in the 17th century.

In the 19th century Harting stated 'a few stragglers occasionally found in thick hedgerows and bean-fields, but nowhere plentiful in the county, even where strictly preserved'. Elsewhere, 267 were shot in Richmond Park in 1839.

20th century

By 1900 they were widespread across the rural areas of London. Large numbers were released on estates for shooting. For example, more than 1,200 were shot at Hillingdon in November 1901. By 1912 a decline had been noted on Wimbledon Common. After the First World War they were kept in semi-captivity in parts of central London and bred in Buckingham Palace Gardens, Holland Park and Kensington Gardens. One was seen flying over Lords during a county cricket match in 1925. In 1928 two escaped from Buckingham Palace Gardens and were later seen calling on a garage roof at Marlborough House; and 30 were hatched out in St James's Park although many were later killed by cats. A Pheasant was once seen entering Russell Square underground station early one morning.

From the late 1920s onwards the population started to increase. A nest was found in Epping Forest in 1929 from where the species had previously been reported as an occasional visitor 10 years earlier. In the 1930s Glegg stated that 'the Pheasant is still preserved in several localities in the county and occasional birds have been observed in various districts, including Inner London'. In about 1930 there was a notable increase in numbers in Gilwell Park and the birds then started nesting regularly. Away from controlled estates, they bred annually in Osterley Park, Colham Green and South Harrow SF. In Bushy Park numbers increased during the early 1930s, and in 1936 about 100 were seen in Kew Gardens.

In the 1940s there were occasional reports of birds seen in built-up areas and a pair bred in a large garden in Hornchurch. There was an increase in numbers in Richmond Park where Pheasants took advantage of the increase in cultivation there during and after the Second World War; in April 1952 about 190 were counted there. The population for most of the 20th century was artificially controlled by releases and heavy shooting in some areas, while in other places they were often left to their own devices. During the 1950s an average of around 1,000 were shot annually on the Panshanger Estate.

There had been a decrease in Richmond Park by 1974. One was seen in Regent's Park on 21 April 1977, the first in Inner London for several years. At the start of the 1990s they were still a common bird in the rural areas with the closest populations to central London at Beddington, Mitcham Common and South Norwood CP. In 1994–95 some 15,000 were released for shooting in Panshanger Park. The second breeding Atlas in 1988–94 showed that they were breeding in 40 per cent of the tetrads in the London Area, an increase of 16 per cent since the first Atlas. The main range expansion was along the Lower Thames Marshes; the vast majority of the population was located outside the suburban area.

21st century

There has been little change to the status of Pheasant in the London Area since 2000 when a minimum of 100 pairs was reported. Although they are typically restricted to the more rural areas, they are occasionally seen in the suburbs, for example in 2009 there were males on territory at Brent Reservoir and Horsenden Hill.

Red-throated Diver   *Gavia stellata*

A scarce winter visitor

Range: breeds northern Europe, Asia and North America, winters at sea in Europe and North America.

Historical

Geological evidence of Red-throated Diver has been found in ancient beds in the Lea Valley.

In the 19th century this was a rare winter visitor with just nine documented records. Two were obtained in 1838, at Darenth in February and on the Thames at Chertsey in the autumn. These were followed by one obtained at Horton Kirby in February 1841; one shot at Brent Reservoir on 9 February 1864; another killed on the Thames near Eel Pie Island sometime prior to 1866; one shot on a reservoir at Clerkenwell in February 1871; two more shot at Brent Reservoir on 29 November 1871 and 27 November 1875; one in Wanstead Park in January 1877; and one shot at Greenford Marsh in February 1891.

20th century

There were no more records until the winter of 1917/18 when one was present for over a month on Batchworth Lake and in 1921 one stayed for over two months in Richmond Park, from 13 February to 27 April. In 1929 one was on Aldenham Reservoir on 17 and 23 March, and there was one in breeding plumage on Staines Reservoir from 2 to 5 May and again on 9 June.

During the 1930s the species began to become more regular. This species seems more prone to being found oiled or in ill health than the other diver species; between 1900 and 1954 there were more than 50 records of the species, 18 of which were found dead. Three were in Inner London, all in Hyde Park and Kensington Gardens: from 14 March to 25 April 1934, 27 January to 2 February 1941, and 9 to 10 February 1948. In 1951 eight were seen between 1 January and 24 February at Stone Marshes, four of which were thought to have been shot. In 1956 there was a large influx caused by severe weather and around 14 were seen in the London Area. The severe winter in early 1963 saw a small influx, including five between 17 January and 2 February, four of which were found dead and the other injured; another was found dead that March.

In 1979 nine were seen between 1 January and 11 April. In 1985 there was a record count of about 17, all but one of which was seen between 22 January and 16 March. Another influx took place the following year when about 19 were seen, all between 6 February and 11 April; this included a small flock on the Thames in the Rainham area, which peaked at seven on 3 March. During the late 1980s and the early 90s very few were seen in the London Area. In 1996, 11 were seen during the year, including eight in February and March, and one on the unusual date of 7 June at Beddington.

21st century

Due to the mild winters between 2000 and 2007 very few were seen, with an average of about two per year. They were most often noted on the large reservoirs in the Lea Valley and in west London with the occasional bird on the Thames. They are also, typically, short-staying birds.

Facts and Figures

The earliest was on 12 September 1953 and the latest was on 9 June 1929, both on Staines Reservoir.
The longest stay was of two months and two weeks, from 13 February to 27 April on the Pen Ponds, Richmond Park, in 1921.
The largest group was of seven birds on the Thames at Rainham on 3 March 1986.

Black-throated Diver *Gavia arctica*

A scarce winter visitor

Range: breeds northern Europe and Asia, winters at sea in Europe.

Historical

This was a very rare visitor in the 19th century with just four records. The first was of an immature taken at Fulham in the winter of 1841; the specimen now resides in the British Museum. Then two were shot at Brent Reservoir, in the winter of 1843 and in January 1893; and one was shot on one of the Walton Reservoirs on 26 March 1881.

20th century

There were three records in the first decade of the century: of two shot at Dartford in about 1907; one in Richmond Park in December 1907 and found dead on 1 January 1908; and one at Caterham on 6 November 1909. From the 1920s they started to be seen more often, no doubt helped by the recent construction of more large reservoirs which these divers frequented.

In the winter of 1937 there was an influx of divers and grebes in the London Area, including at least four Black-throated Divers, which was unprecedented. They arrived between 31 January and 6 February, and most of them stayed only a short while, one remained on Lonsdale Road Reservoir and on the adjacent Thames until 5 June, by which time it had attained full breeding plumage. Another two were seen on the Thames in the 1940s: opposite Isleworth Eyot on 30 November 1941 and in the Barnes area between 9 March and 12 April 1948.

Between 1900 and 1954 there were about 26 records, on average about one every other year; after this they became more frequent. In 1956 there was an influx in severe weather at the beginning of the year and about eight were seen, including five on the Lea Valley reservoirs on 25 February. In 1961 another eight arrived between 28 January and 4 March. The first in Inner London occurred in 1977, in Surrey Docks from 8–12 December. There was an influx in early 1979 due to severe weather with about eight seen between January and May, mostly on the reservoirs.

During the 1980s they were seen almost annually. Another one was seen in Surrey Docks on 31 December 1981. In some winters there were up to three in west London where they toured the reservoirs and gravel pits. In 1985 there were two spring migrants on Girling Reservoir on 4 May. In 1987 a long-staying bird in west London spent most of its time on Wraysbury GP and was in full breeding plumage by the end of its stay on 9 May. None were reported in the London Area between 1989 and 1992 and numbers varied during the rest of the 1990s, with a peak of six in 1996.

21st century

Since 2000 they have become scarce, averaging fewer than one per year. Like the other divers they are most often seen on the large reservoirs and they sometimes overwinter.

Facts and Figures

The longest staying bird was at Walthamstow Reservoirs between 24 September and 22 June 1957; its last date was also the latest-ever record.
The earliest date was on Girling Reservoir on 11 August 1956.

Great Northern Diver *Gavia immer*

An uncommon winter visitor

Range: breeds northern Europe and North America, winters at sea in Europe and North America.

Historical

In the severe winter of 1813/14 two were captured alive at Woolwich but escaped and returned to the Thames. In 1847 Ruegg stated that the species visited the Thamesmead area 'chiefly in autumn and winter'. The only other 19th-century records were of singles shot at Aldenham Reservoir on 26 December 1876 and in October 1884; shot at Erith on 12 February 1876; and at Kew Gardens in 1876.

20th century

The next sighting was at Staines Reservoir, where one wintered in 1905/06. From the late 1920s they became regular and were virtually annual from 1927 to 1939. As in the case of the corresponding increase in numbers of Black-throated Diver around the same time, this appears to be a direct result of the construction of the large reservoirs that these birds favour. Staines was obviously an attractive site as Great Northern Divers were seen there more than anywhere else, including up to three in the winter of 1930/31.

One at Brent Reservoir between 26 November 1949 and 3 January 1950 was the first recorded in the London Area for 10 years. Between 1950 and 1980 they were seen less than annually. During the 1980s about 23 were noted, 16 of which were in west London. In the 1980s and 1990s several overwintered, often remaining until early May. In 1993 a summer-plumaged bird was on Queen Mary Reservoir on 18 July; although a few had been seen in this plumage before they had been long-staying birds and this is the only record of one arriving at this time of year.

21st century

Since 2000 this has become the most common diver in the London Area with an average of six or seven per year. They now regularly overwinter on the west London reservoirs where they are usually present between December and April. They also occur fairly regularly in the Lea Valley but are scarce elsewhere. Occasionally, there are influxes, such as the 10 that arrived during stormy weather in December 2006.

Facts and Figures

The longest stay was five months and four days, between 15 December 1951 and 18 May 1952, on Staines Reservoir. The earliest autumn arrival was on 23 October 1934.
The highest count is of seven birds on King George VI Reservoir between 21 January and the end of February 2010, almost certainly the highest count on any inland water in the UK.

Fulmar *Fulmarus glacialis*

A scarce visitor (39)

Range: breeds coastal Europe, Asia and North America, winters at sea.

20th century

A familiar species of rocky coasts around much of the UK during the breeding season, where it has increased in the first half of the 20th century, the Fulmar is rarely seen in London. However, isolated individuals are occasionally swept off course during harsh weather, and turn up in the capital, often in an injured condition. The first record for the London Area occurred in 1954 when three flew north-east over Crayford towards the Thames on 4 September; this remains the only sighting of more than one bird.

There were four records in the 1960s, three of which were on the Lower Thames Marshes. In 1962 there was one on King George VI and the adjacent Staines Reservoir on 4–5 March; later in the month one was picked up dead under high-tension cables at Swanscombe; these records coincided with a major wreck of this species in eastern England. In 1965, after strong easterly winds, one was found dead at Littlebrook on 21 February; and in 1968 one was picked up at Erith Marshes on 25 February but it died overnight while in care. There were another four in the 1970s: at Bushey on 16 June 1974, just before a severe storm; found freshly dead at Farthing Down on 25 September 1977; and two in 1979, on New Year's Day at Girling Reservoir and at Rainham during strong winds on 27 May.

During the 1980s there was a big increase in sightings and 16 were recorded. In 1981 four were seen within a six-day period, three of them on 27 April when northerly gales brought a number of seabirds into the London Area. There were two in 1983: an injured bird found in a Twickenham garden following a severe thunderstorm on 6 July, and one flying over Woodmansterne in November. In 1986 five were seen between April and September. Only one of these required care, an individual picked up on Hounslow Heath which weighed just over half its usual weight; it was fed and subsequently released. There were three more the following year, including two in Inner London on the Thames foreshore: one was sitting on the steps down to the river at County Hall on 3 February; and an adult was found dead at Battersea on 10 March. In 1988 one flew south over Amwell and the following year one was picked up in Tottenham and released on the coast.

There was a return to a more typical four records in the 1990s, including two more in Inner London. Two occurred in 1991: one, which appeared to be injured, floated down the Thames past Rainham Marshes on 16 February; and bizarrely one seen flying around the Barbican Centre on 10 May, which at one point attempted to land on the ledges of one of the buildings. In 1994 one flew downriver at Tower Bridge on 14 February during a period of strong easterly winds and heavy rain, and in 1996 one was picked up at Banstead in December and taken into care.

21st century

Since the start of the 21st century there have been another eight records. In 2000 just one bird probably accounted for all three sightings on 1 July when it was seen first at Queen Elizabeth II Reservoir and later moved to Walton Reservoirs before flying over Bedfont Lakes CP. Two were found on 1 March 2001: one circled over the Wetland Centre then flew off down the Thames; and another was at Beddington where it became moribund the following day and was taken into care. In 2003 one was seen on the Thames from several sites in the Northfleet area on 16–17 and 19–20 October. There were no further records until 2008 when four were seen, all on 25–26 May: one was at Beddington and three were on the Lower Thames.

Manx Shearwater *Puffinus puffinus*

A scarce visitor (69)

Range: breeds coastal western Europe, winters South Atlantic.

Historical

For a maritime species which mainly breeds off the west coast of Britain and Ireland and wouldn't normally venture inland there are surprisingly a large number of records in the London Area. Most of these occur in autumn, especially during gales and this is when young birds are particularly prone to being blown inland where they are picked up injured or exhausted.

There were seven records in the 19th century, the first being of one found at Epping on 21 September 1866, although a shearwater picked up dead in Epping prior to 1835 was probably a Manx. The next one was at Staines on 30 August 1882, followed by one caught exactly five years later at St Albans; also in 1887, one was found in a Dartford street on 24 September. In 1888 an injured bird was picked up at Haling Park, Croydon, on 6 September. The next was caught alive near Sudbury on 22 September 1891 and the last was one taken at Moat Mount on 5 September 1892.

20th century

Ticehurst listed an undated record of one obtained near Erith. In the first half of the century only two more were seen: one picked up alive at Honor Oak Park in September 1930, which died during the night; and another picked up in Clapham in September 1946.

There were two in September 1953: a badly oiled bird found on a pond on Mitcham Common, which later died while in care; and an apparently sick individual on Barn Elms Reservoirs. In 1958 one in Inner London was seen flying in and landing on the City Road Basin of the Regent's Canal on 5 September; it was exhausted and taken into care.

During the 1960s four were seen, including the first for Staines Reservoir; it was present on 11–12 September 1963; on the latter date it flew over the road where it clipped a power line and crash-landed. It was taken into care and successfully released. In 1967 one was found dead at Brookmans Park on 10 September. In 1968 and 1969 single birds were picked up in different gardens in South Norwood in autumn and released after rehabilitation.

The number of sightings began to increase and six were seen in the 1970s. There were two in 1970: at Old Slade NR in July; and seen flying down the Thames past Swanscombe in mid-September. The other 1970s records were: on Wraysbury Reservoir on 3 September 1974; a sick bird found at Dagenham Reach on 11 September 1976, which later died; one that spent three days on Queen Mary Reservoir in September 1977; and one picked up on a Loughton road in November 1978 and released at King George V Reservoir.

There were 25 in the 1980s, more than in any other decade, the same pattern as with Fulmar. This was probably due to the number of storms during that period. Westerly gales in the first half of September 1980 brought six into the London Area; three of these were picked up and one was found dead. In 1983 there were two: the first June record, which had probably been affected by a thunderstorm; and a more typical record in September. Three were seen in 1984, one of which was picked up in a garden at Rayners Lane and one was found dead at Wraysbury GP. There were two in the first week of September 1985, at Wraysbury Reservoir and seen flying low over wasteland at Hersham. In 1987 one was seen on the Thames at Rainham on 29 May, the first one ever recorded in that month. Less than a month later there was another, this time at Wraysbury GP on 23 June, then two more in the more typical month of September. Six were seen in 1988 between 29 August and 14 September, three of which were picked up and later released; and two more were seen in September 1988.

In the 1990s there were 13, starting with three in 1991. The next year five were picked up exhausted between 23 August and 7 September; all of these were found away from waterbodies and had presumably been attracted to street lights. Singles were seen on the Thames in 1994 and 1997. There were three in the autumn of 1998 between 30 July and 11 September; two of these were picked up and taken into care.

21st century

Since 2000 Manx Shearwaters have become scarcer and only seven have been seen, all in autumn. These were: on Staines Reservoir in 2003 and 2004; picked up in Tolworth and taken into care in 2004; two separate birds on Queen Mother Reservoir in 2007; two in September 2008, one picked up in Paddington and released on the Thames and another on the Lower Thames; one on the Thames at both Northfleet and West Thurrock in October 2009; and at Crayford Marsh on 15 September 2010.

Balearic Shearwater *Puffinus mauretanicus*

A very rare vagrant (1)

Range: breeds Balearic Islands, winters North Atlantic; regular visitor to southern coasts of UK in late summer.

There has been one record in the London Area, on 17 August 1984 when one was present on Island Barn Reservoir. This appears to be the only time this species has been seen alive inland in the UK. It was first found at 10:00 hrs and was observed flying around among a flock of hirundines before settling down onto the reservoir where it spent the rest of the day.

Although this species is frequently encountered in the English Channel in autumn, there had only been one previous inland record, found dead in Kent in about 1865. There were no strong prevailing winds at the time of the Island Barn bird, but on the morning of the 17th there was a light to moderate south-easterly and misty conditions. The most obvious route for this bird to arrive was along the Thames as far as Teddington Lock and then overland to the reservoir. It remained until dusk and presumably departed overnight.

1984: Island Barn Reservoir, one on 17 August

Macaronesian Shearwater *Puffinus baroli*

A very rare vagrant (1)

Range: North Atlantic islands; vagrant to UK waters.

The only record was of one picked up dead at Welling in August 1912. This species was previously known as Little Shearwater and many authorities have split this population into at least two species. Upon examination, the specimen was found to be of the race *baroli*, which breeds in the Azores and Canary Islands, and around Madeira. It remains a rare visitor to the UK, typically seen from headlands in the south-west of England between July and September. This was the second record in the UK, the first being of a bird found dead in Norfolk in 1858.

1912: Welling, one picked up dead on 20 August

Storm Petrel *Hydrobates pelagicus*

A scarce visitor (28)

Range: breeds coastal Europe, winters South Atlantic.

Historical

Apart from during the summer months when they breed on remote islands and coasts they are at home on the sea, enduring all weather conditions so their occasional visits inland are usually a result of very severe weather. The first record for the London Area was of a bird shot at Walthamstow at the end of the 18th century. Like the following species, this was a rare visitor in the 19th century with 10 records.

There were two in 1824: one shot in November while flying over the Thames between Blackfriars and Westminster Bridges (Yarrell also listed this as March 1825); and one found alive in Old Street. In 1835 one was taken on the Thames near Richmond around the end of January (the date of this is also uncertain, being given as March by Bucknill). In 1852 one was shot at West End, Esher, in October. In October 1857 one was 'knocked down with a stick and caught' on the Edgware Road at Paddington; 'it was a wet, windy night, and the bird was much exhausted'. In 1886 two were picked up: an exhausted bird in Hyde Park on 9 December; and another caught near St Albans two days later. In 1888 Belt wrote that one had been shot on the Thames near Brentford a few years previously. One was found alive on the Commercial Road on 9 October 1889. In around 1896 one flew into glass and was killed at Nork Park, Banstead. Remarkably, five of these records occurred in Inner London.

20th century

Surprisingly none were seen in the first half of the century but there were four in the 1950s. The first of these was at Staines Reservoir on 11 November 1950. This was followed by two in 1955: one flew down the Thames at Stone on 15 September; and one was found with a broken wing in Epping Forest on 9 December. In 1959 there was one on King George VI Reservoir on 27 October.

Another four were seen in the 1960s: at Staines Reservoir on 18 November 1960; one found exhausted at Woodford Green on 26 November 1962, which subsequently died; on King George V Reservoir on 25 December 1964; and on Walthamstow Reservoirs on 23 October 1966. No more were seen until 1983 when three were found: on Girling Reservoir on 3 September; an exhausted bird at Dartford Creek on 7 September, which died the next day; and one that dropped slowly out of the sky and landed in an open car boot at Crayford on 25 October and later died.

In 1990 one was found in a garden at Becontree on 8 December and later died in care. In 1992 one flew down the Thames past West Thurrock on 14 July, a particularly early date for a storm-petrel in the London Area. In 1999 there was one at Queen Elizabeth II Reservoir on 10 October.

21st century

Just three have been seen since the beginning of the 21st century. In 2000 probably two different birds were seen on the Thames in the Dartford and Rainham area on 30–31 October; there was a small influx of storm-petrels inland with three Leach's also seen at the same time in the London Area. On 2 December 2003 one spent the morning on Queen Mother Reservoir before being taken by a Peregrine.

Leach's Petrel *Oceanodroma leucorhoa*

A scarce visitor (96)

Range: breeds coastal northern Europe, winters South Atlantic.

Historical

Unlike the previous species which typically migrate down the west coast of Britain, Leach's Petrels regularly migrate down the east coast as well and are therefore more likely to be blown into the London Area during autumn storms.

In the 19th century this was a rare visitor with six documented records, mostly occurring after severe weather. The first record was of a bird caught alive in the London Area in November 1823 although the exact site was unrecorded. This was followed by one shot near Chelsea in October 1827; one picked up exhausted on the High Road between Edgware and Stanmore on 4 January 1850; one shot near the Steam Mill, opposite Bow Creek, in March 1864; one shot on the Thames at Barking in 1874; and another shot in a field at Croxley Green on 26 September 1896.

20th century

In the first half of the century this remained a rare bird in the London Area with six records, four of which were in the 1940s. At Cassiobury Park, one was picked up dead in November 1905 and one was shot at Winchmore Hill in December 1907. In 1942 one was captured at Honor Oak Park station on 13 September. In 1948 one was found dead at Chalfont Park on 24 December and the following year there were two at Staines Reservoir: on 7 January and 13 November.

There was an increase in sightings during the 1950s, with 14 seen during the course of the decade. In the autumn of 1952 there was a large wreck of Leach's in the UK, including nine in the London Area. The first of these was found dead in Regent's Park on 28 October. Of the other 1950s records, one was picked up alive in Romford in 1954 and the others were seen on the west London reservoirs during 1957–59. There were six in the 1960s, including three on Staines Reservoir and one on Queen Mary Reservoir, one seen flying north over Rye Meads and one picked up at Sundridge Park Golf Course, which later died. There was a much wider geographical distribution in the 1970s when 11 sightings included another in central London, in Hyde Park on 13 November 1973. Four in 1977 included the first spring record, at Hilfield Park Reservoir on 8 May.

As in the case of many other seabirds, there was an upturn in records in the 1980s when there were several large storms. The first of the decade's 36 records was of one on King George V Reservoir on 26 April 1981, which appeared during a small wreck of seabirds following gales. In 1983 there was a large wreck in southern England caused by an intense depression tracking across the Atlantic Ocean; 11 were seen in the London Area over the next few days, including two together at Girling, King George V and Queen Mother Reservoirs; another was picked up exhausted in Kings Cross. In October 1987 the Great Storm brought another five into the London Area, including two on Staines Reservoir, and there was a bizarre record of one seen flying along the runway at Heathrow Airport on 1 November. In 1989 there was a national influx on 17 December when four were found on the west London reservoirs.

There was a return to more usual numbers in the 1990s when seven were seen. Three of these were on the south-west London reservoirs, but unusually no birds were on the large west London reservoirs. Two were outside the usual September to December period: flying along the Thames past Rainham Marshes on 29 August 1991 and at Brent Reservoir during very stormy weather on 2 January 1998.

21st century

Another 21 were seen between 2000 and 2009. There were small influxes of four in mid-October 2003, and five during stormy conditions on 7 December 2005, including one in Inner London at Tower Bridge. Elsewhere, there were five on the south-west London reservoirs; five on the Thames between Rainham and Northfleet; three on Queen Mother Reservoir; three at Staines; and singles at four other wetland sites.

Storm-petrel sp.

There are several records of unidentified storm-petrels in the London Area, but the only ones listed here are those in Inner London: on the Thames near Waterloo Bridge on 15 January 1947 and picked up in Battersea Park on 24 November 1949.

Gannet *Morus bassanus*

A scarce visitor (84–85)

Range: breeds coastal Europe, winters at sea.

Historical

This is one of the commonest seabirds around the British coasts and the UK holds a large proportion of the world's breeding population. They are usually only seen inland as a result of adverse weather.

The earliest record was of one taken alive near Dartford in September 1798 and the only 19th-century record was of a bird also captured in a field near Dartford in April 1847.

20th century

In November 1911 one was found stranded on the ground at Stratford after having collided with overhead wires. This was followed by one found in a garden at Kenley in November 1927, and there were three in the 1930s: one picked up on Wandsworth Common in late May 1930 and taken to London Zoo; an adult captured alive on the Grand Union Canal by Harlesden Station and taken to London Zoo on 30 November 1934; and an adult shot at High Barnet early in 1937. There were four in the 1940s: on the Thames at Waterloo Bridge on 28 January 1941; picked up in Epping Forest on 29 May and later released on Wake Valley Pond, where it was last seen on 3 June; picked up in a field at Hayes on 6 August 1948 and later released; and at Kempton Park Reservoir on 26–27 October 1949 and found dead the next day.

As the UK population increased during the 1950s so did the number occurring in the London Area, and another seven were seen. These included two in Inner London: one seen swimming among Mute Swans on the Round Pond in Kensington Gardens on 14 October 1952; and an adult in a Bermondsey garden on 18 January 1957.

Four more were seen in the 1960s, two of which were found dead. The other two were taken into care. One of these was released on King George V Reservoir on 18 March 1960 and was still present a week later but subsequently died. The other was an exhausted adult found in Deptford High Street on 21 January 1965; it was taken to a local police station but unfortunately it is not known what the local constabulary did with it. Of the five seen in the 1970s, four were fly-overs and the fifth, a juvenile, was forced to the ground by two crows at Chidwickbury, near St Albans; it was picked up and released on a nearby lake.

The 1980s was the peak decade. Most of the 18 birds seen were storm-blown. Seven occurred between 24 April and 1 May 1981 when gales brought many seabirds into the London Area; four of these were taken into care and subsequently died. Almost as many were seen during the 1990s and they were seen every year apart from 1994, but there were no influxes. Only four of the decade's 17 birds were taken into care. They were widely recorded across the London Area and included an adult that flew north over Regent's Park on 22 April 1995.

21st century

Since the turn of the century they have been seen annually with a total of 24–25 up to 2010. Their survival rate has improved and only one was picked up. Of the rest, five were seen away from water. There were three in both 2008 and 2009, all in late September or early October. The following year there was an influx on 25 September, with six or seven along the Thames past Chafford Hundred and three over Beddington.

Cormorant *Phalacrocorax carbo*

A common resident and localised breeder

Range: Europe and Africa.

Historical

The fossilised remains of a Cormorant were found at Grays in Upper Pleistocene deposits dating from about 108,000 years ago.

Captive birds were kept on the Thames at Westminster by King James I (1603–25) and by King Charles I. The first sighting of a wild bird came in 1629 when one was seen atop the steeple at Kingston church on 11 July. There was a similar record in July 1793 when one was seen resting on top of St Magdalene's Church, Ludgate Hill, in the City of London, where it was eventually shot.

The only 19th-century records were: killed on the Thames at Swanscombe on 30 April 1848; at St Albans in 1886; one in St James's Park in 1888 that was reported by Pigott 'to have been tame and hungry enough to accept a couple of herrings for breakfast'; the following year there was one by Lambeth Bridge; in 1891 there was one near Hampton Court on 11 August; and there was an undated record from Putney. Some of these may have been escapes from a collection in Richmond Park although they coincided with an increase in records along the north Kent coast.

20th century

Cormorants from the Farne Islands had been brought to St James's Park in 1888 and bred on and off from 1892 onwards, probably up to the beginning of the First World War. Another pair was introduced in 1923 and at least one pair bred annually from 1930 to the start of the Second World War. Their offspring were left full-winged, and some of them were seen on nearby lakes. In 1933 a presumed wild bird mated with a locally hatched one in St James's Park and by 1936, the population of full-winged birds there numbered 11, including two newcomers. Both of the original birds had died by 1943 but Cormorants continued to be seen on the lake for several years afterwards, usually in autumn. A wild bird joined the pinioned birds in St James's Park and became tame enough to take herrings from the keeper's hand. Another pair of pinioned birds was introduced in 1949 but they had to be destroyed later as they ate some of the ducklings that were bred there.

Elsewhere they were rarely recorded during the first two decades. From 1924 they were seen annually until at least 1932. Most of these records were of one or two birds but 27 were on Queen Mary Reservoir on 18 September 1931. Typically they were recorded from the large reservoirs in west London and in the lower Lea Valley, but they were also seen on the Thames and at Highgate Ponds; occasionally they would stay for several weeks on the reservoirs. Two frequented the top of the Palace of Westminster in June 1928 and one roosted nightly on the top of St Paul's Cathedral in August 1931. An early ringing recovery suggested the likeliest source of these birds, having been ringed originally in Rhossilli, Glamorgan, in June 1930 and recovered at Staines Reservoir two months later.

From the late 1940s there was a large increase in the number of sightings and they became much more widespread. They were often seen on the Thames above Kew Bridge, particularly at Syon Park. There was a report that they had bred at the latter site in 1936 but there was no evidence to support this. Cormorants were mostly seen outside the breeding season with only a few still present in summer. By 1949 sightings had increased across the London Area, with groups of up to seven on the Thames and the large reservoirs. Occasional single birds were seen in central London, either on the Thames or flying over.

The increase continued during the 1950s. In 1953 there were 43 at Staines Moor on 15 February and in 1956 up to 70 were seen during the winter at King George VI Reservoir; also that year a bird ringed as a nestling on the Farne Islands in June was recovered at Walthamstow Reservoirs in December. Numbers continued to rise year on year and in 1959 the peak count had risen to 86 at King George VI Reservoir. The first count of more than 100 came in 1968 when 104 were on Walton Reservoirs on 28 January. They continued to increase throughout the 1970s and by 1979 the peak site count had reached 216 at Queen Elizabeth II Reservoir, on 1 December.

At the start of the 1980s the peak counts were fairly similar to those of the previous few years with 206 at Queen Elizabeth II Reservoir and 154 at Walthamstow Reservoirs. They had become very widespread and were often seen in Inner London, particularly on the Thames. The maximum count at Walthamstow topped 200 in 1982 but the following year attempts were made to disperse the roost and only 60 were present at the end of the year; most moved to other sites up the Lea Valley. This action failed to work, though, since by December 1984 there were 232 at Walthamstow. One found dead at Walthamstow had been ringed on Anglesey, the second Welsh-ringed bird to be recovered in the London Area. In 1985 two more Welsh-ringed juveniles were seen in the London Area.

By 1986 there were five main roost sites: at Fishers Green and Walthamstow Reservoirs in the Lea Valley, and at Broadwater Lake, Queen Elizabeth II Reservoir and Wraysbury Reservoir in west London. The peak count at these sites was 1,041 in March of that year. In 1988 the first foreign-ringed Cormorants were seen in the London Area, two originally reared in Denmark; remarkably, one of these regularly wintered on Lake Geneva, Switzerland, and had been seen there on 4 October before being sighted at Queen Elizabeth II Reservoir on 11 December. At the start of the 1990s the highest count was 525 roosting at Queen Elizabeth II Reservoir in November; in Inner London small numbers were seen in the parks and on the Thames, with a peak of 16 in Hyde Park. In 1991 a large feeding flock of 520 was present on Queen Mary Reservoir on 17 February. In 1996 the peak roost count had risen to 780 on Queen Elizabeth II Reservoir in October.

21st century

Since the beginning of the 21st century the wintering population has been around 1,500–2,000 birds. The largest numbers were on Staines Reservoir, including the first four-figure count of 1,018 in September 2003. The only locality of national importance for Cormorants (based on 2009/10 figures) is Walthamstow Reservoirs. They are widely recorded throughout the London Area and small numbers are regular in central London, on the park lakes and on the Thames.

Breeding

The first signs of breeding occurred in 1984 when a single bird built and occupied a nest at Stocker's Lake. This may have been part of an expansion from the Continent, as a tree-nesting colony was established at Abberton Reservoir, Essex, in 1981 and had grown to 46 pairs by 1983. The only other inland breeding colony was in Staffordshire (also started in 1981). In 1984–85 Cormorants were seen nest-building at Broadwater Lake and in 1987 a pair bred there and raised two young. By this time the colony at Abberton exceeded 150 pairs.

There was no further breeding until 1991 when five pairs bred at Walthamstow Reservoirs and raised six young. The following year this new colony had rapidly expanded to 30 pairs and raised 50 young. The islands provided trees for nesting and there was a plentiful supply of fish, so the colony grew, virtually doubling to 58 pairs in 1993; by 1996 it had reached 188 pairs. Also in that year, a pair bred again at Broadwater Lake, and by 1999 there were 11 pairs there.

The breeding population continued to expand during the 21st century. In 2000 there were 242 pairs at Walthamstow and 24 at Broadwater. The following year 15 pairs bred at Holyfield Lake and two pairs attempted to breed at Amwell. The peak year was 2004 when a total of 435 breeding pairs included 360 at Walthamstow, 42 at Broadwater and 27 at Holyfield.

It seems that the London Area was first colonised by continental birds of the race *sinensis*, which nest in trees. Once they had become established, birds of the nominate race *carbo*, which typically nest on cliffs in the UK, also started to breed. In 2000 the breeding colony at Walthamstow consisted of 82 per cent *sinensis*, 13 per cent *carbo* and the rest undetermined. It is not known what the current ratio is between the two races and whether or not they interbreed, but the numbers of *carbo* have increased.

Facts and Figures

The largest flock was of 1,316 birds at Staines Reservoir on 29 September 2006.

Shag *Phalacrocorax aristotelis*

A scarce winter visitor

Range: coastal Europe and North Africa.

Historical

There were only two records in the 19th century, of one shot on the Thames near Wraysbury in 1850 and one killed at Vauxhall in 1865.

20th century

Between 1900 and 1925 the only record was of one shot at Aldenham Reservoir on 21 December 1909. During the period 1926 to 1954 they became more frequent with the majority seen on the Thames, especially on the tidal stretch upriver of Hammersmith. Elsewhere, they were seen on the large reservoirs as well as on smaller waterbodies such as the Serpentine in Hyde Park. Storm-driven birds were occasionally seen away from water, e.g. in 1935 three were picked up in a nursery garden in the Lea Valley on 7 February and taken to London Zoo. Most records were of single birds but there were also a few small groups, the largest being nine in Osterley Park in February 1935. One picked up at Rickmansworth on 12 December 1936 had been ringed earlier that year on Bass Rock.

Although typically occurring in winter, some records appear to have concerned birds on passage since there were several spring records. For example, five were at Ruislip Lido on 24 April 1938. After 1947 there was a break of six years until the next record in February 1953 when a young bird, ringed the previous year on the Farne Islands, was found at Morden. In early 1958 there was a large influx into southern England and 30–40 were seen in the London Area, including seven found dead at Ruxley on 1 February.

During the 1960s there was a series of weather-related influxes. The first of these occurred in early 1960 when about 20 were seen in the London Area, some of which remained and oversummered. In 1962 a notable 'wreck' occurred during strong north-easterly winds in the second week of March. At Walthamstow Reservoirs there were 19 immatures between 10 and 14 March, most of which died within a few days although two remained until June. At Ruxley GP there were 12 on 13 March, half of which were found dead. Elsewhere, up to four were seen at 16 other locations including three in Inner London, one of which was picked up while walking along the road in Knightsbridge and released onto the Serpentine. In 1965, 22 arrived in late November, mostly in the Lea Valley where there were 15 juveniles at Walthamstow Reservoirs and four juveniles at King George V Reservoir; two of these had been ringed on the Isle of May. Six of the Walthamstow birds remained until at least February but five of them had perished by April; the single remaining bird stayed there for over two years and was eventually found dead on 1 January 1968.

In the 1970s there were just a few records annually, apart from in 1974 and 1979 when there were bigger numbers. In the first of those years a flock of up to 15 was at Hilfield Park Reservoir between 29 November and 14 December. In 1979 there were about 20, including up to three on the Thames between Blackfriars and Lambeth Bridge from 31 January to 31 May, and four in the Lea Valley on 23 December.

Stormy weather during the 1980s was responsible for several more influxes. The first occurred in early 1984 with at least 22 birds seen, including seven on the Thames in Inner London, many of which remained throughout the year. Another influx took place in November 1985 with about 23 arriving, including 14 adults at Hilfield Park Reservoir and seven at Queen Mother Reservoir; two of these had been ringed on the Farne Islands. There was a record influx of about 67 in February and March 1988, and the year's total was 87. Peak counts in 1988 were up to 12 on the Thames between Blackfriars and Vauxhall Bridge from March to July, and nine at Wraysbury Reservoir on 26 February. Two remained into 1989, including one in the Lea Valley which was present until 1 October, a stay of 19 months.

In late 1990 there was a small influx early in the year with about a dozen birds seen, and another influx at the end of the year involving at least 18; this included a flock of 10 on the Thames at Rainham on 29 December. The following year even more were seen. About 53 were recorded, including the flock on the Thames at Rainham, which peaked at 30 on 26 January with some remaining until mid-February. There was another large influx in 1993, with about 60 seen, more than half of which arrived during the last three days of January. These included groups of 14 at Wanstead Park and 11 on King George VI Reservoir.

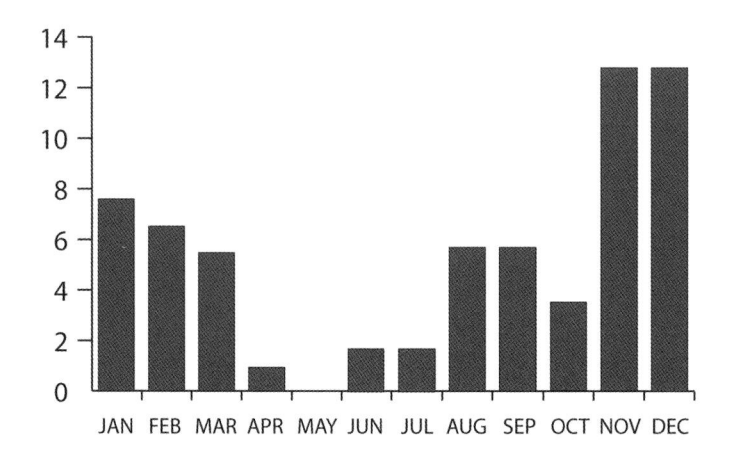

Seasonal occurrence of Shag: 2000–2009.

21st century

Shags breed around much of the British coast on rocky outcrops; in the southeast and East Anglia they are mainly a winter visitor. The UK holds 45 per cent of the global breeding population of this species, but studies have shown a major decline from 1986 to 2011. This has been reflected in the London Area as by the start of the 21st century there had been a decrease in the numbers visiting the capital to about 10 a year. Sightings have continued to decline and none were seen in 2004 for the first time in 50 years. Up to 10 have been seen in subsequent years and are mostly immatures in late autumn and winter.

Facts and Figures

The highest count was of 15 birds at Walthamstow Reservoirs in late November 1965 and at Hilfield Park Reservoir in November and December 1974.

Bittern *Botaurus stellaris*

An uncommon winter visitor, occasional in summer

Range: Europe, Asia and Africa.

Historical

Although there is no documented evidence that Bitterns ever bred in the London Area, they may have bred at Dartford as Latham stated, in 1801, that 'I believe that this bird may be met with in the marshes about us at all seasons, yet is observed most frequent in winter'.

In the 19th century Bittern had declined over much of its former range in England due to shooting for food (it was considered a delicacy) and increased draining of marshes. By the 1880s it was declared extinct in the UK. Bitterns could once be seen on the marshes at Thamesmead before the area was developed for the expansion of Woolwich Arsenal and the construction of Crossness Sewage Works.

Harting knew it to be a rare winter visitor due to a lack of suitable habitat. About 20 were shot in the 19th century with the earliest being in 1832. Elsewhere in the London Area there were records from the Lea Valley and at a dozen sites in south London.

20th century

It remained a rare winter visitor in the first half of the century with only 24 records by 1954. Most of the sightings were between December and March with a few in October and November. During this time there was limited habitat for Bitterns with most records coming from the Thames-side marshes and gravel pits. Occasionally Bitterns turned up in unusual places. For example, in 1921 one was discovered tangled up in a bush in Kew Gardens, and in the severe winter of 1946/47 birds were seen in a Carshalton garden and in the college grounds at Strawberry Hill. The following winter one was found on a main road at Oxted and kept alive on a nearby farm for several months. The Lea Valley occasionally played host to Bitterns and two were seen at Stanstead Abbotts on 14 October 1944. Early migrants were seen at Brent Reservoir on 5 August 1959 and at Rye Meads on 30 August 1957.

In the 1960s they remained fairly rare visitors. There was an influx in the severe winter of 1962/63 with five between 5 January and 13 March; three of these were found dead and another was found at a bus stop in Stoke Newington and taken to London Zoo. Also in 1963, one was seen flying over Dollis Hill on the very early date of 1 August. There were no more records until 1967/68 when one wintered in Cassiobury Park. In 1968/69 up to three wintered at Rye Meads and in November 1969 four were seen on adjacent land. On 7 July 1970 one was briefly at Iver SF before flying off south. In 1974 one remained at Rye Meads throughout the year. The species remained scarce with only a handful seen annually until 1979 when 12 were seen between January and March, including four at Rye Meads.

By the beginning of the 1980s Bitterns had started wintering regularly in the Lea Valley, with single birds at Cheshunt and Rye Meads. Elsewhere, they were occasionally seen at other sites. There was a small influx in the winter of 1981/82 when 11 were seen, including three at Cheshunt. In 1987 one was at Sevenoaks Reserve from 5 to 7 May, a most unusual time of year as most were on their breeding grounds by then.

There had been little change in their status by 1990 with about five seen each winter. In 1992/93 three were regularly seen in a small reedbed in front of a hide (now known as the Bittern Watchpoint) by Seventy Acres Lake, Cheshunt, from the beginning of the year through to early March. Numbers began to increase from 1995 onwards and the following year a record 27 were seen in the London Area, including four at Cheshunt.

21st century

During the first few winters of the 21st century Bitterns were seen mostly in the Lea Valley with several at Amwell, Cheshunt and Rye Meads. They started to winter at the Wetland Centre in 2001/02 when up to three were present. In the winter of 2002/03, 27 birds included seven at Cheshunt. Since then typically 10–20 have been seen in the London Area most winters; the most regular overwintering sites are the Ingrebourne Valley, Lea Valley and the Wetland Centre. Although they have not yet started to oversummer, booming Bitterns were heard at Amwell and Cheshunt in the spring of 2009, raising hopes that they may eventually breed.

Facts and Figures

The earliest migrant was on 7 July 1970 at Iver SF.

Little Bittern *Ixobrychus minutus*

A rare vagrant (18)

Range: breeds Europe, Asia and Australasia, winters Africa.

Historical

The earliest record, of one 'shot near London' in May 1782, ended up in the Leverian Museum. There were seven documented records in the 19th century suggesting it was a scarce summer visitor and may even have bred. A bird shot in 1847 on the River Lea was described as 'a young bird of the year, and from its appearance had probably been bred in the neighbourhood, as there were some remains of the nestling down and it was not fully fledged'. Ruegg stated in 1847 that it was occasionally shot in the Woolwich area. In about 1860 one was shot at Lullingstone. A record of one shot near Broxbourne Station in 1898 appears to relate to the 1884 bird at Carthagena Weir.

20th century

There were no further sightings until 1954 when two were seen: an adult male at Nazeing GP; and an immature present for five days on Pen Ponds in Richmond Park, with probably the same one at Beddington a few days later. In 1956 a pair was seen in overgrown lagoons at Beddington from 18 June to 14 July, with one seen again on 6 and 14 August; breeding was not thought to have taken place. There were single records in the 1960s and 70s.

The mid-1990s was a good time for this species with three seen in successive years. The first of these, an adult female, was found on one of the least-visited ponds on Hampstead Heath, at West Heath, in the early evening of 18 June 1995; it remained there until dark and was not seen the next day. The following year an adult male spent three days on a small pond on Epsom Common, and in May 1997 a male was present for two days at Rye Meads

There were no further records for 15 years when one spent a week on the River Colne by Stocker's Lake in 2012.

1782: near London, shot in May
Pre-1831: Uxbridge Moor
1840: Aldenham Reservoir, male shot

Pre-1843: Brent Reservoir, male shot
1847: Enfield, juvenile shot on the River Lea on 18 September
1860: Lullingstone, shot
1874: Passingford Bridge, shot on 15 September
1875: near Uxbridge, immature shot on the River Colne in autumn
1884: Broxbourne, female shot at Carthagena Weir on 13 October
1954: Nazeing GP, on 30 May
1954: Richmond Park, immature on 20–24 August; also at Beddington on 27–29 August
1956: Beddington SF, two on 18 June–14 July, with one on 6 and 14 August
1961: Weybridge, adult found dead on 22 August
1976: Sevenoaks WR, on 17 October
1995: Hampstead Heath, on 18 June
1996: Epsom Common, on 30 May–1 June
1997: Rye Meads, on 17–18 May
2012: Stocker's Lake, on 10–17 June

Night Heron   *Nycticorax nycticorax*

A rare vagrant (16)

Range: Europe, Asia, Africa and North America; scarce visitor to UK and a BBRC rarity until the end of 2002.

Historical

The first record for the London Area in 1782 was also the first British record; the bird was added to the collection in the Leverian Museum. Harting stated that since that first record it was probable that 'many others have been obtained in this county without any record preserved'. However, only three 19th-century records are documented.

20th century

There were no further records until the 1970s when there were two, at Stocker's Lake and Sevenoaks Wildfowl Reserve. The 1980s witnessed the start of an increase in the number of sightings, beginning with two in the first year of the decade: an adult in spring, and a juvenile found dead beneath overhead wires in autumn. In 1983 an adult in full breeding plumage remained for five days at Wraysbury GP. There was a large influx into the country in the spring of 1987 and an adult spent four days at Sevenoaks. In 1989 an adult or second-summer was seen at Sevenoaks on 28 August and 3 September.

Four were seen in the 1990s, including the fourth to be recorded at Sevenoaks; its nine-day residence made it the longest-staying bird in the London Area. There have been two so far in the 21st century: a first-summer bird at the Wetland Centre where it was present for most of the day on 5 May 2004 and was then seen again briefly three days later as it flew into the main reedbed; and a first-summer at Stocker's Lake on 19 June 2011.

1782: near London, one killed in May
Pre-1843: Brent Reservoir, undated
1883: Plumstead Marshes, immature shot on 3 December
1884: Molesey, immature on an eyot on the River Thames in autumn
1970: Stocker's Lake, adult on 6 September
1978: Sevenoaks WR, immature on 17 July
1980: River Thames near Ham House, adult on 23 May
1980: Walthamstow Reservoirs, juvenile found dead on 11 October
1983: Wraysbury GP, adult on 4–8 May
1987: Sevenoaks WR, adult on 1–4 May
1989: Sevenoaks WR, on 28 August and 3 September

1990: Bookham Common, first-summer on 7 April
1992: The Causeway NR, adult on 9–10 June
1994: Brent Reservoir, first-summer on 19 May
1998: Sevenoaks WR, on 7–15 November
2004: London Wetland Centre, first-summer on 5 and 8 May
2011: Stocker's Lake, first-summer on 19 June

Unacceptable Records

In 1885 Aldridge claimed to have seen five flying over his Norwood garden one evening; in comparison to Grey Heron 'their legs seemed shorter, the wings smaller and the body thicker and the colour...was much darker'. The description is insufficient to accept this record. One was seen feeding in garden ponds in Watford during April and was presumed to be one of the four birds released earlier that month from London Zoo. There is confusion around the actual year for this record as it is listed in the LNHS book as 1935 but in most other references as 1936; the latter is believed to be correct.

In spring 1951 an immature was seen at Brent Reservoir. The record was originally put in square brackets with the agreement of the observer as it was the first modern sighting and was only seen in poor light and not in flight. In 1966 a very tame adult was present at Berrylands SF from December to the following June; it was also seen in New Malden on 11 March. Also in 1967 one was seen in Battersea Park from 10 October to the end of the year. Both of these had escaped from London Zoo.

Squacco Heron *Ardeola ralloides*

A rare vagrant (5)

Range: breeds Europe, Africa and Asia, winters Africa.

There were two in the 19th century, both shot at Brent Reservoir. No more were seen until more than a century later when there was one at Rye Meads in 1979. In 1997 an adult was present all morning on one of the drained basins at Walton Reservoirs before flying north over the Thames towards Stain Hill Reservoir but it was not seen again. Ten years later an adult in full breeding plumage was present on the Southern Marsh at Crossness for 11 days in late spring.

1840: Brent Reservoir, shot
Pre-1866: Brent Reservoir, shot
1979: Rye Meads, on 1 July
1997: Walton Reservoirs, on 17 June; also seen in flight over Stain Hill Reservoir
2007: Crossness, adult on 29 May–8 June

Cattle Egret *Bubulcus ibis*

A rare vagrant (16)

Range: cosmopolitan; range expanding, former rarity in UK.

During the spring of 1992 there was a small influx into Britain and on 3 May a flock of eight was seen flying over Stocker's Lake at 08:00 hrs. They remained in the area, being seen at several sites until the evening of the following day. Most were in breeding plumage and spent a lot of time living up to their name by feeding with cattle on the dairy farms in the vicinity. This was the largest flock ever to be seen in the UK at that time.

In 1995 one roosted at Stocker's Lake for three nights in July; it was believed at the time to be the escaped bird of the eastern race which was touring south-east England at the time. However, several years later it was accepted as a genuine vagrant by the BBRC. Another was seen briefly in Wanstead Park in the spring of 1998.

During the first decade of the 21st century Cattle Egrets were seen more frequently across the UK and even started to breed in the West Country. Although it remained a rarity in the London Area there was an increase in the number of sightings. In 2001 one was present at the Wetland Centre for three and a half hours in the afternoon of 18 August. The following year one wintered in a wetland area at Higher Denham although its presence was not made public at the time. There were three in 2007, including two individuals at Rainham Marshes. In 2008 one spent six days at Girling Reservoir; it was also seen over Walthamstow Reservoirs on 15 April.

1992: Stocker's Lake/Maple Cross and West Hyde, eight adults on 3–4 May
1995: Stocker's Lake, on 20–23 July
1998: Wanstead Park, on 10 May
2001: London Wetland Centre, on 18 August
2002: Higher Denham, from 'early January to early February'
2007: Sevenoaks Wildfowl Reserve, on 31 January
2007: Rainham Marshes, adult on 20–21 May
2007: Rainham Marshes, on 17–23 October
2008: Girling Reservoir, on 12–17 April
2010: Ingrebourne Valley, on 30 July

Unacceptable Records/Escapes

In April 1930 five full-winged birds of the eastern race *coromandus* were released at Foxwarren Park, Cobham.

In 1995 one was seen at various sites from south-west London through central and north London to the Lea Valley between 1 July and 21 September; it was also seen in neighbouring counties between June and August. It was believed that one escaped bird of the *coromandus* race was responsible for all these sightings, including one that roosted at Stocker's Lake as it did not overlap with any other sightings. A ringed bird of the eastern race was in Osterley Park between 12 August and 7 December 2000. In 2003 one in St John's Wood on 15 November had escaped from London Zoo.

Little Egret *Egretta garzetta*

An uncommon resident and localised breeder

Range: Europe, Asia and Africa.

Historical

The remains of one dating back to 1500–1700 were found at Southwark; in 1828 Jennings wrote that they used to breed in Britain and this evidence helps to support that statement. There were no documented sightings in Britain until the early 19th century.

20th century

The earliest record was in 1956 when one was at Old Parkbury on 1–2 July. No more were seen until 1972 when one was at Walthamstow Reservoirs on 29–30 April. In 1985 Little Egret was still a national rarity although by this time it was being recorded annually. That year four were seen in the London Area, three at Staines Reservoir on 27 May and one at Queen Mary GP on 3–6 July. In the summer of 1989 there was an unprecedented influx into the country and this was repeated over the next few years, bringing about a major change of status.

The early 1990s saw a slow increase in the number of sightings in the London Area. In 1991 one was seen in February at various sites in west London and singles were then noted at Beddington on 2 June and at Stocker's Lake on 2 August. The first year with multiple sightings was 1993: six were seen, including one at Charlton GP for 10 days in August during which time it was joined by a second bird. Four were seen in

1994, two in spring and singles in autumn and winter. Another four were seen in 1995, all between 30 July and 31 August, indicative of a post-breeding dispersal. By the mid-1990s the species had started to breed on England's south coast. They very quickly colonised southern England, which may have helped to contribute to the increase in the London Area; by the end of the decade up to 22 were seen annually.

21st century

By 2000 there had been a dramatic increase in the number of sightings and its status in the London Area had changed from being a rarity to an uncommon visitor. Around 66 were seen in 2000, three times as many as the previous year. The vast majority of sightings were between July and September with just a few in winter and spring; they had also started to breed in Essex. By 2002 they were too numerous for annual tallies to be kept and could now be seen at any time of year. They are now fairly widespread in mostly small numbers throughout the year. In the Colne Valley there is a large winter roost at Broadwater Lake of up to 48. Rainham Marshes regularly attracts the most birds during the day. The first Inner London record was at Regent's Park on 28 June 2003.

Breeding

In 2006 one pair bred successfully in the heronry at Walthamstow Reservoirs. They have slowly begun to colonise the London Area; at Walthamstow 10 pairs bred in 2011 and pairs have bred at Wraysbury Gravel Pits since 2007 and at Amwell in 2011.

Unacceptable Records

In 1966 an adult was captured in St James's Park by a member of the public on 12 September and taken to London Zoo where it later died; it was presumed to have been an escape.

Facts and Figures

The highest count was 56 birds at Rainham Marshes on 23 July 2008.

Great White Egret *Ardea alba*

A rare vagrant (17)

Range: cosmopolitan; former rarity in UK.

The first record for the London Area occurred in May 1997 at Brent Reservoir where one was seen briefly in the afternoon before flying off east in the early evening.

During the first decade of the 21st century there was a sudden increase in the number of sightings in the London Area; this followed a large increase in national sightings. In 2000 one was seen at Kempton Park Nature Reserve in October. This was followed by two at each of the following sites: Rainham Marshes, Staines Moor and Walthamstow Reservoirs; some of these were also seen at other sites. The two that occurred in 2007 were the first ones to remain for more than a day. The following year there were two together at Rainham.

1997: Brent Reservoir, on 13 May
2000: Kempton Park NR, on 17 October
2002: Staines Moor, on 6 September
2006: Walthamstow Reservoirs, flew over on 30 April
2006: Rainham Marshes, on 17 July; later at Ingrebourne Valley
2006: Walthamstow Reservoirs, flew over on 11 November
2007: Rainham Marshes, on 19–24 October
2007: Staines Moor, on 22–24 December; also at Queen Mother Reservoir on the 24th and Stanwell Moor on the 26th
2008: Rainham Marshes, on 4 and 7 June, with two on 8 and 2 July and 9 September
2008: Beddington, on 21 September
2008: Holmethorpe SP, on 22 December
2009: Ilford, on 5 April
2009: Rainham Marshes, on 14 April
2009: London Wetland Centre, on 24 September
2010: Holmethorpe SP, on 6 September
2011: Queen Elizabeth II Reservoir, on 28 October

Grey Heron *Ardea cinerea*

A common resident and localised breeder

Range: Europe, Asia and Africa.

Historical

Remains have been found from Bermondsey dating back to 1350–1500 and from Merton dating back to 1350–1450.

The earliest record was in 1523 when they were breeding in the grounds of Fulham Palace which had been leased by the Bishop of London, who took out an action for trespass against a defendant who had been caught taking birds from their nests. There is no later information about this heronry so it is not known when it ceased to be occupied. In 1607 they were probably breeding at Beddington. In 1613 there was a reference to one seen perched on top of St Peter's Church in Cornhill during an outbreak of plague and it was believed that it would cause the sickness to increase. During Henry VIII's reign they frequented the marshy pools in Hyde Park. The only other historical heronries were at Osterley Park, which dated back to the middle of the 16th century, and at Belhus, which was protected so that falconers could hunt the herons but later abandoned after the 'sport' became less fashionable.

In the 19th century Harting described the species as a common resident, most numerous in autumn when "many may be seen in the course of a walk, fishing at our reservoirs and brooks". In south-west London there were a number of heronries close to the Thames between Kew and Sunbury including at Bushy Park, Hampton Court Park and Syon Park. All of these were upstream of the most polluted stretches of the river. Elsewhere, there were heronries at Black Park, Cobham Park, Dulwich Woods, Godstone, Osterley Park, Richmond Park, Walton-on-the-Thames and Wanstead Park. The heronry at Osterley Park was in use for the longest period – at least 300 years – but by 1872 it was no longer in existence. In the 1860s there was also a small heronry near Uxbridge.

20th century

By the 1930s the breeding population across the London Area averaged around 130 pairs; this dropped to 104 pairs by the early 1940s as a result of several cold winters, plus damage and disturbance during the war. Numbers soon recovered and reached 165 pairs in 1945 but following a particularly severe winter in early 1947 dropped back down to 126 pairs. There was a steady increase over the next few years and by 1951 just over 200 pairs were recorded breeding in the London Area.

There appeared to be little change until after the severe winter of 1962/63 when there was a dramatic reduction to only 95 pairs, the lowest reported total since the beginning of the century. In 1964 only 87 pairs were recorded, giving cause for concern; there was a suggestion that toxic chemicals which were thought to be affecting water-birds may have also have been affecting Grey Herons.

There was a gradual recovery with 155 pairs located in 1973. The population continued to increase, reaching 350 pairs by 1990 and 509 pairs by 1998, the highest total ever recorded at that time. During the second breeding Atlas in 1988–94 15 new breeding sites were located that did not exist in the 1968–72 survey.

Details of the colonies are as follows:

East London – the oldest colony was at Wanstead Park which dates back to at least 1834; by the end of the 19th century there were around 50 pairs. During the First World War a few pairs moved to an islet at Walthamstow Reservoirs. Wanstead remained the largest heronry in the London Area in the 1920s with up to 70 pairs which nested in tall elms, but most of the colony transferred to Walthamstow in the 1930s as a direct result of human disturbance. By 1945 there were only six occupied nests at Wanstead while the colony at Walthamstow had increased to 68. The heronry at Wanstead remained at a low level during the 1950s and the last breeding occurred there in 1957.

The Walthamstow colony increased markedly during the 1960s and had reached 84 pairs by 1968. By 1981 it had risen to 128 pairs, making it Britain's second largest heronry. However, after this there was a slow decline and by the end of the 1990s the colony had decreased to 108 pairs.

The only other known heronry in the east was at Parndon. This colony began in 1945 and consisted of around five pairs until it increased to 12–14 pairs in the 1950s. This heronry may have been started by birds from nearby Hunsdon where a colony of eight to nine pairs existed up until 1939. In the early 1960s the Parndon colony decreased and was no longer in use by 1963. The heronry had become totally surrounded by the new town of Harlow, but the site lives on in name as the neighbourhood is now called Heronswood.

South London and the Thames – the heronry at Richmond Park started with a single breeding pair in 1880; by 1909 there were 30–40 pairs and it peaked at 61 pairs in 1939. During the war, some bombs landed among the nesting trees and some of them were cut down and nine of the 40 pairs present deserted. By 1946 only 11 pairs were nesting at Richmond and a similar number bred there until 1951 when the colony increased to 18 pairs. Although they continued to increase over the next two years, reaching 23 pairs in 1953, there was a large fall the following year to only 13 pairs. By the early 1960s the colony was no longer active; however nesting started again in 1994, reaching 25 pairs within five years.

Some of the birds displaced during the war started to prospect at nearby sites; in 1942 one or two pairs bred at Kempton Park and in 1944 several pairs nested at Syon Marsh and on Tagg's Island. A new heronry quickly became established at Kempton Park with 20 pairs breeding there in 1943. Numbers increased annually, probably by birds that formerly bred in Richmond. Heavy predation by Carrion Crows caused the colony to reduce to seven pairs in 1970, however they recovered fairly quickly and by the beginning of the 1980s had reached 50 pairs. They continued to increase and by the end of the 1990s had reached 72 pairs.

Heronries also became established along the Thames at Isleworth Ait (perhaps descendants of the colony which used to breed at Syon Park) with up to 11 pairs in the 1990s and Brentford Aits, with 23 pairs by 1998. At Lonsdale Road Nature Reserve, Barnes, Grey Herons began nesting on recently constructed tern rafts in 1998.

Another colony became established at Burwood Park, near Hersham, which had 15–20 pairs from at least 1925 to 1942. By 1948 it had overtaken Walthamstow to be the London Area's largest heronry with 57 nesting pairs. It peaked in 1951 when there were 93 occupied nests, making it the third largest heronry in the country; for the next few years it remained just slightly below this figure. An isolated colony near the southern border at Gatton Park formed in 1930 when one pair nested. It took a long time for this heronry to become established but by 1948 it had risen to 11 pairs. During the early 1950s it remained fairly constant with 10–12 pairs nesting. A new heronry was established at Chevening and had 10 pairs by 1973.

West London – away from the Thames in west London there was a heronry in Black Park, just on the edge of the London Area; from 1868 to 1925 about 12–30 pairs nested annually but then declined to an occasional breeding pair by the 1930s. At Iver, there was a colony from 1900 to at least 1912 but it had fizzled out by the 1940s. Breeding occurred at Osterley Park in 1950–51 for the first time in some 80 years; there was then a break of 30 years before three pairs bred in 1990. In the Colne Valley, large heronries became established in former gravel pits at Broadwater Farm GP, with around 40 pairs by the 1990s, and Stockers Lake, which peaked at 67 pairs in 1995.

North London – there were several isolated breeding records from the far north of the area around the turn of the century, including: a pair in Hatfield Park in 1880; a pair in Moor Park in 1899; a pair in Grove Park Estate, Watford, in 1901; and there was a small heronry at Beech Hill Park in at least 1903–04. New heronries were established at Verulamium Park, St Albans, since the early 1990s with 12 nests by 1999, and at Brocket Park, with 20 nests by 1997.

Central London – in 1936 a wild bird started displaying with a captive female in the aviary at London Zoo; the bird was released and they built a nest on top of the aviary although this was abandoned after three weeks and the pair moved away. In 1937 Herons regularly visited the sea-lions' pool in London Zoo, presumably to pick up scraps of fish.

In 1968 breeding activity was seen in Regent's Park for the first time when two pairs began building nests but they later deserted. The following year two pairs again attempted to nest there and one succeeded in raising two young. The colony slowly began to increase and by 1973 had risen to five pairs. In 1975 eight pairs reared 20 young, a high productivity rate. By 1979 there were 18 pairs breeding at Regent's Park and it had become the fourth largest heronry in the London Area. By 1991 the colony had increased to 28 breeding pairs. In 1990 a new colony was started at Battersea Park where two pairs bred.

21st century

Since 2000 around 450–520 pairs have been recorded annually, although as not all heronries are counted every year the total population may be much higher. This represents a threefold increase since 1975 when 155 pairs were reported compared to a national increase of 12 per cent between 1970 and 2010. Local factors which may have helped include a cleaner Thames – there are more nests downstream of Teddington and birds can now be seen fishing at low tide in Westminster and Southwark; fish stocking of reservoirs, especially after the 1973 Water Act; development and fish stocking of gravel pits; lack of persecution and the fact that herons have got used to living alongside people. National factors including generally milder winters and a reduction in some of the more harmful pesticides.

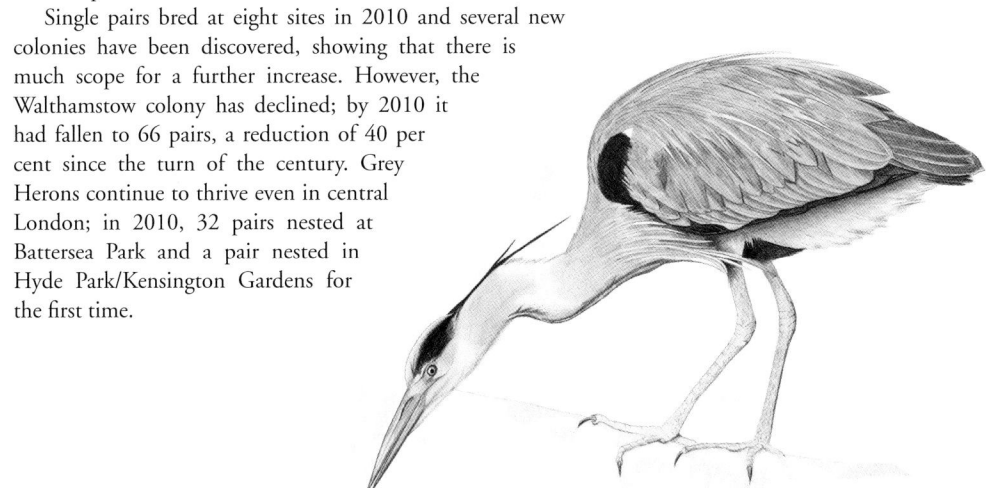

Single pairs bred at eight sites in 2010 and several new colonies have been discovered, showing that there is much scope for a further increase. However, the Walthamstow colony has declined; by 2010 it had fallen to 66 pairs, a reduction of 40 per cent since the turn of the century. Grey Herons continue to thrive even in central London; in 2010, 32 pairs nested at Battersea Park and a pair nested in Hyde Park/Kensington Gardens for the first time.

Ringing

A nestling ringed at Sjaelland in Denmark on 25 May 1911 was recovered in Middlesex on 24 January 1912; and one ringed in Slesvig was found four years later at Mill Hill. As well as foreign-born birds, several that were raised in Cambridgeshire and Kent were discovered in the London Area the following winter. London-bred herons have been recovered in Cambridgeshire, Norfolk and Sussex in their first year.

Facts and Figures

The highest count away from breeding sites was of a roost of 110 birds at West Thurrock on 16 December 1990.

Purple Heron *Ardea purpurea*

A scarce summer visitor (28)

Range: breeds Europe, Asia and North Africa, winters southern Europe and Africa.

Historically this was a very rare bird in the London Area with one record in the 18th century and one undated record from the 19th century. The first bird was shot 'near London' in 1722 and added to the collection in the Leverian Museum and the other was killed near Harrow and came into the possession of a taxidermist in 1829.

Well over a century passed before the next record, when one was seen flying over Fishers Green at midday and again in the evening on 12 May 1956. Another was seen in the Lea Valley just two years later, this time at Walthamstow Reservoirs. Two more were seen towards the end of the 1960s, at Staines and Walthamstow.

The 1970s was the peak decade with nine seen, including four in 1972. All of these were short-staying birds apart from one at Rye Meads, which was present for just over a month. Five more were seen in the 1980s, two of which were immatures that stayed for two days in the autumns of 1985 and 1989.

In the spring of 1990 there was an influx of herons into Britain and one was seen at Dagenham Chase on 27 April. No more were seen until the end of the decade when there were two in 1999: a first-summer at Brent Reservoir for five days in late spring and a juvenile at Beddington in autumn.

In the 21st century none were seen until 2005 when there were two in early autumn: one over Staines Reservoir in the evening of 22 July and one at Beddington on 18 August. Four years later another three were seen, including one which visited two sites in spring.

1722: near London, shot
Pre-1829: killed near Harrow
1956: Fishers Green, on 12 May
1958: Walthamstow Reservoirs, on 25 May
1967: Staines Reservoir, on 19 July; also at Staines Moor on 20 July
1969: Walthamstow Reservoirs, 16 September
1970: Shepperton GP, adult on 2 June
1971: Sewardstone, adult on 16 October
1972: Rainham Marshes, on 1 June
1972: Sevenoaks WR, immature on 17 August
1972: Rye Meads, on 2 September–3 October
1972: Brent Reservoir, on 12 September
1975: Rainham Marshes, adult on 31 May
1976: Sevenoaks WR, on 2 May
1978: Rainham Marshes, adult on 26 May
1982: Beddington SF, on 11 April
1984: Wraysbury GP, adult on 15 May
1985: Dartford Marsh, juvenile on 13–14 September
1987: Rainham Marshes, on 2 May

1989: Amwell GP, immature on 5–6 August
1990: Dagenham Chase, on 27 April
1999: Brent Reservoir, first-summer on 30 May–3 June
1999: Beddington SF, on 22 September
2005: Staines Reservoir, on 22 July
2005: Beddington SF, on 18 August
2009: Walthamstow Reservoirs, on 23–25 April, then at Crossness on 25–27 April
2009: Staines Reservoir, on 26 April
2009: London Wetland Centre, on 1 August

Black Stork *Ciconia nigra*

A rare vagrant (6)

Range: breeds Europe, Asia and Africa, winters southern Asia and Africa.

The earliest record in the London Area was in 1893 at Northolt where an adult male was present for six weeks. It had been eating chickens and was said to be very shy but was eventually shot on a haystack where it roosted; the specimen ended up in Liverpool Museum.

Towards the end of the 20th century four were seen in three years. The first of these was seen flying over three sites in the Lea Valley in the late afternoon of 14 May 1989. There were two the following year: at Tyttenhanger in spring and one drifting over the Oval in Kennington while England played a test match against India. In 1991 one was seen at Epping Forest on 7 July, the third consecutive year that Black Stork had been seen in the London Area. The only record since then was of a juvenile at Sevenoaks Wildfowl Reserve in 2010.

1893: Northolt, present for six weeks and shot on 25 July
1989: Amwell GP, flew over on 14 May; also seen at Waltham Abbey and Cheshunt
1990: Tyttenhanger GP, on 16 April
1990: Kennington, flew over the Oval on 25 August
1991: Epping Forest, on 7 July
2010: Sevenoaks Wildfowl Reserve, on 4 August

Unacceptable Records

On 4 June 2002 one flew towards Amwell from the north, circled over Easneye for 25 minutes then flew off east. Although published in the LBR it has since been established that this bird did not cross the boundary into the London Area.

White Stork *Ciconia ciconia*

An uncommon summer visitor, occasional in winter (30 since 1800)

Range: breeds Europe, Asia and Africa, winters southern Europe and Africa.

Historical

Although they were thought to be breeding in England in the past, the earliest reference for the London Area was in 1738 when Albin recorded that he had seen two at the Duke of Chandos's estate at Edgware; however, these are not listed by Harting or Glegg so it is unclear if they were wild or captive birds. In 1847 Ruegg stated they were present in the 18th century in the Woolwich area. There were no other records until 1866 when one was shot at Rotherhithe in July; a record of a probable White Stork seen flying around St Paul's Cathedral the previous day may have been the same bird. This was the only record in the 19th century.

20th century

Over 70 years elapsed before the next sighting when one flew over Mill Hill in 1938. Unlike the record of one seen on the ground in Richmond Park eight years earlier, which was considered to have been an escape, this particular record was deemed acceptable although there had been an attempt to establish White Storks in 1936 when a number of young birds were released on the Kent marshes. Although they spread out and were seen at various sites on the south coast, none of them were reported again after that autumn.

The first wintering record occurred in 1960/61; although they were rare at this time of year in the country there was no evidence that the bird was an escape as it was unapproachable and could not be traced to any collection. Three were seen in successive years during the 1970s. Another three were seen during the 1980s: two flew straight over and one attempted to land in a tree at Hampstead Garden Suburb before flying off north-east.

In 1993 three were seen circling over Staines Reservoir on 24 April before gaining height and flying off west; they were first seen at Seasalter, Kent, on the previous day and subsequently seen over Bristol on the 27th. In 1996 one was present on Hounslow Heath for at least half an hour in the evening and in 1999 one flew over Brent Reservoir.

21st century

There were just 13 in the 20th century but since 2000 they have been seen almost annually. One toured the London Area in 2003, first being seen at Dartford and Wilmington, and roosting near Hextable on 15 September. The following morning it flew off north and was seen at Orsett. On the 17th it flew over Old Slade Lake and the M4/M25 Junction before roosting on a lamp-post in Harefield. It was last seen the next day at Troy Mill GP before overwintering in Kent. It returned to the London Area in 2004, being seen on a rooftop at South Ockendon on 18 March; over Wraysbury GP, Chertsey, Bookham Common and Leatherhead on 1 April; Fetcham on 2 April; Amwell GP on 9 April and 11 May; Rye Meads on 13 May; Hertford on 15 May; Whipps Cross Common on 18 and 24 May; and finally New Denham where it remained from 30 June to 3 July.

Six were seen in 2006, two separate widely seen birds on 19 March and a flock of four thermalling over Godstone on 6 August.

1866: Rotherhithe, shot on 17 July
1938: Mill Hill, flew over SW on 6 April
1960: Cheshunt, on 3–17 December; also at Wormley on 26 December and 15 January 1961
1976: Ilford, on 25 September; also seen at Hornchurch on 30 September
1977: Sewardstone, on 3 April
1978: Sevenoaks WR, on 30 July; also at Beddington SF/Croydon on 6 August
1982: Dartford Marsh, on 2 May
1983: Hampstead, on 15 April
1986: Rickmansworth, on 6 June
1993: Staines Reservoir, three on 24 April
1996: Hounslow Heath, on 25 April
1999: Brent Reservoir, on 3 May
2000: East Sheen, on 17 June
2001: Berwick Ponds, on 2 May
2003: Hertford, on 27 July; also seen over Hoddesdon cemetery
2003: Various sites on 15–18 September and in 2004 on 18 March–3 July
2004: Addlestone, flew over on 7 September
2005: Reigate, on 17 March
2006: Crossness/Darenth/Sidcup, flew over on 19 March; later seen at Rainham Marshes
2006: Tyttenhanger GP/London Colney, a different bird flew over on 19 March
2006: Godstone, four flew over on 6 August
2007: Orsett Fen, on 8 March
2007: Borough, on 10 June
2008: Maida Vale, on 30 August
2012: Tyttenhanger GP, on 12 May

Unacceptable Records/Escapes

In 1883 one was shot on a rooftop at Hatfield near the end of July and it was believed that it had escaped from Chequers Court, near Tring, Herts. In 1891 three were shot near Hertford but their provenance is unknown and they were not accepted as wild birds by Gladwin and Sage. In 1930 one was seen in Richmond Park on 11 May and thought at the time to probably be an escape. A record of one at West Drayton in 1971 was later withdrawn by the observer. In early 1975 one escaped from Chessington Zoo and was seen flying over there in June. Escapes were seen in the Cobham area in 1979 and 1982. In the winter of 1988/89 one was seen twice in the Rickmansworth area; it was treated as an escape although it was not known to have escaped from captivity. In 1990 there were three records which were all treated as escapes due to the fact that two had escaped from Whipsnade Zoo and were roaming the country at the time. In 2000 two escaped from Bristol Zoo and toured around the country; both were seen over Brent Reservoir on 1 August, and one bearing a colour ring 'AX' was seen at various sites over the next two years.

Glossy Ibis *Plegadis falcinellus*

A rare vagrant (8)

Range: cosmopolitan; vagrant to UK.

There are three records from the 19th century, the first two of which were in south-east London. An additional reference to one 'near Dartford' prior to 1837 relates to the Bexley bird. There were no further sightings until 1974 when one was found at Swanscombe Marsh on 11 April where it remained for just over a month apart from a brief excursion across the river to West Thurrock. In 1977 one spent over four months on Totteridge Common between May and September; although it was suggested at the time that the prolonged length of its stay suggested a captive origin, later records of wild birds have indicated even longer stays in this country.

No more were seen until over 30 years later when one flew over Rainham in 2008. The following autumn there was an exceptional influx into the country with at least 38 seen, one of which was briefly present at Tyttenhanger GP. A similar influx occurred in 2011 and one was seen in the Horton area intermittently in October, often at Arthur Jacob Nature Reserve.

1830: Blendon Hall Park, Bexley, shot in April
Pre-1857: Swanscombe, shot
1881: Balls Park, Hertford, shot on 10 September
1974: Swanscombe Marsh, on 11 April–12 May; also seen at West Thurrock on 5 May
1977: Totteridge Common, on 7 May–22 September
2008: Rainham Marshes, on 21 October
2009: Tyttenhanger GP, on 21 September
2011: Horton, on 22–25 October

(Eurasian) Spoonbill *Platalea leucorodia*

A scarce visitor and historic breeder (51 since 1800)

Range: Europe, Asia and Africa.

Historical

This was first documented in 1523 when it was recorded in Henry VIII's year-book that an action for trespass was brought by the Bishop of London. The defendant was caught taking herons or Spoonbills from their nests in the grounds of Fulham Palace, where both birds were known to be breeding. It is not known for how long after

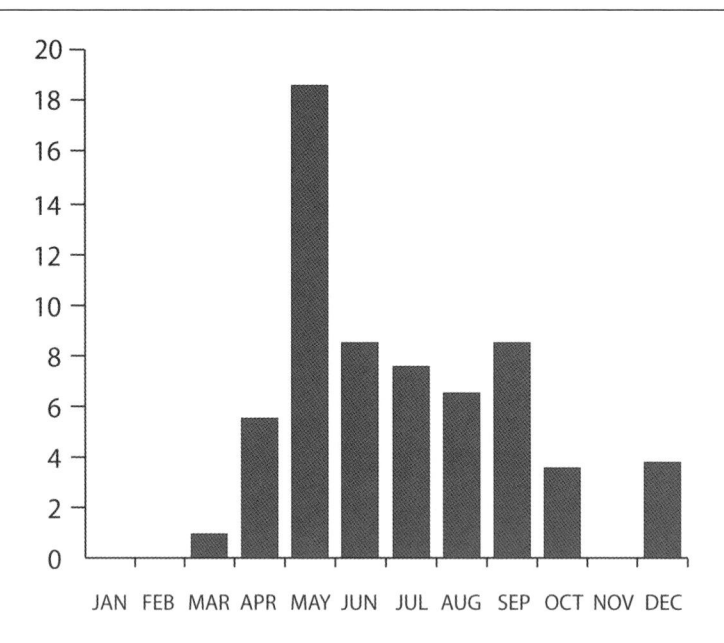

Seasonal occurrence of Spoonbill: all records.

this Spoonbills continued to breed there. The area still exists as Bishop's Park, albeit as a much-changed place. Spoonbills were also known from the Woolwich area until at least 1600.

The next documented records were over 200 years later when there was one at Greenwich prior to 1812; two were shot at Brent Reservoir on October 23rd, 1865; and one flew over a garden in Camberwell in August 1889.

20th century

There was another long absence of over 70 years before the next record in 1961 when one was present at Rainham Marshes on 5 October. In 1969 the same site hosted an immature on 27–28 July, with two from 30 July to 11 August, followed by two adults on 6 September. There were four in the 1970s: an adult at Walthamstow Reservoirs on 15 June 1973; an immature on Staines Moor on 27–28 September 1973; one at Richmond Park on 30 June 1976; and one over Wraysbury GP on 7 May 1977.

The 1980s saw a sudden increase in sightings with 12 seen, commencing in 1980 with one in Romford on 6 April, followed by an adult, which spent two days at Holmethorpe Sand Pits on 12–13 August. In 1982 two adults were present at Beddington for half an hour on 1 June, and the next year one flew over Rainham Marshes on 25 October. There were three in 1986: at Queen Mary GP on 4 July and at Wraysbury GP on 5 July; a different immature at Queen Mary GP and King George VI Reservoir on 23 August; and one at Amwell GP on 10 October. Four occurred in 1989: one on 3 March at Hall Marsh and Bowyers GP; two at Beddington on 25 May; and one on the partly drained Staines Reservoir on 7 October.

Nine were seen during the 1990s. At Kempton Park Reservoir two adults were present on 27 May 1992. The following year an adult flew east over Brent Reservoir on the evening of 25 May, and in 1995 an adult flew over Queen Elizabeth II Reservoir on 2 December. There were three in 1997: at Stanwell Moor on 17–18 April, which was also seen at Perry Oaks and Staines Moor on 18 April; at Fishers Green on 7 June; and over Brent Reservoir on 8 July. In 1998 one was seen briefly at Walthamstow Reservoirs on 28 June, and the next year one was present at the Wetland Centre for less than an hour on the evening of 13 May.

21st century

The number of sightings has increased since 2000 with another 27 seen. The first to be seen in the new millennium were four juveniles at Stanwell Moor on the evening of 11 September 2001. The following year two arrived at Rainham

Marshes on 2 April, one of which remained on and off until the 18th. In 2003 one was seen on six dates at Queen Mother Reservoir between 6 and 25 November. In 2005 two spent the afternoon at Staines Reservoir on 6 September.

Four were seen in 2006: over Beddington and Farthing Downs on 7 May; at Kempton Park NR on 23 May; at Rainham Marshes on 9 June and at Walthamstow Reservoirs on 10 and 17–18 June; and at Redhill on 20 July. In 2007 a record nine were seen, including five individuals at Rainham on various dates between 4 May and 1 August, some of which were also seen at Cornmill Meadows and Dartford Marsh; and single birds at Amwell, Beddington, Rye Meads and the Wetland Centre. There were three in 2008 and two in 2009.

Records Not Accepted

In 1988 one flew over Beddington on 4 May and two flew over Kenley on 31 July. They were accepted as spoonbill species by the Records Committee but in each case the specific identify could not be confirmed; during the year an escaped African Spoonbill *P. alba* was at large in the country and two hybrid Spoonbill x African Spoonbill escaped from a zoo in Surrey in the summer.

Pied-billed Grebe *Podilymbus podiceps*

A very rare vagrant (2)

Range: North America; vagrant to Europe.

The first record for the London Area occurred in January 1997 when an adult was present on South Norwood Lake for two months; during its stay it moulted into breeding plumage and was heard calling. This was also the first one to be seen in south-east England. The following winter it returned in December, albeit a few miles away on Tooting Common Lake, where it stayed until February. This was the twenty-first bird to be seen in Britain and was one of 10 seen in Europe during 1997.

In March 1999 one spent a day on Walton Reservoirs. It left overnight and was refound at Stowbridge, Norfolk, the following day; it may have been the bird that had been present in the previous two winters.

1997: South Norwood Lake, on 26 January–30 March; then at Tooting Bec Common on 5 December–12 February 1998
1999: Walton Reservoirs, adult on 16 March

Little Grebe *Tachybaptus ruficollis*

Common resident and widespread breeder

Range: Europe, Asia and Africa.

Historical

In the 19th century this was mainly a spring and autumn passage migrant, with some winter records and only occasionally breeding. Large numbers were occasionally noted in winter; for example, in 1870 there was a large influx and around 100 were on the Round Pond in Kensington Gardens.

The first breeding was recorded on Chelsea Common in 1805. In Inner London at least three pairs bred in 1867 at Kensington Gardens; a pair bred for the first time in St James's Park in around 1883 and then began to breed there regularly. By the end of the 19th century six or seven pairs were nesting there, a relatively high breeding density for a small lake. This breeding population was not resident and the birds were absent in winter. They nested in Regent's Park during the 1890s but rarely raised any young.

In 1896 a pair attempted to breed in Clissold Park but their nest was repeatedly destroyed by Moorhens and partially eaten by Mute Swans despite the grebes attempting to defend the nest by attacking the swans, both on the water and underwater. Despite these occasional setbacks they continued to expand their breeding range; in 1898 Hudson stated that they were now as well established in the London Area as Moorhen and Woodpigeon.

20th century

During the first quarter of the century breeding was recorded from a number of locations around the London Area, mainly in the outer suburbs and rural areas but with a few pairs in the central parks. In the 1930s Glegg described them as a common resident, becoming more numerous during winter. They could be found breeding in Inner London and the suburbs, and were relatively widespread in the rural areas, for example in Epping Forest where they bred on various ponds. Although the total population was increasing, there were inevitably some local losses. In 1931–32 about 12 pairs bred on Fetcham Mill Pond and up to 60 were present in winter; however, some of this site was later converted to watercress beds and the population decreased.

They returned to breed in St James's Park in 1939 after an absence of 25 years and it was thought that this was due to the reintroduction of sticklebacks into the lake; they continued to breed there irregularly afterwards. However, between 1954 and 1961 none nested in Inner London.

In the first half of the 20th century there was a regular passage through the London Area in autumn. One exhausted bird landed in a garden at Gray's Inn in October 1927 and another was found by a block of flats in Fulham during fog in November 1934. During winter they became more numerous and were seen at many sites where breeding did not occur. The largest numbers were at Brent Reservoir where there was a notable build-up during autumn.

By the beginning of the 1950s the population had increased further, particularly in autumn and winter when large concentrations could be found. Typically they numbered up to 50–60 but occasionally more than 100 could be seen at Brent Reservoir where they peaked at 121 on 11 October 1953. The latter site, along with Ruislip Lido, is a relatively shallow reservoir with natural banks and holds a good population of breeding Little Grebes, which rarely breed on the large concrete-banked reservoirs. They were fairly scarce on the Thames but in November 1959 there was an unprecedented group of 130 on the river at Northfleet and up to 74 were still present in January 1960.

In the early 1970s fewer were seen at Brent Reservoir and the maximum counts never exceeded 40. By the start of the 1980s the breeding population had increased to more than 60 pairs but they were still only infrequent visitors to Inner London. In 1983 a pair bred in St James's Park, the first breeding in any of the Royal Parks since 1945. The second Breeding Atlas Survey showed a range expansion of just over half since the previous survey in 1968–72. By the early 1990s the London Area's breeding population had increased to at least 90 pairs; the largest concentration was 12 pairs at Rye Meads. In Inner London they were seen at various sites but were not breeding.

21st century

At the beginning of the century the breeding population was at an all-time high of 170 pairs, including a record 36 at Rainham Marshes. They have also colonised Inner London and at least five pairs breed usually in the central London parks. The largest count since 2000 was 113 at Rye Meads in 2006.

Facts and Figures

The largest count was 130 birds on the Thames at Northfleet in November 1959.

Great Crested Grebe *Podiceps cristatus*

Common resident, widespread breeder and many depart in winter

Range: Europe, Asia, Africa and Australasia.

Historical

For most of the 19th century this was an occasional winter visitor, not staying to breed in the London Area. The British breeding population was at an all-time low in the middle of the century, at about 42 pairs, due almost entirely to birds being shot for their feathers, which were used in the millinery trade. The earliest record was some years prior to 1824 when there was a large flock on the Thames between Gravesend and Greenwich. The next record was of a young bird shot on the Thames at Penton Hook near Laleham in 1844. There were also four separate records from Brent Reservoir prior to 1866 and on Wimbledon Park Lake in the 1860s.

Legislation was introduced between 1869 and 1880 to protect grebes although the fashion for having feathers in hats did not die out until 1908. The Bird Protection Act of 1880 was soon followed by the London Area's first breeding record in 1886 at Ruislip Lido. Breeding then occurred at Osterley Park in 1896, followed by other pairs in Richmond Park and Gatton Park.

20th century

By 1900 the British population was making a good recovery and breeding numbers had increased to about 300 pairs. This was reflected in the London Area and a gradual colonisation continued throughout the early part of the century. Even as early as 1906, when few pairs bred in the London Area, 80 were counted at Staines in autumn. From 1906 onwards up to three pairs nested annually at Aldenham Reservoir although they rarely succeeded in rearing young; the same year a pair bred at Cheshunt. A pair nested in 1907 at Grovelands Park, where it then continued to breed annually. At Ruislip Lido at least two pairs bred in 1908 and by 1930 eight pairs were breeding there.

Up to the middle of the century they were only known in Inner London as occasional visitors. Breeding was first recorded in the suburbs in 1912 at Stoke Newington Reservoirs. By 1921 they were breeding at 19 sites across the London Area. There was also an unusually high count one spring when 160 were seen at Staines on 20 April 1922. Counts such as this, along with the national censuses, showed that both the post-breeding build-up on the large reservoirs and the spring passage is made up of birds breeding further north.

In the 1930s Glegg stated that at times the species could be very numerous, especially during autumn migration. The largest numbers were found on the reservoirs, particularly in August and September when more than 300 were counted, e.g. there were 330 at Queen Mary Reservoir on 29 September 1934; these peaks declined before the onset of winter. However, the maximum numbers were later in the year at some sites, e.g. at Banbury Reservoir and Barn Elms numbers peaked in November. A census carried out in 1931 found a total of 1,160 pairs across England and Wales. Of these, 68 pairs were in the London Area, and the largest breeding concentration was 11 pairs at Walthamstow Reservoirs.

In the 1940s the colony at Aldenham Reservoir grew to six pairs from 1944 to 1947. In the Lea Valley, four pairs bred on gravel pits between Broxbourne and Waltham Abbey in 1942; they soon became established in the area and by the 1950s up to 15 pairs bred. Another national census was carried out in 1946. Although the overall totals were similar to those in 1931, the breeding population in the London Area had increased by over 50 per cent to 350 birds. However, the following winter was quite severe and this reduced the population by around 20 per cent. A good breeding season in 1947 more than made up for these losses and the numbers in the London Area continued to increase.

The change in population in the 50 years since the beginning of the century was quite dramatic; by 1950 it had reached 550, including 149 breeding pairs, compared to just a handful in 1900. Most of these birds were on gravel pits, which provided suitable breeding habitat, and combined with legal protection allowed Great Crested Grebes to prosper. In 1953 the population had reached a new high of 627. By then the population in the London Area was of national significance, and in several years in the 1950s there were more present in London in June than in any other county.

The 1960s started with a new record count of about 650 on Queen Mary Reservoir on 10 January, and in November 1,179 were counted on the main waters, a doubling of the population in a decade. During the severe winter in 1963 many of the reservoirs froze over; on 13 January the wildfowl count produced a total of 658, including 375 at Queen Elizabeth II Reservoir, but by 26 January seven were found dead there and only 10 remained. Many grebes moved to the Thames. By 2 February there were 400–500 on the river between Kingston and Putney, and it was thought that many had become too weak to leave the area completely; in the Lea Valley 22 were found dead. The weather had a serious effect on the population as the numbers breeding in the London Area were substantially down compared to the previous year. Although the following winter was much milder, the population was still very low and there were far fewer overwintering. By 1967 the breeding population had recovered to about 75 pairs.

At the beginning of the 1970s the breeding population had further increased to about 120 pairs; the largest concentration was 24 pairs at Thorpe Park. In 1972 breeding occurred for the first time in Inner London, at both Kensington Gardens and Regent's Park. A national census was conducted in the summer of 1975 and 816 adults were located at 87 sites in the London Area, more than double the previous count 10 years earlier. The increase in population since the beginning of the century was directly linked to the increase in artificial habitats, as more than 90 per cent were found on gravel pits and reservoirs. There was a large influx in early 1979 due to very cold weather; for example, at Wraysbury Reservoir they increased from 250 to 515 in two days.

By 1980 the breeding population was 172 pairs; there were over 800 birds present in October and November. They were also increasing in Inner London, where by 1983 there were 14 breeding pairs. In 1984 there were 214 breeding pairs, including 32 at Brent Reservoir. By 1988 the breeding population at Brent had risen to 41 pairs although the success rate was low due to fluctuating water levels. Between the two breeding Atlas surveys there was a range expansion of almost 90 per cent; the greatest increase was in the area between the Thames and the Lea Valley.

The breeding population in 1990 had reached a total of 238 pairs. In 1991 about 50 pairs bred at Brent Reservoir, making it second only to Rutland Water for breeding pairs. By 1993 a decline had started and only 194 breeding pairs were recorded; there was also very poor productivity at some sites, including Brent Reservoir, where only two broods were raised due to unseasonable flooding. Of the five nesting pairs in Regent's Park only one was successful, possibly due to predation by intoduced terrapins. In 1994 the population further decreased to 179 pairs and breeding success was again hampered by flooding at some sites; nevertheless this species attempted to breed virtually throughout the year with chicks seen in February and December.

21st century

At the beginning of the 21st century the breeding population was about 178 pairs. The decline in breeding numbers has continued and by 2009 there were an estimated 150 pairs; the largest concentration is 15 pairs at

Brent Reservoir where the breeding population had decreased by two-thirds since 1991. At least five pairs breed in Inner London. Freezing conditions in early 2002 forced most remaining birds onto Queen Mother Reservoir where there were 671 on 13 January, just below the all-time record count.

Facts and Figures

The largest count was 679 birds on Queen Mother Reservoir on 14 January 1997.

Red-necked Grebe *Podiceps grisegena*

Uncommon winter visitor

Range: Europe, Asia and North America.

Historical

There are only three records in the 19th century, the first of which was in Wanstead Park in February 1877. The other two records were of one shot at Farthing Down in April 1890 and one killed at Poynter's, near Cobham, prior to 1900.

20th century

In 1913 one was seen on Highgate Ponds, Hampstead Heath, from 2 to 15 February. This was followed by one at Barn Elms Reservoirs in 1922 and two on the Lea Valley reservoirs in 1924. From then on they became more regular, particularly on the large reservoirs. At Staines Reservoir, for example, there were five between 1926 and 1930. In 1937 there was a small influx into the London Area during the first three months of the year following a severe north-easterly gale and very low temperatures in the Baltic Sea; seven were found, including one on the Round Pond in Kensington Gardens.

Between 1900 and 1954 about 61 were seen, of which 25 were on the reservoirs in the Lea Valley, 13 at Staines and King George VI, and seven at Barn Elms. This helps to explain why the numbers had increased so much as none of these reservoirs existed at the beginning of the 20th century. All of these bar one arrived between September and March, with January being the most frequent month of arrival. The exception was in 1930 when one was found at Staines Reservoir on 13 July.

In 1956 severe weather caused a small influx of at least eight into the London Area. In the severe winter of early 1963 there was a small influx of grebes and divers, but no Red-necked Grebes were seen in the London Area. There was a higher number than usual in 1967, including up to 10 different birds throughout the year on Staines Reservoir, one of which was present on and off during the summer.

More were seen in the 1970s than in any other decade, beginning with 16 in 1970; this included six in the first two months of the year and five spring migrants. In February 1979 there was an exceptional influx into the country due to an easterly blizzard in the middle of the month; in the London Area, 43 were seen in the second half of the month, including 11 in the Lea Valley and a group of five on Staines Reservoir.

There were also higher numbers than average during the 1980s. In 1981/82 three wintered on Girling Reservoir and one summered there in 1982–83. A large influx occurred in 1985 with about 36 recorded, mostly arriving in January and February; the peak counts were seven on Queen Mary Reservoir on 19 January and five on Queen Mother Reservoir on 22 January; one also summered on King George V Reservoir. Smaller influxes occurred in February and March 1986, with about 18, and in early 1987, with about 28 including three on the Thames at Rainham; at the end of the year a further 14 arrived. Towards the end of 1988 there was an influx of about 18 birds, including nine in October, which is extremely early for a mass arrival; the peak count was five on Queen Mother Reservoir on 26 October; six of these then overwintered. The best year of the 1990s was 1996 when about 29 were seen. By 1999 the annual total had dropped to just six birds, the lowest for many years.

21st century

Since 2000 there has been an average of five per year with the lowest annual total being just one in 2005. They are typically seen on the large reservoirs, especially in west London. In 2003 one summered at Queen Elizabeth II and Walton Reservoirs and then overwintered in west London; another one summered on the west London reservoirs in 2008.

Facts and Figures

The largest group was seven birds on Queen Mary Reservoir on 19 January 1985.
A long-staying individual remained at Queen Mother Reservoir from 2 November 1993 until 6 November 1994, a total of 370 days.

Slavonian Grebe *Podiceps auritus*

Uncommon winter visitor

Range: Europe (including small breeding population in Scotland), Asia and North America.

Historical

This was a rare visitor in the 19th century with just four records. The first was documented by Graves in 1821 when he stated that he had received one that was killed on the New River, near Clay Hill, Enfield. There were two in 1870, at Carshalton and at Charlton, followed by one shot on the Thames at East Molesey on 23 December 1876.

20th century

There were no further records until 1917 when one was on the Thames at Richmond on 16 February and later off Chelsea Embankment on two further occasions. From 1924 the species was seen almost annually, and in the 30-odd years up to 1954 about 50 were recorded, including three on the park lakes in Inner London. Most arrived between September and March, and typically stayed for one or two days with the occasional bird remaining for up to two months; they were usually seen on the large concrete-banked reservoirs, notably in the Lea Valley, at Staines and at Barn Elms. In early 1937 there was an influx and nine were seen at five localities, including four together on Mitcham Junction GP on 6 February; the only other count of four during the first half of the century was at Staines Reservoir on 30 November 1930.

There were also two midsummer records, at Walthamstow Reservoirs on 16 June 1934, and in 1948 one in full breeding plumage remained at Poyle GP from 5 April to 1 June. During its stay it attempted to build a nest, then relocated to Staines Reservoir on 6 June; it stayed in the area until 21 September.

Severe weather in February and March 1956 brought many divers and grebes into the London Area and at least nine Slavonian Grebes were noted, including four on Walthamstow Reservoirs from 4 to 17 March. In the winter of 1962/63 the weather was even more severe but there was only a small influx with six arriving between 31 December 1962 and 28 January 1963. In 1971 spring migrants were seen on five reservoirs during the second half of April. In February 1979 there was a large influx of grebes into the country following severe weather and 14 arrived in the London Area.

The 1980s was the peak decade for Slavonian Grebes in the London Area as there were several influxes related to cold weather. The record year was 1985, with 48 birds; about 36 arrived in the first two months of the year and coincided with two separate influxes of Red-necked Grebes, both caused by severe winter weather. The largest count was four at Stone Marsh from 16 to 18 February. Seven arrived following the Great Storm in the autumn of 1987, including four together on Queen Mother Reservoir on 19 October.

Fewer were seen during the 1990s, coinciding with milder winters. However, in 1996 a record 58 were seen, including about 20 at Staines Reservoir during the year with a peak of seven between February and April.

21st century

Since 2000 the annual totals have varied between three and 22 with an average of about 10 a year. Most birds have been seen on the large reservoirs in west and south-west London and in the Lea Valley, but there was also one in Kensington Gardens and St James's Park in December 2002. A pair in breeding plumage spent two weeks at Dagenham Chase in April 2004 and was seen displaying.

Facts and Figures

The largest gathering was seven birds at Staines Reservoir on 4 April 1996.

Black-necked Grebe *Podiceps nigricollis*

An uncommon winter visitor and scarce breeder

Range: Europe, Asia, Africa and North America.

Historical

The earliest record was of one in partial summer plumage that was 'taken about the large ponds at Hampstead' prior to 1743. There were no more until almost a century later when two were shot at Brent Reservoir in 1841; this was the only 19th-century record.

20th century

The only documented records in the first two decades of the century were of singles at Dartford in 1906/07 and at Staines Reservoir from 13 to 26 October 1907. During the 1920s sightings began to increase and the species was seen almost annually from 1922 to 1934. Most birds were on Staines Reservoir; elsewhere, they were noted at Queen Mary Reservoir, Ruislip Lido, Staines Common Pond and even in central London, in Kensington Gardens and Regent's Park.

In 1935 Glegg described it as 'an irregular visitor, which has been recorded in every month of the year except June', although most were seen between August and December. Up to four were seen annually, some of which made prolonged stays of two to three months. This increase continued, and by 1939 small parties had become regular on Staines Reservoir with up to eight seen during the last half of the year. Elsewhere, odd ones were recorded on some of the other reservoirs in autumn and winter. There were few observations during the war as there was no access to many of the reservoirs, but in the autumn of 1945 a flock of 21 was counted at Staines. The autumn peak at Staines varied during the 1950s and 60s from four to 15.

There was little change during the 1970s but by the beginning of the 1980s small numbers had started wintering on Girling Reservoir. There was usually a noticeable build-up in spring at Staines Reservoir but the main peak was after the breeding season; in September 1981 a record 39 were on Staines Reservoir. During the 1990s the autumn build-up at Staines Reservoir decreased and in 1998 only one was seen. The wintering numbers also declined and in 1993/94 no birds were seen on Girling Reservoir, the first time that none had overwintered for many years.

21st century

At the beginning of the 21st century the largest counts were all in the Lea Valley. In 2000 five overwintered on Girling Reservoir; up to eight were there in spring and there was a large build-up towards the end of the year with a peak of 20 in November. There were also up to seven on King George V Reservoir in February and at Walthamstow Reservoirs in September. The numbers have continued to increase on Girling Reservoir, both in autumn and during winter, with peaks of 36 in July 2011 and 28 in January 2007. This is a significant proportion of the UK wintering population of around 130.

Fewer are now seen in west London with the highest count being nine on Staines Reservoir in 2009.

Breeding

Harting referred to a pair that bred on Chelsea Common in June 1805; however, these were later proven to be Little Grebes.

The beginning of the 1990s was a significant time for Black-necked Grebe in the London Area as this was the first year in which they bred. On 9 June a pair was seen displaying at Brent Reservoir and it was assumed to be the same pair that then moved to another reservoir and successfully bred. The following year a breeding plumage adult was present and heard calling on 18 May but it was not seen subsequently. In 1992 a displaying pair was present in April and May but did not remain to breed. There were no further signs of breeding until 1996 when a pair was seen displaying at Brent Reservoir on 29 April but these birds did not stay to breed. In 1997 a pair returned to the original breeding site and was present for two weeks during the spring and was seen displaying; the following year the birds bred successfully. In 1999 four were present during the breeding season but failed to breed.

Since 2000 they have bred annually at the same site although the breeding success has been variable. For example, in 2009 nine broods were hatched but in 2006–07 there were none, although up to 21 adults were present. Additionally, a pair raised two young at Rainham in 2002; the birds returned the following year but did not stay to breed.

Facts and Figures

The largest flock was of 39 birds on Staines Reservoir on 13 September 1981.

Honey Buzzard *Pernis apivorus*

A scarce passage migrant

Range: breeds Europe and Asia, winters Africa.

Historical

The first record was of a bird trapped in Lullingstone Park in 1838. This was followed by one shot at Munden, near Watford, at some time between 1840 and 1850. There was an influx into England in autumn 1881 and two were seen in the London Area, one shot at Dartford and one seen several times in Epping Forest. The only other 19th-century record was of one trapped at Rose Hill, Hoddesdon, on 2 October 1896.

20th century

There were no further records until 1954 when one was seen flying low just inside the London Area boundary at Box Hill on 27 July. In 1959 one flew south at Esher Common on 22 August.

The sightings began to increase during the 1970s when six were seen. The first of these occurred in 1974 when one was seen in Hatfield Park on 1 June. In 1976 there were three: at Dagnam Park on 17 April; over Regent's Park on 22 September, the first Inner London record; and one at Barn Elms Reservoirs on three dates between 16 and 21 November when it was picked up in a weak condition and taken into care where it later died. In 1979 two migrants were seen, over Ewell on 27 May and in Epping Forest near a wasps' nest on 3 September.

There was another large increase in the 1980s with 13 seen. While most of these were in spring and autumn there were also three in June, at Orpington, Oxshott and Rye Meads. All of these were fly-overs apart from one that was feeding on garden lawns in South Croydon on 23 May 1986. Another 14 were seen in the 1990s, including five or six autumn migrants in 1995.

21st century

The year 2000 will be remembered as being the year of the Honey Buzzard invasion. A migrant on 13 August at Limpsfield Chart was notable enough as it was the first in the London Area for four years and only the thirty-eighth record ever. However, by the end of autumn another 188 had been added to the total.

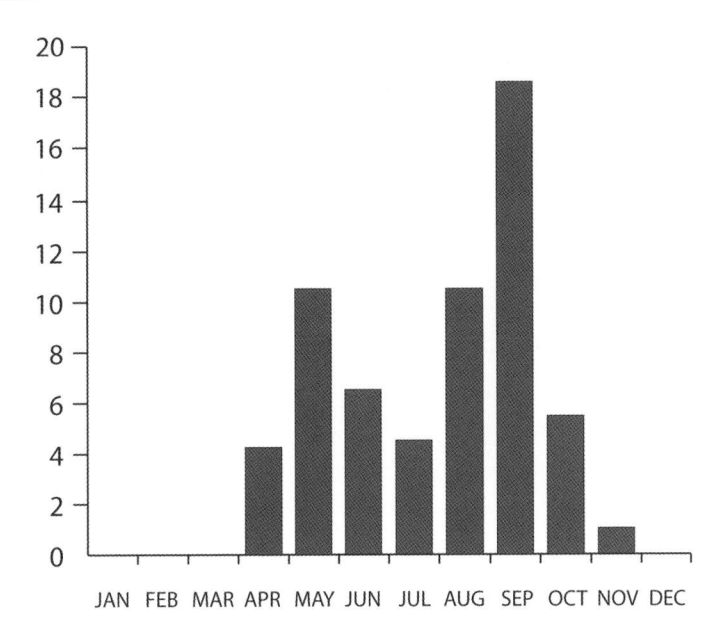

Seasonal occurrence of Honey Buzzard: all records.
NB. The influx in 2000 is excluded from this chart.

The influx began on the East Coast on 19 September when birds that had probably left Scandinavia in good weather then got caught up in strong anti-cyclonic conditions which forced them west where they encountered cloud and rain over Britain. Hundreds, possibly more than 1,000, then arrived in the country and continued their southbound migration, with many passing through the London Area. Two days after the influx began the first two were seen in the London Area. The following day 13 were seen in the afternoon, including a flock of seven over Regent's Park. On 23 September 33 more were seen, all of which were single birds apart from two over West Ham and three over Hampstead Heath. The next day 'only' 12 were seen but on the 24th another 21 passed over and there was an increase in sightings elsewhere in the country, suggesting that another arrival had taken place. Ten or more were then seen daily over the following six days, including 27 on 1 October, the last 'big' day of the invasion. During the following week a few were seen on most days with the last being recorded on the 14th, eight days after the previous one. During the influx they passed through the London Area on a wide front spanning almost the entire width of the recording area but the vast majority flew over the middle of London.

Since then they have become annual, apart from in 2002 when none were seen. Between 2001 and 2009 there were 33, all of which were singles apart from two together over West Bedfont on 12 May 2006. There were eight in 2008, the highest annual total apart from in the 'invasion year'.

Facts and Figures

The earliest migrant was on 12 April 2001 at Beddington.
The latest record was on 21 November 1976 at Barn Elms Reservoirs.

Black Kite *Milvus migrans*

A rare vagrant (10)

Range: Europe, Asia, Africa and Australasia; former rarity in UK.

There have only been ten accepted records of Black Kite in the London Area. The first occurred in 1979 on a rubbish dump at South Ockendon on 19 April; it was still present the next day but flew off south early in the morning. The following year an adult was found freshly dead in the Navestock/Ongar area on 1 October. All subsequent sightings were of birds flying over. One remained for a week in the Weald/Navestock area in May 2009 but was rarely seen during its stay. While 70 per cent of national records are in April and May only four of the ten London Area records occurred during those months.

1979: South Ockendon, on a rubbish dump on 19–20 April
1980: Navestock/Ongar, found dead on 1 October
1986: Hampton Hill/Bushy Park, on 30 April
1994: Stocker's Lake, on 26 June
2003: London Wetland Centre, on 30 August; also seen at Regents Park
2003: Staines Moor, on 29 September
2004: Goodmayes, Ilford, on 25 June
2009: Stoke Newington Reservoirs, on 13 April
2009: Weald Park/Bridge, Bedfords Park and Navestock, on 3–9 May
2011: Beddington SF, on 15 July

Unacceptable Records

One seen perched on the Houses of Parliament in 1906 had escaped from London Zoo. In 1939 several were deliberately released from London Zoo and one was later seen flying over Stepney.

Red Kite *Milvus milvus*

An uncommon resident and passage migrant; historic breeder

Range: Europe and North Africa; reintroduced in parts of UK.

Historical

Red Kites have a long history in the London Area, indeed longer than any other species. The first documented record was in the 2nd century when the Roman author Aelian stated that the kites in London had developed a habit of plucking hair off men's heads for weaving into their nests.

In the 15th and 16th centuries they were abundant in London. They were recorded by Schaschek who visited London in about 1465 and wrote that nowhere had he seen such a number of kites; he noted that it was a capital offence to kill them. In about 1500, in *A Relation of the Island of England*, an unknown Italian author wrote that kites were so tame that they would often take bread and butter out of the hands of little children; it has been assumed that this statement referred to London. In Henry VIII's reign between 1509 and 1547 it was stated that London swarmed with kites, which were attracted by the offal thrown into the streets and were so fearless that they would swoop down in the midst of great crowds of people.

In 1571 Clusius wrote that he had seen large numbers in London throughout the year and considered that the number of kites in Cairo could hardly be greater than the number of kites in London. He described how they picked up and devoured the rubbish thrown into the streets and even into the Thames, adding that 'they flocked thither in multitudes, and displayed great boldness in securing their food'.

Although there has been some speculation that the kites that frequented London may have been Black rather than Red, this notion can be laid to rest by the ornithologist William Turner who in 1544 stated that he knew two sorts of kites, a larger rufous-coloured one which was abundant in England and snatched food from children's hands, and a smaller, blacker one which he had not seen in England but knew from Germany.

The next reference was not until 1639 when Howell poetically described the flight of a kite he had seen on 17 March on a country walk from his home in Holborn. Hamilton stated that they nested in elms surrounding St Giles's Church (near Covent Garden) when it was constructed in 1734. Turner, writing in August 1777, referred to a pair which had bred that year in a garden in Gray's Inn Square and added that he knew of two earlier instances when they had bred in Hyde Park. That was the last known breeding record in London, although it was not successful as the nest was pulled down and the unfledged young were cut open to find out what they had been eating. In 1782 two were seen at Enfield Chase. Breeding also occurred in Swanscombe Park Wood, probably in the late 18th century.

At some point they went into a terminal decline and by the mid-19th century had become rare visitors. Fitter believed that the decline of Red Kites in London was as much a result of the lack of carrion and refuse on which to feed as the increased persecution from gamekeepers and collectors.

Yarrell noted them as being rare in the southern counties of England. They had become extinct in Epping Forest by 1835 due to persecution. In 1840 one was shot near Watford, at Munden. In 1850 Bond saw one over Kingsbury and said it was the only one he had seen in that neighbourhood. Another was shot by Bond about five years later over Grove Park. In 1859 one flew over Piccadilly and in 1893 Jefferies stated that he had seen one over Surbiton but gave no date.

20th century

There were no further sightings until 1948 when one was observed soaring high over Epping Forest on 10 May. The following year, one or two were seen over Baldwin's Hill in Epping Forest on 23 April. In the 1950s the UK population had diminished to about 15 pairs, all in mid-Wales, and they were rare in the rest of the country. The only sightings in the London Area were in the spring of 1958 when single birds were seen near Wraysbury and over Shepherd's Bush.

Just one was seen in the 1960s, over Beddington on 15 August 1966. The next record was 10 years later when one was found freshly dead at Havering on 6 April 1976; it had been ringed as a nestling in Germany four years earlier and was only the second foreign-ringed bird to be recovered in Britain. This helped to confirm the theory that Red Kites seen in spring in southern England were from the Continent rather than from Wales. In 1979 one flew east over Epsom on 6 September.

The 1980s saw the start of an increase in the number of sightings in the London Area and they were seen almost annually. In 1981 one flew over Regent's Park on 7 May – the first sighting in Inner London in over 120 years. There was a notable influx into southern and eastern England in 1988 and nine were seen in the London Area between 21 March and 26 April.

By the start of the 1990s there was a large number of released birds in the Chilterns but they were all wing-tagged and were mainly sedentary. The increase in sightings continued in the London Area but as they were mainly in spring and were of untagged birds they were assumed to be wandering birds from the Continent. The first wing-tagged bird was seen in 1993 and was present in the Epping area from 3 to 20 January. Another three wing-tagged birds were seen the following year.

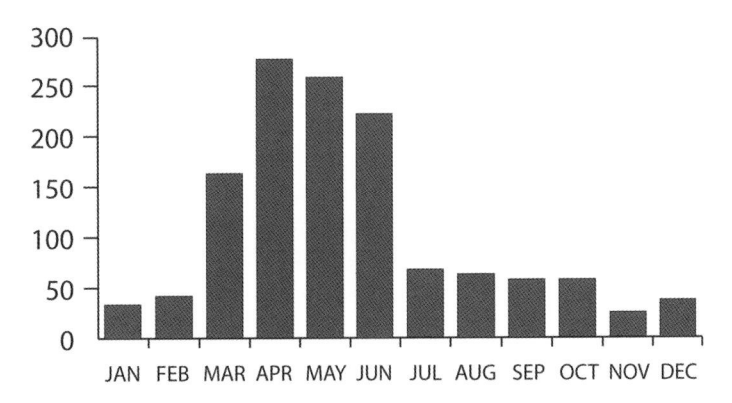

Seasonal occurrence of Red Kite: 2000–2009.

21st century

By the beginning of the 21st century there was a healthy population of over 100 pairs in the countryside on either side of the M40 corridor west of London thanks to the reintroduction programme in the Chilterns. During 2002 the number of sightings in the London Area increased dramatically and some 27 were seen; all but three of these were in the western half of London and most probably originated from the Chilterns population. Sightings have continued to increase every year since then.

In 2004 about 60 were seen, including flocks of four at Croxley Moor and three at Tyttenhanger. The following year a pair bred and raised two young in Hertfordshire, just beyond the London boundary, although they were often seen inside the area; elsewhere over 100 were seen during the year. In 2006 over 200 were seen, including a flock of five over Leyton on 7 June. The breeding pair in Hertfordshire was present all year and was seen mating although no juveniles were seen; another pair was seen in south London in June. A pair bred again at the same Hertfordshire site in 2007–09 and a second pair has been present since 2008. A pair was seen nest-building at a nearby site, within the London boundary, in 2009; although there was no confirmation of breeding, it is surely only a matter of time before Red Kites are again breeding in the London Area.

The peak period is between March and June with about 80 per cent of each year's sightings. Red Kites are now seen annually in Inner London.

Facts and Figures

The largest flock since 1800 was of seven birds at Maple Cross on 2 June 2009.

White-tailed Eagle *Haliaeetus albicilla*

A rare winter vagrant (7)

Range: Europe and Asia; reintroduced in UK.

Historical

This was a rare winter visitor to the coasts of eastern and south-east England. In the 19th century, Harting wrote that 'on some parts of the coast they may be seen frequently'. Apart from the remains of a bird found in an archaeological site in Southwark which was dated to around 300 AD or later, there were no sightings in the London Area until the 19th century.

20th century

There were three records in the 20th century, all within the first three decades. The first of which was an immature male shot in 1906 at Cheverells, near Titsey. Three years later, one was observed soaring over Weald Country Park for half an hour. The third one was seen flying over Navestock in 1928. The small UK population then became extinct until the species was reintroduced into Scotland late in the 20th century; however individuals from the continent were occasionally still seen in winter.

21st century

After an 82-year wait, an immature was seen on a pylon at Orsett Fen in October 2010. Less than two years later, one flew over Rye Meads. Within an increasing population in Scotland it is hoped that sightings in the London Area will become more regular.

Pre-1821: Epping Forest, shot
Pre-1845: Wimbledon Common, shot at Coombe Wood
1906: Cheverells, shot on 12 November
1909: Weald Country Park, on 6 February
1928: Navestock, on 1 February

2010: Orsett Fen, on 5 October
2012: Rye Meads, on 16 February

Other Records

An eagle was shot at Forest Hall, Ongar, in 1839 and another, thought to be a young White-tailed Eagle, was shot in Theobalds Park, Cheshunt, in January 1891. A pair of eagles was reported circling over Westminster in the severe winter of February 1895. Another eagle was seen near the River Roding at Stapleford Tawney/Passingford Bridge by Glegg on 15 December 1929 but he was unable to determine which species it was. A White-tailed Eagle escaped from London Zoo in November 1936 and was later recaptured in St James's Park.

Marsh Harrier *Circus aeruginosus*

An uncommon summer visitor and rare breeder

Range: Europe, Asia and Australasia, winters Europe and Africa.

Historical

In the 19th century a nest containing five eggs was found in an osier bed near the Grand Surrey Canal at Deptford in 1812 and the female was shot. There were no further records until 1892 when one was killed at Shoreham. Four years later there were another two records, of one bird at Headley in January and one killed at Cheam on 5 September.

20th century

Persecution and habitat loss due to drainage led to the extinction of this fine raptor in the UK by the end of the 19th century so it was hardly surprising that there were no records during the first half of the 20th century. There was a slow recovery in the British breeding population but it wasn't until very late in the 20th century that Marsh Harriers really became established in this country.

Four birds were seen in the 1950s; there were two in 1951: at Epping Long Green on 22 September and at East Bedfont on 14 May. The others were at Beddington SF on 5 September 1954; and over King George VI Reservoir, then hunting at Heathrow Airport, on 28 April 1957.

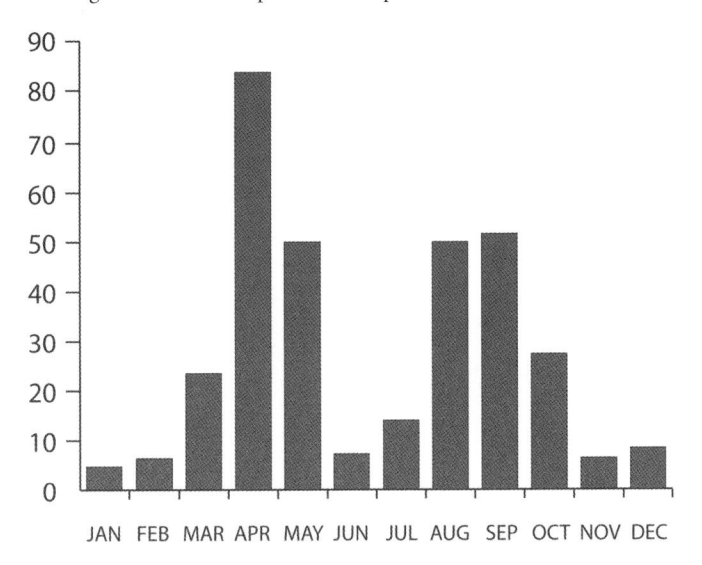

Seasonal occurrence of Marsh Harrier: 2000–2009.

In the 1960s the population in England was extremely low and there were no sightings in the London Area until 1965 when a male was seen on Hampstead Heath on 3 April, a relatively early date for a spring migrant. The only other record that decade was over South Weald on 8 August 1967. No more were seen until 1976 when a female was seen at Beddington on 27 August. After this they were seen annually and the number of sightings slowly began to increase. In 1979 seven were seen, including four at Rainham; all were between April and October apart from one at Swanscombe in February.

In 1981 one remained at Rainham Marshes between 17 August and 14 September, the first long-staying bird for well over a century. Two years later another lingered at Rainham Marshes between 27 July and 17 September. Ten were seen in 1987 and again in 1989; in the latter year all but one was seen in east London, between the Lea Valley and the Thames.

By the start of the 1990s they were being seen on passage with increased frequency. In 1990 there were nine in spring, three in autumn and further signs of oversummering with a female seen regularly during June. At least 25 were seen in 1994, all but five in spring, representing a new yearly record and a change of status to regular passage migrant. The following year a pair of immature birds remained at Rainham Marshes for most of May.

21st century

Since the start of the 21st century the number of sightings has continued to increase, particularly at well-watched sites where observers scan the skies for large raptors; for example, 13 were seen at Beddington in 2000. In 2009 at least 68 were recorded during the year, just over half of which were in spring; there was also an increasing number of winter sightings; at least 20 were at Rainham Marshes, some of which were seen hunting. In 2010 a pair bred in east London, the first breeding record for 200 years.

Hen Harrier *Circus cyaneus*

A scarce winter visitor

Range: Europe, Asia and Americas, some winter Africa; scarce breeder in northern Britain, more widespread in winter, especially on coastal marshes.

Historical

In the early 19th century Hen Harriers probably overwintered regularly on the Thames marshes. Graves wrote that in about 1820 they were often seen in fields at Rolls Meadows by the Old Kent Road at Bermondsey, and Christy stated that they were 'not uncommon about the marshes of Kent and Essex, bordering on London'.

Elsewhere they were a rare visitor with about a dozen records. The first of these birds was shot in Lullingstone Park in 1841 and was followed by one near Harrow between 1846 and 1850; nearby an adult male was shot in Willesden in about 1850 and one was seen several times close to Brent Reservoir in 1862. One was seen over Langley Park prior to 1868 and a female was caught by a bird-catcher as it attempted to carry off a decoy-bird near Hendon Brook on January 12th, 1869. There was one near Tadworth in 1875 and one was shot at New Park, Northaw, on 11 September 1879. On 20 January 1880, a male was picked up dead in Brompton (Glegg) and there was one near Banstead (Wheatley); these may refer to the same bird. In the winter of 1880/81 there was one at Chelsham; one shot at Swanscombe on 30 October, 1888; there were also records from Headley and Walton Heath in the 1890s, and an undated record from near Cobham. Yarrell referred to another one which upon examination was found to have the remains of at least 20 lizards in its stomach.

20th century

In the first half of the century Hen Harriers had become slightly less rare and a total of 13 was recorded up to 1959. Nine of those were in winter and four during migration – one at the end of March and three in October. They were mainly seen in the rural areas and Thames-side marshes, and all but two were seen on one day only; the only suburban record was at Beddington on 7 November 1934. Not included in these figures is a specimen of an adult female in the County Museum in St Albans that was shot at Bricket Wood prior to 1959.

Only six were seen during the 1960s although unidentified harriers seen at three other locations during winter were presumably this species. During the 1970s the number of sightings increased and 19 were seen. By the end of the decade they had started to overwinter and two were present at Rainham Marshes between January and April; another two remained for three weeks in January at St Mary Cray, and two were at Symondshyde from 2 March into April.

Throughout most of the 1980s they regularly wintered on the Lower Thames Marshes with a peak of five in 1987. However, in 1988 they ceased wintering at Rainham, probably because of excessive disturbance, particularly by falconers. Elsewhere they were seen on passage in small numbers, even flying over central London where they were occasionally seen migrating over Regent's Park. During the 1990s they were scarce passage migrants with just occasional visits in winter.

21st century

By the beginning of the 21st century they had become scarce in the London Area. The UK breeding population has continued to decline due to illegal persecution, particularly on keepered grouse moors. Since 2000 there has been an average of four or five each year; most of these were on passage but some have lingered at Rainham, and in the winter of 2006/07 one remained there for about six weeks.

Facts and Figures

The earliest autumn migrant was on 11 August 2005 at Rainham Marshes.

Montagu's Harrier   *Circus pygargus*

A scarce passage migrant (28+)

Range: breeds Europe and Asia, winters Africa; scarce breeder in England.

Montagu's Harrier was only recognised as a species in Britain in 1802 when it was distinguished from Hen Harrier. The first record for the London Area occurred 10 years later when a pair was shot in mid-May in Battersea Fields; the birds were believed to be nesting nearby. They were also regularly recorded in the Woolwich area until the early part of the 19th century and the only other record prior to 1900 was from Coombe Wood, Wimbledon, in the mid-1880s.

In the first half of the 20th century there were three records: of one trapped in spring at Black Park in 1929; the feathers of one were found near a badger's sett in Richmond Park in 1942 and sent to the British Museum for identification (they were found to be 'those of a female or young male harrier, almost certainly Montagu's'); and the following year a pair were seen between Mill Hill and Finchley on 16 April. The only sighting in the 1950s was of a juvenile at Staines Moor in 1953.

The 1960s saw the gradual start of an increase and four were seen, commencing with a rare dark-phase bird at Heathrow Airport for four days in May 1960. The only records in the 1970s were of two in May 1979. Four were seen in the 1980s, three of which were on the Lower Thames Marshes. A female in June 1990 was the first midsummer bird to linger, albeit only for two days, and gives hope that this fine raptor may once again breed in the London Area.

On 2 September 1995 one flew east over Wraysbury GP at 10:30 hrs, and presumably the same bird was then seen over Hampstead Heath an hour later; this was the first one to be seen at more than one site in the London Area.

Since the beginning of the 21st century there have been four records. In 2000 a dark-morph bird flew south over Nore Hill, Chelsham; the second time this colour phase has been seen in the London Area. Two were seen at Rainham Marshes, a subadult in spring 2002 and a ringtail that was present for two days in June 2006. Another two were seen in 2009, including a well-watched bird that spent two days at Rainham Marshes.

1812: Battersea Fields, pair shot
Pre-1847: Woolwich, 'regular'
1880s: Coombe Wood

1929: Black Park, trapped in late April or early May
1942: Richmond Park, feathers of one found on 4 April
1943: near Mill Hill, pair on 16 April
1953: Staines Moor, juvenile on 5 August
1960: Heathrow, male on 18–21 May
1963: Regent's Park, ringtail on 29 August
1967: Clement Street, Sutton at Hone, on 14 and 21 May
1969: Richmond Park, on 11 October
1979: Staines Reservoir, male on 7 May
1979: Maple Cross, male on 18 May
1980: Dartford Marsh, immature male on 25 July
1986: Rainham Marshes, male on 30 May
1987: Dartford Marsh, male on 9 September
1988: Hornchurch CP, male on 2 May
1990: Rainham Marshes, on 15–16 May
1992: Dartford Marsh, on 11 June
1994: Amwell GP, on 7 May
1995: Wraysbury GP, ringtail on 2 September; also at Hampstead Heath
2000: Chelsham, on 12 August
2001: Rainham Marshes, subadult on 28 May
2002: Staines Reservoir, juvenile on 5 August
2006: Rainham Marshes, ringtail on 3–4 June
2009: Rainham Marshes, male on 24–25 August; also at Crayford Marsh
2009: Brent Reservoir, juvenile on 13 September

Unacceptable Records

In 1889 a pair was said to be present in Hornchurch during July; Glegg recorded that it had 'an element of uncertainty' about it and this record is now treated as unacceptable. In 1922 a pair, probably of this species, was said to have been seen at Ashtead on 12 March; in view of the exceptionally early date, this record is also treated as unacceptable.

Goshawk *Accipiter gentilis*

A scarce visitor (39)

Range: Europe, Asia, North Africa and North America.

Historical

The earliest reference was when Edmund Bert of Collier Row trained Goshawks and published a book on falconry in 1619. There were no records of any wild birds until the 19th century when a female was caught at Beddington on an unknown date.

20th century

By 1900 they had become extinct in Britain due to persecution and loss of habitat, and it was not until the middle of the 20th century that this raptor was seen again in the London Area; during the 1950s five were seen. The first of these was in 1955 at Harefield on 20 March and 24 April; this was followed by one at Stone Marshes on 6 November in the same year. In 1957 one was seen soaring over Staines and King George VI Reservoirs on 16 June, and in 1959 two were present at Woldingham on 8–9 August. Unfortunately, this rush of sightings was not sustained and there were only single records in the next two decades: in 1960 an immature was seen hunting Woodpigeons at Beddington on 15 September; and in 1970 one was seen perched and flying in Broxbourne Woods on 25 January.

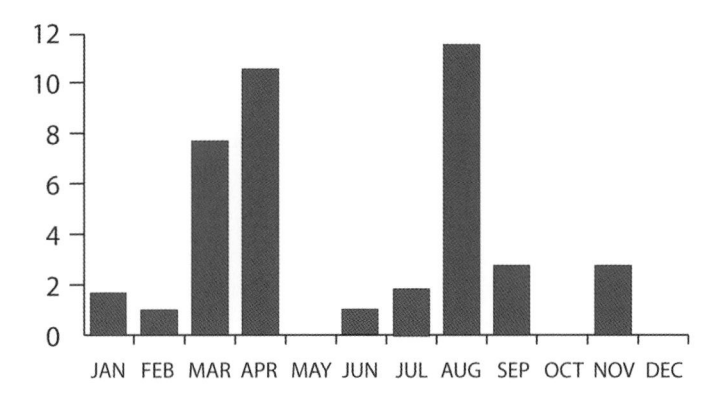

Seasonal occurrence of Goshawk: all records.

There was a gradual increase in sightings during the 1980s with three seen, starting with a female at Epsom on 2 and 8 March 1980. In 1984 an adult male was seen at Bookham Common on 27 August; it was seen soaring over before plummeting into the wood. In 1989 a male was at West Kent Golf Course, Downe, on 23 July. The number of records increased again in the 1990s when 14 were seen. Eight of these were spring migrants, including three over Rye Meads in April 1999. Another four were seen in August; a displaying male was seen over Holmethorpe Sand Pits in January 1993; and one was present at a possible breeding site in north London in March and April 1998.

21st century

As the population increased throughout the country so did the number seen in the London Area. Initially, much of this increase was due to escaped falconers' bird and reintroductions which started to breed in the wild; there is now a healthy and increasing breeding population in the UK. Since 2000 there has been a dramatic rise in sightings in the London Area with 16 seen up to 2009. It is clear that this raptor has established a regular passage over the capital but there have been few signs yet that it will become a breeding bird; however, given the suitability of woods in the rural areas surely this is just a matter of time. In the autumn of 2002 there was a series of sightings involving a family party on the North Downs; these were not thought to have bred in the London Area but probably bred nearby.

Escapes/Unacceptable Records

In 1872 a young male was killed at Hampstead on 3 September; upon examination it was found to be an escape. Bryer claimed to have shot one at Northaw on 6 September 1879 but provided no further details. There have also been several reports of escaped falconers' birds since 1969.

Sparrowhawk *Accipiter nisus*

A common resident and widespread breeder

Range: Europe, Asia and North Africa.

Historical

Remains have been found at Southwark dating back to AD 270–350. In the early 19th century the birds were common in some of the rural districts around London such as at Edgware, Hampstead, Pinner and Stanmore but by the 1860s had become scarcer due to persecution. Towards the end of the century they were only occasionally seen in Richmond Park and on Wimbledon Common but they were still nesting in the Epsom area in 1895.

20th century

By 1900 Sparrowhawk had declined further due to relentless persecution and was only just surviving in the London Area. In woods where game was preserved there was very little chance of successful breeding. For example, at Hayes in south-east London 17 were shot in 1937, yet four pairs still managed to nest and another six were shot by a gamekeeper at Fawkham. This state continued until the Second World War.

In the 1930s Glegg described them as unusual in rural areas and as occasional visitors in the suburbs and central London. He recorded recent breeding from Bushy Park, Harefield Place, Osterley Park, Ruislip, Stanmore and Swakeleys. Elsewhere breeding was reported from Addington, Beckenham and the northern outskirts of Enfield. Up to two pairs bred in Richmond Park in the mid-1930s, and around this time they also bred at Bookham, Headley, Limpsfield and Oxshott.

From the 1940s onwards there was a decrease in game preservation and Sparrowhawks were quick to establish themselves in areas where they were no longer being persecuted. In many of the outer districts in London, observers reported a noticeable increase in Sparrowhawk numbers. Even in areas where they were still being persecuted, for example in the Oxted and Limpsfield area, they managed to find sufficient refuges in which to nest; seven pairs nested in 1946 and 18 pairs bred two years later. Around the same time they also started moving into the suburbs, nesting in Dulwich Woods, Greenwich Park, Richmond Park and Wimbledon Common. In 1941–42 they returned to Hampstead Heath after a long period of scarcity and a pair bred there in 1950.

By the 1950s they were being seen more often in the parks in central London and the inner suburbs, and in 1953 they nested for the first time in Inner London at Holland Park, although the breeding attempt was unsuccessful. However, in other areas they were completely absent; in Barnet, Cuffley, Hatfield and St Albans, Sage stated in 1959 that he had not seen one in the previous ten years. Across the London Area only four breeding pairs were located in 1959 and it was known at the time that there had been a widespread reduction in numbers across the country.

At the start of the 1960s Sparrowhawks were at a very low ebb and only two pairs were known to have bred, although a few other birds were seen in the breeding season. The following year they had become scarcer still and there was only one confirmed breeding record. The decline in the population was reflected nationally and was attributed to pesticides in the food chain, which accumulated in birds of prey. In 1963 there were no confirmed records of successful breeding; a pair nested but deserted, and one or two other pairs was present during the breeding season. The following year there were no signs of breeding at all. They failed to breed again in 1965 and no pairs were seen during the breeding season but there was an increase in the number of sightings compared to the previous year. There was a short-lived recovery in 1966 when one pair bred and another pair probably bred. Single pairs bred in the following two years but in 1969 the nest and eggs of the sole breeding pair were destroyed.

Breeding was confirmed at seven locations during fieldwork for the Breeding Atlas for 1968-72 with another 12 pairs probably breeding, suggesting that even though the population in the London Area was extremely small it wasn't as low as the annual reports suggested. During the Seventies the population began to recover; by 1980 there were 17 pairs reported south of the river, yet none were located north of the Thames. The increase continued and in 1983 six pairs were found in north London; the following year about 50–60 breeding pairs were reported across the London Area, the most for over a century.

By 1990 Sparrowhawk was going from strength to strength and was probably at its highest-ever population with more than 100 breeding pairs. They had not yet colonised Inner London but were being seen there more frequently. One indication of their breeding success was three pairs that each laid seven eggs, all of which fledged. In 1992 a pair almost certainly bred in Regent's Park. During the second Atlas in 1988–94 they were found breeding in almost 40 per cent of the tetrads in the London Area. By 1997 the breeding population had increased to 167 pairs, including two in Inner London.

21st century

Nationally, the population of Sparrowhawks doubled from 1970 to 2010 but the rate of increase in the London Area was even greater. At the start of the 21st century it had overtaken Kestrel to become the most common raptor in the London Area and the breeding population was close to 200 pairs. However the UK population has levelled off in recent years and may have declined slightly and this appears to have also been the case in the London Area. Although it is primarily a woodland nester, it is widely distributed throughout most of the London Area, making use of parks, cemeteries and woodland nature reserves in urban areas; in Inner London it is less common with only a few pairs breeding.

Common Buzzard   *Buteo buteo*

An uncommon partial migrant and localised breeder

Range: Europe and Asia, some winter Africa.

Historical

Very old remains have been found at Southwark dating from the Roman period, and at Merton Priory dating back to 1230–1300.

The earliest record was at Mitcham Common in the winter of 1834/35. In 1838 Torre wrote that one had been obtained near Harrow. Harting stated that he had only ever seen one in Middlesex and referred to just a few other records that he was aware of including one of a bird caught by bird-catchers in Kilburn.

The only documented breeding record from anywhere in the London Area was around Epping Forest where Christie stated in 1903 that they had since ceased to breed. By the middle of the 19th century they were no longer breeding in the Home Counties and were rare in the London Area. In the autumn of 1876 one was killed at Park Wood Ruislip. There were several records from south London: at Woodmansterne in October 1875; Walton Heath in October 1880; Sanderstead on 8 July 1881; Cobham in November 1881; and Dartford Park Wood on 14 June 1886; and undated records from three other locations. They were also considered to be a regular autumn migrant in the Dartford area.

20th century

In the early part of the century they were still very rare but there were signs of an increase. In the autumn of 1909 a pair spent almost three weeks in the Banstead area and from 1912 onwards they were recorded almost annually. At Gilwell Park in Epping Forest one was present for ten days in March 1924 and one remained for four weeks in early 1926. Additional birds were occasionally seen elsewhere over Epping Forest, particularly over the western escarpment. However, they were still being persecuted and this hindered their recolonisation. One was caught in a trap near Uxbridge in 1915; in the Farleigh area one was seen regularly for successive winters but was found shot in 1935; and a pair which had established a territory in the Oxted area was shot by gamekeepers in 1937.

Only a few were seen each year in the 1930s, including one soaring over Westminster on 11 June 1937. In 1939–40 nine were released near Godalming, about 20 kilometres beyond the London boundary, and they eventually formed a small breeding population in Surrey. This may have accounted for many of the sightings in south London in subsequent years; for example, two seen at close quarters at Croham Hurst in August 1943 were thought to have been released birds.

They were also seen passing over the London Area in spring and autumn, even over the most heavily built-up areas. During the 1950s they were seen over Inner London at High Holborn, Regent's Park, Westminster and even the City of London. They remained scarce throughout the 1960s with between one and five records a year. In 1960 there was one in the Lea Valley between Fishers Green and Rye Meads from 16 August to 15 October, the first one to linger in the London Area since 1937.

During the 1970s they were just about seen annually, albeit in very small numbers and typically just flying over. In 1972 one remained in the Cobham area for much of November. Sightings slowly began to increase in the 1980s when up to 11 a year were seen.

At the beginning of the 1990s they were still a scarce but increasing passage migrant and there were the first signs of potential recolonisation when one or two were seen during June and July at a site in east London. In 1993 at least 17 were seen, including some that lingered around the northern edge of the London border. In 1996 at least 37 were seen, with some birds beginning to take up residence in the far north of the area, and two years later a pair bred in the Copped Hall area, the first breeding in the London Area in the 20th century. Three pairs probably bred in 1999 although one pair, at Hogtrough Hill, was of captive origin and was described as being of an eastern race.

21st century

At the start of the 21st century it was a rare breeding bird in the London Area with just two pairs located. Within two years they had begun to establish themselves in many rural areas of north and east London, most likely from the expanding population in Hertfordshire. There were up to a dozen breeding pairs found north of the Thames in 2002, but they were much scarcer south of the river where only one or two pairs were located. By 2006 at least 41 pairs were found across the London Area although Oliver (2007) estimated the breeding population to be 66–93 pairs. Nationally, the breeding population increased by over 400 per cent between 1970 and 2010 and they have continued to increase in the London Area.

The number of non-breeding birds seen has continued to increase at a staggering rate. In 2000 there was a notable increase in sightings coinciding with the large influx of Honey Buzzards in autumn, and it seems that about 20 Common Buzzards migrating from the Continent were caught out by the same weather conditions and had to continue their southward migration through the London Area. This may hint towards the origin of the birds that clearly migrate over the London Area in spring and autumn. The peak month for passage birds is September, followed by April. There were 25 seen over Inner London in 2009 and observations from Tower 42 have shown the species to be a regular migrant over the City.

Facts and Figures

The largest flock was of 17 birds over Brent Reservoir on 15 September 2011.
One bred with a Harris's Hawk in the Trevereaux area in 2009, raising two hybrid young.

Rough-legged Buzzard *Buteo lagopus*

A rare winter visitor (26)

Range: breeds northern Europe, Asia and North America, winters Europe and North America; scarce winter visitor in eastern UK.

The earliest record was of one killed at Eltham prior to 1787. Pennant wrote in 1812 that one was shot near London and the specimen was in the Leverian Museum; in 1821 Latham stated that this was one of just four records in the country that he knew of. There were another seven records in the last quarter of the 19th century including those of two birds shot by Doubleday at Epping on unknown dates. Ticehurst also refers to an undated record from Dartford.

There were just three more in the first half of the 20th century, the first of which were of the remains of one found on a keeper's gibbet in Trent Park in 1911 and identified at the Natural History Museum. After one the following year at Croydon, there was a gap of 35 years before one was seen hunting over ploughed fields at Mill Hill in 1946.

Four were seen in the 1960s, including one at Panshanger Park in April 1963, which may have been present in the area since the previous October. In 1973 there were two seen on consecutive days in late autumn, over Wraysbury Reservoir and Rainham Marshes. The following year there was an exceptionally large national influx in late autumn and three were seen in the London Area, all between 24 October and 10 November. Another two were seen in early 1979; both occurred during a period of severe weather.

Since then they have become extremely rare and only two more have been seen. In 1994 there was a large influx on the East Coast and there was one at Bickley, and in 2007 one was seen flying low over Richmond Park on 27 October.

Pre-1787: Eltham, shot
1812: near London, shot
Pre-1875: Epping, two shot
1876: Walton Heath, shot
1879: Croydon
1881: Chelsham, on 17 February
1881: Ruislip, shot in Park Wood
1892: Bishops Wood, Moor Park, trapped on 19 February
Pre 1909: Dartford
1911: Trent Park, found shot on a keeper's gibbet in October
1912: Croydon, on 21 April

1946: Mill Hill, on 14 December
1963: Panshanger Park, on 15 April
1966: Nyn Park, Potters Bar, on 28 October
1968: Perry Oaks SF, on 15 September
1969: Walthamstow Reservoirs, on 30 November
1973: Wraysbury Reservoir, on 17 October
1973: Rainham Marshes, on 18 October
1974: Joyden's Wood, on 24 October
1974: West Drayton, on 25 October
1974: Sevenoaks, on 10 November
1979: Wraysbury GP, on 1 January
1979: Thorndon Park, on 4 January
1994: Bickley, on 2 November
2007: Richmond Park, on 27 October

Golden Eagle *Aquila chrysaetos*

A very rare vagrant (2)

Range: Europe, Asia, Africa and North America; resident in Scotland with a few in Cumbria.

The first record for the London Area was recorded in 1811 by Graves who illustrated a female that was shot near Brompton about two years earlier; the specimen was in the London Museum. The only other record was of one seen flying in the neighbourhood of Barnet in 1859; Harting stated that the observer was 'well acquainted with the bird'. Golden Eagles were very occasionally recorded in southern England in the 19th century although they were often confused with immature White-tailed Eagles, which were more likely to be encountered. A record of one shot at Claverham Bury Farm, Nazeing, on 15 November 1858 remains unproven.

About 1809: near Brompton, shot
1859: Barnet, one seen

Escapes

In March 1965 one famously escaped from London Zoo and spent 12 days at large in and around Regent's Park; it was seen devouring a Muscovy Duck in the grounds of the American Embassy and also attacked two dogs. It was seen by over a thousand people before eventually being recaptured. It escaped again in December and had another four days of freedom. In 1992/93 an escape from a local zoo was seen in the St Albans area between October and January.

Osprey *Pandion haliaetus*

An uncommon passage migrant

Range: cosmopolitan, European population winters Africa.

Historical

Osprey remains were found in the peat beds of Walthamstow. In the 19th century this was a scarce visitor to the London Area; the earliest record was between 1808 and 1816 when Hayes illustrated one from a specimen shot at Osterley Park. Latham also refers to one shot near Dartford prior to 1821.

A further 23 were seen prior to 1900: on the Thames at Laleham in 1855; at East End Reservoir in Bow in the autumn of 1858; shot at Dartford on 23 September 1859; one shot near Uxbridge on 1 October 1863 that had been observed in the neighbourhood for some days; at Laleham in 1863; a pair frequented an estate near Southgate (now Grovelands Park) for several days in September 1865 and were also seen at Enfield; at Forty-Hill and Whitewebbs in September 1867; at Gatton in August 1868; at Decoy Farm, Hendon, on 24 September 1868; at Tottenham in August 1870; shot in Hatfield Park in September 1880; at Park Wood, Ruislip, in the autumn of 1881; at East Molesey in October 1881; caught at Hayes in 1884; at Richmond Park on 22 September 1889; on the Thames at Barnes on 21 October 1889; shot at Belvedere in May 1898; at Kew and Richmond Park in December 1898; at Painshill Park on about 16 October 1899; and at Kew in 1899. Osprey was also seen near Bromley late in the 19th century.

20th century

Ospreys were heavily persecuted and all but extinct in Scotland by 1916 apart from occasional passage migrants. This was reflected in the London Area and only two migrants were recorded between 1900 and 1950: at Weald Country Park from 11 to 24 October 1903; then one stayed for over six weeks in Panshanger Park from 17 September to 2 November 1930. Breeding began again in Scotland in 1954 at Loch Garten and thanks to major conservation efforts by the RSPB a small breeding population became established.

During the 1950s they were recorded far more frequently and 10 were seen between 1951 and 1953, all in the Colne Valley apart from one at Ruislip. In 1952 one was in the Staines area from 29 April to 1 June, the first occasion one had remained for more than a few days in the London Area during the breeding season. In 1958 one frequented the Chingford Reservoirs between 4 and 25 September; during its stay it was seen to take a Moorhen and a large bat. There was also a winter record, at Moor Mill on 8 January 1959.

Another 14 were seen in the 1960s; while the records were fairly widespread across the London Area there was a clear preference for the river valleys with five along the Lea and others along the Brent, Colne, Cray and Darent. One was observed trying to catch fish at Brent Reservoir on 9 November 1961 despite the presence of 65 sailing boats. Two were seen over central London: at Regent's Park on 19 September 1965 and at Hyde Park on 2 May 1967. The number of sightings continued to increase throughout the 1970s and 80s, with an average of four a year and a maximum of eight in 1976. At Sevenoaks there were five different birds in 1972 and probably six different birds the following year. The first July record occurred in 27 July 1977 at Lullingstone.

In 1979 one spent almost four weeks at Gatton Lake between 17 November and 12 December; on the last day it flew off high to the south-east. Further long-staying autumn migrants were seen in the 1980s: at Claremont Lake, Esher, between 25 September and 13 October 1980; at Amwell between 27 September and 18 October 1981; and in the Lea Valley between 13 September and 7 October 1989. There were also two short-staying birds during the spring of 1988, in the Lea Valley from 1 to 4 May, and at Tyttenhanger GP from 2 to 10 May. During the 1990s migrants were seen more frequently, including 19 in 1996, but there were no long-staying birds.

21st century

The UK breeding population has continued to increase with birds naturally beginning to colonise parts of northern England and Wales as well as the successful breeding of reintroduced birds at Rutland Water in the Midlands. Reflecting this recovery the number of Ospreys seen migrating over the London Area has also risen and in 2008 a record 57 were seen. Most just pass straight over but occasionally one will linger for a short period.

Facts and Figures

The earliest migrant was on 17 March 2005 at Walton Reservoirs.

Lesser Kestrel *Falco naumanni*

A very rare vagrant (2)

Range: breeds southern Europe and Asia, winters Africa; vagrant to UK.

There were only 16 British records up to 2009, two of which were in the London Area. The first was of an adult male present for four days at Rainham Marshes in 1974; it was the first British bird to remain for more than a day. The second was also a male, seen several times over Hampstead Heath on the morning of 31 May 1992 before flying off high to the east.

1974: Rainham Marshes, on 31 July–3 August
1992: Hampstead Heath, on 31 May

Kestrel *Falco tinnunculus*

A common resident and widespread breeder

Range: Europe, Asia and Africa.

Historical

The earliest record was in the 1830s when the species bred in Wimbledon Park. Harting recorded that Kestrels were the most common raptor in Middlesex despite being heavily persecuted by farmers and gamekeepers; they were most abundant in autumn and comparatively few pairs stayed to breed. They bred in Inner London through the 19th century up until 1871, when they last bred on the top of Nelson's Column.

20th century

Up to the start of the Second World War there were relatively few breeding pairs. Although they were still the most common bird of prey they were mainly confined to rural areas, with just occasional records from the built-up areas. Between 1910 and 1915 they were regularly seen around the Houses of Parliament but there was no confirmed breeding. They still suffered from persecution at this time, and later. For example, about 12 were killed in Hayes during 1936 and the remains of 11 were found on a gibbet at Bayfordbury Park in June 1944.

Breeding was confirmed again in Inner London in 1931 when a pair bred in a loft at Paul's School in Hammersmith. In 1936 a pair bred on the Houses of Parliament and birds were present in the breeding season at the Imperial Institute from 1936 to 1944 and may well have bred. Following the Blitz, during the Second World War, London was peppered with bombed sites and Kestrels were quick to take advantage; they were seen around three churches in the City of London and a pair was believed to have bred by Dean's Lane. Nesting was confirmed at the latter site in 1946 and again in 1947. About five pairs nested on buildings in 1948 and again in 1950 (at Battersea, Bayswater, Lambeth, Westminster and in the City of London). Their staple food was the House Sparrow but they also took other small birds, mice and insects.

Elsewhere in the London Area they typically nested in hollow trees or in old crows' nests but also utilised structures such as gasometers, and a pair also bred on Greenwich Power Station. After the late 1940s five or six pairs bred in hollow oak trees in Richmond Park; prior to the war there had only been one confirmed breeding record in the park. Many of these nestlings were ringed, and two of them were recovered elsewhere in the London Area, at Acton and Walton, while one was recovered three years later in Gloucestershire. There is some evidence of migration through the London Area; for example, on 28 March 1947, 10–20 were seen flying south over Dulwich during a 20-minute period in mid-morning.

During the 1950s the breeding population began to decline in central London, possibly due to redevelopment work, and by 1959 there were only two or three breeding pairs. By the early 1960s they had disappeared as a breeding species from central London and were in decline throughout the London Area as well as across the country. In eastern England there was a marked reduction in the population which was attributed to toxic chemicals and was especially noted on arable land. They began to recover during the mid-1960s. In 1967 a full survey was conducted across the London Area and found a total of 98 confirmed breeding pairs and a total population of 142 pairs; the highest population was in Richmond Park where 20 pairs were present. There was still a shortage of Kestrels in the rural areas of east and north London compared to the south and west, where more than 75 per cent of confirmed breeding pairs were located. Also in 1967 the central London population was increasing and three pairs bred, including one pair which bred in a window box on a tower block in Walworth. In the first Atlas in 1968–72 they were found breeding in 377 tetrads across the London Area.

During the 1970s no more than 100 pairs were located annually; however, there was no evidence of a decline and it is likely that the lower figures represented under-recording. The overall population may actually have been increasing as by 1974 there were 12 breeding pairs in central London. By 1992 the breeding population had risen to 15 pairs in central London and, despite the apparently inhospitable habitat, three pairs bred in the City of London. The following year 138 pairs were located across the London Area. The second Atlas in 1988–94 recorded an increase in the number of occupied tetrads of almost 40 per cent, suggesting a significant increase in population since 1968-72.

21st century

There appears to have been a large decrease since the beginning of the 21st century; in 2010 just 56 pairs were reported compared to 139 pairs in 2000. Although this decrease may be partly due to reduced observations, there is some evidence of a genuine decline, for example, in Richmond Park there were only five pairs located in 2010, about one-quarter of the total there in 1967 whilst in Inner London only three pairs were located in 2010 compared to 10 pairs in 2000. This mirrors a decline in the national breeding population of almost a third between 1995 and 2010. The reasons for the decline in the London Area are unknown but may be due to a number of factors such as the development of brownfield sites and the crash in the House Sparrow population which is a key prey species in the built up area.

Facts and Figures

The largest concentration was of 30 birds at Beddington in March 1971.

Red-footed Falcon *Falco vespertinus*

A rare vagrant (15)

Range: breeds eastern Europe and Asia, winters Africa; former rarity in UK.

The first record was at Nunhead in 1892 when an adult female was shot by a man testing rifles who happened to notice an unfamiliar bird. There was a later suggestion that this may have been an escaped bird from a nearby aviary but there was no evidence that this species was present in the collection; the contents of its stomach consisted of beetles and caterpillars. No more were seen until 40 years later when one was seen at Beckenham; this record was overlooked in *The Birds of the London Area* but was documented by Harrison in *The Birds of Kent*.

The next two sightings were both at Sevenoaks, a female in 1962 and a male in 1971; also, in the latter year a female spent two days at Godstone. In September 1976 a female at Amwell was the first autumn record. The only record in the 1980s was of a male near Kings Langley, at Hunton Bridge in 1985; in the late 1980s there were several large influxes into the country but none found their way into the London Area.

The 1990s was the peak decade for this summer visitor, with five records. Two were at Rainham Marshes, a first-summer female in 1990 and a first-summer male the following year. In 1992 there was a large influx into the country resulting in a first-summer female appearing at Banbury Reservoir, and in 1996 a juvenile flew south over Walthamstow Reservoirs. A first-summer male was present at Horton for four days in June 1999.

There were three in the first decade of the 21st century; an immature male at Greenwich in May 2007; a first-year female in the Lea Valley in May 2008; and another female in the Colne Valley in the same month.

1892: Nunhead, shot in late September
1932: Beckenham, on 20 May
1962: Sevenoaks, on 6 June
1971: Sevenoaks, on 29 April
1971: Godstone, on 8–9 May
1976: Amwell, on 9 September
1985: Hunton Bridge, on 20 June
1990: Rainham Marshes, on 31 May–17 June

1991: Rainham Marshes, on 8–17 June
1992: Banbury Reservoir, Walthamstow, on 2 June
1996: Walthamstow Reservoirs, on 14 September
1999: Horton GP, on 11–14 June
2007: Greenwich, on 27 May
2008: Seventy Acres Lake, on 15–21 May; also at Sewardstone on 21 May
2008: Colnbrook, on 31 May

Merlin *Falco columbarius*

An uncommon winter visitor

Range: breeds northern Europe, Asia and North America, some winter Europe and North Africa.

Historical

In the 19th century Merlin appeared to be a fairly regular winter visitor in parts of south London; Blyth stated in 1836 that they were common visitors at Tooting with a few shot every winter, while Ruegg wrote that they were sometimes seen in the Woolwich area. However, Harting considered them to be a rare visitor north of the Thames.

The only documented records are: one taken near Harrow prior to 1838; captured on Hampstead Heath in the winter of 1857; shot at Stonebridge on the River Brent in April 1861; two at Brent Reservoir, in 1864 and 1866; caught in Willesden on 4 October 1867; at the Natural History Museum in South Kensington in January 1886; at Richmond Park in late autumn 1892; on Hampstead Heath in autumn 1895; over Brixton on 4 January 1899; and undated records from Ashtead Woods and Turnham Green.

20th century

In the first half of the century Merlin was an occasional passage migrant and winter visitor with around 50 records, mostly between November and January. The majority of these birds were on the heaths in the south of the London Area. Two were seen in central London, both in Kensington Gardens: a male on 24 August 1936, and one on 22 August 1943. In the middle of the 20th century it was not unusual for Merlins to be seen passing through the London Area in August, whereas now the peak time in autumn is about two months later.

By the 1950s they were most often seen in the Staines district, particularly at Perry Oaks SF. In the far north of the area, one or two were seen in most years, including a pair at Old Parkbury on 3 January 1950; and at the same site there were single birds seen on the unusual dates of 5 July 1953 and 2 June 1956.

During the 1960s and 70s one or two were seen in most years, either on passage or in midwinter. Another two migrants were seen over central London: in Regent's Park on 27 December 1961, and over Primrose Hill on 13 February 1965. They began to be seen more frequently from 1979 when five were recorded, including one circling high over St James's Park on 3 January and one lingering at Queen Mary Reservoir during November.

In the 1980s they started to be seen more often at Rainham Marshes than anywhere else. Between 1944 and 1984 there were 111 records, an analysis of which showed that the peak month was December, and there were also obvious spring and autumn passage increases across the area. By the mid-1980s the number of sightings had increased to about 10 a year. In 1985 a pair wintered in the Staines area and was even seen mating. In 1989/90 two wintered in the Ingrebourne Valley. In 1991 about 20 were seen, the largest-ever total in a single year and double the average of the previous decade; long-staying birds were present at Dagenham Chase and Rainham Marshes.

21st century

In the UK, Merlins breed mainly in the uplands of Scotland, Wales and northern England with many birds wintering on coastal marshlands. The breeding population has recovered since the early 1980s and they are now more frequently encountered in the London Area on passage and during the winter months. Between 2000 and 2007 the annual average was about 10 but in subsequent years sightings have increased with a record 41 in

2010. They are most often seen on the Lower Thames Marshes where they have begun to overwinter, but passage migrants are sometimes noted flying over the built-up areas of London.

Unacceptable Records

There was a claim of breeding at Ongar Wood in 1867 and possibly also in 1857 by an egg dealer who apparently found a nest containing eggs; however, the record was dismissed by Glegg even though Merlins bred further east on the Essex marshes. Merlins are rarely reported as escapes but in 1969 one was seen at Beddington in April – pursued by its owner.

Hobby *Falco subbuteo*

An uncommon summer visitor and localised breeder

Range: breeds Europe and Asia, winters Africa.

Historical

In the mid-19th century Hobby was an uncommon breeder in the London Area. It was known to have bred at Debden in 1829 and 1835; near Harrow two or three times between 1831 and 1838; in Epping Forest in 1846–47; in Pinner sometime between 1846 and 1850, and again in 1861; at Belhus Park in 1879; and in Moor Park in 1881. In south-east London breeding occurred in the Darenth Valley in 1841 and at Woolwich on an unrecorded date.

Apart from these isolated breeding records they were fairly uncommon visitors. Blyth stated that two or three were usually seen in late summer at Tooting. Harting refers to them as being seen infrequently and records the capture of birds at Chiswick, Hampstead, Harrow, Kilburn, Kingsbury and Primrose Hill. A Hobby was shot on Hampstead Heath in August 1864 while pursuing a Swallow. Some years afterwards one was seen attempting to catch a Sand Martin over Earlsfield station. They were also seen at a number of sites in rural south London and in Langley Park.

20th century

They were much rarer in the early part of the century and Glegg described them as an occasional visitor, being recorded only in four years between 1900 and 1935. There was a small breeding population in Surrey and in 1937, after a gap of 56 years, a pair bred again in the London Area, just inside the boundary. The birds bred again the following year in the same nest but unfortunately the tree where they nested was felled before the breeding season in 1939, and although one or two Hobbies were seen in the area there were no signs of nesting then or in subsequent years.

In 1946 four pairs nested in south London and during the next few years pairs summered every year although breeding was not always successful. Elsewhere in the London Area there was no evidence of breeding and they were recorded as passage migrants in small numbers at many other sites in the 1940s and during most of the 1950s. In 1949 one was watched trying to catch bats at dusk at South Weald Park on 8 October. In 1959 they once again bred in the London Area and raised two young; their prey included Great Spotted Woodpecker, House Martin, Greenfinch, Goldfinch, Budgerigar and probably Crossbill.

In 1960–61 two pairs summered, one of which probably bred, but there were no further breeding records until 1968 when two pairs bred successfully. The following year three pairs bred but this small breeding population took a backward step as no breeding pairs were located in 1970–72. Three pairs bred again in 1976–77 but none were found after 1978 and it appeared that Hobbies were extinct in the London Area. However, in 1983 a pair bred successfully in north London and at least one other pair summered; this was the start of a successful colonisation, and by 1985 the breeding population had increased to eight pairs, six of which bred successfully.

By 1990 the London Area's summer population was still increasing and the breeding population had reached 14 pairs. The expansion continued so that by 1993 35 pairs were located. The second Atlas in 1988–94 showed probable breeding in 84 tetrads, compared with just five in the first breeding survey. In 1999, 20 pairs were proven to have bred, the highest total of confirmed breeding in the London Area.

21st century

There has been a gradual increase in the breeding population since the start of the 21st century. From a minimum of 25 pairs in 2000 it had increased to 33 pairs by 2006. Breeding is not confined to the rural areas as many pairs nest in the suburbs, occasionally in small parks. In 2001 a pair bred in Inner London for the first time, at Kennington, and another pair attempted to breed the following year in South Hampstead but was evicted by Carrion Crows. Another pair bred successfully in Inner London in 2011.

Facts and Figures

The earliest migrant was on 21 March 2000 at the London Wetland Centre.
The latest migrant was on 24 October 2001 at Wimbledon Common.
Unusual prey items caught in 2003 included Sanderling and Pectoral Sandpiper.

Gyrfalcon *Falco rusticolus*

A very rare vagrant (2)

Range: northern Europe, Asia and North America; vagrant to UK.

There have only been two records in the London Area, in successive years in the early 1970s. The first was of a bird at Queen Mary Reservoir in March 1971 and this was followed by a light-phase bird that was found at Rushett Farm, Chessington, on 14 March the following year. The latter remained for six days; there were three other records in the UK in March 1972.

1971: Queen Mary Reservoir, on 6 March
1972: Chessington, on 14–19 March

Peregrine *Falco peregrinus*

An uncommon resident and localised breeder

Range: cosmopolitan.

Historical

There is some evidence that Peregrines once inhabited central London in the 18th century. Several authors wrote that a pair frequented the top of St Paul's Cathedral for many years where it was supposed that the birds had a nest, but there was no actual proof that they bred there. By 1826 several Peregrines had taken to wintering on some of London's churches, including Westminster Abbey.

Harting stated that they were formerly not uncommon in winter and early spring, but by the 1860s the species had become only an occasional visitor between August and March, and was not recorded every year. In 1859 one was caught in Hatfield Park in June. The only dated records since then were: one bird at East Molesey on 26 April 1876; a pair shot in Harrow in December 1888; one found wounded on Stratford Marshes in January 1889; and one near Wimbledon in about 1890.

20th century

In the first half of the century Peregrine was a scarce winter visitor. Apart from one that overwintered at Aldenham Reservoir most others were seen on one day only; almost half of these were seen in the inner suburbs and central London parks. At St Paul's Cathedral, a pair was seen in 1921, and in March 1931 one spent five days in the area. Peregrines were also recorded in Westminster, once even perching on the Houses of Parliament. In 1937 a party of four was seen at Queen Mary Reservoir in July; one can only speculate whether these were a locally bred family party but there was no evidence that a breeding pair was resident in the area. One wintered in the Staines area in 1948/49 and 1949/50.

They remained scarce during the 1950s and 60s and were typically seen just a few times each year. During the 1970s the national population declined due to continued persecution and the effects of pesticides which thinned their eggshells and only 11 were seen in the London Area. By the start of the 1980s they had become rare in the London Area and were not recorded every year; it was also thought that some of these sightings related to escaped birds. There was a very slow recovery in numbers. Two seen together at two localities in south-east London in July 1984 were thought to be escapes.

During the 1990s the status of Peregrine changed from being a scarce visitor to a resident breeding bird. In the early 1990s one wintered in the Lea Valley and in 1993, 10 were seen, mostly on passage in autumn. The following year there was a large increase in sightings with up to 19 seen, including at least one bird which had taken up residence in central London. In 1995 at least two birds were present for most of the year along the Thames in central London. Another pair was present for most of 1996 at Littlebrook Power Station and possibly nested there the following year. In 1998–99 a pair bred by the Thames at Silvertown, the first confirmed breeding record in the London Area.

21st century

At the start of the new millennium Peregrines were on the verge of colonising the London Area. In 2000 two pairs bred, including a pair at Battersea Power Station that was seen throughout the year in central London, mainly near the Thames; there were also several birds in residence around the west London reservoirs. The following year two or three pairs bred along the Thames and in 2002 there were four breeding pairs. A pair bred on a tower block in a south London suburb, well away from the Thames, in 2003.

By 2004 they were seen regularly in central London and even over St Paul's Cathedral, perhaps giving some credence to historical sightings there. Six pairs were located in 2005 and within five years the breeding population had increased to at least 23 pairs and is continuing to expand. Peregrines have colonised many other cities across the UK; the abundance of tower blocks and other high structures for nesting and lookout sites, along with a plentiful supply of pigeons and other prey, has given them the opportunity to make a home in the London Area. Peregrines can often be seen on some of London's iconic landmark buildings such as Westminster Abbey, the Houses of Parliament and Tate Modern; there is even a web-cam on the nest site at Charing Cross Hospital in Hammersmith.

Water Rail *Rallus aquaticus*

An uncommon winter visitor and localised breeder

Range: Europe, Asia and North Africa.

Historical

In the 19th century this was a regular and widespread winter visitor with just one known breeding record, at a small pond near Kingsbury where several young were caught one summer. They probably bred regularly in parts of south-east London; they were thought to breed in the upper Darenth Valley; they were seen at Northfleet in June; and a pair was present every summer at Beckenham up until 1893.

20th century

At the beginning of the century they frequently bred in osier beds along the River Colne in the Staines district and they were a fairly common resident on Dartford Marshes although no nests were found. At Buckhurst Hill, a pair was present in spring in 1906 and a pair possibly nested in the Roding Valley in 1915.

In the 1930s Glegg described them as 'a local winter resident in limited numbers'. The only place he knew where they regularly frequented was along the River Colne between Uxbridge and Harefield. In 1935, a pair bred in Cassiobury Park and the following year breeding was suspected at Chigwell SF. In 1949 a pair summered at Moor Mill, Radlett.

The only confirmed breeding record in the 1950s was at Beddington in 1955, although they were sometimes reported during the breeding season, particularly in Cassiobury Park. They bred at Rye Meads in 1960. There were sporadic summering records during the 1970s and breeding occurred in Richmond Park and at Sevenoaks, and they probably also bred at Broxbourne GP, Bushy Park and Thorndon Park. The highest winter count was at least 11 at Maple Cross SF in 1971, where they congregated around an effluent pipe.

At the start of the 1980s only two breeding pairs were known. A count of at least 13 at Cheshunt GP in 1984 was a record at the time. Wintering numbers began to increase and 69 were located in January 1987. By the beginning of the 1990s there were six known breeding pairs, concentrated at two sites: Dagenham Chase and the Ingrebourne Valley. Breeding numbers fluctuated during the decade and by 1999 eight breeding pairs were reported, of which at least five bred successfully.

21st century

At the start of the 21st century around 115 were located wintering in the London Area, double the number of a decade earlier; the actual figure was probably much higher as they can be fairly secretive. For example, at Brent Reservoir 22 were located one January by use of a tape lure, compared to six seen the previous year. Wintering numbers appear to have increased as there were at least 157 present in the winter of 2002/03.

The breeding population has also expanded, from 14 pairs in 2002 to 22 pairs by 2006, including 10 in the Ingrebourne Valley and six at Rainham Marshes. A pair also bred at the Wetland Centre in 2001.

Inner London

Water Rails are occasionally seen in Inner London, either on passage or during cold weather, and have been found in odd places such as a school playground in Holborn in January 1924; a coal bunker at the Houses of Parliament in December 1936; the roof of a government laboratory in Aldwych on 27 October 1954; the inside of a drainpipe in Battersea on 21 March 1958; and on a window sill on the Guildhall on 3 April 1967. A similar event occurred in 1979 when one spent all day on the window sill of a bank on Fenchurch Street on 15 November; it was still present the following day but flew off when an attempt was made to catch it. Additionally, one was caught by a cat in Buckingham Palace Gardens in December 1966 and released unharmed.

Facts and Figures

The highest count was 52 birds at the London Wetland Centre on 15 December 2001.

Spotted Crake *Porzana porzana*

A scarce passage migrant; historic breeder

Range: breeds Europe and Asia, winters southern Europe, Asia and Africa; scarce in UK.

Historical

In 1813 Graves wrote that Spotted Crake was more common within a few miles of London than anywhere else in the UK and that the species bred at Rolls's Meadows at Bermondsey. A pair was shot during the summer at Dartford Marshes prior to 1824. In 1842 Meyer stated that they were occasional visitors to the Chertsey district. Harting knew of 12 that had been shot in Middlesex by 1866, adding that there were probably others he had not heard about, so they occurred fairly regularly across the London Area.

Dated records include two shot at Laleham in 1857; two shot on the Thames at Chiswick in both 1862 and 1863; three shot on Hackney Marshes in October 1863; and two found dead by the railway at St Albans in October 1880. In 1891 a pair bred in the Lea Valley at Dobb's Weir and raised five young. There may well have been a regular breeding population in the Lea Valley as two were taken at Leabridge in 1890 and nine were shot near Waltham Abbey in 1896.

20th century

In the early years of the 20th century they were regular visitors. In 1903 Christy wrote that 'it is occasionally met with in the soft low-lying parts of the [Epping] Forest, such as would be attractive to Snipe'. In 1907 Davis stated that he had received several specimens, mainly from the Sidcup area. Over the next 10 years there were another four records, including one of a bird that died after flying into telegraph wires at St Albans on 7 September 1910.

With the decline of the British breeding population there was a corresponding decline in the numbers passing through the London Area after the First World War and no more were reported until the 1950s when three were seen, the first of which was at Epsom SF on 11 December 1955. During the 1960s there was an increase in the number of sightings, which began with one near Denham on 23 December 1962; in the New Year this was joined by several more and as the severe winter continued numbers increased to five although two were found dead on 24 March 1963. In 1963–65 Spotted Crakes summered at Beddington and may have bred. In 1965 four were found at Cassiobury Park on 4 January and at least one remained until 4 March. They were almost annual visitors over the next two decades; although most of these were on passage, there were a few winter records and one remained at Cassiobury Park for three weeks in the spring of 1973. There was also one found by Waterloo Bridge on 29 September 1982 and taken into care.

They were particularly rare during the 1990s and only two were seen: one spent a week at Beddington in September 1996 and one was present at the Wetland Centre for eight days in October 1999.

21st century

Since 2000 they have been seen almost annually with 10 records up until 2009. The first of these was on 3 May 2002 at Waterworks NR, the first spring record for almost 30 years. All the other records were in autumn at Beddington, Cornmill Meadows, Ingrebourne Valley, Rainham Marshes and the Wetland Centre apart from one during winter at Warren Gorge, which was heard calling from 2 to 13 January 2009.

Little Crake   *Porzana parva*

A very rare vagrant (1)

Range: breeds Europe and Asia, winters Africa; vagrant in UK.

The only record is of one shot on the banks of the Thames near Chelsea in 1812. This was the fifth record in the UK.

1812: near Chelsea, shot in about May

Unacceptable Records

One was purchased from a London poulterer in May 1812. There was no proof that this bird was killed in London but it is interesting to compare the date to the above record. There was also a record of one at Chalk Hill House in 1863, but no further details are known about this and Glegg did not accept it.

Baillon's Crake   *Porzana pusilla*

A very rare vagrant (4)

Range: breeds Europe, Asia, Africa and Australasia, European population winters Africa; vagrant in UK.

There are four records for the London Area, three of which were in the 19th century. The first was of an adult female captured in a meadow between Mitcham and Carshalton. Later the same year an immature was caught by a dog in a thick reedy ditch in Dagenham that is now part of the Ford motor works area; the specimen is in the Essex Field Club collection. This was a particularly good year for this species as there were another five records in the UK in 1874. The third bird was shot on flooded marshes around Cheshunt.

There were no further records for more than 120 years when a juvenile spent 16 days at Rainham Marshes in the autumn of 2012. Earlier that year there had been an influx into the UK and several territorial birds were heard calling. Although none were located at Rainham, this juvenile may well have been raised elsewhere in the country.

1874: between Mitcham and Carshalton, caught in May
1874: Dagenham Gulf, caught on 3 October
1891: Waltham/Cheshunt, shot on 24 October
2012: Rainham Marshes RSPB, on 7–22 September

Little/Baillon's Crake *Porzana parva/pusilla*

A crake was flushed from an overgrown dyke at Beddington SF in November 1955. It was an adult male and was thought to have been a Little Crake but the views were not conclusive.

Corncrake *Crex crex*

A scarce passage migrant; former breeder

Range: breeds Europe and Asia, winters Africa; breeds on Scottish islands, reintroduced to other parts of UK.

Historical

The Corncrake is one of the species whose status has changed most dramatically in recent times. In the 19th century it was a common summer visitor, breeding widely across the outer districts of London. In 1835 Jesse stated that it was far from scarce around Hampton and in 1847 Ruegg wrote about the Woolwich area stating that 'we are seldom without a few'.

According to Harting their numbers varied from year to year; around Kingsbury he recorded few in 1861 yet the following year they were present in 'great numbers' and on 24 May he heard five calling at the same time. He added that most left in about the end of July as nearly all of the farms in Middlesex were grass-farms which left very little shelter once the hay was cut. There were occasional winter records; Harting knew of one shot in January and he also noted that they could sometimes be seen for sale in the Christmas markets in London.

After the middle of the century they began to decrease in some areas and in 1875 Wharton stated that he had not heard it once during that spring to the north-west of London. In 1877 Leach stated that it used to be very common around Hampstead and Hendon but he had not heard one for two years. In 1878 Harting noted that they had become scarce. During the 1880s they bred in Epping Forest and at Orsett and in south London they could still be heard regularly at Tooting and around Norwood; in the 1890s they were resident throughout the summer on Streatham Common.

In 1891 they bred in Harlesden and in 1893 Swann wrote that they could be heard every spring around Harrow. Further evidence of a decline was noted by Read in 1896 when he stated that although they could still be heard every year in the Lower Brent Valley, they were not as common as they used to be. They used to occur regularly at Winchmore Hill but had disappeared from there by about 1897. By then all areas of London were reporting decreasing numbers; the decline in parts of south London may have been due to the introduction of the mowing machine in 1892–95 to the Warlingham district on the North Downs.

20th century

In the early years of the century, there were a few pairs breeding in the Bromley area and at Lewisham and West Wickham in around 1901. In June 1903 the nests of all the ground-nesting birds were destroyed by a great flood, yet the following year there were more than usual numbers of Corncrakes and they were said to be fairly common throughout the Dartford district. They were still common around Staines in 1906 and the following year Kendall stated that the Corncrake was a regular visitor to the rural areas around Willesden and was fairly common in the meadows. In 1909 Macklin wrote that they were 'generally distributed but nowhere plentiful' in parts of south-east London.

They were still present at Aldenham in 1910 and they bred in 1911–12 at Caterham. In about 1911 they were common in parts of east London and about the same time they were still being reported in small numbers at Godstone, Oxted and Warlingham. However, they started to disappear from some areas and were last recorded on Hampstead Heath in 1908 and Wimbledon in 1912.

The decline in numbers was not restricted to the urban areas of London; Boyd Watt reported that between 1904 and 1911 he had only heard them twice in the course of many outings in Middlesex, Hertfordshire and Buckinghamshire. Breeding was last reported at Haileybury, Hertford and Watford in 1913; at Radlett and Rickmansworth in 1916; Theydon Bois in 1917; and Beddington in 1918. At a Corncrake Inquiry that was held in 1914 it was stated that only occasional pairs in Middlesex had been noted during the previous 10 years.

During the 1920s and 30s there were still isolated breeding records. Up to 1926 they were still being heard regularly at Mill Hill, Northolt, and between Eastcote and Roxeth but thereafter there were only occasional reports. A pair possibly bred in the Lea Valley in 1927 and they probably last bred in north-west London at Mill Hill, Northwood and Ruislip in 1928. Two pairs summered near East Lodge in 1931. Breeding continued on the North Downs at Limpsfield to at least 1928 and at Epsom up to 1934; and records of possible breeding came from Addington, Chigwell, Headley and Oxhey. In 1938 they were fairly numerous at Hatfield but had disappeared from Leatherhead by 1939. The last known breeding pair was at Westerham in 1941; although two were present there the following year on 6 August, there was no sign of any breeding activity.

Occasional migrants were still reported during autumn in Inner London in the 1930s and 40s: at Leicester Square on 29 November 1930; Haymarket on 21 October 1931; St James's Park on 1 November 1932 and 5 December 1933; and picked up dead in Red Lion Court, off Fleet Street, in October 1943 after a very foggy night. One heard calling from long grass on Primrose Hill on 22 August 1938 at 22:45 hrs was probably the last one heard in Inner London. Elsewhere, they continued to be heard during most of the 1940s with calling birds recorded at Oxhey in 1943–45; Ravensbourne and Westerham in 1944; Epping Long Green and Oxshott in 1947; and Headley in 1948. The first year that no calling birds were heard in the London Area was 1949 when the only record was of one picked up dead in a garden in Carshalton.

In 1951 none were recorded for the first time ever, although migrants were seen in most subsequent years. The 1950s ended with two calling birds at Ware; these were the first ones in the breeding season for 11 years. During the 1960s they remained a scare passage migrant with the occasional one heard calling. In 1967 one near St Thomas's Hospital in Lambeth on 6 September was the last one seen in Inner London.

Remarkably, a pair was discovered breeding near Cuffley during fieldwork for the Hertfordshire Breeding Atlas in 1967–73. After one was found dead at Abridge in September 1970, there were no further records until 1976 when four were seen: at Navestock on 11 August; at Dagnam Park on 13 August; one retrieved by a dog on Harlow Common in August and released unharmed; and at Swanscombe Marsh on 12 September. The following year one was heard calling at dawn at North Ockenden on 13 June and an immature was found dead at Abridge on 15 October.

By the start of the 1980s they had become rare and only three were seen: one flushed from a crop field at Chessington Farm in early October 1980; on Walthamstow Marsh on 28–29 September 1983; and trapped and ringed at Cole Green on 24 September 1985. Only two were seen in the 1990s: in 1992 one was caught by a cat at Feltham on 15 May and taken into care; and in 1995 one was seen at Amwell GP in the evening of 7 September and again the following morning.

Despite a recovery in the breeding population in Scotland and a reintroduction programme in eastern England they have remained an extremely rare visitor in the 21st century. The only record was of one at the Wetland Centre on 30 August 2010, which met an unfortunate end when it was taken by a Grey Heron.

Facts and Figures

The earliest migrant was on 26 March 1867 in Middlesex.
The latest migrant was on 5 December 1933 in St James's Park.

Moorhen *Gallinula chloropus*

A very common resident and widespread breeder

Range: cosmopolitan.

Historical

In the 19th century Harting described the Moorhen as a 'common resident, found all along our brooks and at quiet pond sides'. Breeding occurred in Hyde Park and Regent's Park from at least the 1850s and they had become so habituated that they fed with the ducks on bread. In 1860 the introduction of boating on the Serpentine and the resulting disturbance caused them to abandon the site. They first bred in St James's Park in around 1881. In the early 1890s they were still summer visitors but in 1898 Hudson stated that they could be found all year round in some of the parks; in St James's Park a dozen could be seen feeding together, more than in any other London park.

20th century

At the beginning of the century they could be found breeding in the central parks where they were much tamer than they were in the rural districts. They also bred regularly in many other parks, such as Dulwich Park, Hampstead Heath, Kew Gardens, Springfield Park, Tooting Bec Common, Walthamstow Park and Wanstead Park.

There had been no change to their status by Glegg's time, when he wrote that they were a 'very common resident wherever there is a suitable sheet of water', adding that in winter they could be seen in numbers. They were recorded from ponds, canals, streams and sewage farms but were rarely seen on the large concrete-banked reservoirs such as Staines. In 1930, at the end of the breeding season, there were 80–100 in St James's Park.

In 1949 the breeding population in St James's Park was three or four pairs and these were joined by immigrants in winter. This increase during the winter months was also observed elsewhere and occasionally led to Moorhens being recorded in odd places – such as the fountains in Trafalgar Square in November 1946 and a group of 10 on the Thames by Waterloo Bridge in February 1945. By the 1950s they were much more numerous as a breeding species than they had been 50 years previously. Breeding was reported on almost every small pond in Epping Forest and on the Thames at Staines.

During the very cold winter of 1962/63 many dead and dying Moorhens and Coots were seen lying on the ground at Brent Reservoir. At Weald Park only three were seen during the breeding season in 1963 compared to about 20 in the previous summer. In 1967 the breeding population in central London was at least 26 pairs.

By the early 1990s the largest breeding populations were 40 pairs on the River Wandle between Butterhill and Morden Hall Park, and 31 pairs at Brent Reservoir. The second breeding Atlas for the years 1988–94 found breeding in 570 tetrads, two-thirds of the London Area total, and an increase of 15 per cent since the first breeding Atlas. In 1997 there were 56 territories along the Wandle, an increase of more than 300 per cent since 1983.

21st century

Considerable effort has gone into improving London's tributary rivers in recent years, including naturalising banks and planting of waterside vegetation, which is likely to improve the habitat for Moorhens. It remains a common and widespread breeding bird, present in almost all wetland areas, from park lakes in central London to the Thames marshes.

Facts and Figures

The largest concentration was of 239 birds at the Wetland Centre on 12 February 2006.
The highest count was 282 birds along the River Wandle between Hackbridge and Morden Hall Park in November 2002.
A first-winter bird ringed at Rye Meads on 26 November 1967 was recovered the following year on Drejo Island, Denmark, on 21 July.
Hybrids with Coots were seen at Lamorbey Park, Sidcup, in May 1983, and at Hampermill Lakes in February and March 1984.

Coot *Fulica atra*

A very common resident and widespread breeder

Range: Europe, Asia, Africa and Australasia.

Historical

Remains have been found at Southwark dating back to Roman Britain. In the 19th century this was an uncommon resident but it was more numerous during winter when it was more common than Moorhen. In particularly cold weather there were influxes when groups of more than 20 could be found; however; when the freshwater pools and rivers froze the Coots departed to the coast. Harting knew of two occasions when one was seen on the Serpentine in Hyde Park, suggesting that it was particularly rare in central London at the time.

20th century

In the early 1920s they were still a relatively scarce visitor to the central parks and suburbs but by the end of the decade they had become a more regular winter visitor with a few pairs breeding on the park lakes. After 1925 the numbers breeding in the London Area increased, probably as a result of new habitats that had been created. Glegg stated that it was a widespread breeding bird on lakes, ponds and gravel pits although it shunned the large concrete-banked reservoirs. At some sites it was fairly common, such as at Osterley Park where 15 pairs bred in 1931.

In winter it was found in much larger numbers, including on the large reservoirs. In 1907 Kendall stated that Coots were only occasionally seen on Brent Reservoir but by the middle of the 20th century large numbers were wintering there. Glegg estimated that more than 1,000 wintered in Middlesex alone during the early 1930s, and he credited the new large reservoirs for an increase in wintering numbers. In 1938 a total of 1,930 was counted on 17 December. The highest count during the 1930s was about 500 at Cheshunt Marsh in January 1939.

In the early 20th century, Coots wintered in St James's Park but did not breed so eggs were brought from nests in Richmond Park and placed in Moorhen nests. Pinioned birds were introduced and a pair of these bred in 1926. After that breeding numbers increased and a thriving colony was created. There was so much competition for nesting places that some pairs even built their nests on open banks and in flowerbeds, though they were rarely successful. By the 1930s a few pairs bred occasionally in the other central London parks and flocks of a dozen or more could be found on the Serpentine in hard weather. In the 1950s they successfully colonised Battersea Park and Regent's Park.

By the late 1950s the wintering population had significantly increased and the first flock of more than 1,000 was seen at Hilfield Park Reservoir in December 1957. Also in 1957, a survey across the London Area produced a total of 419 breeding pairs; the highest count came in the Lea Valley where there were 61 pairs. They suffered significant mortality during the severe winter of 1962/63 when many waterbodies froze over; at least 40 were found dead at the Lea Valley reservoirs and further losses were reported from several other sites. In subsequent years both breeding and wintering populations began to recover.

Wintering numbers continued to increase, with a high of 4,600 counted in November 1982. The highest counts in the London Area are typically in October/November and in 1990 the total count from the top 15 sites was in excess of 8,200. The largest breeding population was at the Cheshunt GP complex where there were 90 pairs. The second breeding Atlas found an increase of 60 per cent in the breeding distribution across the London Area. They were located throughout the London Area apart from most of the North Downs and farmland in north and east London. By 1999 there were at least 50 breeding pairs in Inner London.

21st century

At the beginning of the 21st century there were two areas of national importance for Coots in the London Area: the gravel pits in the Lea Valley and at Wraysbury; the peak counts at these sites in the winter of 1999/2000 were 3,559 and 869, respectively. However, by 2009/10 the Lea Valley alone retained this designation. The largest breeding population is on Cheshunt Gravel Pits where 92 pairs bred in 2010.

Facts and Figures

The largest count on a single water-body was 1,800 at King George V Reservoir in October 1990.

Crane *Grus grus*

A rare vagrant (20+ since 1900)

Range: breeds northern Europe and Asia, winters Europe and North Africa; small resident population in eastern UK.

The remains of two Cranes were found in Borough that were dated to AD 70–100. Later remains may relate to birds taken for food.

In 1847 Ruegg, writing about the marshes at Woolwich, stated that 'this swampy territory must have been, some half-century since, the chosen abode of the Crane'. In 1872 Palin wrote that Cranes were once common around Aveley prior to the area being drained.

The first 20th century record was of a flock heard flying over Kensington during a foggy night in May 1924. There were no more records until 1973 when one was seen Greenford; the next four sightings were also of single birds

Since 2000 there has been a dramatic increase in the number of sightings which may be linked to the increasing population in Norfolk. In Inner London a flock of six flew north-west over Regent's Park in 2002 and one flew over Lords cricket ground in 2006. In 2010–11 there were three records from Beddington, including one of a juvenile which remained for three days during very cold weather. With a reintroduction scheme just starting on the Somerset Levels and natural colonisation in East Anglia, the outlook for Cranes is very positive and hopefully sightings will continue to increase in the London Area.

1924: Kensington, flock heard at night on 8 May
1973: Greenford, on 3 June
1982: Rainham Marshes, on 15 May
1987: Withybridge, on 11 January
1990: Crayford, on 6 April
1997: Collier Row, on 8 June
2002: Regent's Park, six flew over on 3 September
2003: Fairlop Waters, on 13 October
2006: Lords Cricket Ground, on 12 May
2010: Beddington, on 2 May
2010: Beddington, juvenile on 5–7 December
2011: Beddington, two on 16 April
2012: Romford, on 6 May

Unacceptable Records

In 1799 a flock was seen at a great height over Hackney, and in 1848 or 1849 two birds flew over Hyde Park in spring; neither of these records was sufficiently documented and they were not considered acceptable by Glegg. In 1932 one was seen by the River Colne on 9 and 11 May and it was believed that this was one of the full-winged birds from Woburn Park. In 1947 one escaped from London Zoo and was later seen flying over London. In 1957 a juvenile at Radlett on 4 and 6 July was considered to be an escape as it was tame and allowed a close approach. It is very likely that this was the same immature bird seen at Rainham on 19 July, which was previously accepted as a wild bird. A crane flying over Alexandra Palace on 19 July 1975 was thought by the BBRC to possibly be a Sarus Crane (*Grus antigone*). In 2009 three adults were seen at Tyttenhanger on 22 April, where they roosted overnight and left early the follow morning; they had been ringed and were traced to a release scheme in Norfolk.

Great Bustard *Otis tarda*

A very rare vagrant (1)

Range: Europe, Asia and North Africa; rare vagrant to UK, also reintroduced.

The only accepted record dates back to the 19th century when a female in poor condition was shot at Feltham. Three days later there was a flock of seven in Devon, suggesting that there had been a small influx into the country. The specimen was later exhibited at a meeting of the Zoological Society of London.

Once widespread in the UK, this species became extinct in this country when the last resident bird was shot in 1832. As a contribution to the conservation of this globally threatened species, a programme has been started to try to re-establish a small population on Salisbury Plain. A record of one that flew over Queen Mother Reservoir on 17 November 2011 was probably from this Wiltshire reintroduction programme and suggests that if a breeding population can be established further sightings in the London Area can be expected.

1870: Feltham, shot on 28 December

Unacceptable Records

There was a claim of one seen at Hatfield Broad Oak on 31 October 1899 during severe weather. Christy considered the record as unsubstantiated; it is probably also beyond the London Area boundary.

Oystercatcher *Haematopus ostralegus*

An uncommon passage migrant and scarce breeder

Range: Europe and Asia, some winter Africa.

Historical

Oystercatcher was a rare visitor in the 19th century with just 11 records. Four of those were from Brent Reservoir, one prior to 1843; one shot in July 1859; three on 12 April 1866; and one shot in 1886. All the other sightings were also from the western half of the London Area with none on the Lower Thames Marshes.

20th century

There was a gap of almost 30 years before the next sighting, of one that flew along the Thames past Barn Elms on 30 August 1927. From then on they were seen on a near-annual basis. The dated records up to the 1930s showed that nearly all records were of passage migrants, with three in April and five in autumn between July and September; there were also a few in winter, including one that flew over the City of London on 12 February 1929 during very cold weather.

Up to 1954 there were at least 60 records, almost one-third of which were in August. From the mid-1950s they became more regular and larger flocks were seen, including nine at Barn Elms on 19 March 1953 and 18 at Perry Oaks in August 1956. There was a succession of sightings in Inner London during the 1960s, including one on a mooring float on the Thames opposite the Houses of Parliament on 28 November 1968. In 1969 two flocks totalling 22 birds flew up the Thames past Plumstead Marshes on 21 December.

By the start of the 1970s Oystercatcher was predominantly a passage migrant with occasional records during summer and winter; the highest count was a flock of 14 that flew up the Thames past Swanscombe on 25 August 1970. There were several large flocks during the 1980s and 90s, including 18 at Dartford Marsh on 9 August 1980; 16 at Rainham Marshes on 3 May 1980; 18 at Amwell and seven at Fishers Green on 2 January 1982, when there was an influx during severe weather; and 16 at West Thurrock on 28 December 1998.

21st century

They are regularly seen in small numbers on the Lower Thames Marshes during most of the year except in winter when they are largely absent. Elsewhere they are mainly a passage migrant with around 60 seen annually, mostly at the reservoirs.

Breeding

Oystercatchers bred for the first time in 1971 on the Lower Thames Marshes: one pair nested at Rainham but the eggs did not hatch, and across the river a pair bred at Swanscombe. In 1973 a pair bred at Rainham Marshes and pairs summered at Crossness in 1975 and at Dartford Marsh in 1976. They bred at both Rainham Marshes and Swanscombe Marsh in 1977. From 1978 to 1980 several birds were present during the breeding season on the Lower Thames Marshes but there was no proof of breeding.

In 1983 a pair bred again at Rainham Marshes, the first proved breeding there for six years. The breeding population began to increase and by 1989 there were four breeding pairs on the Lower Thames Marshes and a pair attempted to breed in the Colne Valley at Moorhall GP; the birds were present between May and July, and built a nest but no eggs were laid. This was the first breeding attempt away from the Thames marshes.

By the early 1990s there was a small population of about three pairs on the Lower Thames but breeding success was fairly low. In 1994 a pair was present for a week in late May at Tyttenhanger GP and was seen displaying but did not breed. By 1998 there were at least six breeding pairs on the Lower Thames Marshes. Since 2000 the breeding population has fluctuated from none in 2004 to a high of 12 pairs in 2005 and very few appear to have bred successfully. A pair has also attempted to breed on several occasions at Tyttenhanger. In 2009 five pairs summered away from the Thames, including a pair that may have bred at Walthamstow Reservoirs.

Facts and Figures

The largest count was 20 birds at Crossness on 17 January 2009.

Black-winged Stilt *Himantopus himantopus*

A rare vagrant (11)

Range: breeds Europe and Asia, winters Africa; vagrant to UK, occasionally breeds.

The first record was from Brent Reservoir in the autumn of 1918 when two were seen on the muddy foreshore; the reservoir was drained at the time and had attracted many waders. The autumn of 1955 produced two more records: one circled over Beddington SF for five minutes but did not land, and two were present at Epsom SF. The only record in the 1960s was at Kempton Park Reservoir in July.

No more were seen until 1984 when a pair arrived at Chandler's Cross, on 7 May; the birds then relocated to Perry Oaks SF the following day but flew off west at 05:30 hrs on the morning of the 9th. In September 1997 a first-summer bird was present for 11 days at Rainham Marshes and is the longest-staying bird. The following year an adult spent two days at Park Street GP. The only one seen so far in the 21st century was a male at Rainham Marshes in 2010.

1918: Brent Reservoir, two in September
1955: Beddington SF, on 17 August
1955: Epsom SF, two on 9 September
1968: Kempton Park Reservoir, on 6 July
1984: Chandler's Cross, Watford, two on 7 May; then seen at Perry Oaks SF on 8–9 May
1997: Rainham Marshes, on 10–20 September
1998: Park Street GP, on 27–28 May
2010: Rainham Marshes, on 18 April

Avocet *Recurvirostra avosetta*

An uncommon winter visitor and rare breeder

Range: Europe, Asia and Africa.

Historical

At the beginning of the 19th century this was a regular summer visitor to marshes in England where it used to breed. The extensive drainage of these marshes, along with an increase in the number of people shooting birds, caused the Avocet to become an extreme rarity in Britain by the middle of the century. The first documented record in the London Area was of one shot at Brent Reservoir in May 1854. The only other 19th-century record was of two shot at the same site on 30 August 1897.

20th century

Up to 1930 the only record was of one flying over Hampstead Heath on 10 August 1909. There were single records in both the 1930s and 40s, one at Brooklands SF from 13 to 16 June 1932 and two at Mayesbrook Park GP on 28 March 1949.

In the 1940s they began to recolonise England, first establishing a breeding population in Suffolk leading to a gradual increase in the number of sightings in the London Area. The first of these was of an overwintering individual along the Thames by Littlebrook Power Station and at Stone between 22 December 1952 and 24 January 1953. This was followed by five spring migrants over the next five years. No more were seen until 1967 when one flew over Staines Reservoir on 9 January. The following year there was one at Rainham Marshes on 21 April. In 1969 two or three flocks were seen in spring, 12 birds alighted briefly at Holmethorpe before flying off north-east on 16 March and may have accounted for the flock of 11 the same day at Sevenoaks; there were also three at Rainham Marshes a week later.

The 1970s began with the first record for Inner London – three birds that flew around the Serpentine on 13 March. In subsequent years they were seen almost annually, usually in spring or autumn; most records were from the Lower Thames Marshes, Staines Reservoir and the Lea Valley. The 1970s finished with a flourish; after two spring records, Avocets were seen on the Lower Thames Marshes, mainly at Rainham, on many dates in autumn between 30 July and 8 September, with a peak of 15 on 19 August.

At the start of the 1980s Avocet was a scarce spring migrant with occasional records in autumn and winter. However, in 1981 there were two records in June, two at Hilfield Park Reservoir and one at Rainham Marshes. There was a good spring passage in 1983, including a flock of 22 at Amwell on 25 April, a record count at the time. On 2 December 1984 there was a large passage through south-east England and two flocks were seen in the London Area, 13 birds at Staines Reservoir and seven birds at Island Barn Reservoir.

By the beginning of the 1990s Avocets were becoming more frequent in the London Area in line with the increase in the UK breeding population. In 1990 there was a flock of 14 on 1 June at Staines Reservoir. In 1992 they again visited Inner London when two circled the Round Pond in Kensington Gardens on 24 May. By the mid-1990s there was a large wintering flock on the Thames just beyond the London border and three of these birds were seen at Dartford Marsh in December 1995.

21st century

Since 2000 Avocet has become a regular migrant through the London Area, particularly along the Thames Valley, both on the Lower Thames Marshes and on the reservoirs in west London. They have also increased along the Lower Thames in winter, reflecting the 75 per cent increase in the UK winter population between 1998/89 and 2008/09. There is a large wintering population on the Thames not far from the London boundary as well as a healthy breeding population on the Essex and North Kent marshes.

Breeding

In 1996 a pair unexpectedly bred in the London Area for the first time; the birds arrived at Perry Oaks in spring and remained there until 2 June when they moved to Kempton Park Reservoir where they successfully raised

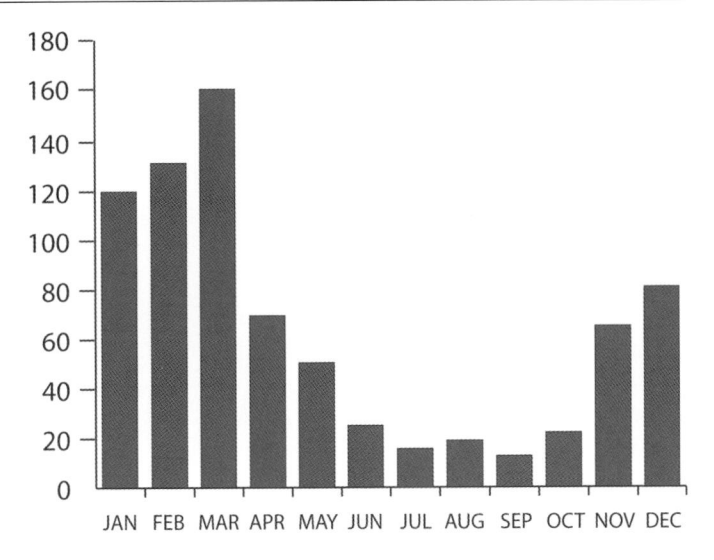

Seasonal occurrence of Avocet: 2000–2009.

one young. The same pair returned in 1997 and was seen at several sites before spending a month at Walton Reservoirs although the birds did not attempt to breed.

In 2005 three pairs nested at Rainham Marshes, one of which successfully raised two young; the breeding area was at a large lagoon away from the RSPB reserve. Breeding was attempted at two sites in 2006: a pair hatched four young at the Wetland Centre but none survived; and four pairs nested at Rainham, one of which raised two young, but the others failed when the lagoon dried out. In 2007 a pair arrived in the Ingrebourne Valley on 13 May and made a nest scrape but did not stay to breed. In 2008 a pair attempted to breed at the Wetland Centre; the birds laid three eggs but the nest was later abandoned.

Facts and Figures

The largest flock was of 82 birds at West Thurrock Marshes on 5 February 2012.
The largest flock away from the Lower Thames was of 15 birds at Walton Reservoirs on 22 November 2002.

Stone-curlew   *Burhinus oedicnemus*

A scarce passage migrant; historic breeder

Range: breeds Europe, Asia and North Africa, winters southern Europe and Africa.

Historical

In 1823 Latham wrote that Stone-curlews were 'not uncommon' in the Darent Valley. There were 12 other records in the London Area in the 19th century, listed below. The 1848 record of a male shot in a fallow field near Bushey Heath was originally documented by Harting but was not counted by Glegg as it was deemed not to be in Middlesex; however, it is still within the London Area. On the western border of the London Area they were occasionally shot in Langley Park.

Pre-1823: bred in Darent Valley
Pre-1842: Chertsey Meads
1848: Bushey Heath, shot in April
1867: Feltham, killed in March
Pre-1868: Langley Park, shot at various times

1870: Hounslow, shot in autumn
1875: Hounslow, present for about three weeks in September
1883: Epping Forest, on 21 April
1889: Hackney Marshes, shot in December
Pre-1893: Epping Forest, one taken
1897: Haste Hill, Ruislip, on 24 May
Pre-1900: Kilburn, one caught alive by a bird-catcher
Pre-1900: Willesden, one caught alive by a bird-catcher

20th century

In the first half of the century Stone-curlew was considered to be a regular but rare migrant in the London Area with around 25 records away from the North Downs. In the latter area, between Woldingham, Tatsfield and Chelsham, they were heard on a number of occasions in spring; although there was suitable habitat they were not considered to be breeding there as the area was too disturbed.

All records were of single birds apart from a group of three seen near Drayton on 4 October 1929. They were usually seen or heard at open spaces like farmland, old sewage farms and parkland, such as at Richmond Park, from where there were four records, and also at Chingford Plain, Hainault Forest and Wimbledon Common. Of the total number of records, only five occurred in autumn: singles in July, August and November, and two in October. The August record was rather bizarre as the bird was picked up from a children's sandpit in Grange Park, Kilburn.

During the 1950s there was a gradual increase in the number of records with a total of 14. In 1956 an unseasonable bird flew low over Barn Elms Reservoirs on 27 December. Eight more were seen in the 1960s, including a group of three that flew south over Rye Meads on 12 July 1961 and the first one in Inner London occurred on 8 May 1964 when one flew south over Regent's Park. There was a gradual decrease in sightings during the 1970s as the UK population began to decline. In 1977 another one was seen in Inner London, at Surrey Docks on 10 June. The following year one remained at Rainham Marshes from 2 to 6 October.

In 1980 one was trapped at Upminster on 23 September; it had previously been ringed near Icklingham, Suffolk, on 1 June as a nestling. In 1982 a pair was seen at one site in rural north London on 24–25 April, with one remaining until 1 May. They remained scarce during the 1990s and just four more were seen.

21st century

Conservation measures have helped the UK population to increase although they are still largely restricted to two small populations in the East Anglian Brecklands and Salisbury Plain. There has been a corresponding increase of migrants seen in the London Area; most have been recorded in spring, some of which have remained for a day before continuing their migration. One or two are seen in most years but there were four in both 2007 and 2008.

Facts and Figures

The earliest migrant was on 28 February 2000 at Sewardstone.
The latest migrant was on 27 December 1956 at Barn Elms Reservoirs.
Migrants have been identified flying over the London Area at night by their calls on five occasions in March and three in April.

Cream-coloured Courser *Cursorius cursor*

A very rare vagrant (1)

Range: Africa and Middle East; vagrant to UK.

The only record occurred in the 19th century when one was shot at Temple Mill Marshes, on the east side of the River Lea at Stratford. Hudson relayed the account stating that 'a working man came full of excitement to the White House to say that he had just seen a strange bird, looking like a piece of whity-brown paper blowing about on the marsh; whereupon the late Mr George Beresford took down his gun, went out, and secured the wanderer'. The specimen was later mounted and went on view in the inn.

1858: Stratford, one shot on 19 October

Collared Pratincole *Glareola pratincola*

A very rare vagrant (2)

Range: breeds Europe, Asia and Africa, winters Africa; vagrant in UK.

The first reference was in 1847 when Ruegg wrote that 'a friend of mine once saw a pair flying over the marshes [at Woolwich]'. Unfortunately, no further details were given so this record cannot be accepted. The first confirmed record occurred in June 1983 when one was found on the drained north basin of Staines Reservoir. It was then watched for three hours as it hunted over the reservoir and Staines Moor; at about 12:30 hrs it flew off high to the north-east and was not seen again. However, one was seen later that day on the Ouse Washes and it is conceivable that it was the same bird.

The second record came in 2005 when an adult spent four days on the new RSPB reserve at Rainham Marshes. Although the reserve had not been officially opened, access was arranged to allow many people to see this rare wader.

1983: Staines Reservoirs, on 19 June
2005: Rainham Marshes, on 2–5 July

Pratincole sp.

An immature pratincole was seen at Barn Elms Reservoirs on 8 and 11 September 1948. Unfortunately, the underwing colour could not be determined so it was not assigned to a particular species. Another unidentified pratincole was seen flying over the North Downs on 14 September 1971, between Belmont and Banstead.

Little Ringed Plover *Charadrius dubius*

An uncommon summer visitor and localised breeder

Range: breeds Europe, Asia and North Africa, winters Africa.

Historical

This was a very rare visitor in the 19th century with just three records, all of which were from Brent Reservoir. The first two were of juveniles shot in 1864, on 20 and 30 August. The third was seen the following year on 29 April.

20th century

There were no more records for almost 80 years. By then a pair had bred for the first time in England in 1938 at Tring Reservoirs, Hertfordshire, about 19 kilometres beyond the London boundary. There was no further breeding recorded until 1944 when two pairs bred at Tring and one pair bred in London, at a gravel pit near Ashford. The same year, one was seen at Brent Reservoir on 3 and 9 August.

The following year a pair bred again at the same site near Ashford and a second pair bred a couple of miles away near Shepperton; these were the only breeding records in the UK in 1945; two were also seen near Radlett in June but were not thought to have bred. In autumn birds were again seen on passage at Brent Reservoir. In 1946 five or six pairs summered in west London. Up to 13 pairs bred in 1947, including three at King George VI Reservoir and four at Girling Reservoir, which was under construction at the time. By the summer of 1948 King George VI Reservoir had been filled but six pairs summered at Girling Reservoir and a further seven or eight pairs bred at various gravel pits.

Breeding continued at Girling Reservoir until 1951 when it was filled. By 1953, just 10 years after the first breeding record, 18 pairs were reported in the London Area, of which at least nine were known to have bred; one of these pairs bred on a small patch of stony ground at the end of a runway at Heathrow for the second consecutive year. However, there was no room for further expansion as all the suitable sites were occupied. Once the two large reservoirs had been filled, Little Ringed Plovers bred almost exclusively at gravel pits, sometimes for only one or two years since once the gravel had been extracted, the pits soon became overgrown and were no longer suitable for breeding. Little Ringed Plovers readily used working gravel pits despite the excavating machinery and associated disturbance. In 1957 a total of 30–32 pairs were located including six at Rye Meads. At the beginning of the colonisation, Little Ringed Plovers arrived at the end of April or in early May, but by the mid-1950s they were arriving about a month earlier with some back on site by the end of March.

By the beginning of the 1960s 37 pairs were reported summering and the total remained similar for the rest of the decade. In 1963 the first one to be recorded in Inner London was seen on the grass in Archbishop's Park, Lambeth, on 18 July. In the early 1970s a post-breeding build-up became a feature at Rainham Marshes; in 1972 the peak was 17 but by 1979 this had increased to 75. In 1973 the total number of pairs found was 72, of which 39 were proven to have bred; the large number that year can partly be attributed to the efforts of a national survey; also that year a pair bred in Inner London for the first time, at Surrey Docks.

During the 1980s, the number of pairs reported each year was around 50-65. In 1984 there were two breeding pairs in central London on opposite sides of the Thames. By the start of the 1990s, 75 pairs were found, of which 53 were proven to be breeding; the numbers were boosted by the draining of the north basin of Staines Reservoir where 15 pairs bred. Between the two breeding Atlases there was little change in the distribution, with most pairs found along the Lower Thames Marshes, in the Lea Valley and around the gravel pits and reservoirs in west and south-west London. After the end of the second atlas survey, there was a fall in numbers reported to the LBR. This may signify reduced recording effort although it could also reflect loss of suitable breeding habitat.

21st century

The increase in this delightful wader during the 20th century is one of the conservation success stories in the UK, remarkable in the way the species makes use of man-made habitats in gravel pits and construction sites, quickly exploiting the opportunity as new habitats become suitable. For example, the recently created London Wetland Centre has provided good habitat for this species and eight pairs bred in 2005. Since 2000 the number of breeding pairs reported has varied between 15 and 55, representing around 1.5-5 per cent of the British breeding population. There are no longer post-breeding gatherings at Rainham and the largest concentration in autumn was 21 at Cornmill Meadows on 1 July 2000.

Facts and Figures

The earliest migrant was on 27 February 2010 at Staines Reservoir.
The latest migrant was on 4 October 2004 at Staines Reservoir.
The highest count was 91 birds at Rainham Marshes on 31 July 1980.

Ringed Plover *Charadrius hiaticula*

An uncommon passage migrant, winter visitor and scarce breeder

Range: breeds northern Europe, Asia and North America, winters coastal Europe and Africa.

Historical

In the 19th century this was only known as a regular passage migrant, appearing in small flocks in early May, and again from July onwards in flocks of six to 20 or more. The earliest reference was in 1842 when Meyer wrote that it had been seen on the Thames at Chertsey. There are also records from Brent Reservoir, Brixton, Headley Heath, Redhill and West Molesey.

20th century

The first breeding record took place in 1901 at Enfield SF. The following year some birds were again present but they did not stay to breed. In 1909 Ticehurst stated that they could be found in winter along the Thames from the docks all the way out to the east. The first sighting in Inner London was on the Thames at Millbank in March 1917.

In the 1930s Glegg stated that it was seen in spring between March and May, although not recorded annually; and in autumn from mid-July, with most being seen in August and September. This passage sometimes continued through to early winter. The largest flocks recorded were in 1927 at Staines Reservoir, when the water level was low, with two to three dozen in May and 25 in August. In 1926 a juvenile was shot at Langley SF on 2 August and upon examination was shown to be of the Arctic race *tundrae*; this was the first time this race had been identified in the London Area.

Between 1937 and 1940, flocks of up to 30 or more were seen in winter at Romford SF. By the 1940s it was mainly known as a passage migrant with the occasional record in winter. Ringed Plovers were quick to take advantage of new habitats, and during the construction of King George VI Reservoir up to 20 were seen on passage from 1947 to 1949 on the partially flooded shingle bed. Nearby, they also were regularly seen at Perry Oaks SF from the mid-1940s onwards; 45–60 were seen between 20 August and 10 September 1950; and up to 51 were recorded there in early June 1952. Sometimes they were also seen at the adjacent Heathrow Airport which was under construction at the time.

In the Lea Valley, particularly at King George V Reservoir, they were regular on passage and flocks of 30 or more could be seen. Elsewhere, the highest count during the early 1950s was 92 at Stone Marshes in August 1954. During cold weather in 1956 there was a flock of 57 at Beddington on 2 February. There was a large passage in the autumn of 1956 with about 100 at Perry Oaks and 87 at West Thurrock. In 1959 up to 100 were seen at Swanscombe Marshes in November.

In the late 1950s they returned to breed in the London Area. In 1957 a pair bred successfully at Rye Meads and a pair nested at Stone Marshes, although no young were seen; the following year a pair summered at Stone Marshes. In 1959 three pairs bred on the Lower Thames Marshes and in subsequent years three or four pairs then bred annually at Swanscombe Marshes.

During the 1960s there was a significant rise in the numbers seen in the London Area. In 1961 about 120 were present at Swanscombe Marshes on 1 January and there were about 300 there on 17 August. The following year the numbers at Swanscombe had increased again and a new record high of about 400 was seen on 22 August. The breeding population on the Lower Thames Marshes fluctuated between one and seven pairs during the 1960s. Elsewhere, a pair was present during the breeding season at Walton Reservoirs in 1961 and may have bred. In 1967 pairs bred at Nazeing GP and Stanwell Moor, and in 1969 a pair bred at Bedfont GP.

There was a gradual increase in breeding numbers during the 1970s and by 1977 about 15 pairs were reported, including one which bred successfully at Heathrow Airport in a disused builder's yard. The highest counts in the 1970s were 250 along the Thames at Purfleet and Rainham between August and November. At the start of the 1980s a pair bred in Surrey Docks, the first breeding record in Inner London. By 1985 the breeding population had risen to a new high with 31 pairs recorded. In 1987 there were 114 at Perry Oaks on 26 August, a particularly high count away from the Thames marshes.

The breeding population in the London Area was temporarily increased in 1990 when 10 pairs bred on the drained north basin of Staines Reservoir. The breeding population was centred on the Thames marshes and the Lea Valley with an isolated population of four pairs at Hersham GP. In 1994, 34 pairs bred or attempted to breed, half of which were away from the Thames marshes. There was a marked increase in the breeding population between the first two breeding Atlases; in 1968–72 they were found in 12 tetrads and during 1988–94 they were in 45. The actual distribution was very similar to Little Ringed Plover's with the majority of pairs found along the Lower Thames Marshes, in the Lea Valley and around the waterbodies of south-west London.

In February 1992 a coordinated count found a total of 362 at low tide along the Lower Thames. Due to the closure of the ash lagoons at West Thurrock in 1994 there was no high-tide roost on the Lower Thames Marshes and the numbers declined dramatically to a high of just 70 in winter; there were actually more at both Perry Oaks SF and Staines Reservoir in autumn.

21st century

Since 2000 there has been a large decrease in breeding numbers and by 2004 the population appeared to be halved with only 13 pairs reported; in 2008 this figure had reduced further to just six pairs. The peak count on the Lower

Thames was about 250 at Rainham in November 2008; away from the river there were 120 on the drained Queen Elizabeth II Reservoir on 23 August 2007. Small numbers are often encountered on passage at the large reservoirs.

Facts and Figures

The largest flock recorded was of about 400 birds at Swanscombe Marshes on 2 August 1962.

Killdeer *Charadrius vociferus*

A very rare vagrant (1)

Range: North America; vagrant to UK.

The only record occurred in 1984 when one, probably a first-winter bird, was seen briefly flying with a flock of Lapwings on the afternoon of 31 January at Beddington. It reappeared the following morning and remained until early afternoon when it was seen to fly off north; it was then relocated outside the London Area, in Berkshire, on 25 February.

1984: Beddington SF, on 31 January–1 February

Unacceptable Records

A record of one found dead in South Kensington on 20 December 1938 was not accepted.

Kentish Plover *Charadrius alexandrinus*

A rare vagrant (24)

Range: Europe, Asia and Africa; rare passage migrant in UK.

The first record was in the spring of 1878 on the River Mole near East Molesey. In the first half of the 20th century just two more were seen, at Staines Reservoir in 1915 and on the adjacent King George VI Reservoir in 1947. There was then a succession of records with five seen in six years between 1958 and 1963; of these, three were in April.

The 1970s was the peak decade, beginning with one at Rainham Marshes in 1970. Two years later another spring migrant was seen on the Lower Thames Marshes and in 1974 one stayed at Rainham Marshes for 10 days in autumn. Between 1976 and 1979 another four were seen, including two at Staines Reservoir in the spring of 1978.

The early 1980s continued in a similar vein, commencing with two at Perry Oaks in consecutive years. In 1983 there was an unprecedented series of records at the partially drained Staines Reservoir, where four different birds were seen between 10 April and 12 June; there was also one at Rainham Marshes in April.

Since then there have only been two more records, both in the spring of 1999, at Perry Oaks and Sevenoaks. Of the total of 24 records, 13 were in the Staines area, six were on the Lower Thames Marshes, three were in the Lea Valley and two others were elsewhere.

Spring records outnumber autumn records by a factor of nearly four to one: with an impressive 11 records in April and five in May (the 1878 record was undated), then just two in June followed by singles in July and August with three in September.

1878: near East Molesey, in spring
1915: Staines Reservoir, on 21 April
1947: King George VI Reservoir, female on 3 September
1958: King George VI Reservoir, female on 1 April
1959: Walthamstow Reservoirs, female on 19 April
1960: Perry Oaks SF, female on 3 April
1962: Rainham Marshes, male on 1 July

1963: Girling Reservoir, on 7 September; also seen at Ponders End SF
1970: Rainham Marshes, female on 18 April
1972: Swanscombe, on 6 May
1974: Rainham Marshes, on 2–11 September
1976: Girling Reservoir, male on 24 April
1978: Staines Reservoir, female on 9 April
1978: Staines Reservoir, female on 29 May
1979: Rainham Marshes, on 19 and 25 August
1980: Perry Oaks SF, male on 1 June
1981: Perry Oaks SF, on 19 May
1983: Staines Reservoir, female on 10 April
1983: Staines Reservoir, female on 22 April
1983: Rainham Marshes, female on 24 April
1983: Staines Reservoir, female on 16 May
1983: Staines Reservoir, male on 12 June
1999: Sevenoaks WR, female on 24 April
1999: Perry Oaks SF, female on 2 May

Dotterel *Charadrius morinellus*

A rare passage migrant (74+)

Range: breeds Europe and Asia, winters Africa.

Dotterel was a very rare migrant in the 19th century with just four records. However, considering the relative lack of observers and this species' tendency to stay for short periods in open fields, most probably went unrecorded. The first record was of five that were killed on Nunhead Hill in November 1817. Dotterel used to be seen in spring and autumn in an area known as Hungry Down, which was a bare open tract of land between Edgware and Kingsbury; there were two records from farms there. In 1871 a flock of 14 was seen in east London at Forest Gate.

No more were seen until almost 80 years later when 10 were discovered at the newly constructed Heathrow Airport on 26 August 1950. The following day, seven were still present and this reduced to five birds by the day after, one of which remained until 30 August. Five years later, a flock of four flew over Perry Oaks SF on 1 August.

The only record in the 1960s was of a single bird at Girling Reservoir on 16 September 1962, the latest date one has been seen in the London Area. The next three records were all from Staines Reservoir: two singles in autumn and a group of three in the spring of 1992. The only sightings since were in May 1994 when a trip of 16 arrived at Bowmansgreen Farm near London Colney during a day of exceptional wader migration, with heavy showers and a strong easterly wind that grounded many migrants. There was a similar occurrence in May 2012 when a flock of 15 spent the day at Canons Farm.

1817: Nunhead Hill, five killed in November
1856: Fox Mead Farm, Kingsbury, one shot in September
1858: Burnt Oak Farm, Kingsbury, a small flock in April
1871: Forest Gate, 14 at the end of August
1950: Heathrow Airport, 10 on 26 August, declined to one on the 30th
1955: Perry Oaks SF, four on 1 August
1962: Girling Reservoir, on 16 September
1977: Staines Reservoir, on 11 September
1985: Staines Reservoir, on 29 August
1992: Staines Reservoir, three on 29 April
1994: London Colney, 16 on 7 May
2012: Canons Farm, Banstead, 15 on 4 May

American Golden Plover *Pluvialis dominica*

A very rare vagrant (1)

Range: Americas; vagrant to UK.

The only record occurred in September 1995 when an adult in partial summer plumage was found in the evening at King George VI Reservoir.

1995: King George VI Reservoir, on 2 September

Pacific Golden Plover *Pluvialis fulva*

A very rare vagrant (2)

Range: breeds Asia, winters East Africa to Australasia; vagrant to UK.

The first Pacific Golden Plover to be seen in the London Area was also the first record for the UK. The bird was shot at Epsom Race Course in November 1870 and the specimen is in the Charterhouse Museum. Apart from one other record in Essex in 1896 there were no other sightings in the UK for more than a century. Since then the number of national records has slowly increased and another bird was seen in the London Area in 2010. It was first seen flying over Beddington and a short while later landed at the Wetland Centre where it remained for several hours before the breeding Lapwings harassed it and it flew off west.

1870: Epsom, shot on 12 November
2010: Beddington/London Wetland Centre, on 2 May

Golden Plover *Pluvialis apricaria*

An uncommon winter visitor

Range: breeds northern Europe and Asia, winters Europe and North Africa.

Historical

Harting wrote that the Golden Plover 'visits us in flocks towards the end of autumn, and is, perhaps, most numerous about the end of November'. At that time he regularly found flocks on fallow fields between Kingsbury and Stanmore; for example, he counted nearly 200 there on 23 November 1861. They were also known to be present on the marshes at Woolwich in the middle of the 19th century. A flock of 100 or more regularly wintered at Epsom and they were common on Staines Moor until around 1880. Towards the end of the 19th century there was evidence of a decline in numbers. Around Harrow, 'great flights' were reported between 1846 and 1850, but in 1891 Barrett-Hamilton stated that they had become less numerous.

20th century

In 1901 Lodge stated that although they still wintered in flocks of 100 or more in the Lea Valley, there had been a decline in numbers since the 1880s. In 1902 Vaughan described Nazeingwood Common as a great haunt of Golden Plover and from there, and in other areas east of the Lea Valley, flocks occasionally numbering up to 400 could be found on spring passage in March; some of these were of the northern subspecies *altifrons*.

From 1910 onwards flocks of 30 or more were often seen at Beddington on the irrigated grass fields – one of the few places where they could be found regularly south of the Thames. Up to 50 were seen annually at Warlingham between 1907 and 1932 but the land was then built upon. At Walton Heath, similar-sized flocks were often noted between 1905 and 1924.

By the 1930s the decline had continued and it was no longer a regular winter visitor; Glegg attributed this decline to suburbanisation. However, large numbers were occasionally still reported: on 25 March 1928 there were 170 near Stanwell; 400–500 were seen by the junction of the Great West and Bath roads on 5 December 1931; and there were 300–400 at Thornwood, near Epping Forest, on 22 December 1934. Two winters later 500 were seen in the Radlett and Shenley area on 3 January 1937, and a similar number wintered there in subsequent years, peaking at about 600 on 3 January 1950. In east London more than 400 were seen at Abridge on 2 December 1938. Very few have been seen in Inner London; the earliest sighting was in 1939 when about 40 flew west over Lord's cricket ground on 29 December.

In 1940 about 250, most of which were in breeding plumage, were seen at Romford SF on 3 April, and a similar number was at Waltham Marsh on 17 April. After the Second World War the wintering numbers increased; flocks of up to 300 wintered at Fairlop aerodrome, and in December 1950 about 1,000 were counted there. In the winter of 1948/49 there were about 100 at Northfleet and a similar number was seen at Rainham; these were the only large flocks recorded from the Thames-side marshes. During the 1950s large wintering flocks began to be recorded in the Lea Valley; for example, in January 1956 there were about 350 at Hoddesdon SF. Towards the northern boundary, flocks of up to 500 were noted around Radlett and another 300–600 were regular at Shenley.

During the 1960s there were fewer wintering in the London Area and the highest count was 350 at Heathrow Airport from 25 November to 2 December 1960. By the middle of the 1960s the large flocks around Shenley had reduced to less than 100 birds. At the start of the 1970s the only wintering flock of over 100 was at Staines Moor where the maximum count was 235 in January. By 1973 they had ceased wintering in the Shenley area and the highest count was 500 at Symondshyde on 15 March.

Numbers increased at Fairlop during the cold winters in the 1980s with a peak count of 1,500 in 1987. At the start of the 1990s there were several sites with a regular large wintering population, notably at Fairlop and in the London Colney area; elsewhere, there was a particularly large count of 1,500 at Stanborough on 30 December. At Romford there were 1,200 in December 1991 and the following year there were 3,000 at Fairlop in December. In February 1995, 1,000 birds were present at four sites: Fairlop, London Colney, Rye Meads and Tyttenhanger. At the end of the year there was a notable cold-weather movement with a flock of 360 flying west over Hampstead Heath.

21st century

In January 2000 there were about 8,000 wintering birds; the largest flock was of 2,500–3,000 birds at Sewardstone. Over the next few winters the wintering population was fairly similar with the largest flocks numbering 3,000 birds at Great Amwell in 2002 and 3,500 at Beech Farm in 2003. However, since then there have been fewer birds wintering in the London Area with less than 2,000 in 2009/10. The largest flocks are often recorded on passage in March.

Facts and Figures

The largest flock was of 3,500 birds at Beech Farm on 8 February 2003.

Grey Plover *Pluvialis squatarola*

An uncommon passage migrant

Range: breeds Arctic Circle, winters coastal Europe and North Africa.

Historical

The earliest record was in September 1766 when two were shot in St George's Fields (just to the north of Hyde Park) and, being considered a great rarity, were sent as a present to a noble duke. They were equally rare in the 19th century and the only record was of a flock seen in the spring of 1890 on Stratford Marshes.

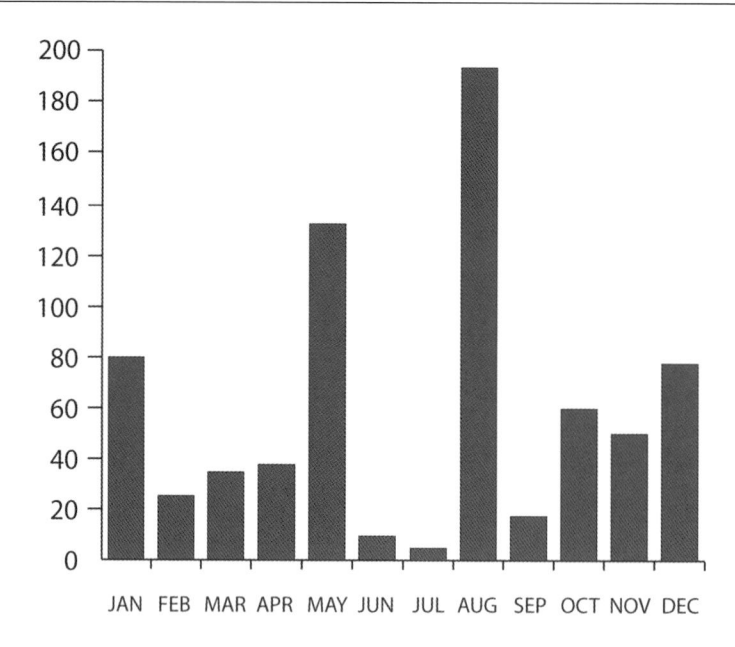

Seasonal occurrence of Grey Plover: 2000–2009.

20th century

They remained rare up to 1930 with the only documented record being of six at Staines Reservoir on 9 April 1922. Between 1930 and 1954 the status of Grey Plover changed dramatically and there were more than 40 records, divided equally between spring, autumn and early winter. Most of these were surprisingly away from the Thames-side marshes, on reservoirs, sewage farms and gravel pits; there was one on the Thames foreshore near Chiswick Eyot on 1 January 1942.

By the mid-1950s they had become a regular migrant through the London Area but only in very small numbers. In 1956 seven were seen at Perry Oaks on 28 October, and in 1958 five were seen flying down the Thames at Dartford on 26 January. In 1963 up to five wintered at Swanscombe Marshes and one flew over Regent's Park on 3 March, seemingly the first record in Inner London.

By the start of the 1970s Grey Plover had become an annual but scarce migrant, mainly seen on passage but with the occasional record in winter. It continued to be seen more frequently, with about 50 seen in 1980, including a flock of 32 which flew south over Queen Mary Reservoir on 9 August. In 1982 there was a good passage at Staines Reservoir where one of the basins was drained; birds were present there on many dates during October to December, with a peak of 20 on 5 November. A similar passage took place there the following spring between 19 March and 8 June with a peak of 10 on 6 May. There were three high counts in the late 1980s: 25 at Dartford Marsh in February 1986; 17 flew over King George V Reservoir on 14 September 1988; and 17 flew over Staines Reservoir on 2 April 1989.

In 1990 an exceptionally high number was seen on passage, notably on 23 September when 50 were present at seven sites, mostly on the Lower Thames Marshes and along the Lea Valley. In the winters of 1990/91 and 1991/92 they wintered on the Lower Thames Marshes, with peaks at the West Thurrock roost of 16 the first year and 30 the second; they were also seen feeding at several sites. At the beginning of 1992, large numbers were still present at West Thurrock with 35 on 20 January. In the autumn of 1992 a flock of 30 flew south down the Lea Valley on 30 August on a day of large wader passage. In early 1996 large numbers were present on the Lower Thames, with a peak of 58 at Rainham Marshes on 22 February.

21st century

Grey Plover continue to winter on the Lower Thames but numbers are now generally small; a few birds are occasionally encountered elsewhere on migration. However, there have been three very large flocks: 170 that flew high over Limpsfield Chart on the evening of 8 August 2004; 120 over Staines Reservoir and 185 over Wormwood Scrubs, both on 23 August 2010.

Facts and Figures

The largest flock recorded was about 185 birds over Wormwood Scrubs on 23 August 2010.

Sociable Plover *Vanellus gregarius*

A very rare vagrant (2)

Range: breeds central Asia, winters Middle East; vagrant to UK.

There have been two records in the London Area, both on the Lower Thames Marshes. The first was of an adult at Dartford Marsh in 1985, which associated with a group of up to 100 Lapwings and was also seen on the adjacent Crayford Marsh on occasions. By the end of its stay it had attained full breeding plumage.

The second record came 20 years later when a first-winter bird was found at the new Rainham Marshes RSPB reserve, where it remained for over two weeks apart from a brief visit to Dartford Marsh, across the Thames.

1985: Dartford Marsh, on 8 March–13 April
2005: Rainham Marshes, on 4–20 December; also at Dartford Marsh on 9 December

White-tailed Plover *Vanellus leucurus*

A very rare vagrant (1)

Range: breeds Asia, winters Africa and Asia; vagrant to UK.

There has only been one London Area record of this nationally very rare wader, only the seventh to be seen in Britain. This particular individual toured the UK and north-west Europe in 2010, starting at Seaforth, Lancashire, on 27–28 May; it was then seen in Holland the following day before disappearing for almost six weeks. It spent a day at Rainham Marshes before relocating to Slimbridge, Gloucestershire, on 9 July where it stayed for two days. The next day it was back in the south-east of the country, this time at Dungeness where it remained until the 21st. The following day it was found across the Channel in France.

2010: Rainham Marshes, on 7 July

Lapwing *Vanellus vanellus*

A common winter visitor and localised breeder

Range: Europe and Asia, some winter North Africa.

Historical

Harting stated that 'this species is resident with us throughout the year, but receives an increase to its numbers in spring and autumn, when large flocks are frequently seen on the fallows and wastes'. He noted how they used to breed by Kingsbury Reservoir but moved to a quieter area about 2 miles away after their nests were continually plundered for their eggs, which were a delicacy at the time – 'plover's eggs'. They were clearly a common and widespread breeding bird in the outer districts of London as Harting once found more than 20 nests in a day near Elstree. They were probably widespread across the London Area but breeding was only documented at Chertsey, Kempton Park, Kenton and Wembley.

20th century

By the 1920s the effects of suburbanisation had started to impact on Lapwing as fields were built over for new housing. By 1922 they had ceased breeding in the Harrow district and further losses were reported from around Mitcham Common in 1926, Eastcote and Rayners Lane in 1927, Perivale in 1931 and New Malden in 1934.

Glegg noted that it was 'a common resident, whose numbers become much increased in winter', adding that it bred on unfrequented fields and some sewage farms. During autumn a considerable migration was recorded with flocks of up to 500 being seen. In winter it was quite widespread, typically seen in small flocks but sometimes of up to 200 birds. Occasionally large post-breeding flocks were noted on Staines Moor numbering several hundred, and on 25 December 1931 at least 2,000 were seen in fields by the junction of the Great West and Bath roads. Glegg stated that they had been seen flying over central London on many occasions but rarely landed.

During and after the Second World War, there were considerable changes of land use, which impacted on the species. Some farmland areas, such as at Parndon and between Northwood and Harefield, were drained, causing Lapwings to abandon these areas completely. Pastureland in several outer north-west London districts was converted to arable land, providing winter feeding but being less suitable for breeding. Additional habitat was provided by the temporary conversion of parkland to arable land, and flocks made use of these areas for feeding as well as breeding; Bushy Park, Hampstead Heath, Hampton Court Park, Mitcham Fair Green, Richmond Park and Tottenham Marsh were converted in this way. This had the short-term effect of increasing the Lapwing population during the 1940s, but this trend was reversed once the parkland was reclaimed in the following decade. An example of this was at Richmond Park where Lapwings had not bred since 1834. After some of the park was ploughed, a pair nested in 1943 and then bred regularly; by 1950 five or six pairs were breeding. After the harvest in 1951, most of the stubble fields were left to revert to grassland and only one pair nested the following year, the eggs failing to hatch. When the park was cultivated flocks of up to 150 were seen on the stubble in winter. Along the Lower Thames, about 3,500 were present at Northfleet on 19 December 1948.

Further decreases were reported during the 1950s and Lapwings were now absent from some apparently suitable areas. Within 16 kilometres of the centre of London, they only bred at two strongholds: Beddington and Hendon Aerodrome. However, beyond the suburbs they were still a widespread breeding bird. A census carried out in 1957 found about 220 breeding pairs. The population on the Lower Thames Marshes had seriously declined after the great floods of early 1953 subjected the marshes to increased salinity. Research showed that the largest gatherings occurred in winter around the full moon when small flocks congregated to feed during the night; about 2,000 gathered at Beddington on one such occasion in 1953.

During the extremely severe winter in 1962/63 many Lapwings left the area completely in December and did not start to return until early March. In 1973–74 a pair bred at Surrey Docks, the first modern breeding record in Inner London; the birds summered there again in 1975 but did not breed. In the 1980s the largest wintering flock was of 3,000 at Colney Heath; up to 2,500 were present on the drained Staines Reservoir in December 1985.

At the start of the 1990s there was a fairly stable breeding population of about 150 pairs. The highest concentrations were 19 pairs at Rammey Marsh, 14 at Rainham Marshes and 11 at Beddington. Twelve sites hosted winter flocks of at least 1,000 birds, with the biggest count being at Stanborough where there were 4,500 on 17 November 1990. At the beginning of 1991 at least 20,000 were present across the London Area; however, heavy snow in February caused almost all of these to move on. In early 1994 there were 4,000 on one of the drained basins at Staines Reservoir; on the Lower Thames there were similar numbers at Aveley in November.

Somewhat surprisingly there was an increase in the breeding population between the two Atlases; during the 1988–94 survey the birds were found breeding in 193 tetrads, an increase of about 8 per cent. Although some of this may be due to increased coverage, a stable population in the London Area contrasted sharply with the national decline of 40 per cent between 1974 and 1999. The bulk of the breeding population was located in the more rural areas although there was sufficient habitat along the Thames and Lea for Lapwings to breed in suburbia. The largest wintering flock in 1994/95 was of 5,000 birds at Maple Cross.

21st century

Since 2000 there have been increases in the breeding populations at several sites, notably at Beddington, the Wetland Centre and Rainham Marshes. Between 2005 and 2010 about 65-150 breeding pairs have been reported annually in the LBR; given that some pairs will always nest undetected the breeding population is likely to be around 100-200 pairs; most of these are located on farmland beyond the suburbs. The Lapwing is a Red listed Species of Conservation Concern in the UK having declined by 56 per cent from 1970 to 2010, due mainly to changes in agricultural practice. The largest wintering population is 2,000–5,000 at Rainham.

Movements

There were several large cold-weather movements in the 1970s. The first was on 4 March 1970 when 10,500 birds flew over Kingston in half an hour and at least 4,000 flew over Worcester Park in 90 minutes. At the end of the same year a total of about 17,000 flew over London in the last week of December. In 1972 about 14,000 flew south over Wraysbury Reservoir on 30 January.

Ringing Recoveries

During the summer there is an immigration of continental birds, mainly juveniles, into south-east England from June onwards, and some of these pass through the London Area. A young Lapwing ringed at Brandenburg, Germany, in May 1934 was recovered at Hertford in April 1938, and one ringed in northern Bohemia in May 1942 was recovered in north London in the following August.

Facts and Figures

The largest wintering flock was of about 5,000 birds at Maple Cross in 1994/95 and a similar number at Rainham Marshes in 1999/2000.

Knot   *Calidris canutus*

An uncommon passage migrant

Range: breeds Arctic Circle, winters coastal western Europe, Africa and Australasia.

Historical

This was a very rare migrant in the 19th century with just five records. The first two were at Brent Reservoir; in June 1844, Bond wrote that one had recently occurred there and on 23 February 1870 a flock was seen. Migrants were seen at West Molesey in the spring of 1874 and the autumn of 1877. On 11 August 1895 Bowdler Sharpe heard 'a large migration' of Knot calling as they flew over his garden in Chiswick; he stated that they were too high to see and that he had previously heard them passing over London at night from the top of Primrose Hill.

20th century

Between 1900 and 1930 it remained a rare bird in the London Area, occasionally seen on the banks of the Thames between Greenhithe and Dartford, and there was only one record away from the river, of two at Staines Reservoir on 10 September 1927.

From the mid-1930s they began to be seen more frequently. Most of the records were of passage migrants, especially in September and October, but there were also a few winter records. Like many other waders migrating through the London Area, they were recorded on the reservoirs, at sewage farms and on the Thames. The largest flocks were seen in winter: 10 birds flew over Staines Reservoir on 16 November 1935; and in the 1950s there were two flocks of 10 each on the Thames near Dartford. During severe weather in 1956, a flock of 23 was seen at Beddington on 2 February and about 20 were at Stone Marshes two days later.

In 1960 Knot roosted at Swanscombe Marsh at high tide between August and December; the peak count was about 250 on 31 October and about 200 remained for much of November. In subsequent years only much

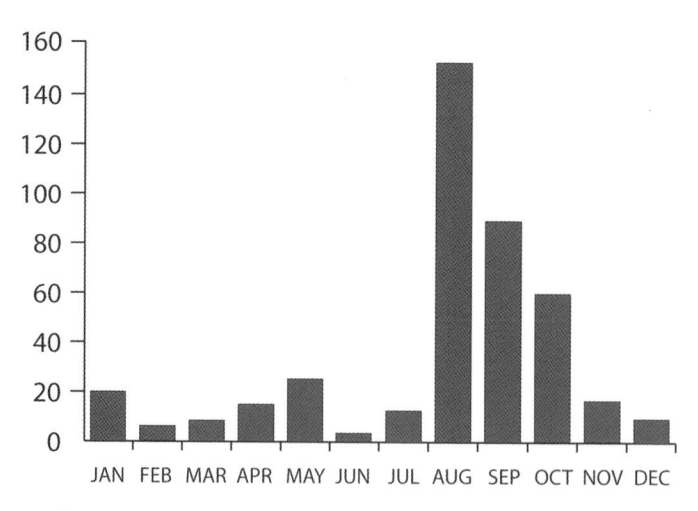

Seasonal occurrence of Knot: 2000–2009.

smaller numbers were seen at Swanscombe. In 1962 there was a good passage in the autumn through Perry Oaks SF, with a peak of 24 on 16 September. By the start of the 1970s there was a small flock wintering at Dartford but on 15 February a flock of over 250 was seen. In 1973 a flock of 24 flew south over Wraysbury Reservoir on 29 August. By the end of the 1980s they had ceased wintering on the Lower Thames Marshes and had become erratic visitors elsewhere in the London Area. In 1992 a flock of 22 appeared out of thick fog at Walthamstow Reservoirs on 2 February.

21st century

Since 2000 Knot has continued to be an irregular visitor, typically in small numbers. Large flocks are occasionally seen, such as the 45 that flew over Regent's Park on 6 October 2001 and the 90 that flew over Queen Mother Reservoir on 19 August 2007.

Facts and Figures

The highest-ever count was about 250 birds at Swanscombe Marshes on 31 October 1960 and at Dartford Marsh on 15 February 1970.

Sanderling *Calidris alba*

An uncommon passage migrant

Range: breeds Arctic Circle, winters coasts on all continents.

Historical

This was a rare migrant in the 19th century, at least away from the Thames. One or two were shot near Putney Bridge and a pair was shot at West Molesey in the spring of 1870. The only records away from the river all come from Brent Reservoir; in 1844 Bond stated that one had been shot there and two others were shot in autumn about 10 years later.

20th century

Glegg stated that one had been reported on the Serpentine in Hyde Park on 30 July 1900 but this was considered doubtful. No more were seen until 1928 when one was at Staines Reservoir on 4 September and another was seen eight days later at Brent Reservoir. From then on one or two were seen in most years (apart from during

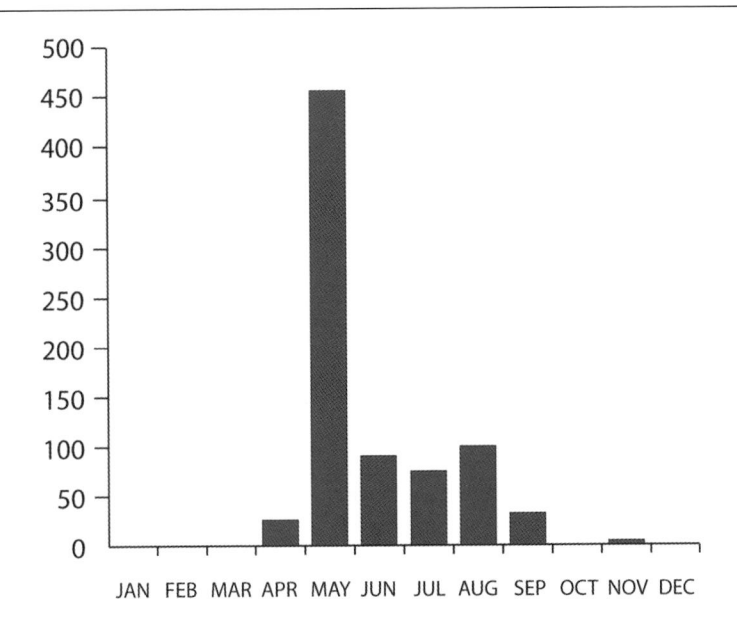

Seasonal occurrence of Sanderling: 2000–2009.

the Second World War) until the 1950s. They were typically found feeding along the water's edge on the large concrete-banked reservoirs; there were also six at Watford SF on 13 September 1936. In 1949 up to three were present on one of the drained basins at Barn Elms between March and May. As well as the reservoirs they were regular at Perry Oaks and were occasionally found at some of the other sewage farms. The only site on the Thames where they occurred was Stone Marshes.

In the spring of 1950 there was a particularly strong passage in west London and a flock remained at Staines and King George VI Reservoirs for a week, peaking at 16 on 20 May. In 1953 a new high count was reached when there were 17 at Perry Oaks on 16 May. In 1971 a flock of 22 was at Wraysbury Reservoir on 16 May. The only definite one to have occurred in Inner London was in 1979 when one was seen on the edge of the Round Pond in Kensington Gardens during a snowstorm on 15 February.

In 1980 a flock of 26 was at Rainham Marshes on 29 March. There was an exceptional passage of around 45–50 birds in the spring of 1994; most were seen on Staines Reservoir where there were peaks of at least 22 birds on 7 and 16 May; on the latter date there were also eight or nine at Queen Mother Reservoir.

21st century

Since 2000 Sanderling has mostly been seen on passage, typically in small numbers during May. In favourable weather conditions, they can be more numerous, e.g. there was a good passage in 2000 with 43 migrants in spring, and at least 42 passed through Queen Elizabeth II Reservoir during May 2007.

Western Sandpiper *Calidris mauri*

A very rare vagrant (1)

Range: Americas; rare vagrant to UK.

The only London Area record was in July 1973 when an adult was present at Rainham Marshes for three days; this was just the second record in Britain, the first being on Fair Isle in 1956.

1973: Rainham Marshes, one on 21–23 July

Little Stint *Calidris minuta*

An uncommon passage migrant

Range: breeds Arctic Circle, winters southern Europe, Asia and Africa; passage migrant in UK.

Historical

Harting wrote that this species was 'occasionally found at our reservoirs during the periodical migrations in spring and autumn, but more commonly at the latter season, when the little flocks which arrive here consist chiefly of young birds'. There were 16 records in the 19th century from Brent Reservoir; the highest numbers were five in August 1862, and four in autumn 1864 and on 10 June 1865. Elsewhere, the only known record was of several killed at Battersea in 1869.

20th century

In the early years of the century one shot on the Thames near Dartford in 1903 was the last one to be killed in London. There were no further records until 1925 when two were seen at Langley SF on 12 September. In 1929 one stayed at Brent Reservoir from 27 September to 2 October; this was the first one to be seen on more than one day.

From 1930 onwards one or two were seen virtually every year in autumn, typically in September or October. The first spring migrants of the 20th century appeared in 1939 when there were two at Brooklands SF from 31 May to 1 June and another at Staines Reservoir on 4 June. In subsequent years they were occasionally noted in spring. They were recorded on passage at a number of localities, typically at the large reservoirs and sewage farms, and along the Thames. Most were seen at Perry Oaks SF; there was a particularly good passage there during the autumn of 1950, with up to five seen virtually daily throughout September and 15–17 on the first four days of October, then up to four daily until the 27th.

The first winter record was at Dartford Marshes on 11 and 19 December 1954. In 1957 there was an exceptional influx in late autumn, with about 40 at Perry Oaks on 28–29 September; elsewhere, there were peak counts of 14 at Swanscombe Marshes on 6 October, up to 12 at Rye Meads on 24–30 September and smaller numbers at five other sites.

There was another large passage in the autumn of 1960 when 42 were present at Swanscombe Marsh on 18 September. In the winters of 1961/62 and 1962/63 up to two overwintered at Swanscombe Marsh. One was seen at Swanscombe on 21 January 1965, the fourth consecutive year that at least one was present during winter. In the autumn of 1967 there was a large passage in September when the peak counts were 33 at Rainham Marshes and 30 at Perry Oaks. In 1969 up to six were present at Rainham during the first two months of the year and seven in March.

By the start of the 1970s they had ceased wintering along the Thames. In 1976 there was a maximum of 32 at Rainham on 2 October; exceptionally, one was seen on the Round Pond in Kensington Gardens on 14 May, the first one recorded in Inner London. In 1977 up to five wintered at Rainham. In 1978 there was a very large passage in autumn, the peak counts being 50 at Rainham Marshes on 9 and 17 September, and 39 at Perry Oaks on 10 September. One wintered at Rainham Marshes in 1981/82.

In the autumn of 1996 an exceptional influx began on 8 September when 10 were seen on Wraysbury Reservoir; small numbers began arriving elsewhere from mid-month onwards and there was a large influx from the 21 to the 25th, including peak counts of 76 on Wraysbury Reservoir and 48 at Beddington.

21st century

Between 2000 and 2010 there were far fewer Little Stints compared with previous decades; the largest flock comprised 11 birds at Staines Reservoir on 2 September 2004. Only four were seen in 2006, the lowest annual total for over 60 years. In 2001 an unseasonable migrant spent two days at Tyttenhanger GP on 25–26 June.

Facts and Figures

The largest flock was of 76 birds on Wraysbury Reservoir on 24 September 1996.

Temminck's Stint *Calidris temminckii*

A scarce passage migrant

Range: breeds northern Europe and Asia, winters southern Europe, Asia and Africa.

Historical

A rare species in the 19th century with eight records, two in spring and six in autumn, all from Brent Reservoir. The first was in 1839 when Bond saw two adults in the spring and at least five juveniles in autumn. No more were seen until 1859 when one was shot in May and another was shot on 31 August 1861. The next records were in 1866 when four were shot on 15 September; one was shot on 4 September 1869; two were seen on 4 October 1871; and one was shot on 29 August 1872.

20th century

In the first half of the century there were only two records, the first of which was in 1936 – over 60 years since the previous one – at Brooklands SF on 29 May. The other was on Staines Moor on 27 April 1949; this was the first April record.

Between 1950 and 1979 they were seen more frequently, with 40 recorded. Most were seen at sewage farms, including groups of three at Perry Oaks on 21 September 1957, at Rainham Marshes from 26–29 August 1970 and at Staines Reservoir on 16, 22 and 31 May 1978. One also wintered at Rainham from 4 November 1970 to 19 April 1971.

The peak decade was the 1980s with a total of 43 birds, 10 of which were in 1987. Several small flocks included three at Broxbourne GP on 11 May 1980 and five at Kempton Park Reservoir from 17 to 19 May 1987. During the 1990s only 22 were seen, about half the total of the previous decade. They were seen at Beddington in five of the 10 years. The only flock was of four birds at Hersham GP on 19 May 1999. Apart from two at Rainham and one at Sevenoaks all the rest were in the western half of London.

21st century

Between 2000 and 2004 there were 23, but in the following five years only six more occurred. During the latter period there was also a reduction in the numbers seen nationally and it may be that a lack of easterly winds during their migration accounts for this decline. There were nine in 2004, six of which were in May. In 2008 five were seen at Rainham Marshes in the period 11–16 May.

Facts and Figures

The largest flock was of five birds at Kempton Park NR from 17 to 19 May 1987.

White-rumped Sandpiper *Calidris fuscicollis*

A very rare vagrant (3)

Range: Americas; vagrant to UK.

The first record for the London Area was discovered by Harting in the collection of his friend Mr Dresser who had recently purchased the specimen. When Harting examined it he found on the back of it a small label on which was written 'Schinz's Sandpiper. Shot by Mr Goodair at Kingsbury Reservoir, 1856'. Although the exact date it was shot was unrecorded, the specimen was in winter plumage. Schinz's Sandpiper (Buonaparte) was the name in use at the time for White-rumped Sandpiper as opposed to *Tringa schinzii*, which was known to Harting as a variety of Dunlin. There were only four other British records of White-rumped Sandpiper prior to this one.

The second record occurred over a century later when an adult was present at Perry Oaks SF for two weeks in 1984 and then again at Staines Reservoir after disappearing for six days; there had been an influx into the country with a total of 23 birds, nine of which were between 14 July and 7 August. In 1995 an adult was at Walton Reservoirs in late August.

1856: Brent Reservoir, shot
1984: Perry Oaks SF, one on 30 July–12 August, also at Staines Reservoir on 18 August
1995: Walton Reservoirs, on 30 August

Baird's Sandpiper *Calidris bairdii*

A very rare vagrant (3)

Range: Americas; vagrant to UK.

The first record was in 1950 when one was present at Perry Oaks SF for six days in September. It occurred during a period of strong westerly and south-westerly gales, which also brought a number of Pectoral Sandpipers to the UK. This was only the second national record of Baird's Sandpiper, following one shot on St Kilda in 1911. By the time of the second London Area record in 1977 the species was not quite so rare in the UK, but only a few had been seen inland. The third occurred in October 1982 at Staines Reservoir, this bird remaining until the following spring. It was the first individual to have overwintered in the Western Palearctic; during the course of its stay it moulted from juvenile into winter plumage.

1950: Perry Oaks SF, one on 17–22 September
1977: Rainham Marshes, one on 3 September
1982: Staines Reservoir, on 14 October–24 April 1983

Pectoral Sandpiper *Calidris melanotos*

A scarce passage migrant (77)

Range: Asia and Americas; vagrant to UK.

Historical

The first record was in 1846 when one was shot at Brent Reservoir. The specimen came to light in 1891 when it was sold along with the Rev. Frere's bird collection.

20th century

No more were recorded until over a century later when one was discovered at Perry Oaks SF in the autumn of 1950. It remained there from 16 to 29 September and was part of a small influx of this species into the UK following a period of gales that swept across the Atlantic.

From then on the number of Pectoral Sandpipers arriving in the UK, and to a lesser extent the London Area, increased substantially. During the 1950s there were a further seven records, all between late August and the end of September; six of these were at sewage farms. In 1960 three separate birds were seen at Perry Oaks between 29 July and 3 September. The London Area's first winter record was in 1964 when one was present at Harefield Moor on 29 December; it was found dead on 2 January 1965.

The peak decade, as it was for several other scarce waders, was the 1980s when 24 were seen, eight of which were at Perry Oaks. There was a national influx in the autumn of 1984 when six were seen in the London Area. In 1989 the first-ever spring Pectoral Sandpiper was seen in the London Area, at Beddington on 12–14 May. Ten were seen during the 1990s, three of which were at Beddington, which had begun to take over from Perry Oaks as the most regular site for this species. The only June record was at Kempton Park Reservoir on 1 June 1996.

21st century

Between 2000 and 2010 there were a further 23 birds. Of these, eight were at Rainham Marshes and six at Staines Reservoir. All arrived between late August and the end of October apart from a late one at Ruislip Lido on 8 November 2003. Five were seen in both 2002 and 2003. With the vast majority of all birds arriving in autumn, most have been aged as juvenile; two birds have been seen together on five occasions, most recently at Rainham Marshes in September 2002 and at Staines Reservoir in September 2004.

Sharp-tailed Sandpiper *Calidris acuminata*

A very rare vagrant (1)

Range: breeds Siberia, winters Australasia; vagrant to UK.

The only record occurred in 1976 when an adult was found at Staines Reservoir, which was drained at the time and attracted a large number of passage waders.

1976: Staines Reservoir, on 6 August

Records No Longer Accepted

A record of one at Staines Reservoir on 28 September 1966 was originally accepted by the British Birds Rarities Committee but later withdrawn by the observers following a review of all the records.

Curlew Sandpiper *Calidris ferruginea*

An uncommon passage migrant

Range: breeds Siberia, winters Africa, Asia and Australasia; passage migrant in UK.

Historical

The earliest reference was in 1801 when Latham stated that one had been taken near Greenwich Marshes. In the 19th century this was a scarce migrant with a total of 14 birds, all at Brent Reservoir: one prior to 1843; two juveniles on 2 September 1844; one shot 'some years later'; one shot on 17 September 1864; two shot on 2 September 1865; two prior to 1866; one on 1 May 1868; one on 27 August and one on 10 September 1870; one in May 1871; and one on 11 September 1873.

20th century

There was a 50-year gap before the next record, at Staines Reservoir on 7 September 1927. This was followed by two at Brent Reservoir from 12 to 17 September 1928. From the early-1930s through to the end of the 1940s small numbers were seen in most years apart from during the war when none were recorded. The most unusual location was Richmond Park where there was one on 23 October 1938 – also a particularly late date. The largest group was four at Brooklands SF on 2 June 1939.

They were more frequently seen from the 1950s onwards. In 1950 there was a particularly good passage at Perry Oaks SF between August and mid-October, peaking at 15–17 on 19–24 September. Perry Oaks remained the best place to see Curlew Sandpiper during the next few years apart from in 1953 when up to 26 were present at Stone Marshes.

From 1959 through to the mid-1960s the largest numbers were seen on the Lower Thames Marshes. In 1959 there was a large influx and the highest count was 40 at Swanscombe in September. There were several autumn influxes during the 1960s. The first of these was in 1960 when about 30 were at Swanscombe Marsh, and 36 were at Brent Reservoir on 28 September. In 1963 the peak counts were 68 at Swanscombe Marsh on

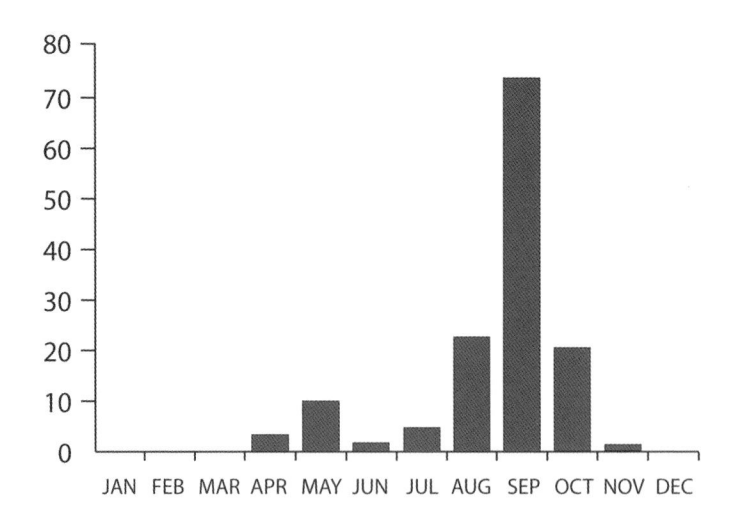

Seasonal occurrence of Curlew Sandpiper: 2000–2009.

9 September, 38 at Rainham Marshes on 7 September and about 20 at Perry Oaks on 1–3 September. In the autumn of 1969 there was a particularly heavy passage across Britain but although large numbers were seen in the London Area they were mainly confined to two sites: at Perry Oaks the peak count was 39 on 29 August and 1 September, while at Rainham the maximum count was 70 on 31 August and 1 September.

This pattern of occasional influxes continued during the 1970s with the highest counts being along the Lower Thames and in west London at Perry Oaks and Staines Reservoir. In 1975 one injured bird remained at Queen Mary Reservoir from 6 to 30 December. During the 1980s and 90s there were far fewer seen and the only large count was in 1985 when there were 36 at Staines Reservoir on 10 September. In the autumn of 1988 there was a large influx on the East Coast but few reached the London Area.

21st century

Since 2000 they have become much scarcer on passage in the London Area. As is the case with some other waders this is probably due to a lack of easterly winds during their migration. They are seen annually in ones or twos, typically on the Lower Thames Marshes or on the west London reservoirs. There was a small influx in 2004 with up to eight at Staines Reservoir and groups of four seen flying over Beddington and along the Thames at Northfleet. None were seen in 2006, the first blank year since 1974. They are exceptionally rare in winter and the only one was at Rainham Marshes on 24 and 26 February 2012.

Facts and Figures

The largest flock was of 70 birds at Rainham Marshes on 31 August and 1 September 1969.
The earliest migrant was on 11 April 2003 at Walton Reservoirs.

Purple Sandpiper   *Calidris maritima*

A scarce winter visitor (47)

Range: breeds northern Europe and North America, winters coastal Europe.

Historical

The only 19th-century record was of one shot on the Thames near Molesey on 9 January 1871.

20th century

No more were seen until the 1930s when three occurred, all in November: one at Barn Elms Reservoir and two at Staines Reservoir. There were two in the 1950s: at Beddington SF in the very cold winter of 1956, along with Dunlin and Knot, on 2 February; and at Queen Mary Reservoir on 1 November 1959.

The frequency of sightings increased during the 1960s when a total of 13 were seen; 10 of these were on the west London reservoirs, including a group of three at Staines Reservoir on 28 December 1962. The 1970s began with one that was trapped and ringed at Dartford on 2 January 1970, then remained for at least three weeks; this was followed by 10 more, most of which were seen between late October and December.

Only two were seen during the 1980s: at King George V Reservoir on 18 August 1981; and on the Thames at Teddington Lock on 9–10 September 1984. There were six in 1991: three appeared in February after heavy snowfall in the country, including the first record for Inner London, on the Thames at Chelsea Embankment on the 7th; and there was one on the Thames at Rainham Marshes from 7 December intermittently into January 1992 when it moved just outside the London Area to East Tilbury, Essex. The only other one in the 1990s was at Thorney CP on 19 October 1997.

21st century

Since 2000 they have been seen almost annually and have been recorded in every month between July and December. All the records were from the reservoirs in west and south-west London with three at Queen Mother, three at Staines and singles at Island Barn and Queen Elizabeth II.

Dunlin *Calidris alpina*

An uncommon passage migrant; winter visitor on the Lower Thames

Range: northern Europe, Asia and North America, winters Europe and Africa.

Historical

In the 19th century Dunlin was known as a regular passage migrant, being one of the first spring waders to arrive. Harting wrote that 'it frequents large sheets of water, such as the Reservoirs of Kingsbury, Ruislip and Elstree', adding that it 'has more rarely been found here in winter'.

Migrants were seen over the Dulwich area on three occasions between 1876 and 1890, and there are also records from Cobham, East Molesey and Mitcham. However, its status on the Lower Thames Marshes went unrecorded.

20th century

In the first three decades there were occasional records during the migration seasons along the Thames between Dagenham and Thurrock, and from reservoirs such as Barn Elms, Staines and Walthamstow. In winter there were scattered records across the London Area, e.g. up to six were seen at Barn Elms and Putney in the winter of 1907/08, and Ticehurst stated in 1909 that they could be found along the Kent coast right into the London Area. In 1921 Horn wrote that they were regular at Thurrock Marsh and especially numerous during cold weather. In August 1927 there were at least 40 at Staines Reservoir when the low water levels exposed some mud.

In the 1930s Glegg stated that Dunlin was mainly a regular passage migrant in the London Area. It had been noted in every month apart from June and was most frequently seen in August, particularly at the large reservoirs, at Staines Moor, along the Thames and in central London. Occasionally, large flocks were seen during the spring passage. During a period of severe cold weather in December 1939 a flock of 85 was recorded at Romford SF. In the autumn of 1948 there was a good passage in the Staines area with 30–40 seen regularly between July and mid-October, peaking at 70 on the site where King George VI Reservoir was being constructed.

In the 1950s large numbers started to be seen at Stone Marshes in winter; in February 1951 about 50 roosted on a jetty at high tide and on 7 February 1954 the first three-figure count was made there – 270. Elsewhere, they were frequently seen on passage, especially at Perry Oaks SF; the maximum count there was 66 on 2 May 1953. In 1956 there were 130 at West Thurrock Marsh on 11 December, and in 1959 a new high count was made when about 500 were at Swanscombe Marsh on 29 November.

In 1960 the numbers roosting at Swanscombe Marsh at high tide began to build up in late autumn; there were about 300 on 9 October and they continued to increase until they peaked at about 2,500 on 20 November. Numbers then began to recede and about 2,000 were seen roosting on the opposite side of the Thames at Grays in mid-December; by the end of the year, only a few hundred remained in the area. It was thought that the main feeding area for these birds was at Cliffe Marshes, a few miles beyond the London Area boundary in Kent. In subsequent years up to 1,800 wintered on the Lower Thames Marshes.

The 1970s began with a record number on the Thames marshes with peak counts of more than 4,000 at Dartford, up to 1,500 at Swanscombe and about 1,000 at Woolwich in January. A simultaneous count at sites on the Lower Thames produced a total of 7,455 in December 1972. In December 1973 the estimated total on the Lower Thames Marshes was about 10,000; the peak count for an individual site was 6,000 at Purfleet. There was a particularly strong passage on the Lower Thames Marshes in the spring of 1974 with 500 at Purfleet on 2 May; this coincided with larger numbers further downstream, beyond the London Area boundary. In 1978 about 20 flew over Regent's Park on 7 December, the largest flock seen in Inner London.

In 1983 there was a good movement recorded at Staines Reservoir on a wet day in spring when the numbers built up from about 90 in the morning of 17 April to 237 by the afternoon. Although fewer birds had been wintering on the Lower Thames, large numbers returned in 1984 when up to 3,000 roosted on the ash lagoons at West Thurrock; in 1988 there were 7,000 at the West Thurrock roost. The roost count continued to increase, reaching 10,000 in December 1991 and peaking at 22,000 in February 1992. The following winter there were up to 15,000 but by December 1993 the peak count had fallen to 2,500; it was thought that this was due to the cessation of pumping at the power station and the drying out of the ash lagoons where the waders roosted at high tide. In 1994 a low-tide count along the river at all sites found a total of almost 6,000. A new roost was formed in December 1996 when 4,000 were discovered at Dagenham.

21st century

Since 2000 large numbers have continued to winter on the Lower Thames Marshes; the peak counts were in January 2003 when up to 5,000 were at Tilbury Docks and there were 4,000 at West Thurrock. Away from the river they are regularly seen at the reservoirs in small numbers on passage and occasionally in winter.

Facts and Figures

The highest-ever count was 22,000 birds at West Thurrock in February 1992.
Although most ringing recoveries from the Thames Estuary are from around the Baltic Sea, one ringed at Littlebrook in January 1974 was controlled in northern Norway in July 1975.
One picked up dead in the grounds of the Natural History Museum in South Kensington on 29 December 1938 was identified as belonging to the race *alpina*.
During the Second World War, a flock was heard low over the rooftops in Mill Hill as it evaded gunfire aimed at a low-flying aircraft on 18 October 1943.
The largest flock away from the river was of 290 birds on King George VI Reservoir in November 1991.

Broad-billed Sandpiper   *Limicola falcinellus*

A very rare vagrant (1)

Range: breeds northern Europe, winters Africa; vagrant to UK.

The only record was in the Lea Valley in 1958. The bird was first seen at Rye Meads on 27 July, then after an absence of 11 days was relocated at Broxbourne SF, where it remained for 12 days.

1958: Rye Meads, on 27 July; also at Broxbourne on 8–19 August

Buff-breasted Sandpiper   *Tryngites subruficollis*

A rare vagrant (5)

Range: Americas; rare visitor to UK, mostly in south-west.

There have been five records in the London Area, three of which were in the Staines area. The first bird was found at Perry Oaks SF on 18 October 1953 and was then seen on a further four days up to 27 October and again on 3 November. In September 1977 one was found at Staines Reservoir, where it remained for over two weeks.

There were two in the autumn of 1981: a long-staying bird at Perry Oaks in September and another at Barn Elms Reservoirs on 12 October; the latter flew off south-east and was then seen at Bough Beech Reservoir, Kent, the following day. The only adult to be seen in the London Area was at Rainham Marshes in July 1986.

1953: Perry Oaks SF, intermittently on 18 October–3 November
1977: Staines Reservoir, on 14 September–2 October
1981: Perry Oaks SF, on 12–25 September
1981: Barn Elms Reservoirs, on 12 October
1986: Rainham Marshes, on 21 July

Ruff   *Philomachus pugnax*

An uncommon passage migrant

Range: breeds northern Europe and Asia, winters Europe and Africa.

Historical

The earliest record was of one shot at Greenwich on 5 August 1785. In the 19th century Harting referred to the Ruff as 'an uncertain visitant, appearing occasionally during the vernal and autumnal migrations in May and August'. In 1831 one was shot in Bushy Park. Harting knew of about a dozen shot at Brent Reservoir between 1838 and 1866, as well as a flock of five which flew over there on 21 May 1866. In the 1840s two were shot at Munden House near Watford. There were another four records up to 1900, including two more at Brent. Between 1860 and 1864 a flock was present on Turnham Common and was seen lekking; the nearest breeding area to the London Area at this time was in Norfolk.

20th century

In the early part of the century, one at Lower Edmonton SF in November 1912 was the first to be seen in the London Area for 41 years; in September 1915 two were shot at the same site. In the 1920s three were seen near Langley in September 1924 and one was shot on a ploughed field at Colnbrook on 5 December 1925. After that, one or two were seen in most years.

Glegg stated that it was more frequently seen during autumn than spring; there was also the occasional winter record. During the 1930s and 40s it had become a regular passage migrant in small numbers, stopping off at many of the reservoirs and sewage farms as well as marshy areas in the river valleys and even on ploughed fields. From 1947 onwards it was found intermittently throughout autumn at Perry Oaks SF or on one of the nearby reservoirs; typically either single birds or small groups were found, but up to 14 were present in March 1950. Spring migrants were fairly scarce until 1949 when 20 were found at Wraysbury GP on 10 March.

During the 1950s larger numbers began to be seen. In 1954 there was an unprecedented influx in the London Area; between 20 and 24 March the flock at Perry Oaks peaked at 115. Ruff were seen at nine locations during April with a total of 119 present on 8 April. From the mid-1950s onwards they began to overwinter in the London Area. In the winter of 1954/55 two or three remained at Perry Oaks and the following year up to 20 overwintered there, with another six at West Thurrock. In 1959 up to 27 were present at the start of the year and there were several summer records, including some in full breeding plumage.

The wintering flock in the Staines area averaged about 30 during the early 1960s. In the exceptionally cold weather at the start of 1963 there were fewer birds wintering in the London Area with a peak of just 13 at Perry Oaks in March and five at Swanscombe. By the mid-1960s they had ceased wintering on the Lower Thames Marshes, while the wintering flock at Perry Oaks had decreased to 15. There was a regular wintering flock on the Thames marshes in Kent, and between 1969 and 1973 it moved back into the London Area at Dartford; typically 50–75 birds were seen but the peak count was 130 between Dartford and Swanscombe in 1971. Also in 1971, the peak winter count in the Perry Oaks/Staines Moor area was 42. There was a large drop in wintering numbers between 1973 and 1974 with just 17 along the Thames. Numbers continued to fluctuate in subsequent years with the maximum count never exceeding 55. They were first seen in Inner London in 1975 when there were three in Surrey Docks on 7 February.

At the beginning of the 1980s up to 50 were still wintering at Rainham and there were up to 23 in the Staines area. By 1984 the wintering numbers had decreased to a maximum of 10 at Rainham and 17 in the Staines area.

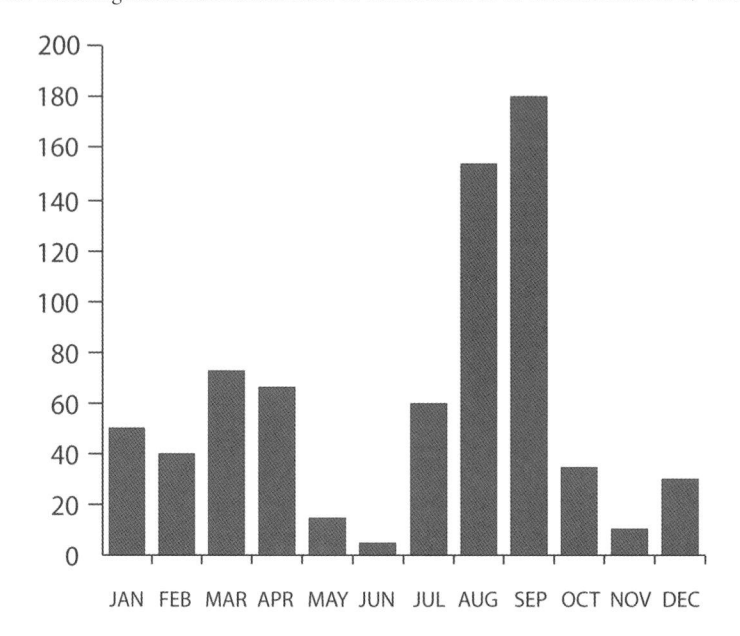

Seasonal occurrence of Ruff: 2000–2009.

In 1989 two breeding plumage males were present at Rainham Marshes in June and a juvenile was seen there later in the month and into July; it was possible that breeding took place.

In 1990 there were wintering flocks of up to 12 at Crossness and five birds on the drained north basin of Staines Reservoir; up to 15 wintered at Walton Reservoir in 1990/91. In 1992 a pair summered at Rainham Marshes and may have bred. By the middle of the decade the wintering numbers had greatly decreased; there had also been a large reduction in the numbers seen on passage. In 1998 a male summered at Rainham.

21st century

By the start of the 21st century they had almost ceased overwintering in the London Area, although one or two were present at Rainham Marshes during most winters. Since then they have become an uncommon passage migrant with only occasional flocks seen, such as one of 25 birds at Rainham on 22 March 2002; 21 at Staines Reservoir on 11 August 2004; and 14 at Queen Elizabeth II Reservoir on 26 February 2003.

Facts and Figures

The largest flock was of 120 birds at Dartford and Erith Marshes on 1 March 1970.

Jack Snipe *Lymnocryptes minimus*

An uncommon winter visitor

Range: breeds northern Europe and Siberia, winters Europe and North Africa.

Historical

During the 19th century Harting typically recorded these returning in the first week of October and he knew several locations where he could expect to see them. They were known from Golders Green and the lower part of Hampstead Heath, where Bond and Harting would go to shoot Common and Jack Snipe; on one occasion Harting flushed six from about half an acre of marsh. In south London they were known from Banstead, Epsom Common, Mitcham, Upper Norwood and Wimbledon Common; at the last site they were seen less frequently after draining took place in 1874.

In central London one was picked up on Horse Guards Parade in October 1869 and another was picked up in January 1894 on the premises of the Bank of England.

20th century

At the beginning of the century they wintered in good numbers on the marshes on both sides of the Thames, including 16 at Greenhithe. Away from the river they were scarce but were reported from Chingford, Edmonton, Epping, Gatton Park, Haileybury, Ilford and Walthamstow.

In the 1930s Glegg stated that it was 'a winter resident in small numbers'; he commented that although it was scarce in comparison to Common Snipe, it was not as unusual as it appeared to observers as it was often shot rather than seen. At Moor Mill, one summered in 1948 and one was seen the following year on 9 July.

In the 1950s Beddington attracted the highest numbers; at the start of a very cold spell 42 were seen on 31 January 1954; at around the same time, 40 were at Epsom SF on 23 January. The species was mostly recorded at sewage farms; elsewhere, the birds wintered in small numbers at Ruislip Lido, Staines Moor and in the Lea Valley and were occasionally encountered at other sites. In the cold winter of 1956 a total of almost 200 was counted at five sewage farms, including 73 at Epsom and 57 at Beddington. There was a smaller influx during the severe winter of 1962/63, with peak counts of 47 at Elmers End and 20 at Beddington.

In the early 1970s they regularly wintered on the Lower Thames Marshes, including up to 25 at Swanscombe, 15 at Rainham, and 10 at Littlebrook and Dartford. During the 1980s fewer were seen with a maximum of 11 at both Hersham SF and Swanscombe Marsh. In 1984 one summered at Stanstead Abbotts GP and was even seen displaying. A spell of freezing weather in 1991 brought an influx into the London Area with a total of 22 at

10 sites on 10 February. Another cold weather-related influx occurred in January 1996 with a peak count of 12 at Swanscombe Marsh.

21st century

Since the beginning of the 21st century the largest numbers have occurred between 2000 and 2002; the maximum count was 32 at Rainham Marshes on 10 October 2002. Since then, wintering numbers have been quite low, with no large counts. They are most regularly seen at Beddington, Rainham and the Wetland Centre.

Facts and Figures

The largest count was 73 at Epsom SF on 11 February 1956.
The earliest migrant was on 15 September 2002 at Rye Meads RSPB.
The latest migrant was on 28 May 1992 at Dagenham Chase NR.

Snipe *Gallinago gallinago*

A common winter visitor and former breeder

Range: breeds Europe and Asia, winters Europe and Africa.

Historical

The earliest reference was in the 1720s when they were shot in open fields around Hanover Square. They were also shot around St Martin's-in-the-Fields prior to 1815; at Five Fields, now Belgrave Square, prior to 1824; on the site of what is now Sloane Street in the 19th century; and on marshes where Vauxhall Bridge now stands.

After these central London sites were developed they were mainly winter visitors to the outer districts of London with birds occasionally found as close to the city as Hampstead Heath. Harting stated that they were 'tolerably common, and may be found at the brooks, ponds and wide ditches, as well as upon any marshy ground'. In 1878 and 1881 there was a flock of 400 at Norwood in south London.

20th century

In the 1930s Glegg commented that they were resident but much more numerous during winter when they could be found at many sites, particularly at sewage farms; he believed it was the creation of these that had led to an increase of Snipe wintering in the London Area as they provided suitable feeding areas. Flocks of 100–200 were regular at the sewage farms at Beddington, Elmers End, South Harrow and Watford. Similar numbers were occasionally reported at other sites and in 1945 there were 500 at Watford SF on 14 December.

By the 1950s up to 400 were regularly wintering at Beddington SF and on 24 January 1954 about 600 were counted there, and there was a sudden influx in December 1955 when the numbers reached about 1,500. Elsewhere, up to 500 wintered at Elmers End SF in 1959. In early 1963 there was a huge influx at Beddington during very severe weather and about 3,000 were present on 6 January.

The wintering population had declined by the beginning of the 1970s and no site held over 200 birds until 1977 when there were 300 at Sevenoaks in December. The decrease continued during the 1980s apart from in two years when the drained north basin of Staines Reservoir provided excellent habitat and up to 500 wintered there. There were also about 500 at Dartford Marsh in January 1988. The peak counts decreased again during the 1990s when the winters were fairly mild, apart from in January 1997 when a cold-weather influx brought 400 to Swanscombe Marsh.

21st century

Since 2000 there have been far fewer wintering Snipe in the London Area and the maximum count was 205 at Rainham Marshes on 10 October 2002. Most of the other large flocks have also been on the Lower Thames Marshes. Elsewhere, they are a widespread passage and winter visitor in small numbers and are very occasionally seen in the parks in central London.

Breeding

At the start of the 19th century it is likely that Snipe were fairly common breeders in the extensive marshes that bordered the Thames. They probably continued to breed there until most of the marshes were drained and developed. In 1812 many nests were found in the osier beds bordering on the Surrey Canal in the area that is now South Bermondsey. In the Hainault Forest area, breeding occurred until at least 1878. Snipe also bred in south London at Epsom Common, Godstone, Leatherhead, Mitcham and Wimbledon, and probably also in Richmond Park prior to draining in about 1860.

In the early 20th century they began to breed at more inland sites as suitable habitat was created, particularly at sewage farms. During the 1930s they bred in small numbers at various locations including the Colne, Darenth and Lea Valleys, Aldenham Reservoir, Haileybury, Romford and sewage farms at Beddington, Edgware and South Harrow. Up to six pairs nested near Cobham until 1938 when the land was drained. In 1941 they bred in the Lea Valley at Broxbourne and Wormley, and also at Great Parndon where they were decreasing due to drainage. A few pairs bred annually on Staines Moor.

In 1955 the breeding population peaked at about 20 pairs. However, after that it rapidly declined and just two years later there were only five breeding pairs. After the cold winter in 1963 there were no breeding records from anywhere in the London Area. There was a very slow recovery and one pair bred in 1964, followed by two pairs in 1965. By 1969 there were four breeding pairs, and displaying males at three other sites.

During the 1970s the known breeding population was typically just one or two pairs a year, mostly on the Lower Thames Marshes, apart from in 1972 when breeding pairs or displaying birds were seen at eight sites. Away from the Thames breeding occurred in the Colne Valley and displaying birds were seen at Brent Reservoir in 1975. There was very little change during the 1980s but the breeding success was low and a pair that bred at Bell Lane GP in 1986 was the first confirmed breeding since 1979.

Between 1990 and 1993 the annual breeding population was about four pairs; this increased to six in 1994. Breeding took place in most years during the 1990s at Rye Meads. Snipe have not bred in the London Area in the 21st century although pairs were present in 2000 at Rainham and Rye Meads, and again at the latter site in 2002. In subsequent years there were only occasional sightings in summer.

Great Snipe *Gallinago media*

A rare vagrant (19)

Range: breeds Europe and Asia, winters Africa.

In the 19th century this was a regular autumn migrant in small numbers in the eastern counties of England but relatively rare in the London Area. Most records were of single birds that were shot, apart from two at Chertsey Meads which were flushed and then seen flying north over the Thames. There were four records from Brent Reservoir, which was regularly visited by shooters at the time. In 1847 Ruegg stated that it was rare in Woolwich but gave no further details.

In the first half of the 20th century the only record was of one in Feltham; the observer, writing in February 1913, stated that it was seen almost daily for a few weeks although it is not known whether it was present at the time of writing or if it related to a previous year. There were no further records until 1959 when two were seen in the Lea Valley.

These were followed by a single record in each of the last four decades of the 20th century. In the winter of 1962/63 one was present near Denham for two months although it was only seen on five dates. In September 1975 one was present at Sevenoaks GP for almost two weeks. In 1989 one spent two days on the edge of the lake at Dagenham Chase; this was the first June record in the UK for over 30 years.

There was an influx into the country during the autumn of 1996 when 16 were seen, including one at Rye Meads which was seen by a large number of people during an organised flush.

Pre-1838: near Harrow, shot on unknown date
1842: Brent Reservoir, shot
Pre-1847: Woolwich
Pre-1850: Chertsey Meads, two in autumn
1851: Hendon Fields, Brent Reservoir, shot in November
1856: Brent Reservoir, shot in late autumn
Pre-1866: Bushy Heath, shot
Pre-1866: Edmonton, shot
1866: Brent Reservoir, shot in August
1889: Forest Gate, shot in November
1897: near Waltham Abbey, shot on 27 February
Pre-1913: Feltham, one for several weeks
1952: Moor Mill, Radlett, on 22 March
1959: Ponders End SF/Girling Reservoir, on 12–17 September
1959: Rye Meads, on 26 December
1962: Denham, on 23 December–17 February 1963
1975: Sevenoaks Wildfowl Reserve, on 3–15 September
1989: Dagenham Chase, on 1–2 June
1996: Rye Meads, on 30 September–1 October

Long-billed Dowitcher *Limnodromus scolopaceus*

A very rare vagrant (2)

Range: Americas; vagrant to UK.

The first confirmed record of this species was in 1977 when one was found on 1 October at Staines Reservoir; it remained for just over two weeks. Ten years later another one was seen in the Staines area; it was first found on King George VI Reservoir on 4 October. Despite its long stay of over five weeks it was seen by relatively few people due to its unpredictable nature; it repeatedly flew off from the reservoir to feed at either Staines Moor or Perry Oaks SF.

1977: Staines Reservoir, on 1–16 October
1987: King George VI Reservoir/Staines Moor/Perry Oaks SF, on 4 October–10 November

Dowitcher sp. *Limnodromus* sp.

There are two 19th-century records of dowitchers shot in the London Area at a time when they were known as Red-breasted Snipe. They were not split into two species until 1932 so if the specimens cannot be traced no old records can be assigned to either species. The Battersea specimen was in Bond's collection but was sold and its whereabouts are unknown. The other specimen was later identified as a Short-billed Dowitcher.

About 1849: Battersea, shot on the banks of the Thames

Woodcock
Scolopax rusticola

An uncommon winter visitor and localised breeder

Range: Europe and Asia.

Historical

The earliest reference was in the 1720s when General Oglethorpe frequently shot them in open fields on the site of what is now Hanover Square.

In the 19th century Harting described them as 'formerly not uncommon in winter, but now scarce' which he put down to increased cultivation, hedge trimming and the clearing of ditches. He added that they were more frequently seen in the wooded neighbourhoods of Edgwarebury and Stanmore, and that Wembley Park used to be famous for them. In Bishops Wood, near Northwood, up to eight had been seen in one day; this was clearly a regular site as an area adjacent is still called Woodcock Hill. In November 1879, 14 were seen in Oxhey Woods and on 13 November 1880, Harting witnessed a 'very remarkable flight' of about 50 in a wood near Enfield.

Breeding was first recorded in 1839 by Yarrell who had been given an egg taken near Ruislip; he also recorded breeding at Streatham. In 1850 young were taken from a nest at Kenwood, Hampstead, and also in Highgate. At Park Wood, Ruislip, breeding was regular in the 1880s but ceased before the turn of the century after the undergrowth was removed.

20th century

In the early part of the century breeding was recorded from a number of sites in the rural areas. During winter they were present in large numbers at regular haunts. For example, on one day in 1916, 10 were seen and six shot on Stanmore Common, while at Thorndon up to 70 were seen in the winter of 1913/14. In the winter of 1926/27 a group of six spent several days at Dormer's Well in Southall.

Glegg knew Woodcock as a winter visitor and passage migrant that had bred on occasions. It was most frequently recorded during October. Glegg added that a considerable proportion of records had come from central London. At Copse Wood, 41 were shot in 1936.

In the 1950s territorial birds were reported from a number of locations, mostly on the North Downs. The only location north of the Thames where they bred regularly was Broxbourne Woods. On one estate near Hertford about 30 were shot every winter. On 6 December 1952 a flock of six flew east at Stone Marshes in thick fog. In 1956 one was grounded by a firework on 5 November and found in Southwark; it was later released in St James's Park. On two occasions birds flew into the arms of passers-by in the Royal Parks!

During the 1960s at least 10 pairs were thought to be breeding in Broxbourne Woods and a minimum of six were roding at Limpsfield Chart. In 1969 they probably bred in Richmond Park. There was an influx in December 1981 following heavy snow, with counts of six at Berwick Ponds and five at both Bromley Common and Darenth.

Because of their mostly nocturnal habits it is particularly difficult to ascertain their breeding population. The two breeding Atlases found virtually no change in their distribution with breeding activity recorded in 48 tetrads from 1988-94. Only three of these areas were within Greater London with most birds being found in the large wooded areas and heaths at the northern and southern extremities. Cold weather at the start of 1997 brought an influx of at least 80 during January.

21st century

Since 2000 about a dozen roding birds have been reported annually but the breeding population may be double that as it is a difficult bird to accurately census. Outside the breeding season about 50–100 are seen most years with peaks in March and November. Migrants are frequently recorded in central London.

Facts and Figures

Unseasonable birds were seen in central London in Hyde Park on 19 June 1968; in Regent's Park on 1 June 1985; and in Kensington Gardens on 8 July 1996.

Black-tailed Godwit *Limosa limosa*

An uncommon passage migrant; winter visitor on the Lower Thames

Range: breeds Europe and Asia, winters Europe, Asia and Africa.

Historical

This was a rare breeding bird in England in the 19th century and a very rare passage migrant in the London Area. There were just four records, three of which were of birds seen or shot in spring at Brent Reservoir prior to 1843; the other was killed in the spring of 1895 on Epsom Common.

20th century

In the first two decades there were only two records: one bird shot at Edmonton SF in 1911 and a flock of seven on the mud at Surrey Docks in 1917. Between 1927 and 1944 there were another eight records.

After 1944 they were seen almost annually. This was a reflection of the general increase of this species as a passage migrant in south-east England. The spring passage at this time was from mid-March to late May, and in autumn from the end of June to the end of August. The largest flock was of 13 birds at Elmers End SF on 15 March 1947. The first winter records occurred in 1950 when one remained at Perry Oaks from 7 October into early December when it was shot; and there were two at Dartford and Stone Marshes on 16 December.

By the start of the 1970s they were still quite scarce with only a few records a year during the spring and autumn passage, but in subsequent years several large flocks were seen. In July 1970 up to 16 were present at Rainham Marshes, and in the autumns of 1977–78 there was a good passage through Staines Reservoir when the south basin was drained, peaking at 22 on 7 August. In 1985 a flock of 36 flew north over Brent Reservoir on 29 April and in 1989 there were 21 at Staines Reservoir on 27 July. At the beginning of the 1990s passage birds were being seen in increasing numbers, especially at Staines Reservoir; the peak count was 23 on 4 August 1994.

In 1997 a flock of 26 was at West Thurrock on 21 December, heralding the start of a large wintering flock on the Lower Thames Marshes; by 1 February there were a record 144 at Grays. Large numbers returned in 1998/99 with 82 at West Thurrock in November.

21st century

In 2000 there was a distinct lack of wintering birds compared to previous years. However, they returned in 2001/02 when up to 50 were on the Lower Thames Marshes. In 2002/03 up to 180 wintered at Swanscombe and a count between West Thurrock and Grays produced a total of 245 on 10 December. Since 2005 there have been increasing numbers wintering at Rainham Marshes; in December 2005 there were 300, within three years the peak count was 500 and by 2011/12 this had increased to more than 600. Some of this increase is due to climate changes in Iceland where the race islandica breeds which has significantly increased. Essex estuaries hold 26 per cent of the British population and 5 per cent of the Icelandic population, so clearly this has boosted wintering totals along the River Thames in east London.

Elsewhere, they are only encountered on passage, usually in small numbers, but larger flocks are occasionally seen. In 2003 two flocks totalling 56 birds were at the Wetland Centre on 15 August; and in 2004 there was a good autumn passage through King George V and Staines Reservoirs, with a peak of 50 at the latter on 10 August. One that flew up the Thames past Westminster on 15 July 2009 was the first record for Inner London.

Facts and Figures

The largest flock was of 620 birds at Rainham Marshes on 24 January 2012.
Colour-ringed birds of both the nominate race and the Icelandic race were seen in 2004.

Bar-tailed Godwit *Limosa lapponica*

An uncommon passage migrant

Range: breeds Arctic Circle, winters coastal western Europe, Middle East and Africa.

Historical

This species was much more common in England than Black-tailed Godwit in the 19th century; Harting noted it as 'a scarce and uncertain visitant'. It was more plentiful in milder winters on coasts but rare inland. There were only six records for the London Area, all from Brent Reservoir, and they were all of spring passage migrants. The first was recorded by Bond in 1843 and this was followed by one shot in May 1851; four on 29 April 1863; at least one on 29 April 1864 and in May 1866 (there were six on one of these dates); and one shot in May 1872.

20th century

There was a gap of over 40 years before the next record, when a flock was heard flying over Theydon Bois in the autumn of 1916. By 1954 there had been a further 18 records, three of which were in winter; the remainder were roughly equally split between spring and autumn. The three winter records were at Staines Reservoir on 24 January 1933; in Richmond Park on 21 February 1936; and at Stone in February 1954. The spring migrants were recorded between 12 April and 11 May, and in autumn between 7 July and October, with most being in September. The only passage migrants on the Thames were two at West Thurrock on 12 April 1947 and one at Stone Marshes in September 1947; all of the rest were recorded at reservoirs and sewage farms. During very cold weather in February 1956 a flock of 15–20 was seen at Queen Mary Reservoir.

In the 1960s just small numbers were seen in spring and autumn with occasional records in winter, and this pattern continued into the 1970s. In 1974 about 40 flew north-east over Kensington Gardens on 23 April, and in 1978 there was a good spring passage at Staines Reservoir with a peak of 44 on 30 April.

In 1980 there was a large spring passage of some 360 birds between 24 April 11 May, including a flock of about 120 north over Walthamstow Reservoirs on 3 May; the same day three smaller flocks totalling 95 birds were seen elsewhere flying up the Lea Valley; and about 80 flew over Staines Reservoir on 26 April. In 1992 there was an exceptional passage on 30 August when more than 120 birds in three flocks flew south along the Lea Valley, the largest of which comprised at least 50 birds over Girling Reservoir. In 1999 a flock of about 150 flew north-west over Amwell on 30 August. Winter records are rare but in February 1991 there were four records of up to four birds associated with severe weather.

21st century

Bar-tailed Godwits are more frequently seen in spring than in autumn as the chart on page 183 shows. At Brent Reservoir, 29 April is known as 'Bar-wit Day' as this is typically when spring migrants are seen there, although the peak day in the London Area can be anytime in the last week of April or the first week of May. In autumn 2004 a

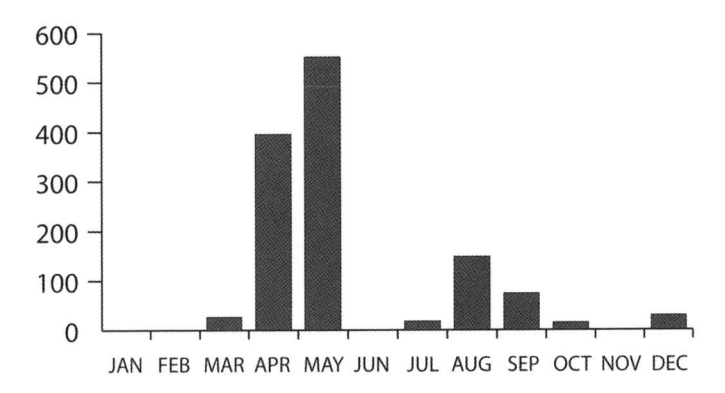

Seasonal occurrence of Bar-tailed Godwit: 2000–2009.

flock of 100–110 flew over Staines Reservoir on 10 August. Occasionally almost none are seen on passage and in 2005 there were only 27 all year. Yet two years later, in the spring of 2007, there was a large passage, particularly on the Lower Thames Marshes, where perhaps several hundred were seen and the single highest count was 73 on 2 May.

The principal reason that this species occurs in the London Area in spring, and also why it fluctuates quite widely from year to year, is because it merely mirrors the spring up-Channel easterly passage of birds that winter in west Africa and are migrating to their breeding grounds in Siberia. The volume of passage varies annually, often down to local weather conditions, so this is reflected in the London Area.

Facts and Figures

The largest flock was of about 150 birds over Amwell on 30 August 1999.

Whimbrel *Numenius phaeopus*

An uncommon passage migrant

Range: breeds northern Europe and Asia, winters southern Europe, Africa and Indian Ocean; passage migrant in UK.

Historical

This was a rare visitor in the 19th century with few documented records. Harting listed three, all of which were of birds shot at Brent Reservoir: in spring 1850; in autumn 1859; and on 10 May 1866. Elsewhere, there was one at West Molesey on 8 October 1879; five over Walton Heath on 14 May 1890; and 20–30 on a ploughed field at Epsom on 29 September 1894. There were also seven records from Dulwich between 1875 and 1900.

20th century

Between 1900 and 1954 there were around 100 records, about half of which were in the last 10 years of that period. In spring, most birds were seen in May with just a few in April. There were two records in June and autumn migrants were recorded between July and early September, with a peak in the second half of August. Most records were of single birds but small parties of up to seven were seen on occasions and there were also two large flocks: about 50 birds flying south over Colley Hill, Reigate, on 24 August 1946, and 20–30 over Wandsworth on 23 August 1953. The first documented records in Inner London were in 1934 when one flew over Campden Hill on 22 May and another flew over Kensington on 28 September.

In the spring of 1957 there were two large flocks in the Lea Valley: 28 birds at Sewardstone on 25 April and 20 at King George V Reservoir on 23 April. During the 1970s the largest flocks were all in the autumn: 28 birds at Barn Elms Reservoirs on 6 August and 19 over Staines Reservoir on 22 August 1972; and two flocks totalling about 50 flew over Dulwich Golf Course on 18 August 1973. There was a large increase in the number recorded in the 1980s. Flocks of 10 or more were seen more frequently and the highest counts were a flock of 75 over Northfleet on 5 August 1981; three flocks totalling 62 at Staines Reservoir on 6 May 1984; and a flock of at least 75 south over Beddington on 5 September 1988. Additionally, 18 flew over Stone Marsh on 22 June 1986; midsummer records are very rare in the London Area.

In 1991 there was a particularly strong passage at Beddington with 131 birds recorded between 28 April and 8 May, including 121 on the last date. Although they are occasionally seen flying over Inner London they have only once been seen on the ground: in 1993 three were on the football pitches in Regent's Park early on the morning of 30 April.

21st century

Since 2000 the average annual total has been 150; the lowest was 51 in 2001 and the highest was 281 in 2009. The largest counts in autumn were 41 over Beddington on 19 August 2005 and three flocks totalling 41 over Staines Reservoir on 21 July 2007; while the peak passage in spring was on 23 April 2012 when a flock of 33 flew over Wanstead Flats.

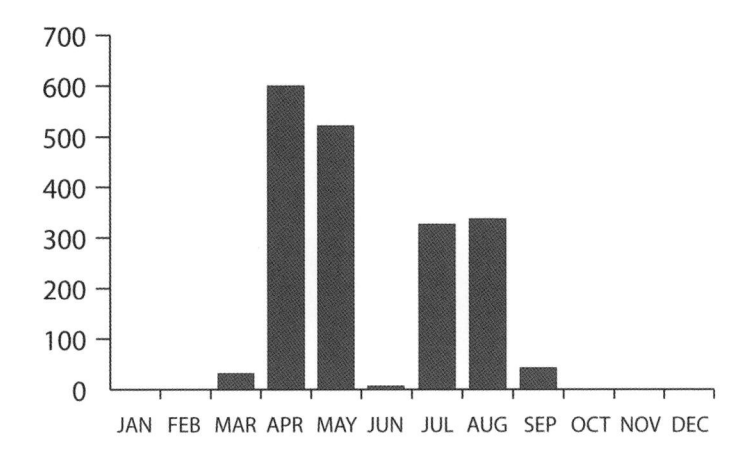

Seasonal occurrence of Whimbrel: 2000–2009.

Facts and Figures

The largest flock recorded was of 75 birds over Northfleet on 5 August 1981 and over Beddington on 5 September 1988.
The earliest date was 9 March 1998 at Silvertown.
The latest date was 30 November 1971 at Hilfield Park Reservoir.

Curlew *Numenius arquata*

An uncommon passage migrant; winter visitor on the Lower Thames

Range: Europe and Asia, winters Europe, Africa and Asia.

Historical

In the mid-19th century, Harting described the Curlew as 'occasionally observed during the periods of migration in spring and autumn' but gave no specific details of numbers or locations. They were also recorded on Hackney Marshes and around Epping Forest. In south London they were seen at Chertsey during a winter prior to 1840; one was shot at Thames Ditton on 10 December 1880; two flew over Walton-on-Thames on 18 January 1892; and a flock of five or six was at Epsom Downs in early October 1894.

20th century

Glegg stated that 'the Curlew is an irregular visitor', adding that it was most frequently recorded in spring and autumn, especially in August. He noted that single birds and small parties had been seen, although some observers had witnessed large groups passing overhead during migration, some of which were over central London. In the early hours of 21 August 1936 a flock of 25 flew over Barn Elms. Migrants were seen or heard over London at almost every time of day or night and in the late 1920s they were heard calling at night every autumn as they flew over north-west London, especially in late August. During the cold winter of early 1947, almost 70 were recorded in the London Area from eight locations. In 1949 a flock of 19 landed briefly on King George VI Reservoir on 20 August. There was no evidence of any change of status during the first half of the 20th century.

By the 1950s it was considered to be a frequent passage migrant; in winter it was much scarcer and only found in a few localities along the Lower Thames Marshes. Most migrants were recorded flying over or stopping for short periods but occasionally stayed for longer periods on the Thames marshes. In 1956, during severe

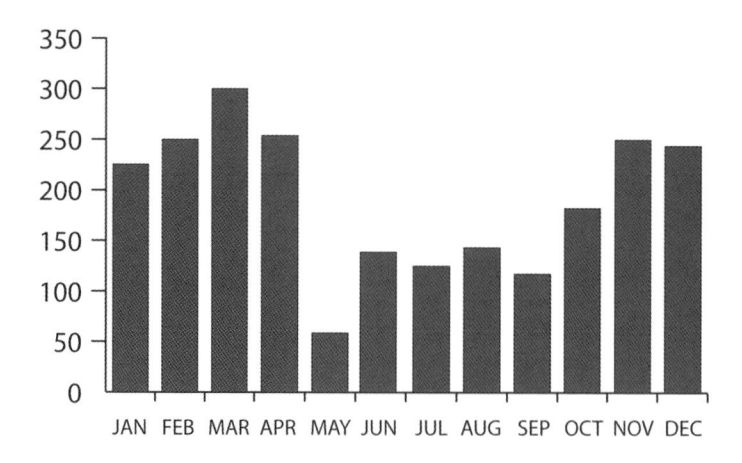

Seasonal occurrence of Curlew: 2000–2009.

weather, a flock of 54 was present at Beddington on 31 January and 1 February. In 1961 about 200 flew over Barnehurst in two hours in the afternoon of 17 September. In 1962 about 50 flew over Ashtead at 23:00 hrs on 4 October. There was an influx during the severe winter of early 1963 when they were recorded from 15 locations in January; more were seen the following month along the Thames, with peaks of 45 at Stone on 3 February and 38 at Rainham Marshes on 12 February.

In the winter of 1972/73 large numbers were found on the Lower Thames with peaks of 43 at Swanscombe in November and 35 at Purfleet in December; although such numbers were not repeated in 1973/74, small flocks were present in subsequent winters. An analysis of migrant Curlews away from the river from 1970 to 1973 revealed that the peak month was August, as might be expected, but the second highest number was in June. In 1978 a flock of 20 flew north over Surrey Docks on 29 June.

By the start of the 1980s Curlews had ceased wintering on the Lower Thames. In 1980 there was a bizarre record of one that remained in Brompton Cemetery for three days from 12 to 14 November; as it ran for cover rather than flying when approached it may have been injured. By 1990 the number of sightings had reduced and the species was less frequently seen than Whimbrel. However, during winter there was a small population of up to 12 on the Lower Thames Marshes, most often at Rainham. In February 1991 there was an influx due to severe weather with a maximum of 33 at Rainham Marshes. After 1991/92 only a few were present in winter along the Thames. In 1996 a flock of 24 flew north-east over the Lea Valley on 12 April.

21st century

Curlew remained scarce in winter until 2002/03 when there were more than usual on the Lower Thames, with a peak of 50 at West Thurrock on 6 March. Since then about 30 have wintered along the Lower Thames Marshes; away from the river they are uncommon, usually seen on passage in small groups or singly. Although the peak passage for many migrant waders is in May, this is the month with the fewest records for Curlew.

Facts and Figures

The largest-ever passage was of about 200 birds over Barnehurst in two hours on 17 September 1961.
The largest flock was of 54 birds at Beddington on 31 January 1956.

Common Sandpiper   *Actitis hypoleucos*

A common passage migrant and uncommon winter visitor; former breeder

Range: Europe and Asia, winters Europe, Africa, southern Asia and Australasia.

Historical

This was a common summer migrant in the 19th century, generally arriving in about the first week in May and returning at the end of July or the beginning of August, and remaining until the middle of September.

20th century

By the 1930s spring migrants were arriving earlier, with most being recorded in April rather than May. In autumn, most birds were noted from the end of July to mid-September with a few in October and one late bird on 2 November 1929. The largest number seen was 35 at Romford SF on 13 August 1938.

Migrants were seen in most years in central London in the Royal Parks; prior to 1925 they were usually recorded in spring. Since then, they have been recorded with increasing regularity in autumn when they are seen more frequently than in spring.

By the middle of the 20th century they were recorded very frequently on passage, and autumn concentrations of up to 30 could be found at some reservoirs and sewage farms and on the saltings along the Thames at Northfleet and Swanscombe. Peak counts continued to increase with 40 at Swanscombe Marshes on 15 August 1961; 57 at Girling Reservoir on 29 July 1963; 86 at Perry Oaks SF on 6 August 1978; and 100 at Rainham Marshes on 5 August 1982.

21st century

This species breeds mainly by streams, lochs and reservoirs in upland areas of Britain, and numbers are thought to have declined in recent years. They are widely seen on passage at waterside locations across the London Area, though since the start of the 21st century numbers have generally been lower than in the past However, there were at least 50 at King George V Reservoir in August 2012.

Wintering

The only winter record in the 19th century was at Feltham on 11 January 1894. There were no more until one was at Wallington in February 1914. The first one to overwinter was at Walthamstow Reservoirs in 1927/28; from then on the species was recorded almost every winter, typically at reservoirs and gravel pits, and along the Thames. At Stone Marshes, up to three wintered in 1949/50 and in the following winter up to five were present between November and March.

By the start of the 1970s small numbers wintered on the Thames marshes, at Perry Oaks and at a few reservoirs. In 1991 up to six wintered at Rainham Marshes. At the beginning of the 21st century at least 10 wintered in the London Area. The wintering population has slowly increased since then with 16 present in 2008/09, most of which were on the Lower Thames Marshes or the large reservoirs.

Breeding

The only confirmed breeding in the 19th century was in the Colne Valley. Although Harting stated that they had bred on the banks of the line of reservoirs near Chiswick, Glegg did not accept this record due to a lack of supporting evidence.

In 1910 a pair bred on an estate at Hertford and hatched four young. Breeding was claimed nearby at Haileybury in 1912 and 1913 but in both cases the evidence was unconvincing.

Between 1950 and 1956 a pair summered at Old Parkbury GP. In the first four years of this period breeding was not confirmed although the bird's behaviour suggested that they may have bred. However, breeding was finally confirmed in 1954 when fledged young were seen with the two adults at the end of June; this was repeated in 1955–56.

Summering records included pairs at both Nazeing GP and Rye Meads in 1959; a pair was seen mating at Holmethorpe Sand Pits in 1962; two pairs remained throughout the breeding season at Hornchurch Chase in 1965; and another pair was seen at West Hyde on 20 June 1965.

Breeding was confirmed again in 1967 when a pair bred successfully at West Hyde and another pair nested nearby at Maple Cross, although the nest was later destroyed. Two pairs summered and were seen displaying in 1969 and a pair held territory the following summer and may have bred. In 1982 a pair attempted to nest at Lonsdale Road Reservoir but abandoned the attempt due to disturbance. In 1988 a pair summered at Amwell and may have bred. Displaying birds were seen at Amwell in June of the following year but they did not breed.

Since 2000 no breeding attempts have been reported, although single birds are occasionally present during June.

Facts and Figures

The largest ever count was 100 birds at Rainham Marshes on 5 August 1982.
The largest flock in Inner London was of 10 birds on the Thames by Lambeth Bridge on 11 May 1960.

Spotted Sandpiper *Actitis macularius*

A very rare vagrant (4)

Range: Americas; vagrant to UK.

The specimen of the first record was among a collection sold in April 1852 and purchased by Bond; the specimen is now in the British Museum. It was said to have been taken at Brent Reservoir and Harting stated that Bond 'traced it to the person who had received it in the flesh and stuffed it and fully satisfied himself of the truth of the statement'.

The next record was in September 1956 when a summer-plumaged adult spent a day at Hilfield Park Reservoir. The only two other records were of birds seen in consecutive years in the late 1980s. A summer-plumage bird was present for a few hours at Barn Elms Reservoirs in 1988 and in the autumn of the following year one remained for 12 days on King George V Reservoir; it was often seen feeding with a small group of Common Sandpipers.

Pre-1852: Brent Reservoir, shot
1956: Hilfield Park Reservoir, on 2 September
1988: Barn Elms Reservoir, on 16 May
1989: King George V Reservoir, on 25 September–6 October

Green Sandpiper *Tringa ochropus*

A common passage migrant and uncommon winter visitor

Range: breeds Europe and Asia, winters Europe, Asia and Africa.

Historical

This was a bit of an enigma in the 19th century; Harting conceded that the Green Sandpiper was one of a few species 'about which we have still something to learn'. He personally did not consider that it was the rarity that it was reputed to be, rather that it was 'a regular visitant to this country in spring and autumn'. He noted that it appeared in late April 'and may then be found, singly or in pairs, along the brook sides, where it frequents the little bays of mud and sand that are often formed where the stream winds', generally departing in the third week of May.

He recorded that it must have a very short breeding season as returning migrants were seen from the middle of July and would then be seen until early November in groups of five or six. Harting observed that 'as the autumn advances, they leave the open sheets of water, such as reservoirs and large ponds, where they may be first seen on their return, and betake themselves to the more sheltered brooks'. During this time they would often be shot for sport and food, being 'a good substitute for Snipe'. He also knew of a few rare occasions when one had been shot in winter. The earliest dated record was in August 1832 when one was shot at Epping.

20th century

By the beginning of the century they were being seen more frequently and in greater numbers. In April 1908, 10 were at Bushey Lodge Farm and by 1910 they were wintering regularly in the Roding and Colne Valleys.

In the 1930s Glegg stated that this was a double-passage migrant but could also be seen in virtually any month of the year. Typically, most sightings were of single birds or small parties. He added that it was infrequently seen in spring but was regular in autumn, with most birds occurring in August. It was most often encountered at sewage farms and Glegg believed that sightings had increased because of the creation of these.

Migrants were occasionally heard flying over after dark and in 1947 two were seen flying over Hyde Park at dusk on 3 October, the first record for Inner London. By the 1950s large groups started to be recorded, e.g. there were 20 beside Watford SF in August 1951; this area of waterlogged tussocky grassy was a particular favourite with this species and up to 12 could be found regularly in all months apart from June. Then there were 21 at Rye Meads on 26 July 1957.

In 1970 exceptional numbers were recorded in autumn at Rainham, with a peak of about 50 on 5 August; during the rest of the decade similar numbers were seen annually at Rainham. By the start of the 1980s more were wintering in the London Area, with records from 26 sites; in December 1989 there were 13 at Rye Meads and in 1993 up to 14 wintered at Beddington.

21st century

In the first decade of the 21st century the wintering population gradually increased from about 20 to 50 in 2008/09. Most of these birds were at Beddington and in the Lea Valley. However, the passage numbers have declined, especially at Rainham; the peak counts were at Beddington with 17 in spring 2000 and 42 in autumn 2002. Small numbers are seen at most wetland localities on passage, particularly in autumn.

Facts and Figures

The highest-ever count was 70 birds at Rainham on 23 July 1979.

Solitary Sandpiper *Tringa solitaria*

A very rare vagrant (4)

Range: Americas; vagrant to UK.

The first record was in autumn 1967 when one was seen along the River Stort between Roydon and Rye Meads; it was trapped and ringed. There were two in the mid-1970s, at Rainham Marshes for six days in September 1974, and three years later at Perry Oaks on 22 July. The fourth was 25 years later, when an adult remained for 12 days at Rye Meads.

1967: River Stort between Roydon and Rye Meads, on 24 September–9 October
1974: Rainham Marshes, on 1–6 September
1977: Perry Oaks SF, on July 22nd
2002: Rye Meads, on 13–24 September

Spotted Redshank *Tringa erythropus*

An uncommon passage migrant

Range: breeds northern Europe and Asia, winters Europe and Africa.

Historical

This was a rare visitor in the 19th century with only five records, all from north-west London. Five birds were shot at Brent Reservoir: in June 1841, August 1859, two on 25 May 1865 and at least one in August 1881. Additionally, one was shot on the River Brent at Stonebridge in either 1848 or 1849.

20th century

There were no further records until 1928 when an immature was at Staines Reservoir on 23 September. In the 1930s there were two at Brooklands SF on 8 May 1937 and three in the autumn of 1938, one at Barn Elms Reservoirs on 1 September and two at Brent Reservoir on 11 September.

After the Second World War they were seen more frequently but had not yet become annual visitors. Following singles in 1945 and 1947, there were four records in the autumn of 1949, including of a group of five at King George VI Reservoir on 14 August. During the 1950s and 60s they were seen almost every year, mainly on sewage farms and reservoirs in autumn; the favoured site was Perry Oaks. In 1969 there was an influx of more than 30 in autumn; the peak count was 11 at Rainham Marshes from 28 to 30 August.

In both 1976 and 1978 there was a good autumn passage but on opposite sides of the London Area. In the first year they were present almost daily at Staines Reservoir between 9 August and 4 October with a peak of eight on 13 August. In contrast, in 1978 they were regular at Rainham on many dates between 23 July and 21 October, the peak being eight in mid-September. There were a few short-staying winter birds and in 1984/85 one stayed at Staines Reservoir from 13 November to 3 January. There were two in Inner London during the 1990s, on the Thames at Chelsea Reach on 6 May 1990 and in Regent's Park on 24 July 1999. An analysis of records between 1983 and 1992 showed that the peak passage was in the last week of August.

21st century

Since the beginning of the 21st century they have become much scarcer, with an average of fewer than 10 a year. The only exception was in 2002 when there were two reasonably sized flocks at the Wetland Centre: 12 birds on 7 May and eight on 20 November. One overwintered in 2011/12 at Crayford Marshes.

Facts and Figures

The largest flock was of 12 birds at the London Wetland Centre on 7 May 2002.

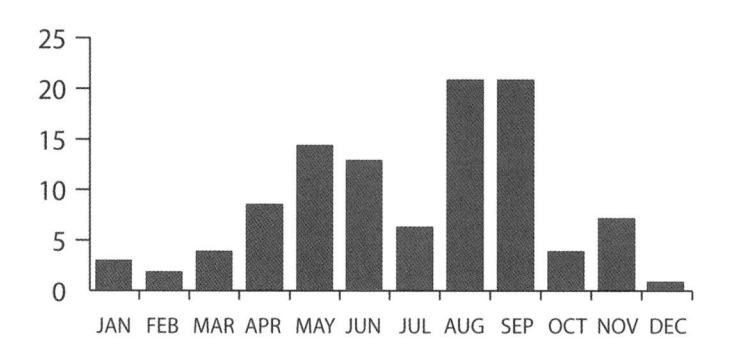

Seasonal occurrence of Spotted Redshank: 2000–2009.

Greenshank *Tringa nebularia*

A common passage migrant

Range: breeds northern Europe and Asia, winters coastal Europe, Africa, Asia and Australasia.

Historical

Harting wrote that this was 'a rare and uncertain visitant in spring and autumn' and listed a number of records, most of birds which were at Brent Reservoir, including a flock of seven on 1 August 1863. The only known records from south London were of one shot on the Thames at Walton in June prior to 1842; one over Brixton on 6 September 1880; and one killed at Mitcham in the autumn of 1895.

20th century

In the first quarter of the century Greenshank remained a relatively rare passage wader with only about five records, some of which were of birds shot at Edmonton SF. By the mid-1930s it was being seen much more regularly on passage, notably in the Lea Valley, in the Colne Valley, along the Thames and at Brent Reservoir. Glegg described it as 'an irregular passage-migrant in small numbers', adding that it was very scarce in spring. In autumn it was seen more frequently, although not on an annual basis. The highest count was 11 at Brent Reservoir on 2 September 1928.

In the 1950s it was regular in the Staines area, mainly at Perry Oaks SF, and also on the reservoirs and at Staines Moor; in some years the species was seen almost daily in autumn. In 1957 there were 14 at Rye Meads on 18 September. The autumn passage across the London Area during the 1950s was much longer than Glegg recorded two decades earlier; it started in July, peaked in August and continued through to October. The spring passage was more established than during Glegg's time, and took place throughout April and the first three weeks of May. Records in midsummer and winter were exceptional.

There was very little change during the 1960s and early 70s; the highest count was of a flock of 26 which flew over Brent Reservoir just after dawn on 8 August 1967. One in Regent's Park on 10 August 1968 was the first one in Inner London. In 1976 the drained Staines Reservoir provided excellent habitat for migrant waders and counts of more than 20 were regular between 20 August and 12 September, the peak being at least 29 on 31 August. The following year the maximum count was again at Staines, about 20 on 3 September.

Greenshank were particularly numerous in 1980 in both spring and autumn. At least 53 arrived at Rainham Marshes just before a thunderstorm on 26 July. There was a good passage through Staines Reservoir in the autumn of 1985, peaking at 66 on 5 September.

21st century

Greenshank numbers have declined considerably since the start of the 21st century although they are still recorded annually at most wetland sites. The largest flocks were of 35 birds over the Chingford Reservoirs on 26 August 2000 and at Rainham Marshes on 26 April 2009. One wintered at Tyttenhanger GP from 2000/01 to 2003/04, and another wintered at Rainham in 2002/03 and 2003/04.

Facts and Figures

The earliest migrant was on 13 March 1991, at Dartford Marsh.
The largest count was 66 birds at Staines Reservoir on 5 September 1985.

Lesser Yellowlegs   *Tringa flavipes*

A rare vagrant (8)

Range: Americas; vagrant to UK.

In the autumn of 1953 many American waders crossed the Atlantic and were discovered in the UK, including the first two records of Lesser Yellowlegs in the London Area. Both were at Perry Oaks SF, which was London's wader hotspot at the time. The first was found on 30 August and remained until 5 September; the following day it was discovered 4 miles away at Ham Fields SF, just outside the London boundary. It stayed there for several months and returned to the London Area on many occasions; it was seen at Staines Moor on five dates in September and October, and again at Perry Oaks on nine dates between September and November and finally on 9 December. During this time, a second bird was present at Perry Oaks on 25–26 September; this bird was also on the causeway of Staines Reservoir at dusk on the second day.

Perry Oaks also claimed the third record in 1962, of one present from 5 to 9 September; it was also seen at Staines Reservoir during its stay. There was one record in the 1970s, at Denham for three weeks in late autumn. In 1984 a juvenile was present at Beddington for a week at the end of September. The drained north basin of Staines Reservoir attracted only one rare wader in 1990, an adult Lesser Yellowlegs which was present between 21 August and 8 September; this bird also visited Perry Oaks on 25 August. In 1998 an adult was at Rainham Marshes for two days in August, and finally in 2002 an immature was present at Amwell for 11 days in October.

1953: Perry Oaks SF, intermittently on 30 August–9 December; also at Staines Moor
1953: Perry Oaks SF, on 25–26 September; also at Staines Reservoir
1962: Perry Oaks SF, on 5–9 September; also at Staines Reservoir
1977: Denham, on 15 October–5 November
1984: Beddington SF, on 24–30 September
1990: Staines Reservoir, on 21 August–8 September; also at Perry Oaks SF on 25 August
1998: Rainham Marshes, on 26–27 August
2002: Amwell GP, on 18–28 October

Marsh Sandpiper   *Tringa stagnatilis*

A very rare vagrant (3)

Range: breeds eastern Europe and Asia, winters Africa and Asia; vagrant to UK.

The first record was in the autumn of 1963 when one was found at Swanscombe Marsh on 26 August; it remained in the area for eight days, also being seen at Aveley and Rainham Marshes. Three years later another arrived on exactly the same date, at Chigwell SF. The last record was in the spring of 1984 in the Lea Valley; it was first located at Broxbourne GP and seen later in the day at Holyfield Marsh GP.

1963: Swanscombe Marsh, one on 26 August–3 September; also seen at Rainham Marshes on 26–28 August
1966: Chigwell SF, one on 26 August
1984: Broxbourne GP, one on 28 April; also seen at Holyfield Marsh

Wood Sandpiper   *Tringa glareola*

An uncommon passage migrant

Range: breeds northern Europe and Asia, winters Africa, Asia and Australasia.

Historical

This was a scarcely recorded migrant in the 19th century with at least 17 records, all but four of which were at Brent Reservoir. The earliest reference was in 1840 when Meyer illustrated one that had been shot on Ditton Marsh. The first one at Brent was noted by Bond in 1843, and between 1863 and 1878 another 12 were recorded there, mostly in autumn. Elsewhere, one was shot in Tottenham in July 1859; there was one at East Molesey on 19 September 1871; and another was shot on Hackney Marsh on 31 August 1885.

20th century

There were no further records until 1927, when one was seen at Staines Reservoir on 17 August, followed by two the following day. During the 1930s there were a further five records, three at Brooklands SF and two at Staines Moor.

In 1946 Perry Oaks SF was 'discovered' by Lord Hurcomb and on 4 May he found a Wood Sandpiper there. Perry Oaks soon became the place to see the species and in subsequent years up to three were often seen between July and September. In 1952 exceptional numbers were recorded there, up to 10 in July and 17 in August. Elsewhere in the London Area that autumn one or two were reported from a further nine sites. They were rarely recorded in spring but in 1955 seven were seen between 8 and 29 May, including four on the last date. In 1956 there was another large influx; at least 18 in spring included up to five at Perry Oaks from 19 to 21 May. In the summer of 1957 one was present at Rye Meads on 18 and 30 June.

In the autumn of 1960 several were seen later than the usual period of passage: at Romford SF one was present throughout October and up to 6 November with two there on 23 October; and another was at Hoddesdon SF on 11 October. In 1963 the highest autumn count was six at Rainham Marshes, the first time that more were seen on the Thames marshes than at a sewage farm. The first Inner London record occurred in 1969 when one was seen in Regent's Park on 7 August.

In 1985 a very late migrant was seen at Hogsmill SF from 30 November to 8 December; it was then seen at Beddington from 14 to 16 December. Higher numbers were seen on passage in 1987: there were 13 in spring and about 61 in autumn, including flocks of 18 at Rammey Marsh on 21 July and 14 at Perry Oaks on 24 July. In 1988 one was seen displaying at Beddington on 8 May, and another at Wraysbury GP from 13 to 19 May was seen in song flight on the last date. There was a large spring passage in 1989, between 12 April and 2 June with 20 birds including six at Amwell on 21 May. At the start of 1990 one wintered at Crossness between 14 January and 11 April. Then in 1997 a flock of 13 was seen at Tyttenhanger and Hilfield Park Reservoirs on 27 August.

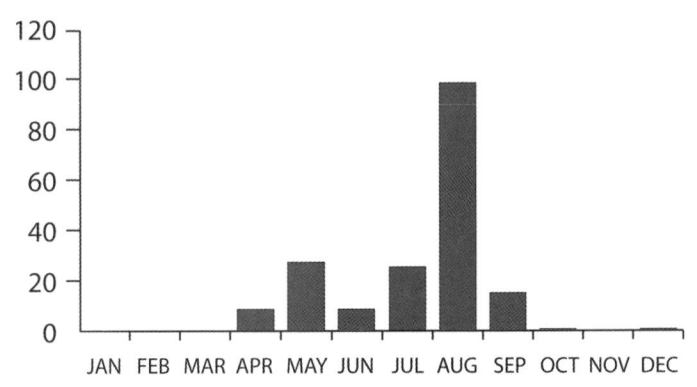

Seasonal occurrence of Wood Sandpiper: 2000–2009.

21st century

Since 2000 there has been an average of 25 a year with a low of 13 and a high of 45. The peak count was 10 at Staines Reservoir on 21 August 2004. The species is most regularly seen at Beddington, Brent Reservoir, the Lea Valley, Rainham Marshes and the Wetland Centre.

Facts and Figures

The earliest migrant was on 9 April 1960 at Barn Elms Reservoirs.
The latest bird was on 16 December 1985 at Beddington.
The largest flock was of 18 birds at Rammey Marsh on 21 July 1987.

Redshank *Tringa totanus*

An uncommon passage migrant, winter visitor on the Lower Thames and localised breeder

Range: Europe and Asia, winters Europe and Africa.

Historical

This species was described by Harting as 'an uncertain visitant, appearing occasionally at the periodical migrations in spring and autumn'. He knew of about six records from Brent Reservoir and one from Hampton so it appears to have been fairly scarce. In south London the only known records are: one shot on Epsom Common in the autumn of 1882; two shot at Mitcham in the autumn of 1895; and a flock over Brixton on 8 December 1889.

20th century

By the 1930s they had been recorded in every month of the year but were not resident. Glegg stated that they were never numerous and were most commonly observed in spring and autumn. The breeding pairs would arrive from March onwards and had generally departed by the end of August. The Lower Thames Marshes was the only regular wintering area in the 1930s.

Wintering numbers began to increase and by the start of the 1960s large numbers were wintering on the Lower Thames Marshes; in 1961 there were about 130 at both Swanscombe and Littlebrook. The following year the peak count had increased to about 250 at Littlebrook. There was an influx during the severe winter in early 1963 and small groups were widespread across the London Area, including a flock between Hammersmith and Barnes which peaked at 11 on 27 January; on the Lower Thames there were about 300 at Aveley Marsh from 12 to 15 February and about 150 at Dartford on 23 February.

In 1971 a massive count of 958 was made from a boat as it followed the Thames from Woolwich to Swanscombe on 16 January. In 1974 about 800 were counted at Purfleet in January. During the 1970s and into the early 1980s up to 26 were regularly seen in central London at the decommissioned Surrey Docks. By 1980 there were still large numbers wintering along the Lower Thames with a peak of 440 at West Thurrock. The numbers using the high-tide roost at West Thurrock rapidly increased during the 1980s with 1,000 in 1983/84, 2,000 in 1984/85 and 3,000 in November 1989.

21st century

By the beginning of the 21st century the wintering population on the Lower Thames Marshes had declined to about 400 birds. Elsewhere, Redshank occur mainly on passage in small numbers.

Breeding

From 1865 onwards there was an increase in the breeding population of Redshank in England and this eventually led to them breeding in the London Area. Breeding was recorded for the first time in the early part of the century although it is likely that it went unrecorded on the Thames marshes. Between about 1902 and 1912 the species bred annually near Enfield Chase, and in 1903 Christy stated that it was breeding as far up the Thames as Dagenham.

On the east side of London two to three pairs started nesting in the Lea Valley at Broxbourne and Nazeingwood Common from 1907 and a pair bred at Edmonton SF in 1910, leading to a gradual colonisation in the Lea Valley. Just to the west of the Lea Valley up to 10 pairs bred at Haileybury in 1915. In the Mardyke Valley around 10 pairs bred in marshy fields at East Horndon from 1908 to at least 1921; and a few pairs bred in the Roding Valley from 1908 onwards.

Elsewhere, they began to spread around the London Area. Two pairs bred in the Watford area in 1906–07; breeding occurred regularly at Beddington from 1910; a pair bred at Radlett in 1915–16 and again in 1919; a few pairs bred on marshy fields where Kent Brook met the River Eden until the land was drained in 1921; and two or three pairs bred in the Darenth Valley between Lullingstone and Westerham in about 1919. They were first recorded in Inner London on 3 February 1917 when four were seen flying around Victoria station during cold weather.

The breeding population had encircled London by 1930 but the species could only be found in the outer suburbs at some sewage farms. The breeding stronghold was the Lower Thames Marshes; for example, there were 20 pairs at Swanscombe in 1934 and at least 16 pairs bred at Erith Marshes in 1938.

The destruction of breeding habitat due to development resulted in the loss of several colonies; the East Horndon colony died out in 1939; wartime drainage led to the loss of a small colony near Harefield; South Harrow SF was converted to allotments in 1942; and in the Roding Valley drainage work reduced the number of pairs from seven or eight in 1937 to two or three in 1942, and then none after 1948. However, Redshank were quick to take advantage of new habitats and up to 10 pairs bred from 1946 to 1950 at the site where Girling Reservoir was being constructed.

By the 1950s they had become a well-established breeding bird in the London Area, particularly on the Thames-side marshes where at least 80 pairs bred, three-quarters of which were on the south side. They were also breeding in some of the other river valleys and on the remaining sewage farms, notably at both Beddington and Hoddesdon, where about six pairs bred. There was then a gradual decrease in breeding numbers and by 1963 the largest population was 11 pairs at Rainham. By 1967 numbers had decreased further to about 27 pairs, of which only seven pairs were by the Thames; the decline continued and breeding pairs had fallen to about 16 in the early 1970s.

The population recovered, though, and by 1980 had increased to about 30 pairs, including eight at both Dartford and Rainham Marshes. The species continued to prosper and by the early 1990s had reached at least 61 pairs, with 13–14 at both Rainham and West Thurrock; three pairs also attempted to breed at Beddington, the first for a decade. There was an increase in the breeding distribution of 45 per cent between the two breeding Atlases; the most notable increases were in the western half of the London Area.

By the beginning of the 21st century the breeding population had decreased to about 30 known pairs with the largest concentration being at least 16 pairs at Rainham. Redshank quickly colonised the London Wetland Centre after it opened and by 2008 there were six breeding pairs. A small number of pairs still breed annually (although not always successfully) at Staines Moor, Stanwell Moor, Horton and Wraysbury GPs.

Facts and Figures

The largest count was 3,000 birds at West Thurrock on 25 November 1989.
The largest group away from the Thames was of 80 birds at Staines Reservoir on 26 March 1983.
Birds of the race *robusta*, which breeds in Iceland, Scotland and the Faeroe Islands, may pass through the London Area regularly; two of these were shot at Sevenoaks on 10 April 1943.

Turnstone *Arenaria interpres*

An uncommon passage migrant; winter visitor on the Lower Thames

Range: breeds northern Europe, Asia and North America, winters coastal Europe, Africa and Indian Ocean.

Historical

This species was rarely recorded in the London Area in the 19th century and there were just three records, all of which were at Brent Reservoir: one in late August 1865 for a few days; one on 15 September 1866; and two shot in autumn many years prior to 1866. There is also a record of one flying over Dulwich on 2 July 1893, although this may have escaped from a nearby collection.

In 1909 Ticehurst stated that they were occasionally seen on the Thames as far up as Dartford; however, no details are known about the status of Turnstone from the Thames-side marshes in the 19th century.

20th century

Away from the river, the first record of the 20th century was from Queen Mary Reservoir on 20 August 1931. Following this, regular watching at reservoirs and sewage farms showed that Turnstone was a scarce passage migrant through the London Area.

Between 1933 and 1944 there were occasional sightings, and after the Second World War the species was seen regularly, in May and early June, then from July through to October. Usually, records were of one or two birds but the occasional small party was also seen. One of the favoured places was Perry Oaks SF where groups of up to 10 were recorded. The only record in Inner London around this time was of a lame bird seen on a bombed site in Chelsea, not far from the Thames, on 28 January 1950.

At the beginning of the cold winter of 1962/63 six were seen in a field at Elmers End SF on 28 December. In 1967 a flock of six was seen at Sevenoaks on 14 January. They first began wintering on the Lower Thames Marshes in 1980 when three were present at Broadness. Turnstone were also recorded along the Thames during migration; in 1980 a flock of 17 flew upriver past Dartford on 9 August, and in 1986 there were 19 at West Thurrock on 14 May. An analysis of all records between 1950 and 1987 showed that the peak passage was at the end of April and in early May.

During the 1990s small flocks were occasionally present on the Lower Thames Marshes in winter, with peaks of 22 at Grays on 24 January 1999 and at least 20 at Swanscombe in the winter of 1994/95. The highest counts on passage were 21 at Swanscombe on 6 May 1991; away from the Thames, a flock of 14 flew south over Staines Reservoir on 14 August 1991.

21st century

Although more typically encountered foraging amongst pebbles on stony beaches, a few migrants stop off in the London Area each year during spring and autumn; they are most often encountered along the Lower Thames Marshes and on the concrete banks of the west London reservoirs as well as other wetland sites such as Beddington and the London Wetland Centre. The largest flock of migrants was of 15 birds at Swanscombe Marsh on 8 May 2000. Wintering numbers along the Lower Thames have varied annually from two to a maximum of 23.

Facts and Figures

The largest flock was 23 at Swanscombe Marsh on 4 March 2001.

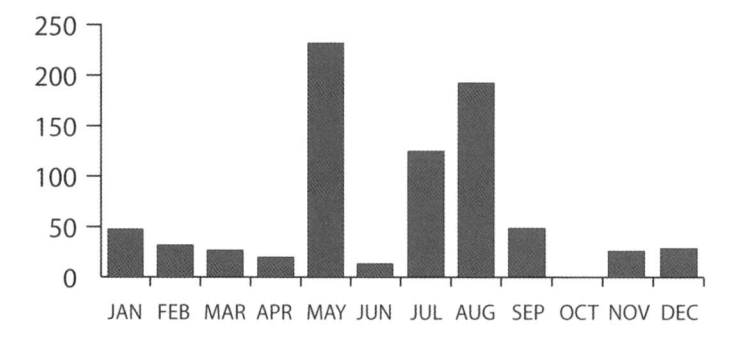

Seasonal occurrence of Turnstone: 2000–2009.

Wilson's Phalarope *Phalaropus tricolor*

A very rare vagrant (2)

Range: Americas; vagrant to UK.

There have only been two records, both at Staines Reservoir in September. The first was in 1983 when a moulting juvenile was present for almost two weeks; it arrived after a long period of westerly winds. In 1997 an immature was present on the partially drained north basin for nine days; it had almost completed its moult from juvenile to first-winter plumage.

1983: Staines Reservoir, on 14–26 September
1997: Staines Reservoir, on 5–17 September

Red-necked Phalarope *Phalaropus lobatus*

A rare passage migrant (25)

Range: breeds northern Europe, Asia and North America, winters at sea in southern hemisphere; rare breeding bird on Scottish islands.

Historical

The earliest record dates back to 1850 when one in summer plumage was shot 'whilst running between the rails near the station at Stratford' (Christy). The only other 19th-century record was of one shot on Nutfield Marsh in about 1885.

20th century

In 1930 one was possibly seen by Glegg at Staines Reservoir on 28 September; however, as its identity was disputed by other observers he decided not to count it and listed it only in square brackets. The next confirmed record was in 1949 when one was seen at King George VI Reservoir in August. Ten years later one was present at Rye Meads for 10 days in October; it was joined by a Grey Phalarope for part of its stay.

The species remained fairly scarce, with just two in the 1960s and three in the 1970s, including one in summer at Rainham Marshes on 18 June 1978. During the 1980s it started to become a more frequent visitor with another six records; all were in autumn apart from a summer-plumage female at Staines Reservoir in early June. The Great Storm was responsible for the only record in 1987; the bird was found on King George VI Reservoir on 17 October and remained there or on the adjacent Staines Reservoir until 8 November; its three-week residence made it the London Area's longest-staying bird.

The peak decade was the 1990s when eight were seen: six in the Colne Valley and two in the Lea Valley. They included a breeding-plumage female on Stanwell Moor in May and an adult male at Staines Reservoir in June and July.

21st century

There have been four since 2000, all of which were juveniles that arrived in September.

1850: Stratford, shot
1885: Nutfield Marsh, shot
1949: King George VI Reservoir, on 1–2 August
1959: Rye Meads, on 18–27 October
1960: Girling Reservoir, on 1 October; also seen at Ponders End SF
1968: Perry Oaks SF, on 10 August
1970: Godstone SP, at least one on 16 August

1974: King George V Reservoir, on 28–30 September
1978: Rainham Marshes, on 18 June
1981: King George V Reservoir, on 26 September–2 October
1983: Staines Reservoir, on 3–4 June
1984: King George V Reservoir, on 16–17 August
1986: Kempton Park NR, on 27 August
1987: King George VI Reservoir/Staines Reservoir, on 17 October–8 November
1988: Staines Reservoirs, on 14 August
1992: Perry Oaks SF/Staines Reservoir, on 30–31 July
1993: Staines Reservoir, on 14–16 September
1994: Staines Reservoir, on 29 June–2 July
1995: Stocker's Lake, on 18 September
1996: Stanwell Moor, on 9–11 May
1998: King George VI Reservoir, on 3–4 October; also seen at Staines Reservoir
1999: Girling Reservoir, on 25 August
1999: King George V Reservoir, on 20–25 September
2001: Queen Mother Reservoir, on 3–4 September
2002: Girling Reservoir, on 16 September
2003: Walton Reservoir/Queen Elizabeth II Reservoir, on 26 September–4 October
2005: Staines Reservoir, on 18–21 September

Grey Phalarope *Phalaropus fulicarius*

A scarce passage migrant, typically in autumn

Range: breeds northern Europe, Asia and North America, winters at sea in southern hemisphere; passage migrant in UK.

Historical

This was a scarce autumn migrant in the 19th century with about 20 records. Most of these birds were on the Thames, where there were 10 records, the first bird being shot at Battersea in November 1824. This was followed by two near Shepperton Lock, in December 1840 and on 2 December 1841; one shot at Swanscombe in the autumn of 1845; another shot on the River Wandle at Carshalton in mid-November 1851; one at Blackwall in November 1862; near Waterloo Bridge in September during a massive immigration in the autumn of 1866; at Erith in 1866; at Chiswick in October 1870; and near Barnes in October 1896.

Another six were shot at Brent Reservoir between 1841 and 1892. In 1847 Ruegg, writing about the Woolwich marshes, stated that 'during severe weather, we frequently obtain specimens'. Elsewhere, there was one at Hackney Marshes on 20 September 1866; one shot on the reservoir at Barnes in October 1870; killed on a brook near Redhill on 16 October 1877; killed on Wanstead Flats on 21 November 1888; shot at West Molesey on 25 October 1891; on Stratford Marshes in 1891; and undated records at Poynter's, near Cobham and at North Weald.

20th century

In the first half of the century there were about 18 records, all of which were between mid-August and December with the majority in September. The earliest of these birds were shot at Edmonton on 18 November 1905 and picked up dead near Rickmansworth in October 1909. In the years that followed, most were seen on the large reservoirs, often after gales, including a total of 14 at Barn Elms and Staines Reservoirs. Elsewhere, in 1950 there was one at King George VI Reservoir from 18 to 24 September and two at Brent Reservoir from 8 to 14 September, which all occurred after very strong westerly winds brought an influx of this species into the UK. In Inner London a phalarope on the Thames by Tower Bridge on 2 September 1951 was thought to be this species.

Between 1955 and 1961 there were almost as many as there had been in the first half of the 20th century. Part of the increase was due to a very large influx into the UK during the autumn of 1957 when five were seen in the London Area. Between 14 and 27 October 1959 another five were seen. After 1961 they became scarce again and only 10 more were recorded up to the end of the 1970s; three of these occurred in September 1974, including the first definite record in Inner London, at Surrey Docks on 11–12 September. In 1964 one was seen on the relatively early date of 18 August at Rainham Marshes.

The peak decade was the 1980s when 31 were seen. Eight of these were in 1981, including the only spring record when one in virtually full breeding plumage was on Staines Reservoir from 26 April to 1 May. In 1987 at least 15 were blown into the London Area as a result of the Great Storm on 15–16 October, which deposited many seabirds inland. Most were found in the first two days after the storm, including flocks of seven on King George VI Reservoir and three on King George V Reservoir; all had departed before the end of the month.

The first wintering record was of a bird in 1990/91 on King George V Reservoir from 16 December to 11 January. During the rest of the 1990s another seven were seen, including on the Thames at West Thurrock from 26 to 28 December 1999.

21st century

Between 2000 and 2009 there were 14, 11 of these being on the large reservoirs. Two others were on the Thames in the West Thurrock area in November and December 2006, one of them remaining until 21 April 2007. The other record concerned a bird on a small pond in Hampton Court Park from 13 to 15 November 2008.

Facts and Figures

The largest gathering was of seven birds on King George VI Reservoir on 17 October 1987.

Pomarine Skua *Stercorarius pomarinus*

A scarce passage migrant (40)

Range: breeds Arctic Circle, winters Atlantic Ocean.

Historical

This was a rare visitor in the 19th century with four records: in 1828 Yarrell stated that one had been shot on Hackney Marshes; one was shot in Harrow in 1841; one was shot at Brent Reservoir prior to 1843; and one was found dead at Wembley Park prior to 1866.

Harting noted that early in the winter of 1837/38, many Pomarine Skuas were received in the London markets for sale and among them were eight or 10 birds that had been taken alive; however, it is not known whether any of these were captured in the London Area.

20th century

There were no further records until almost 90 years later when a dark-phase adult was seen at Banbury Reservoir in 1954 for a week at the start of winter. It fed mainly on dead birds and was seen to fly away strongly on 4 December.

The next one was in 1970 which coincided with increased sightings of all the skuas in the London Area. In 1972 one remained at Wraysbury Reservoir from 26 November until 8 December. There were three in 1976, two at Queen Mary Reservoir in October and one at Girling Reservoir from 11 to 22 December that fed mainly on dead gulls beneath the power lines.

There were 12 in the second half of the 1980s, most of which were seen in the winter of 1985/86 when many overwintered in the North Sea and at least eight were in the outer Thames Estuary. Four were seen in November and another four arrived in the New Year, including two which remained on the Thames at Rainham for several weeks. In 1987 another winter bird was seen, at Stone Marsh on 18 January, then one arrived in the aftermath of the Great Storm on 17 October. After one in the Lea Valley in 1988, no more were seen for the next six years.

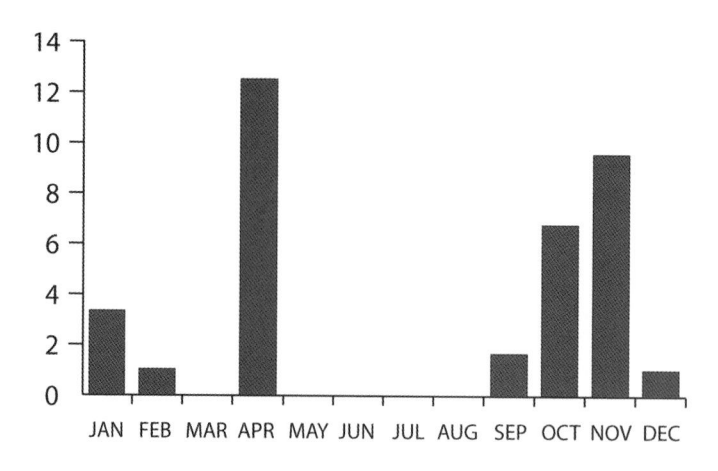

Seasonal occurrence of Pomarine Skua: all records.

They were then seen in three consecutive years in the mid-1990s, followed by another midwinter record at Walthamstow Reservoirs in January 1998.

21st century

Since 2000 they have been seen on four occasions, all in west London. In 2003 an unprecedented flock of 13 adults circled over Queen Mother Reservoir for four minutes on the evening of 25 April before flying off north. Although they arrived during the peak time of passage on the south coast, this is the largest inland flock seen in England and the birds were presumably taking an overland shortcut towards The Wash. Three were seen in November 2007.

Arctic Skua *Stercorarius parasiticus*

A scarce passage migrant

Range: breeds northern Europe, Asia and North America, winters at sea.

Historical

This was a rare visitor in the 19th century with just five records of eight birds: shot on the Thames at Battersea in September 1824; shot at High Ongar in 1837/38; four immatures shot at Brent Reservoir in autumn 1842; shot on the Thames at Greenwich Reach in autumn 1862; and shot at Langleybury in 1882.

20th century

It was over 30 years later before the next one was recorded, over Hyde Park on 16 May 1916. This was notable not only for being in Inner London, but also for being the first spring record.

In the 1920s and 30s there were another four records, two of which were in Inner London. One bird overwintered at Surrey Docks from December 1920 to January 1921. The following year one was seen at Wimbledon on 28 August and again on 8 September; it was not known if it was present for the intervening 10 days. One was seen in St James's Park on 18 September 1935 and another was present at King George V Reservoir on 21 and 28 August 1938.

Arctic Skuas began to be seen more often with five in the 1940s, four of which were in spring; the other was present at Staines Reservoir from 9 to 14 October 1949. There were three during the 1950s, including a late-spring migrant that flew north over Old Parkbury GP on 10 June 1956. They were seen more frequently during the 1960s; all were in autumn apart from one at Queen Mary Reservoir on 18 May 1965. An early returning migrant was seen on 21 July 1961 at Staines Reservoir, and in 1963 a group of six flew south-west over Sutton on 28 September.

By the start of the 1970s they had become a near-annual passage migrant, mostly seen in autumn. During the 1970s there was a total of 46, including several flocks. There was a small influx in the autumn of 1971 when nine were seen between 29 August and 30 September. In the autumn of 1972 two lingered, one at the Chingford reservoirs from 26 August to 9 September, and an immature on the Thames from 2 to 30 September between Purfleet and Swanscombe that spent most of its time on the outflow at West Thurrock Power Station. In 1974 one flew over Kensington Gardens on 29 October. In 1978 one flew north-east over Staines Reservoir on the unusual date of 17 June.

During the 1980s the species was seen annually, with a total of 55. An early autumn migrant was at Island Barn Reservoir on 12 July 1986. Only one was seen after the Great Storm in 1987, over Croham Hurst Golf Course on 16 October. The most productive decade was the 1990s with a total of 71 birds. More than one-third of these were seen in the autumn of 1992; at Staines Reservoir 22 flew over in five groups (including two groups of seven) in three hours on the afternoon of 29 August. In the autumn of 1997 a flock of eight flew over Walthamstow Reservoirs on 30 August.

An analysis of all records between 1945 and 1993 showed that the peak period was from the last week of August to the last week of September.

21st century

The best-ever year was in 2000 when 40 were seen. This included a flock of six adults over Beddington on 4 May, the largest spring flock. That autumn 26 were seen on 16 September: five over Brent Reservoir and 21 at Queen Elizabeth II Reservoir. Between 2001 and 2008 another 37 were seen, the largest flock being of five birds over Queen Mother Reservoir on 23 September 2004; almost all of these were in autumn, with five in spring and one in December.

Facts and Figures

The largest flock was of 21 birds on Queen Elizabeth II Reservoir on 16 September 2000.
The earliest migrant was on 8 April 1984 at Staines Reservoir.

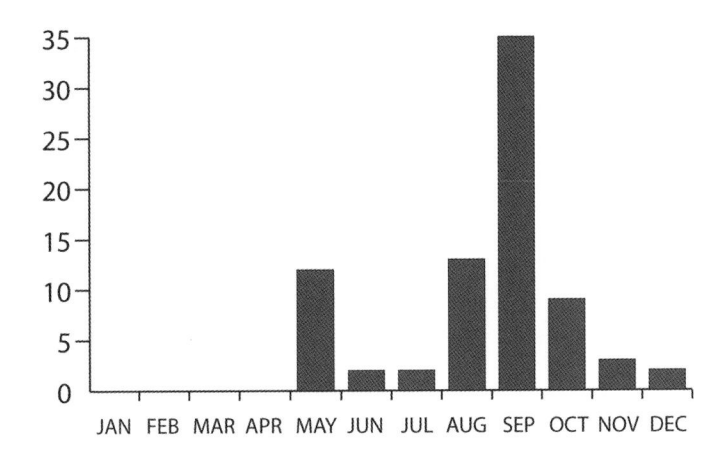

Seasonal occurrence of Arctic Skua: 2000–2009.

Long-tailed Skua *Stercorarius longicaudus*

A rare passage migrant (16)

Range: breeds northern Europe, Asia and North America, winters at sea in southern hemisphere.

The first record for the London Area was of a juvenile found dead at Queen Mary Reservoir in 1966 and identified by staff in the Natural History Museum. It may have been one of the two juvenile skuas present there from 17 to 22 September and which were originally thought to have been Arctic Skuas. The first one to be seen alive was at Wraysbury Reservoir in August 1978.

In 1985 three were seen in autumn: a confiding juvenile which spent five days at Staines Reservoir and was then refound on Queen Mary Reservoir; and two which flew south-west over Barn Elms Reservoirs in October, one of them an adult in breeding plumage. These last two were seen during calm weather and appeared to be migrating through London airspace as opposed to the juvenile, which arrived following strong winds. There was a good coastal passage in the autumn of 1988 and several were seen inland, including an immature at Hilfield Park Reservoir.

In the autumn of 1990 a juvenile was present at King George V Reservoir for six days in September. There was a near-repeat performance three years later when another juvenile arrived on Girling Reservoir on 9 September; it then moved to King George V Reservoir the following day, where it remained for a fortnight. Two immatures were seen in 1995 on the west London reservoirs.

Although this remains a rare visitor to the London Area it has been seen more frequently since the beginning of the 21st century, with six seen up to 2009, all in west London.

Of the total, 11 were on the west London reservoirs, with two in the Lea Valley, two together at Barn Elms and one at Hilfield Park. All have been recorded between 17 August and 8 October.

1966: Queen Mary Reservoir, found dead on 26 September
1978: Wraysbury Reservoir, on 30 August
1985: Staines Reservoir, on 6–10 September; also at Queen Mary Reservoir on 14 September
1985: Barn Elms Reservoirs, two on 8 October
1988: Hilfield Park Reservoirs, on 16–17 September
1990: King George V Reservoir, on 23–28 September
1993: King George V Reservoir, on 9–23 September; also at Girling Reservoir on 8 September
1995: Wraysbury Reservoir, on 30 August
1995: King George VI Reservoir, on 6 September
2003: Queen Mother Reservoir, on 17 August
2004: Staines Reservoir, on 9 September
2006: Queen Mother Reservoir, on 9–17 September
2007: Queen Mother Reservoir, on 21 August
2008: Queen Mother Reservoir, on 24 September
2009: King George VI Reservoir, on 20 September

Records No Longer Accepted

In September 1937 one was found dead at Easneye, near Ware. This record was published with the caveat that it was uncertain whether it was inside the London Area boundary; however, it is virtually certain to be outside the area so is no longer counted.

Great Skua *Stercorarius skua*

A scarce passage migrant (45)

Range: breeds northern Europe, winters Atlantic Ocean; breeds Scottish islands, passage migrant on UK coasts.

Historical

The earliest record was in 1790 when Latham referred to one that had been killed at Greenwich, the specimen of which was in his museum.

20th century

In the first half of the century, there were four records of five birds. The first was of two seen chasing Black-headed Gulls over the Thames by Waterloo Bridge in 1915; this is still the only record in Inner London. In 1928 an immature was observed chasing a Carrion Crow at Mill Hill; so far this is the only July record. The next two were both during winter, at Staines Reservoir for nine days in February 1931 and over Kingsbury in January 1942. There were no further records until the 1970s, which saw the start of an increase in sightings beginning with an exhausted bird that was picked up at Littlebrook and taken into care; two more were seen in the middle of the decade.

There was an even larger increase in the 1980s with 20 seen. These included one at Sevenoaks for five days in January 1984. Seven arrived after the Great Storm in October 1987; during the first day two flew over Queen Mother Reservoir, three flew over Wraysbury Reservoir and one was picked up at a bus stop at Muswell Hill and taken into care. The next day the last bird was released at Queen Mother Reservoir and a second bird was seen there later. Two years later eight were seen on 9 September: six drifted south over King George V Reservoir and two were present for an hour at Queen Mary Reservoir before flying off north-east.

During the 1990s, 12 more were seen, but unlike in the previous decade most were seen on the Lower Thames, including a flock of four that flew upriver past Dartford Marsh on 15 September 1994.

21st century

Since 2000 there have been seven more records. None were seen until 2003 when one was on the Thames off Swanscombe. The following year one was present on Queen Mother Reservoir for five days in September before being found dead. There were two in 2007: one flew up the Thames past Northfleet and Rainham, and another – the first December record – was seen on King George V Reservoir where it killed a Black-headed Gull. In 2009 one remained on the Thames in the Rainham area from 29 December to 1 January 2010.

Facts and Figures

The largest flock was of six birds at King George V Reservoir on 9 September 1989.

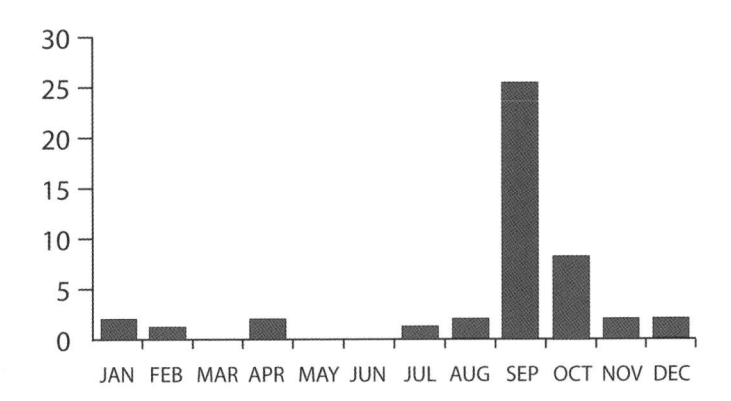

Seasonal occurrence of Great Skua: all records.

Sabine's Gull *Xema sabini*

A scarce passage migrant, typically in autumn

Range: breeds northern Europe, Asia and North America, winters off coasts in southern Africa and South America.

This was a very rare species in the UK in the 19th century; the first bird in the London Area in 1862 on the Thames at Blackwall was only the fourth to have been seen in England. There were no further London Area records until almost 90 years later when, following sustained westerly gales, two birds were seen at Staines Reservoir in September 1950; several others were present in the UK at the same time. No more were seen until the 1970s when two were on the Thames at West Thurrock.

In 1987 there was a massive arrival of seabirds in the country. The remnants of Hurricane Floyd crossed the Atlantic and intensified over the Bay of Biscay, where many Sabine's Gulls congregate in autumn, then headed across the Channel and inland over southern England and through the London Area in the early hours of 16 October. Gusts of wind exceeded 90 mph. The first Sabine's Gull was seen at 09:00 hrs on the 16th when two adults flew west along the Thames at Hammersmith Bridge. Later in the morning an adult was seen at Wraysbury Reservoir; by the afternoon a further 19 had been seen, making the daily total 22 at 11 localities. With the chaotic aftermath of the storm (and it being a weekday), many more went unrecorded.

The following day was a Saturday and it soon became apparent that there had been an unprecedented arrival of Sabine's Gulls – a maximum of 52 were seen throughout the day, including at least seven adults on the Thames at West Thurrock and five at Barking. Most, however, were seen on the reservoirs, particularly in west London. There were five at Queen Mary and Wraysbury, four at Island Barn and King George VI, and three at Queen Mother and Staines. Early on the morning of the 18th three flew east past Barking, then at least nine were present for much of the day; elsewhere, many remained on the reservoirs but a few others were seen flying through and probably 33 were seen in total.

There was a large drop in numbers on the 19th and 'only' 14 were seen, including up to five at Barking. Between 20 and 24 October an average of 10 a day were seen, the highest counts being five in a flooded field by Girling Reservoir on the 21st and up to three at Colnbrook where the last was seen on the 28th. Only one was seen in Inner London, on the Thames by Hungerford Bridge on the 16th. Around 300 were recorded in England during this period, at least 64 of these in the London Area.

Since 1987 Sabine's Gulls have been seen far more frequently than before the Great Storm, with occurrences in 10 of the following 20 years. Out of the 14 seen during this period, five were on the Thames and four were on the west London reservoirs. An immature arrived at Walthamstow Reservoirs in 1989 during easterly winds when there was a large passage of Little Gulls through the London Area. An adult was present for about three hours on the morning of 5 August 1994 at Queen Mother Reservoir; this is the earliest date one has been seen in the London Area (several were on the East Coast at the time). A juvenile was present on the Thames in 1998, ranging between Crossness and Silvertown; its three-week stay made it the London Area's longest-staying individual.

In 2000 an adult arrived on the Thames at Dartford Marsh during a day of considerable passage through the London Area, which included 40 Arctic Skuas. In 2007 three juveniles arrived separately at Beddington within 20 minutes of each other during the morning of 19 October.

1862: Blackwall, immature shot on the Thames in early September
1950: Staines Reservoir, adult on 11–27 September and immature on 18–27 September
1975: West Thurrock, first-winter on 7 September
1978: West Thurrock, adult on 1 October
1987: across London, 64–84 between 16 and 28 October
1988: West Thurrock, juvenile on 23–29 September; also seen at Greenhithe
1988: Hilfield Park Reservoir, juvenile on 23–24 September
1989: Walthamstow Reservoirs, immature on 10 September
1993: Rainham, juvenile on 4 September
1994: Queen Mother Reservoir, adult on 5 August

1997: Wraysbury Reservoir, juvenile on 14–16 September
1997: Dartford, juvenile on 23 September
1998: Barking Bay area, juvenile on 30 August–19 September
1999: Queen Mother Reservoir/Horton Fields, on 29 September–2 October
2000: Dartford Marsh, adult on 16 September
2003: King George VI Reservoir, adult on 12 August
2007: Beddington, three juveniles on 19 October
2009: Rainham, on 7 October; also at Crossness
2011: King George V Reservoir, on 10–13 September

Kittiwake *Rissa tridactyla*

An uncommon visitor

Range: breeds coastal northern and western Europe, winters at sea.

Historical

This was an infrequently recorded species in the 19th century, with at least 18 records. The earliest was in 1837 when one was captured alive on the Thames near Esher on 19 January with another taken nearby on 22 February. There were several records from Brent Reservoir and from various other locations across the London Area including some in central London: up to four on the Thames at Battersea in 1885; one on the Serpentine in Hyde Park in 1891; and a few first-winter birds on the Thames at Embankment among thousands of Black-headed Gulls in February 1895. This was the start of the period when they became more frequent.

20th century

By the beginning of the century they had increased in such numbers that they were considered to be regular in some areas of London. In 1900 Bucknill described them as being 'of regular occurrence' on the Thames and less frequently seen on some of the reservoirs to the south of the river. Ticehurst stated in 1909 that 'in autumn and spring many pass up the Thames as far as Dartford'. However, Glegg considered Kittiwake to be rare in Middlesex until the late 1920s when there was a small increase in the number of records with eight birds between 1926 and 1932, all in winter; five were on the west London reservoirs, there were two on the Thames and one near Hounslow Heath.

Numbers remained fairly static up to 1953 with about four or five records a year although there were nine years when there was a complete absence of sightings. There were several in central London: on the Thames opposite County Hall on 23 April 1941; in St James's Park on 20 November 1951 and 11 January 1952; and at Waterloo on 21 November 1953.

In 1954 there was a large increase in the number of records and there was a notable passage in December when 15 flew through Barn Elms Reservoirs on the afternoon of the 24th during a large movement of other gulls; a further five flew through the following morning. In 1956, during very cold weather, a flock of 34 was seen at King George V Reservoir on 4 March. In 1959 there was a very large movement through the London Area on 22 February when a total of 187 was seen, including groups of about 60 and 50 flying west over Leatherhead Watercress Beds and 52 flying south-west over King George V Reservoir.

Kittiwakes were largely unaffected by the severe winter in early 1963 with the only record being of a single bird on the Thames at Swanscombe. In March 1966 one spent about nine days in the penguin enclosure in London Zoo. In 1969 a flock of 55 was seen briefly at Staines Reservoir on 10 November before flying off in a westerly direction; a little later another 30 flew over in the same direction. During the 1970s numbers continued to increase and several large groups were seen, including 65 adults north-east over Kew Gardens during very cold weather on 2 January 1979.

There was a massive influx in 1985 when about 250 were seen, mostly on the spring passage and including about 150 on 28 April; at West Thurrock about 100 flew upriver on that date, including a flock of 56. Some of these birds must have followed the Thames to Staines Reservoir as 33 were present there on the same day.

Another large movement occurred in 1986 when 124 adults flew west in three flocks over Barn Elms Reservoirs on 25 March. However, only 13 were driven into the London Area by the Great Storm in October 1987.

1993 was a record year with some 1,129 seen, almost all on one day; in late January there were two days of storms in the North Sea and these drove hundreds of Kittiwakes into the Thames Estuary where they then followed the river into the London Area on the 25th. The first flock was a modest 20 over Tyttenhanger but in the early afternoon 238 flew south-west over Barn Elms Reservoirs in an hour and a quarter. There were additional counts of 70–80 at Mortlake and 120 at Staines but it was at Queen Mother Reservoir where the largest numbers were seen; they began to build up from mid-afternoon and by dusk 642 were present, contributing to the day's massive total of 1,100. Another huge movement took place during February 1999 with some 900–1,024 passing through the London Area due to strong north-westerly winds; on 7 February 337 flew west in three flocks.

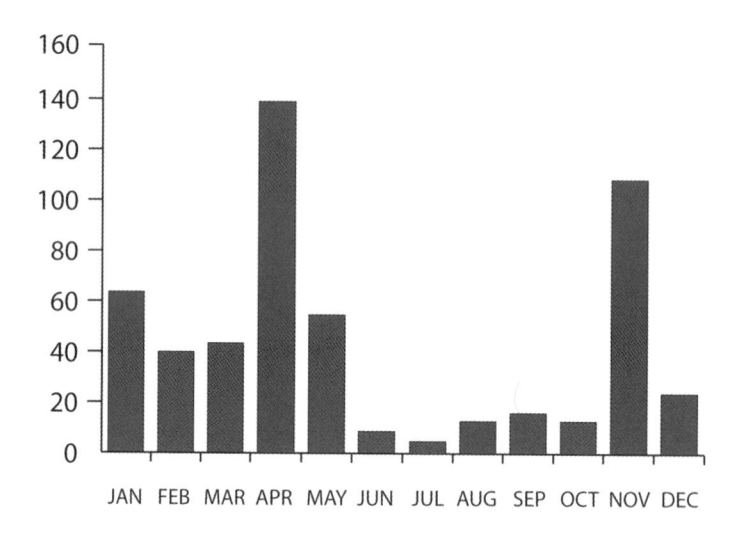

Seasonal occurrence of Kittiwake: 2000–2009.

21st century

Since the beginning of the 21st century the status of this coastal gull is best described as erratic; typically small numbers are seen annually but there are occasional large influxes. During the first decade there was an average of 53 a year, ranging from 15 to 125. The two largest groups were both on Queen Mother Reservoir: about 85 on 4 April 2005; and 40 on 9 November 2007. Elsewhere, 37 flew over Queen Elizabeth II Reservoir on 29 January 2003.

Unacceptable Records

Harting listed two records of birds shot at Aldenham Reservoir in May 1850 and spring 1858; however, these were later rejected by Glegg as were a number of subsequent records, notably in 1895, as he believed that many of them related to Common Gull.

Facts and Figures

The largest-ever concentration was of 642 birds at Queen Mother Reservoir on 25 January 1993.

Bonaparte's Gull *Chroicocephalus philadelphia*

A very rare vagrant (2)

Range: Americas; vagrant to UK.

2012:   Crossness/Barking Bay, first-winter on 19–29 May
2012:   Crossness, first-summer on 26 and 29 May

Amazingly, the only two records both occurred in the same location and in the same month, although despite the overlapping dates the birds were never seen together. The first bird was absent on 25–26 and 28 May.

Unacceptable Records

A first-winter at Barn Elms Reservoirs on 29 January 1983 was accepted by the BBRC but is no longer considered acceptable by the LNHS Rarities Committee.

Black-headed Gull *Chroicocephalus ridibundus*

A very common winter visitor and localised breeder

Range: breeds Europe and Asia, winters Europe, Africa and Middle East.

Historical

In 1738 Albin stated that they 'are very frequent in the River Thames near Gravesend' which is just outside the London Area boundary. Harting wrote that 'of the few species of Gulls which visit us in spring and autumn, the Black-headed Gull is certainly the commonest'. He saw and shot several at Brent Reservoir, although only in small numbers.

Towards the end of the 19th century a change was detected in their status, particularly in winter when they had previously been absent. They were first seen in good numbers on the Thames in the severe winter of 1880/81 when lumps of ice were present on the river. They subsequently returned every winter along the Thames in central London. In 1887/88 several hundred were noted at Putney, 120 were near Waterloo Bridge and they were seen as far up the river as Kingston. Further influxes occurred in the three consecutive winters between 1890/91 and 1892/93.

When they arrived in central London they were regularly shot and this practice did not stop until the winter of 1892/93 when the police effectively ended this by fining the gunmen for the offence of discharging firearms to the public danger. From this point onwards, the public started feeding the gulls instead. Hudson related how hundreds of working men and boys would flock to the river in their lunch-break to feed the gulls. The winter of 1894/95 was one of the coldest on record and thousands of Black-headed Gulls were on the Thames around London Bridge. They were literally starving and people used to feed them from the Embankment and bridges; one enterprising chap started selling sprats to the public to feed to the gulls, and sometimes they would even perch on his head when he was feeding them. The following three winters were relatively mild but the gulls continued to visit to take advantage of the food that was offered to them. It seems that the Thames offered an easy route into the London Area for Black-headed Gulls where there was a variety of food sources available in winter and it is this that led them to becoming regular winter visitors.

During 1894/95 they spread away from the Thames into the parks and open spaces of central London and they continued to explore these areas in later, milder winters. St James's Park soon became their favoured haunt away from the Thames as there were plenty of handouts for them and they would spend all day on the lake and roost on the river at night. They were first seen using trees as perches in the park in 1898 and it is thought they developed this habit there due to the lack of artificial perches as there are few records from elsewhere of this behaviour. They started to be seen regularly over Tooting and by 1898 they were often seen over Brixton. By the end of the century they had made their way up the Lea Valley just to the north of Walthamstow.

20th century

By 1900 they were regular in some of the parks in central London and had begun to spread out further across the London Area following the river valleys. In February 1901 large numbers were seen in Enfield on flooded fields and the following year they were present at Watford SF. In 1903 large flocks were seen in the Twickenham area, both on the Thames and in nearby fields. By 1906 they had become the most numerous gull on Staines Reservoir and could also be found in the area on fields following the plough. In the arable parts of Woolwich they were seen feeding with Rooks in 1909. From around this time they started to return earlier to the London Area, arriving as early as the first week in August. In January 1914 there was an exceptional count of 10,000 on the river between Westminster and Blackfriars compared to the usual 1,000–2,000.

Black-headed Gulls were slow to spread into many districts away from the river valleys, particularly south of the Thames. It was not until the mid- to late 1920s that they were regularly recorded at Coulsdon, between Sutton and Banstead, and in the Warlingham and Chelsham district. By 1928 immature Black-headed Gulls had started to frequent the Thames during the summer months.

In the 1930s they were mostly observed on the Thames and, to a lesser extent, on the reservoirs, but could be found anywhere where there was a feeding opportunity, such as in parks, sewage farms, rubbish dumps, sports grounds, gravel pits and open fields. Glegg estimated the population of Black-headed Gulls on the Middlesex stretch of the Thames to be about 2,000 in winter. The first arrivals were noted in early July and would gradually increase until the middle of September when the main wintering population had all arrived. These would then stay until early March when the numbers started to decrease.

They were first noted roosting in large numbers on Queen Mary Reservoir in the 1920s. Glegg stated that 'every evening towards sunset thousands of gulls, mostly of this species, migrate up the Thames to roosting-places, as at the large reservoirs, where they pass the night, returning to their feeding-grounds at dawn. The numbers of this species in the county, in winter, are much greater during the night than during the day'. Glegg thought that many of these birds flew in from the Thames Estuary and were also joined by those gulls which fed in the parks during the day. One unusual roosting site was a roof in Notting Hill where seven were observed in December 1945.

By the mid-1950s it was estimated that 85,000–100,000 Black-headed Gulls were roosting on the London Area reservoirs in midwinter. In December 1963 a survey of roosting gulls produced a total of about 165,000 Black-headed Gulls on the reservoirs; the majority of these, about 92,500, were on Queen Elizabeth II Reservoir. A repeat census in 1968 found an increased total of about 192,000 and the following year around 100,000 roosted on Queen Elizabeth II Reservoir.

In 1966 about 150 oiled birds were seen at Sevenoaks and 10 were found dead in September; they were thought to have been contaminated by an oil spill in the River Medway. The highest counts during the day were typically along the Thames, about 20,000 were at Dartford Marsh in February 1972 and 10,000 were feeding on the tip at Rainham in December 1978.

A survey of gull roosts in January 1983 found a total of over 236,000 Black-headed Gulls; the highest count was 79,000 on Queen Elizabeth II Reservoir. While most of the major roosts are occupied throughout July to April, a roost of up to 5,000 at Brent Reservoir is only occupied between July and September, after which time the birds move to Hilfield Park Reservoir and the west London reservoirs.

21st century

By 2000 this species had become one of the most common wintering birds and could be found just about anywhere from central London out to farmland in the rural areas. Although large numbers still roost on the reservoirs, particularly in the Lea Valley and in west London, there are far fewer than in previous decades; this may be due to the closure of landfill sites. The gull-roost survey in 2003–06 found a total of over 56,000 on three reservoirs.

Breeding

They did not breed in the London Area in the 19th century although they were a common breeder further out in the Thames Estuary.

The first breeding colony in the London Area was discovered at Perry Oaks SF in 1946 although it was believed to have originated in 1941 or 1942. Because of its secure location in a bed of sewage sludge, the colony grew quickly; in 1946 there were 50–60 nests but the following year this had increased to about 300 and it remained at this level for the next few years. Subsequently, the size fluctuated between 150 and 250 nests.

A second, but much smaller, colony was established on a gravel pit at Maple Cross and six pairs bred in 1961. By 1963 the Perry Oaks colony had been reduced to just five nests and none was seen at Maple Cross. The decline at Perry Oaks continued and by 1965 no pairs attempted to breed in the London Area. There were no further breeding attempts until 1981 and 1986 when a pair nested at Perry Oaks.

In 1998 four pairs bred on Staines Reservoir and a pair attempted to breed at Perry Oaks. By 2000 the small colony on the tern rafts at Staines Reservoir had increased to 14 pairs. Within three years it had expanded to about 50 pairs and by 2007 it had more than doubled to 118 pairs due to the introduction of a new nesting raft on the reservoir. In 2009 there were up to 160 breeding pairs there and one pair attempted to nest on the adjacent King George VI Reservoir. A pair also attempted breeding at Thorpe Park in 2011.

In 2008 a pair bred at Rye Meads, the first breeding record in the Lea Valley; the following year eight pairs attempted to breed there and another pair attempted to breed on Seventy Acres Lake.

Ringing Recoveries

In the early 20th century ringing recoveries showed that most of the Black-headed Gulls that wintered in the London Area had originated overseas, in Estonia, Germany and Sweden. Later recoveries were made from Czechoslovakia, Denmark, Finland, Holland, Lithuania, Poland and Switzerland. Ringing also showed that many Black-headed Gulls stayed loyal to their chosen wintering location and that there was little interchange between sites. An immature ringed at Molesey in December 1936 was recovered at Watford 19 years later in December 1955.

Facts and Figures

Leucistic birds are seen most years and melanistic birds were seen at Hilfield Park Reservoir in December 1992 and at Queen Mother Reservoir in February and March 1995.

Little Gull *Hydrocoloeus minutus*

An uncommon passage migrant

Range: breeds eastern Europe and Asia, winters Europe and Atlantic Ocean; passage migrant in UK.

Historical

The first London Area and British record was of an immature shot on the Thames near Chelsea on an unknown date and was described from the specimen by Montagu in his *Ornithological Dictionary* published in 1813.

There are seven dated records in the 19th century: shot at Greenhithe in September 1860; an adult in nearly full summer plumage shot on the Thames in Blackwall Reach in the early spring of 1863; shot at Cornhill on 26 February 1870; two shot at Brent Reservoir in August 1871; at East Molesey in the autumn of 1890; near Cobham on 13 January 1893; and on the Thames at Charing Cross on 15 February 1895. Undated records include four birds that were shot out of a flock of six at Battersea and one from Walton-on-Thames.

20th century

There were no further records until 1928 when an immature was seen at Staines Reservoir on 6 October. From then they started to be seen more regularly. Between 1928 and 1954 there were at least 50 records, all but two of which were of birds seen on the reservoirs. During this period it was mainly a late-autumn and winter visitor; there were a few in August and one in July but the species was surprisingly rare in spring with just three records in early May. The largest group was five birds at Queen Mary Reservoir on 6 May 1936, which accompanied seven Black Terns; these species have often been seen together in recent springs.

From the late 1950s onwards there was a further increase in numbers and they became a regular passage bird, predominantly in August and September with a smaller spring passage. This tied in with increased numbers in Essex and Kent, particularly from the mid-1960s. In the early 1970s the number of Little Gulls seen flying along the English Channel increased substantially and more were seen in the London Area on passage. In 1974 there was an exceptional influx in May; the peak counts were 17 at Staines Reservoir, 16 at King George V Reservoir and 14 at Swanscombe. In 1975 two adults summered at Wraysbury GP between 10 June and 23 July, and there was a large autumn passage with a peak count of 20 at Staines Reservoir. In 1978 a second-summer bird summered at Staines Reservoir between May and August.

During the 1980s there were several years of exceptional passage, with most birds seen on the large reservoirs. In 1984 at least 159 were seen on passage, mostly in spring, with a peak of 45 at Staines Reservoir on 2 May; during this time none were seen in east London. In 1987 there was a massive total of 407, compared to the five-year average of 84. Out of this total, about 252 arrived following the Great Storm in October, the peak count being 50 on the Thames at West Thurrock. There was another large influx in 1989 with around 307 seen, 40 per cent of which were at Staines Reservoir.

By the beginning of the 1990s they had become a regular passage migrant with just a few seen in winter. In 1995 there was another large influx in spring; the first large groups were on 28 April with 75 at Staines Reservoir and 65 at King George V Reservoir; the following day there was a flock of 186, including 151 adults in summer plumage at Staines Reservoir.

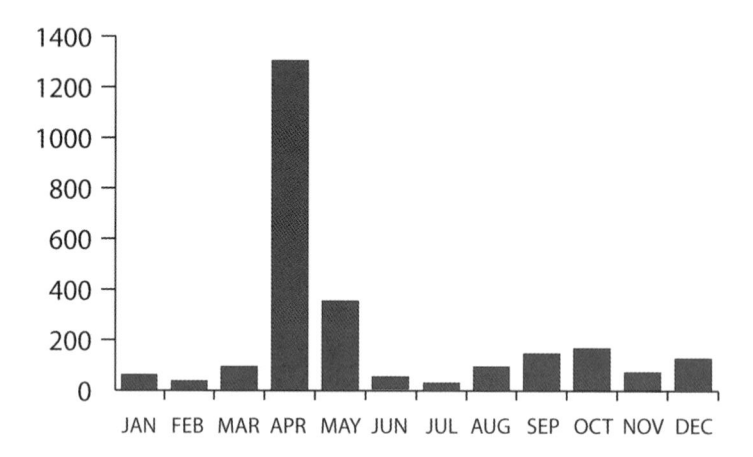

Seasonal occurrence of Little Gull: 2000–2009.

21st century

The passage of Little Gulls through the London Area depends on weather conditions and the number seen fluctuates from year to year. Most of the large flocks are seen on the reservoirs, especially in the Lea and Colne Valleys, suggesting that they are taking an overland route. There are occasional influxes, mostly in spring between mid-April and mid-May. For example, in 2003 there was a large passage between 15 and 20 April with peak flocks of 68 at Staines Reservoir and 67 at King George V Reservoir. In 2004 a flock of 61 was seen at Rye Meads on 1 May. They are scarce away from the river valleys and large reservoirs but there have been a few in Inner London, including one that spent a week on the Round Pond in September 2004.

Facts and Figures

The largest flock was of 186 birds at Staines Reservoir on 29 April 1995.

Laughing Gull *Larus atricilla*

A very rare vagrant (2)

Range: Americas; vagrant to UK.

The first record occurred in September 2006 when a second-winter bird roosted at Hilfield Park Reservoir on two consecutive nights; it was not located during the day. There had been a national influx in 2005 with 52 seen and a further 20 in 2006. Another of a similar age was found in the roost at Amwell just two months later. Although accepted as a separate individual, the same bird may have accounted for both sightings.

2006: Hilfield Park Reservoir, second-winter on 19–20 September
2006: Amwell GP, adult or second-winter on 26–27 November

Franklin's Gull *Larus pipixcan*

A very rare vagrant (1)

Range: Americas; vagrant to UK.

The only record occurred in 2000 when a second-summer bird was present in the Crossness area for four days. It had previously been tracked across the country at Weymouth, Dorset, from February to early March, then at Cheddar Reservoir, Somerset, in mid-March and in Bristol on 10–11 April before being found in the London Area. It spent most of its stay on the southern side of the Thames but occasionally crossed over into Barking Bay.

2000: Crossness, on 13–16 April; also seen at Barking Bay

Mediterranean Gull *Larus melanocephalus*

An uncommon winter visitor

Range: breeds Europe and Middle East, winters Europe.

Historical

The first record for the London Area and the UK was in January 1866 when an immature was shot at Barking Creek; the specimen is in the British Museum.

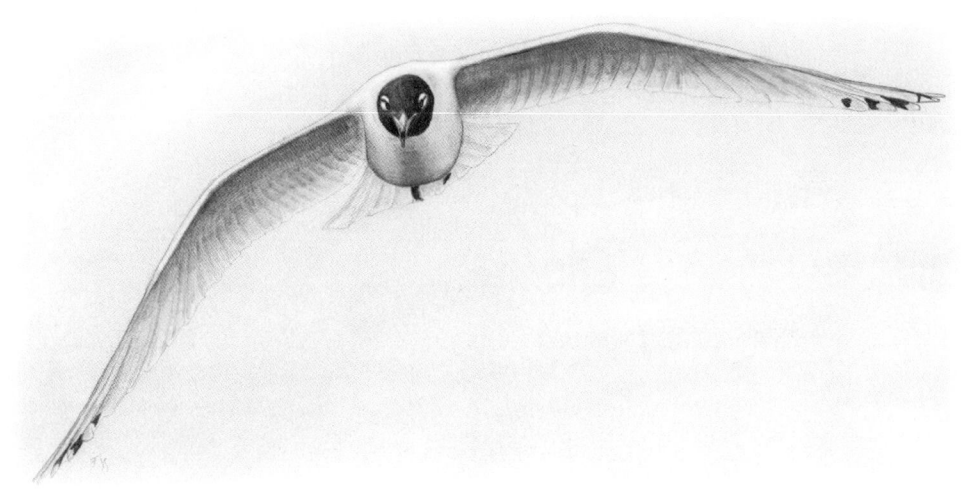

20th century

No more were seen in the London Area until 1957 when an immature was present at Barn Elms Reservoirs on 19–20 September; there had been a small influx into southern England. The mid-1960s saw the start of an upturn in the number of sightings in the London Area. They continued to increase during the 1970s and between 1957 and 1979 there was a total of 31 records. The first overwintering individual was in 1976/77 on Epsom Common and it returned for three subsequent winters.

By the start of the 1980s Mediterranean Gull had become an annual visitor and in 1984 an adult was seen in July for the first time, meaning that the species had been seen in every month of the year. It continued to be seen more frequently almost every year but was unpredictable until one began wintering at Croxley Green, Watford, from 1986 to 1987; this individual returned every year until 1991/92.

Sightings had increased so much by the start of the 1990s that it had become virtually impossible to know how many different birds were seen each year. Around 80 were seen in 1995, representing another large increase; for the first time several groups were seen, with six at Erith Reach on 1 February and five at both Dartford Marsh and Girling Reservoir.

21st century

By the start of the 21st century it had become a regular winter visitor on the Lower Thames Marshes and in the Lea Valley but could be seen almost anywhere, including in the parks in central London. Although it can be seen in any month of the year, most are seen between November and March with some individuals overwintering. Most sightings are of one or two birds and small groups are unusual during the day, although several can sometimes be found roosting on the same reservoir. However, in 2009 a flock of eight juveniles flew along the Thames past Greenwich.

Facts and Figures

The largest flock was of eight birds at Greenwich on 20 July 2009.
In 1993 one or two hybrid Mediterranean Gull x Common Gull were seen on Hampstead Heath on 31 December.

Common Gull *Larus canus*

A common winter visitor

Range: Europe, Asia and North America, also winters North Africa.

Historical

In the first half of the 19th century this was a very rare bird in the London Area with only three known occurrences: in the winter of 1831 one was shot on the River Mole opposite Hampton Court; in 1838 Torre wrote that it had been seen around Harrow; and in 1843 Bond included it on his list of birds observed at Brent Reservoir.

Harting stated that they were rarer in southern England than Kittiwake, and considered them as accidental visitors in the London Area. He listed the following records: one killed in winter at Hampton Court; half-a-dozen shot on the Thames at Blackwall in September; three shot at Brent Reservoir; and a flock of about 20 in Bushy Park in October 1865. In 1888 they were recorded on the Thames at Kingston and by that time they had become occasional winter visitors to the Thames in central London.

20th century

By the turn of the century they had followed the Black-headed Gull by using the Thames as a route into the London Area and had increased in numbers, although on a much smaller scale. They became fairly common on the Thames at Dartford and several had taken to overwintering on Staines Reservoir; and by 1909 they had also become more numerous on the Thames at Putney. There was only a slow increase in numbers during the early to mid-1920s, particularly away from the Thames, but by the end of the decade there had been a noticeable increase in central London, especially during periods of cold weather.

By the 1930s they had become common winter residents, occasionally being seen during the summer. Glegg showed that this species made up about 10 per cent of all the gulls in the London Area; he noted that most of them were seen on the Thames and reservoirs but they could also be found at other open areas. Although some wintering birds started to arrive in July, mainly on the Thames, the main numbers were not seen until September and these would then remain until about the third week of March. During winter they joined the other species of gull in flying up the Thames to roost on the reservoirs.

On 5 March 1940, of the estimated total of 37,000 gulls roosting at Queen Mary Reservoir, almost half were probably Common Gulls; it was thought that these large numbers were mainly passing through the London Area en route to their breeding grounds. In 1958 a coordinated roost count at five reservoirs on 8 March produced a total of 22,500 Common Gulls, most of which were on Queen Mary Reservoir. A census in 1963 found a total of just over 20,000 roosting birds and five years later a repeat census produced a total of about 31,000. A large passage was noted on 17 February 1973 when about 5,000 flew east in 30 minutes at Swanscombe. The next gull-roost survey in January 1983 found 24,000, half of which were on Girling Reservoir. Ten years later there were only 9,200, a decrease of 15,000 since the previous census.

Studies of Common Gulls in the London Area found that there were two distinct feeding habits. In most areas a few would consort with flocks of Black-headed Gulls and harry them for scraps of food. In other areas they would gather to feed in large single-species flocks of at least 50 birds (but sometimes 200–300 strong) in open areas such as downland, playing fields and farmland.

21st century

Common Gulls are a common and widespread overwintering bird, sometimes in flocks of several hundred, although they are vastly outnumbered by Black-headed Gulls. In 2006 an estimated 10,000 roosted on Queen Mary Reservoir in December and in the 2003–06 roost survey there were 7,000 on Girling Reservoir.

Ringing Recoveries

There were two ringing recoveries in the 1920s and 30s, of birds originally ringed in Denmark and Sweden. Further ringing recoveries over the next two decades confirmed that the majority of wintering birds in the London Area bred in the Baltic area and there were no ringing recoveries of British-breeding Common Gulls.

Facts and Figures

The largest non-roosting flock was of 10,000 birds at Dartford Marsh on 10 January 1997.

Ring-billed Gull *Larus delawarensis*

A rare winter visitor (22)

Range: Americas; rare visitor to UK.

This gull first arrived in the UK in 1973 and six years later one was found at Rainham Marshes. By then they had started to become more regularly seen in Britain and three more arrived in the London Area over the next eight years, all of which were present on one day only. The first long-staying bird occurred in February 1991 when an adult was found in Uxbridge, on Rockingham Recreation Ground; it was seen there every morning for the next month. It returned to the same area in the following four winters and was last seen on 5 March 1995. Another 11 were seen during the 1990s across the London Area, including two different birds at Ponders End Lake in 1993.

In 1996 an adult at Charlton on 23 November was the first appearance of the long-staying bird that later frequented the Greenwich and Isle of Dogs area. This individual spent 14 winters in the London Area and was only ever seen around the Thames. In addition to the long-staying adult on the Thames, six were seen between 2001 and 2004, including the only one in Inner London; this spent two days in Kensington Gardens and was considered to be a different individual from the one seen in Richmond Park a few days earlier.

1979: Rainham Marshes, on 25 February
1982: Staines Reservoir, on 28 November
1985: Barn Elms Reservoirs, on 19 April
1987: Rainham Marshes, on 6 December
1991: Uxbridge, on 2 February–3 March; overwintered again until 1994/1995 and also roosted at Queen Mother Reservoir on 27 November 1994
1993: Ponders End Lake, on 24 January and 11 February; also seen at Enfield on 13 February–7 March
1993: Ponders End Lake, on 8–11 April
1995: Barn Elms Reservoirs, on 15 January
1995: Beddington SF, on 6 February
1996: Brent Reservoir, on 23 March
1996: Charlton, on 23 November, overwintered every year to 2009/10 at various locations along the River Thames
1997: Hampton Court Park, on 2 February; also seen at Bushy Park on 6 February and 5–6 March
1997: Rainham Marshes on 8 February
1997: Kempton Park NR, from 14–17 March
1998: Rainham Marshes, on 21 February
1999: Barking Bay/Beckton/Crossness, on 20 November–3 January 2000
2001: King George V Reservoir, on 25 January
2002: Richmond Park, on 9 February
2002: Kensington Gardens, on 13–14 February
2002: Stratford, on 18 March
2003: London Wetland Centre, on 9 November
2004: Queen Mother Reservoir, on 11, 21, 24 and 26 April

Lesser Black-backed Gull *Larus fuscus*

A common resident and localised breeder

Range: breeds northern Europe and Asia, winters Europe and Africa.

Historical

This was a scarce bird in the 19th century; Harting wrote that it was 'only met with in severe weather, or after there has been a prevalence of east wind'. He knew of only three records, all of which were from Brent Reservoir: one prior to 1843, an undated autumn record and one shot in October 1865.

By the end of the century it was being recorded on the Thames. In 1896 Rushen wrote that he had often seen it on the river off Blackfriars and Lambeth the previous year.

20th century

Between 1900 and 1925 there were fewer than 30 records; most were in spring although there were one or two in almost every other month. An increase began in the late summer of 1925 when up to 12 were at Barn Elms Reservoirs. Two years later the peak count there was 57 in August and there were more than 100 nearby on the Thames. This became a regular autumn occurrence in subsequent years with higher numbers and longer stays being recorded.

In the 1930s Glegg described this as mainly 'an autumn passage-migrant in considerable numbers'. The first would be recorded in July and the main passage was in August and September, with some birds still being seen into November; only single figure counts were made during the winter months. Most records were from the Thames and its environs, with up to 277 birds in the Barn Elms area and up to 400 at Wyke SF in the Lower Brent Valley. Aside from this passage, much smaller numbers could be seen at any other time of year, usually single birds but occasionally small flocks; a group of more than 70 at Staines Reservoir in June 1934 was exceptional.

From 1936 they started to overwinter regularly at some sites. In the east of London, they remained rare until 1938 when there was a large influx in autumn on the Lea Valley reservoirs. Around the same time they began to explore other areas and flocks started to be seen on playing fields and golf courses. This range expansion continued during the

1940s and they were also recorded at many of the smaller reservoirs away from the Thames, as well as at most sewage farms and other open spaces. Some immature birds had started to oversummer by the 1940s, especially on the Thames.

There was little further change in the status by the start of the 1950s except that numbers had increased in winter. They were still most common in autumn and in the late 1950s. In the autumn of 1955 there was a very pronounced migration through Hilfield Park Reservoir with a minimum of 50 birds recorded every week from the end of July through to the end of October and a peak of 608 during the second week of September; it was estimated that at least 5,400 birds passed through during the entire autumn. Despite there being a gull roost at Hilfield all these birds flew to Staines Reservoir at dusk to roost. In 1956 the maximum roost counts were of 7,600 birds at Girling Reservoir on 20 August and 3,000 at Island Barn Reservoir on 6 September.

They continued to increase and by the 1960s there were up to 7,000 roosting on the reservoirs in autumn and about 400 at Beddington in winter. In 1971 the peak count had reached about 1,000 at Harefield Moor. The 1983 gull-roost survey found a total of 14,000; this was 15 per cent higher than the count from a similar survey in 1969 and this species was the only gull to have increased since then.

21st century

At the start of the 21st century the wintering status of this species had remained unchanged, although the numbers present in summer had significantly increased over the previous few decades. These gulls can be found throughout the London Area with the largest numbers on landfill sites and on the reservoirs. The gull-roost survey in 2003–06 found at least 8,600 roosting in the London Area.

Breeding

The first signs of breeding were in 1966 when a pair was present in St James's Park between March and May; the birds were seen to display but no nesting attempt was made. In 1968 a pair was present in London Zoo from March to April and was seen displaying but did not breed. The following year a pair remained in Regent's Park and attempted to breed; a pair was also present in Hyde Park and Kensington Gardens during the summer but did not breed.

The first confirmed breeding occurred in 1982 when a pair raised two young at Lords cricket ground. Single pairs bred again in the Euston/St Pancras area in 1987–88, and in 1989 a pair bred again on St Pancras Church and another pair bred at Kings Cross. By 1990 there were at least five pairs breeding in a small colony at Kings Cross.

In 1991 pairs were present at three other sites in central London. Three years later the breeding population had increased to at least 29 pairs, 23 of which were at Kings Cross. In 1997 a new colony of seven pairs was formed at Beckton, on jetties by the Thames.

At the start of the 21st century the breeding population had expanded to at least 100 pairs with the largest colonies being along the Thames with 76 pairs at Beckton and 20 pairs at Billingsgate Market. The following year the population increased by about half to 144–150 breeding pairs, including 13 pairs in central London. In subsequent years they have begun to colonise other areas, including 40 pairs at Highams Park Industrial Estate and 36 pairs at Walthamstow Reservoirs.

Migration

Studies at Barn Elms showed that there was often a difference between the migration routes used on spring and autumn passage, and also in the timing of the peak numbers recorded in autumn; whilst most peak counts were in August, in some years the peaks varied between late July and late October. At Staines Reservoir the peak numbers were usually in late July.

Meanwhile, on the Thames in central London further studies showed that they were the most common gull between early April and early October. During the 1930s various observers commented on the passage of Lesser Black-backs through the London Area. The general consensus was that there was an inland route from the Wash south towards London and through the Brent and Colne Valleys, as well as another route following the Thames. There was also some evidence of a broader front with small numbers seen migrating across many areas of London. The only long-distance ringing recovery in the first half of the 20th century was of a nestling ringed at Walney Island, Lancashire, in June 1933 and recovered at Queen Mary Reservoir in January 1935.

'Scandinavian Lesser Black-backed Gull' — *L. f. intermedius*

Birds of the Scandinavian race have been recorded in most years since the 1930s. The first ones were three at Queen Mary Reservoir on 18 January 1930. Their numbers slowly increased and at Hilfield Park Reservoir there were about 20 on 12 September 1955.

In 1975 they were seen in every month apart from June and some birds overwintered; an analysis at Sevenoaks showed that they outnumbered the British race in April and August. At the time, all these records were erroneously reported as being of the nominate race *fuscus*. In 1979 there were 1,450 Lesser Black-backs at Wraysbury Reservoir on 24 August, of which 70 per cent were considered to be *intermedius*.

Herring Gull — *Larus argentatus*

A common resident and localised breeder

Range: Europe and northern Asia.

Historical

Harting stated that this was 'an accidental visitant; no regular migration in spring and autumn taking place', adding that the few he had seen were all immature birds and were shot on the Thames and reservoirs, mostly in autumn. Towards the end of the 19th century larger numbers started to be recorded. In the severe winter of 1895 some were seen with the many gulls that came up the Thames into London. The following year a flock of more than 100 was seen at Alperton in January; they fed on manure that had been deposited in a field.

20th century

Early in the century they were regular winter visitors along the Thames and in the docks but were also increasingly being seen away from the river. In 1907 Kendall described them as being not uncommon in the Willesden area, noting that it was mainly immature birds that were present and fed with groups of Black-headed Gulls. Pinioned Herring Gulls were kept in St James's Park around this time and they attracted wild birds onto the lake.

During the 1920s and 30s there was a noticeable increase in numbers. By 1931 more than 100 were seen going to roost over Barnes, and the following year in excess of 200 were at Staines in February. Numbers had increased to about 700 at Staines by October 1935 and three months later there were more than 1,000 there.

Herring Gulls continued to spread further out during the 1930s and started to make use of sewage farms for feeding; they were first recorded at Beddington in 1930 and at both Epsom and Watford in 1933. Glegg described the Herring Gull as 'a common and increasing winter resident, but generally in smaller numbers than the Common Gull'. He added that flocks numbering hundreds had sometimes been seen but thought that these were probably migrating birds. As was the case with most gulls at the time, the greatest numbers were found on the Thames and the neighbouring reservoirs, with smaller numbers seen elsewhere. They were mainly recorded between September and March.

The increase continued and by the early 1950s about 10,000 were roosting on the reservoirs. During the day large numbers were present along the Thames; for example there were almost 700 between Barnes and Woolwich in January 1950, and flocks of several hundred were seen at other favoured feeding and resting areas. Ringing recoveries in the first half of the 20th century showed that the majority of overwintering birds originated in Scandinavia. In January 1958 about 15,000 roosted on the Chingford Reservoirs and another 4,000 roosted at Hilfield Park Reservoir. A coordinated roost count in December 1963 produced a total of about 28,000 Herring Gulls on the reservoirs.

The gull-roost survey in January 1983 could find only 14,000, a massive decline of almost 70 per cent since the previous survey in 1969. In October 1985 about 7,500 roosted at Wraysbury Reservoir in October. There were no roost counts in the 1990s.

21st century

Herring Gulls can be found throughout the year at most sites, although the population is greatest during the winter months when Scandinavian birds of the race *argentatus* are present. The largest counts have been 6,940 at Beddington in 2009 and 8,000 roosting on Queen Mary Reservoir during the gull-roost survey in 2003/04–2005/06.

Breeding

In 1961–62 a pair attempted to nest at London Zoo and breeding was confirmed in 1963. The following year the same pair bred again and another pair built a nest in St James's Park. In 1965 three pairs summered in Regent's Park and a pair was present in St James's Park. From 1966 to 1969 several pairs summered in the parks and at least one pair bred annually. In 1968 a pair also bred at Lords cricket ground and subsequently bred there annually into the 1980s. The total breeding population barely increased and by 1980 there were still only four breeding pairs in central London.

The 1990s saw the biggest increase in breeding numbers. In 1990 at least six pairs were breeding in a colony in Kings Cross and one pair probably bred near St James's Park. In 1994 the breeding population had risen to 18 pairs, including the first ones outside Inner London at Greenwich and Poplar Docks. By 1998 there were at least 33 breeding pairs, including at least 15 at Billingsgate Docks and 12 at Beckton, and the following year the breeding population had increased to 53 pairs.

At the start of the 21st century the breeding population was continuing to increase, particularly at the two main colonies where 40 pairs bred at Billingsgate Market and 34 pairs bred at Beckton. Elsewhere at least 21 pairs bred at six other sites, including a burgeoning colony of at least eight pairs at Paddington station. They have since spread out from the Thames and central London and a few small colonies can now be found on rooftops in the suburbs.

Facts and Figures

The largest non-roost count was 6,940 birds at Beddington SF in October 2009.

Yellow-legged Gull *Larus michahellis*

An uncommon winter visitor

Range: southern Europe and North Africa; regular post-breeding visitor to southern UK.

20th century

Large-gull taxonomy and subspecific identification were very much in their infancy in the 1960s and 70s, just at the time when this bird first began arriving in the UK. At first these 'Herring Gulls' with yellow legs were thought to be of the race *omissus*; the first occurrence in the London Area was at Barn Elms Reservoirs on 10 October 1968; subsequent records in the London Area were also believed to be of birds of this race. By the early 1970s it had been determined that Herring Gulls of the *michahellis* subspecies had become regular visitors to Kent and at least one at Sevenoaks in September 1976 was considered to be this subspecies. The following year similar birds were seen at seven sites, including several on the Thames.

In 1984 there was a large increase in the number of sightings of these gulls, particularly on or by the Thames; at West Thurrock the peak count was 16 on 2 September; and at Barn Elms Reservoirs there were up to 15 in August and September. This was an unprecedented influx at the time with most being seen between June and October and a few remaining in winter.

The following year there were sightings at 24 locations, more than double the previous year; the peak month was August when at least 40 were seen. They continued to increase each year with peak counts of 60 at Littlebrook in 1986, and 70 at Stone Marsh on 22 July 1987. The rubbish tip at Stone Marsh quickly became their favoured location, with a peak of 135 in October 1989; far fewer were seen after autumn, with only 26 present in December.

The first ringing recovery was made in 1990 when a bird retrapped at London City Airport on 27 August was found to have been ringed originally in Sardinia on 3 June 1988. In 1990 there were fewer Yellow-legged Gulls than usual on the Lower Thames, especially on the south side, following the closure of the tip at Stone. However, the following year large numbers were attracted to the tip at Rainham with 220 present in August (Dennis). In 1992 up to 200 were again present at Rainham; away from the Lower Thames the maximum count was 12 at Queen Mary Reservoir. There was a slow decrease in the numbers on the Thames and by 1998 the peak count was just 50 at Rainham Marshes but in west London there were 97 at Staines Reservoir in August.

21st century

By the beginning of the 21st century they had been known in the London Area for about 30 years and had increased to such an extent that they were regular visitors that could be seen at any time of year. There is still a large post-breeding peak in July and August, most apparent along the Thames. Numbers are not as high as they were in the late 1980s, however, probably due to the closure of landfill sites. The highest counts are usually on the west London reservoirs, with the peak being 101 on King George VI Reservoir on 15 August 2008.

Facts and Figures

The largest count was 220 birds at Rainham Marshes in August 1991.

Caspian Gull *Larus cachinnans*

An uncommon winter visitor

Range: eastern Europe and Asia; scarce visitor to UK.

20th century

This elegant and striking gull achieved full taxonomic species status in 2007 when it was recognised as such by the BOU; previously it had been considered to be a subspecies of Herring Gull. In Europe it has recently extended its breeding range north and westwards from the Caspian Sea area and now breeds in Poland and eastern Germany. They used to winter no closer than the Middle East but since the breeding range has extended, some birds now winter in Western Europe.

It was first identified in Britain in 1995 at Mucking Tip, Essex and the following year up to eight were seen. In the early days the identification criteria were not fully understood and there was a lot of confusion with other large gulls.

The first record for the London Area was initially thought to be that of an adult which wintered on King George V Dock in 1998/99 and again in subsequent winters. This was very quickly followed by two more the same winter. These were accepted by the BBRC but doubt has recently been cast on the identification of the first two and they are currently being reviewed. Assuming those records will no longer stand, the first confirmed Caspian Gull for the London Area was an adult along the Thames at Twickenham which was first seen on 31 December 1998 and remained around Eel Pie Island until at least 17 January 1999.

21st century

By the beginning of the 21st century it was recognised that this was a regular, albeit very scarce, visitor to southeast England, particularly in the Thames Estuary. It has become an annual visitor to the London Area, typically seen on the Lower Thames Marshes, as well as at Beddington and the west London reservoirs; most occur between October and February.

Iceland Gull *Larus glaucoides*

A scarce winter visitor

Range: breeds Arctic Circle, winters northern Europe and North America.

20th century

Between 1939 and 1945 there were about 10 reported Iceland Gulls in the London Area. Of these only one can definitely be confirmed: an adult in Hyde Park and Kensington Gardens, which was present from 20 February to 10 March 1942; during the course of its stay it was seen to kill two pigeons. All the other reports were considered by Fitter (1949) probably to relate to Glaucous Gull, particularly those seen during the winter of 1941/42 when there was a considerable influx of Glaucous Gulls down the East Coast but no Iceland Gulls were recorded outside the London Area.

The second confirmed record was of an immature at Barn Elms and Lonsdale Road Reservoirs on 11–12 April 1955. Following this they were seen annually until 1964; all of these were between January and April. From 1965 to 1982 only three more were seen, in 1968, 1970 and 1980; two of these were at Sevenoaks and the third was at nearby Ruxley GP.

During the 1980s Iceland Gulls started to become more regular and were recorded annually between 1983 and 1988; the majority of these were on the west London reservoirs with six in 1985 being the best year. The peak decade was the 1990s when an average of just over four birds per year was seen; a record 11 were seen in 1998.

21st century

By the start of the 21st century Iceland Gull had become a scarce annual winter visitor with most sightings between February and April. Eight were seen in both 2001 and 2005 but there were none in 2006, showing the unpredictable nature of this gull. Most of the sightings come from three locations: Beddington, Rainham Marshes and the roost at Queen Mother Reservoir. There was an exceptional influx into the country during the winter of 2011/12; many more than usual were seen in the London Area, including at least 10 different birds at Beddington, one of which remained into June.

Unacceptable Records

The following were published in LBRs but were later considered by Fitter to be unproven: at Barn Elms Reservoirs on 15 April 1939; at Staines Reservoir from 25 December 1939 to 16 January 1940 and again from 11 to 18 February; at Lonsdale Road Reservoir on 15 December 1941; by Hammersmith Bridge from 26 December 1941 to 18 January 1942, and again from 7 February to 23 March; in Kensington Gardens on 8 March, which was thought to have been the Hammersmith bird as it was not seen there on that particular day; at Hammersmith from 5 to 31 December 1942; an older bird at Hammersmith from 2 January to 28 February 1943 was thought probably to have been the previous winter's bird; on the Thames and at Lonsdale Road Reservoir from 17 to 25 November 1943; and at Brent Reservoir on 18 December 1943.

Facts and Figures

The earliest migrant was on 24 October 2007 at Amwell GP.
The latest migrant was on 29 May 2009 at Beddington SF.

'Kumlien's Gull' *L. g. kumlieni*

Although there is still a taxonomic debate about the validity of this subspecies, the following three records have been accepted. Another record of one at Girling Reservoir on 10 February 2003 was not proven conclusively.

1997: Rainham Marshes, adult on 13 December
2001: Langley, adult briefly before flying to roost at Queen Mother Reservoir on 29 March
2007: Queen Mother Reservoir, one roosted on 15 February

Glaucous-winged Gull *Larus glaucescens*

A very rare vagrant (1)

Range: western North America; vagrant to UK.

This individual was the first to be recorded in the UK and was originally found at Hempsted, Gloucestershire, on 15 December 2006 where it was trapped and ringed. It reappeared on the Towy Estuary, Carmarthenshire, in February 2007 and returned to the landfill site at Hempsted in March before making its last appearance in the UK at Beddington.

2007:   Beddington SF, third-summer on 18 April

[Slaty-backed Gull] *Larus schistisagus*

A very rare vagrant (1)

Range: eastern Asia; very rare vagrant to Europe.

A record of one in 2011 is currently being assessed by the BBRC/BOU and would constitute the first British record. It was also seen at various sites in Essex, mostly in the Pitsea area, between its appearances at Rainham.

2011:   Rainham Marshes, on 13–14 January and 16–17, 20 and 24–26 February

Glaucous Gull *Larus hyperboreus*

An uncommon winter visitor

Range: breeds Iceland and Arctic Circle, winters northern and western Europe.

Historical

This was first recorded by Hamilton in 1879 on his list of wild birds seen in London.

20th century

In 1902 Pigott stated that one had been seen circling over 'the ornamental water' (probably referring to St James's Park) by a careful observer who had become familiar with the species in Iceland; the date of this sighting was not recorded. The next record was of an adult in St James's Park on 20 March 1915. This was followed by one on Staines Reservoir from 9 to 13 February 1934.

In the 1940s they started to be seen more often in the London Area. One on the Round Pond in Kensington Gardens on 25 December 1941 coincided with an unusually large influx of this species along the east coast of Britain, and two more were seen later in the winter. One was seen the next winter and then again on the Thames in the Barnes area during the following five winters. Additionally, another adult was seen in St James's Park on 21–22 November 1949.

At Swanscombe Marshes, an immature was seen on the unseasonable date of 30 July 1950 the first record outside the usual period of November to March. Apart from that isolated summer record, the 1950s continued in much the same way with one or two almost every winter apart from in 1958 when there were five. They were seen annually during the 1960s apart from in 1967. During the severe winter in 1962/63 there were three between January and March. In 1966 an immature was present at Queen Mary and Staines Reservoirs between 30 March and 28 May.

In the 1970s there was a shift away from the west London reservoirs to the Lower Thames Marshes, particularly Rainham. The peak year was 1979 when probably 10 were recorded, seven between January and

March and three in December. At least 80 were seen during the 1980s, mainly as a result of a series of cold winters. In early 1986 there was a cold air flow from Siberia, which resulted in the coldest February since 1947, and this was responsible for a large influx early in the year with 15–18 seen; there were six more at the end of the year. Most were in the Lea Valley or on the west London reservoirs, but there were also sightings at Kensal Green Cemetery, Stoke Newington Reservoirs, West Hyde and Wimbledon Common.

During the 1990s there was an average of just over five a year compared to eight in the previous decade. Although they were usually recorded more frequently than Iceland Gull, there was only one in 1994 compared to six of the latter species.

21st century

Since 2000 the annual total has decreased to an average of four, ranging from 12 in 2000 to just one in 2004. The favoured locations are the Lea Valley, Rainham Marshes, Beddington and the west London reservoirs.

Records Not Accepted

There was a 19th-century record of an immature bird purchased in London in a fresh state by Mr Bartlett during the winter of 1838. Harting noted that 'this winter was a very severe one, and numbers of rare wild fowl visited us from the north' and thought that it was not improbable that the bird in question was killed in London. However, as no further details are available as to exactly where it was shot this record cannot stand as the first for the London Area.

There are many records of birds that were published in LBRs as Glaucous/Iceland Gull as they were not specifically identified. These are not included in this book along with a further dozen records that were originally reported as Iceland Gull. See that section for more details.

Great Black-backed Gull *Larus marinus*

An uncommon winter visitor and rare breeder

Range: northern and western Europe.

Historical

In the 19th century this species bred in the Thames Estuary and would occasionally wander upriver into the London Area. The earliest record was of one shot on the Thames at Putney during cold weather in 1841; this is illustrated in Yarrell's book of British birds. The only other documented records are from Brent Reservoir: one shot many years prior to 1866; one in March 1862; two in January 1864; and one in July 1867. There were three records on the Thames in central London: at Lambeth in 1879; at Temple Gardens on 5 November 1895; and three at Blackfriars Bridge on 4 February 1896. In the far north of the London Area a flock was seen over Balls Park, Hertfordshire, on 30 August 1890, one of which was later shot at Bayford.

20th century

Between 1900 and 1921 there were six further records; after this the species started to be seen on a more regular basis, although still in relatively small numbers. Most were on the Thames with the rest coming from the large reservoirs and a few others elsewhere.

During the 1930s the species became a regular winter visitor; the first group of more than 10 was seen in 1931 along the Thames near Greenwich. It continued to increase in the London Area and by the end of the 1940s could be seen at almost any time of year on the Thames except in June and July. One particular bird had to be shot in 1940 as it had developed the habit of killing and eating pinioned ducks in St James's Park. In 1948 large numbers were seen at Hampton Filter Beds, peaking at 188 on 31 January.

By the mid-1950s there were roosts of several hundred on the reservoirs near Hampton; up to 500 roosted at Walton Reservoirs. They were usually seen on the Thames and at rubbish dumps as well as some sewage farms, but were rarely recorded on buildings except at Billingsgate Fish Market where they would compete with other gulls for scraps of fish. They could be found at any time of year although they were still uncommon between May and mid-July; the peak numbers were recorded in December and January. When Hilfield Park Reservoir was completed it was soon used as a roost despite being 17 kilometres from the Thames. In 1956–57 up to 25 roosted there; by January 1958 the roost had expanded to about 200 and on 6 February 1960 numbers had reached more than 300.

During the early 1960s the species continued to increase across the London Area. In December 1961 there were 850–1,000 at Walton Reservoirs and up to about 700 on the Lea Valley reservoirs. By 1965 about 1,500 were roosting on Queen Elizabeth II Reservoir in winter. A census of roosting gulls in 1968 found 5,865 Great Black-backs compared to 6,325 in 1963. This was the only gull whose numbers decreased over the five-year period. Ringing recoveries have given clues to some of these gulls' origins: two birds ringed as chicks in Russia and one in Norway were recovered in the London Area in 1962.

The decrease continued from the mid-1960s. The peak count in 1971 was about 800 at Walton Reservoirs in January. In 1980 the total wintering population was thought typically to be below 1,000 birds: 800 at Rainham on 1 January and a similar number roosted on Queen Elizabeth II Reservoir on 20 January. A count of roosting gulls in January 1983 produced a total of just 728, a decline of 76 per cent since the previous count in 1969. The peak count in the 1990s was 400 at a pre-roost gathering at Nazeing GP in 1994.

21st century

Since 2000 this large gull can frequently be seen in winter along the Lower Thames, in the Lea Valley and on the west London reservoirs but it is relatively uncommon elsewhere. The peak counts have mainly been at Rainham Marshes where they increased from 150 in 2000 to 1,029 in 2005. The roost counts have decreased with only 61 being recorded in the gull-roost survey in 2003–06, but this total only covered Greater London, not the London recording area. There were higher roost counts in 2009 for the entire area, with 170 at Amwell and 137 on Queen Mother Reservoir.

Breeding

Between 2007 and 2012 a pair was seen at Brent Reservoir during the breeding season and exhibited breeding behaviour, including mating. It is possible that the birds nested in a gull colony around Staples Corner during this period but no juveniles were ever seen on the reservoir. The first confirmed breeding was in 2008 when a pair bred on a barge moored on the Thames off Wandsworth Park; the single youngster did not survive to fledging. A pair probably bred at Barking Bay in 2009 and a pair bred successfully on the Isle of Dogs in 2010.

Facts and Figures

The highest ever count was about 1,500 birds at Queen Elizabeth II Reservoir on 10 January 1965.

Sooty Tern *Onychoprion fuscatus*

A very rare vagrant (1)

Range: breeds in tropics, winters at sea; vagrant to UK.

The only record was in August 1971 when an immature was seen by two observers at Staines Reservoir. Despite it being a highly pelagic species, there have been similar records elsewhere in Britain: a first-summer found in Northamptonshire in May 1980 and two shot inland in the 19th century.

1971: Staines Reservoir, on 18 August

Bridled Tern *Onychoprion anaethetus*

A very rare vagrant (1)

Range: breeds North Africa and tropics, winters at sea; vagrant to UK.

This was yet another example of a bird that found its way into the London Area along the Thames but could so easily have gone unnoticed as it was only present for half an hour one June afternoon. Fortunately, that part of the river at West Thurrock and Swanscombe was under scrutiny because of the presence of a White-winged Black Tern, and much to the amazement of those present a Bridled Tern flew into view at 15:50 hrs. After it flew off it was later relocated at Hanningfield Reservoir, Essex, where it remained until dusk. This was the thirteenth British record.

1991: West Thurrock, adult on the Thames on 2 June

Little Tern *Sternula albifrons*

An uncommon passage migrant

Range: breeds Europe and Asia, winters Middle East and Africa.

Historical

In the 19th century Harting described this as an irregular passage migrant in spring and autumn and the least common tern seen in the London Area. It was first recorded by Bond in 1843 when he included it on his list of birds from Brent Reservoir; in 1896 Read stated that there were specimens of this species also taken at Brent Reservoir in the collection at the Welsh Harp Inn. Elsewhere, the only known records are of one at Ruislip Lido on 27 April 1884; two on the Thames at Petersham on 28 August 1895; and two shot at Ashtead on 18 August 1899.

20th century

In 1909 Ticehurst stated that Little Terns could be seen as far up the Thames as Dartford in autumn. The only records up to 1920 are of single birds at Putney on 11 August 1911 and on Highgate Ponds, Hampstead Heath, on 30 April 1913.

In the 30 years between 1925 and 1954 there were about 50 records, showing that this species was a scarce passage migrant. During this period there were 19 records in spring, all between 21 April and 25 May. The autumn migration spanned a longer period, from 4 August to 22 October, although more than half of these records were from September. Although rarely recorded on the Thames, more than half the records came from the reservoirs in the Thames Valley, especially Barn Elms and Staines; however, there was the occasional overland record. Most records were of single birds but there were also occasional small flocks, such as 11 at Staines on 3 September 1954, seven at Barn Elms Reservoirs on 10 May 1947, and six at both Brent Reservoir on 23 August 1950 and Hamper Mill on 22 September 1935. The only record in Inner London was of a bird on the Round Pond in Kensington Gardens in September 1932.

From the 1960s onwards it became an annual visitor, mostly in small numbers but with the occasional larger flock. In 1987 flocks of 19 and 14 flew over Queen Mary Reservoir on 20 August. In 1992 a flock of 18 was at Queen Mary Reservoir on 11 September. An analysis of all records between 1945 and 1983 showed the peak passage was during the first week of May.

21st century

Since 2000 the species has become much scarcer and in the 10 years up to 2009 there was an average of just 19 birds a year; the highest annual total in this period was 41 in 2008. Apart from a flock of 15 that flew past East India Dock Basin on 12 September 2000 all records were of single birds or very small groups. Little Tern is an erratic visitor in the London Area, mainly occurring on the reservoirs; although most birds are seen on passage there have been some midsummer sightings; few birds linger for more than an hour.

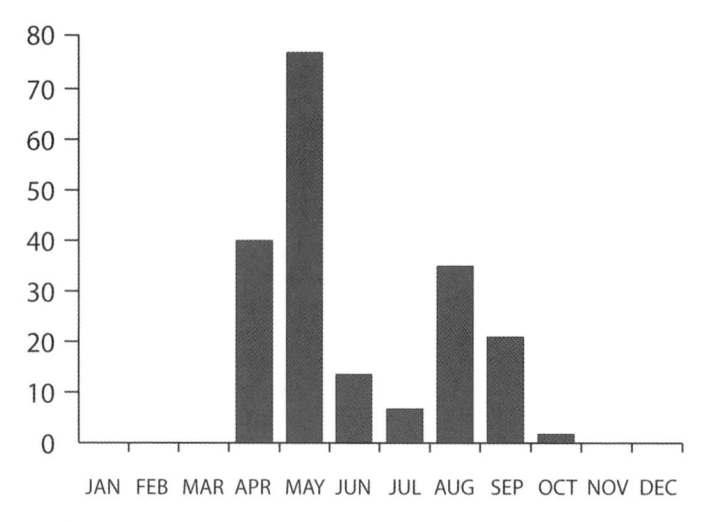

Seasonal occurrence of Little Tern: 2000–2009.

Facts and Figures

The earliest spring sighting was on 12 April 1991 at Shepperton GP.
The latest sighting was on 25 October 1974 at Barn Elms Reservoirs.
The largest flock was of 19 birds at Queen Mary Reservoir on 20 August 1987.

Gull-billed Tern *Gelochelidon nilotica*

A very rare vagrant (1)

Range: cosmopolitan, European population winters Africa; vagrant to UK, mostly coastal.

The only record occurred in 1980 when one appeared at Barn Elms Reservoirs after a thunderstorm. This was the first record of one in an inland county in Britain.

1980: Barn Elms Reservoirs, on 26 July

Caspian Tern *Hydroprogne caspia*

A very rare vagrant (3)

Range: cosmopolitan, European population winters Africa; vagrant to UK.

There have been three in the London Area, all in July and August. The first was in 1961, just over two weeks after the capital's first-ever Whiskered Tern. It was seen at Walthamstow Reservoirs and was presumed to be the same bird seen later at King George V Reservoir. It is likely that this same bird was also responsible for all the British sightings between these two dates (Langney Point, East Sussex, on 14 July; Minsmere, Suffolk, on 16 July; and Abberton Reservoir, Essex, on 6 August). Caspian Tern was a very rare bird in Britain at the time – prior to this record there had only been a dozen since the beginning of the 20th century. The second and third records were both birds seen seen flying over Staines Reservoir.

1961: Walthamstow Reservoirs, adult on 9 July, then at King George V Reservoir on 26–27 August

1979: Staines Reservoir, on 29 July
1982: Staines Reservoir, on 11 August

Whiskered Tern *Chlidonias hybrida*

A very rare vagrant (4)

Range: breeds Europe, Asia, Africa and Australasia; European population winters Africa and Asia; rare passage migrant in UK.

Whiskered Tern has been seen on four occasions, all of which were in west London at King George VI and Staines Reservoirs. The first was in 1961 when one was present for two days in late June. Three years later there was a very similar record when one was present for three days in late June. In 1977 a second-year bird was seen on 17 August and there was then a long gap of 28 years until the fourth record in 2005, when a second-summer bird remained for four days. There had been an influx of nine birds into Britain in May 2005 and the London Area individual was also seen in Berkshire and Hampshire.

1961: Staines Reservoir, one on 24–25 June
1964: King George VI Reservoir, one on 22–24 June
1977: Staines Reservoir, second-year on 17 August
2005: King George VI/Staines Reservoir, second-summer on 22–25 May

Black Tern *Chlidonias niger*

An uncommon passage migrant

Range: breeds Europe, Asia and North America, European population winters Africa; passage migrant in UK.

Historical

In the 19th century Harting stated that 'next to the Common Tern this is the commonest of the four species which visit us, and is the earliest to appear in the spring'. This suggests that they were encountered in most years in small numbers although there are very few documented records to back this up.

The only dated records are: of two shot at Chertsey in May 1849; two or three at Aldenham Reservoir on 21 August 1871; and on the Thames at Hampton on 12 May 1877. Additionally, six were shot in the Lea Valley at Broxbourne prior to 1886 and there was one at Gatton prior to 1890.

20th century

In the first quarter of the century there were only 10 records, seven of which were in spring. From 1926 onwards they were seen annually in the London Area, except during the Second World War when access to many reservoirs was barred.

In the 1930s Glegg described it as a passage migrant in small numbers. In spring it was more frequently seen in May than April and there had been three records in June. In autumn it was mainly seen in August and to a lesser extent in September, and had been noted in October on three occasions. Most records came from the large reservoirs and the Thames; typically sightings were of single birds or small parties, the largest gathering being 21 at Staines Reservoir in autumn. In 1935 there was one in Inner London at Hyde Park on 27 May.

From the mid-1940s to the mid-1950s they were seen more frequently, and there were several years during which there were notable influxes during both spring and autumn. The first of these was in 1946 when they were seen at a number of gravel pits and reservoirs in north London, as well as on the Thames, between 30 April and 26 May. The peak occurred during strong north-easterly winds in the second week in May when there were up to 43 at Staines Reservoir. In 1950 there was an exceptional passage through the London Area during spring; on

14 May 153 were counted at nine localities, including 47 at Staines Reservoir. Two days later there were 121 at Staines; the following day there were 63 at King George V Reservoir.

From the late 1950s onwards they were seen in larger numbers. In the autumn of 1958 about 150 were at Queen Mary Reservoir on 27 August. In 1959 there was a large influx on 23 May and at least 150 were seen throughout the London Area, including a single flock of about 100 at Rye Meads. An influx on 13 May 1961 involved up to 260 birds, including flocks of 127 at Staines Reservoir and 80 at Queen Mary Reservoir. In the autumn of 1963 there was a notable passage on 30 August when five flocks totalling 108 birds flew west up the Thames past Swanscombe. In the autumn of 1969 there were at least 190 at Staines Reservoir on 5 August.

In 1974 there was an exceptional passage on 15 September in southern England with at least 635 in the London Area, mainly flying straight through; the highest counts were 182 at Staines Reservoir and 138 at Wraysbury Reservoir. The majority of these passed through west London with only 55 seen in the Lea Valley and none further east. In 1979 there was another good autumn passage, mainly along the Thames at West Thurrock, where more than 100 were seen on 8 August and 110 on the 26th. Yet another large passage occurred the following year when a total of 777 was seen, mostly during the autumn passage and peaking on 20 September when there were 300 at West Thurrock. Further large flocks occurred in autumn in 1985, 1989 and 1990, the biggest comprising 170 birds at Staines Reservoir on 19 August 1985.

In the autumn of 1992 there was a spectacular and unprecedented passage in south-east England with thousands of birds arriving from the Continent. For example, on 6 September around 10,000 flew past Dungeness, Kent. Some of these made their way up the Thames into the London Area, and in August and September about 2,500 bird/days were logged. The first wave of migrants appeared on 19 August then on 6 September there was a sudden massive influx. The biggest counts were 300 at West Thurrock, 265 at Barn Elms Reservoirs, 200 at Barking and 140 at Staines Reservoir; 600–1,000 were seen in the London Area during the day. There were further influxes over the next week, notably on 13 September with 216 at Barn Elms and 100 at West Thurrock. There were no further large influxes until 1998 when almost 800 were seen in autumn, the peak being 100 at Queen Mother Reservoir on 1 September.

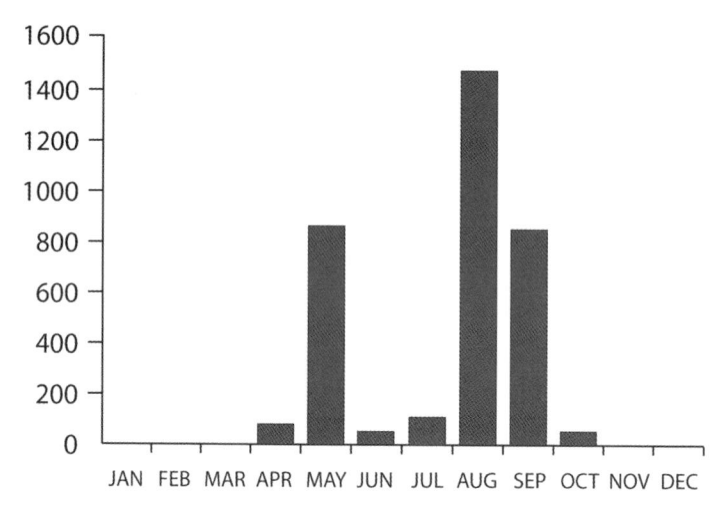

Seasonal occurrence of Black Tern: 2000–2009.

21st century

Black Tern passage through the London Area has continued to be unpredictable since 2000, although there have been fewer big influxes. The largest influx was in the autumn of 2001 when around 930 were seen; the peak count was 85 at King George VI Reservoir on 23 August. They are most often seen on the large reservoirs, particularly in west London as well as on the Thames.

Migration

There have been a number of studies into the migration of Black Terns and their occurrence in the UK. These have shown that spring influxes occur when easterly or north-easterly winds off the coast of France follow an increase in temperature in France or Spain during the passage period. It is unclear what route these birds use into the London Area as there is no evidence of them entering the Thames Estuary. However, as large flocks are often recorded on reservoirs near the Thames, such as at Barn Elms and Staines, Black Terns may be flying along the course of the river but at too great a height to see. They may then drop in to feed on the reservoirs as the Thames narrows. In autumn two routes appear to be taken, an inland one from the north-west and another coming up the Thames, although the birds rarely venture further upstream than Barking Bay before flying back out on the receding tide.

In spring, birds have been recorded between 10 April and early June. Occasional birds are seen in late June and early July with some early autumn migrants noted later in July and early August. The main autumn migration period starts in late August and continues throughout September into early October, and there is a small number of November records.

Facts and Figures

The earliest migrant was on 10 April 1960 at Staines Reservoir.
The latest migrant was on 24 November 1968 at Queen Mary Reservoir.
The largest flock was of 300 birds at West Thurrock on 20 September 1980 and 6 September 1992.

White-winged Black Tern *Chlidonias leucopterus*

A scarce passage migrant (40)

Range: breeds eastern Europe and Asia, European population winters Africa; scarce passage migrant in UK.

Historical

The only 19th-century record was in May 1883 when one was shot at Brent Reservoir. This record was reviewed in 1935 by Glegg who decided not to accept it on the grounds that the specimen had not been examined by an experienced ornithologist; however, as it was shot by the gamekeeper of the reservoir who was familiar with many of the birds there, the record should still stand.

20th century

There were no records in the first half of the 20th century but there was a dramatic change of status in the second half when there were 28 records. The upturn began in the 1960s when six birds were seen, all on the west London reservoirs. The first of these was an adult in the spring of 1961 at Queen Mary Reservoir at a time when a large number of Black Terns arrived in the London Area. The next five were all in autumn on the west London reservoirs.

The 1970s saw a large increase in the number of sightings, beginning with six in 1970 including two together on Queen Mary Reservoir on 3 May. Between 1973 and 1978 another eight were seen, seven of which were on Queen Mary Reservoir; all of these were in autumn and were immatures apart from one in 1978. In September 1979 an immature was present on the West Thurrock outflow for three days; this was the first one on the Thames and ended a long sequence of sightings from the reservoirs. The next four were also on the Thames.

Another five were seen during the rest of the 1990s, four on the reservoirs and one on the Thames at Woolwich. Two of these were adults in breeding plumage: at Woolwich on 4-5 July 1996 and at Brent Reservoir on 27 July 1996.

21st century

Up to 2010 another nine have been seen, one in May and the rest between July and October. There were four on the reservoirs, three on the Thames, one at the London Wetland Centre and the first record in Inner London: a juvenile at Hyde Park and Kensington Gardens from 26 September-4th October 2010.

Sandwich Tern *Sterna sandvicensis*

An uncommon passage migrant

Range: breeds coastal Europe, Americas and Caribbean, European population winters southern Europe and African coasts.

Historical

This species could have been named 'Greenwich Tern' as one was shot on the Thames at Greenwich in about 1784, the same time that the first specimen was obtained in the UK, at Sandwich. The only known record in the 19th century was in 1820 when Donovan illustrated one that had been shot in Chelsea Road.

20th century

Up to 1930 there were only three more records: over Hyde Park on 2 September 1908; in the Lea Valley on 3 May 1913; and two on the Thames at Barnes on 28 August 1926.

During the 1930s they started to be seen on a more regular basis and were seen in five years between 1931 and 1936. Glegg put this down to the fact that they were breeding in increasing numbers on the East Coast of England. From the mid-1930s to the mid-1950s they were recorded almost annually. Most of the records were of birds seen on the large reservoirs with very few on the Thames. During spring they were seen between 25 April and 28 May; apart from a handful of records in July, the main passage in autumn was between 20 August and 10 October with 40 records in this period. The only later one was in 1934 when an individual remained at Staines Reservoir between 19 October and 4 November. Most of these records were of single birds or small groups and there were only three double-figure flocks.

From 1950 they were seen annually, mostly on the large reservoirs or on the Thames. In the autumn of 1964 the total passage through Staines Reservoir was 87 birds, all of which were seen flying west; the peak count was 38 on 24 September. In 1968 a flock was heard migrating at night over Potters Bar on 12 September at 23:00 hrs; three days later a flock of about 90 flew over Barn Elms Reservoirs in heavy rain. More nocturnal migrants were heard at both Chislehurst and Walton-on-Thames in August 1972, and in 1974 a flock of 30–50 was seen flying over Shirley after dark at 21:15 hrs but illuminated by street lights. There were several large flocks during the autumn of 1973, including 42 over Brent Reservoir on 9 September and 47 over Broxbourne GP the next day. In 1975 at least 40 flew over Queen Mary Reservoir on 30 August.

21st century

The number of Sandwich Terns occurring in the London Area has decreased significantly since the start of the 21st century. Between 2000 and 2004 an average of just over 160 birds was seen each year, the highest-ever average over a five-year period. This included a record 218 in 2001, of which all but nine were in autumn; the highest count

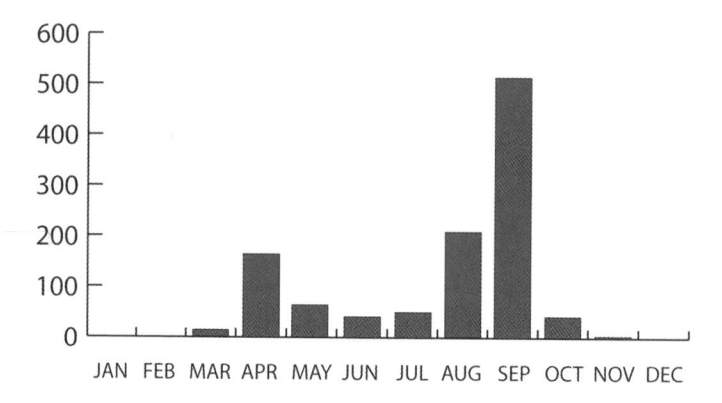

Seasonal occurrence of Sandwich Tern: 2000–2009.

was 55 at the Wetland Centre on 29 September. However, between 2005 and 2009 the average annual total more than halved to just 71.

Facts and Figures

The earliest migrants were two on 12 March 2003 in Regent's Park.
The latest date was 4 November 1934 at Staines Reservoir.
The largest flock was of 90 birds at Barn Elms Reservoirs on 15 September 1968.

Common Tern — *Sterna hirundo*

A common summer visitor and localised breeder

Range: breeds Europe, Asia, North Africa and Americas, European population mostly winters at sea off Africa.

Historical

This was a regular and common passage migrant in the 19th century, arriving in the first week of May, generally on an east wind, and sometimes remaining for several days. The first record was of two dozen on the Grand Surrey Canal, Bermondsey, in 1812. The next dated record was of one taken in Bushy Park in 1831. At Gatton, flocks were said sometimes to occur after storms.

20th century

During the early part of the century there was little change in the pattern of occurrence. In 1906 Hudson wrote that Common Terns were not infrequently recorded on the Thames at Kew in spring, and in 1909 Ticehurst stated that occasional birds were seen on the Thames in spring and that during August and September they could be seen coming up the river on the tide. In Inner London they had been seen in both Hyde Park and Kensington Gardens, the first sighting involving about 20 terns believed to be of this species that flew east over the Serpentine on 12 May 1906.

In 1925 two were seen in winter, on the ground at close range at Tooting Bec Common on 16 December. The next day the same observer saw a flock of about a dozen terns which he believed to be of this species flying high over West Norwood. A further winter record came from Walthamstow Reservoirs in 1932 when one was seen on 9 December; one found dead at King George V Reservoir on 11 February 1933 may have been the same bird.

Glegg noted that in spring most birds went through during May, with the rest in April. In autumn the species was recorded between July and October, with the peak passage being in August. Most sightings were of single birds or small parties, with the largest group being 30–36 at Barn Elms Reservoirs. In the 1940s and 50s the species was described as a regular passage migrant, particularly on the Thames and larger reservoirs, with occasional influxes.

Large groups of terns were rarely specifically identified and the following details could apply to either Common or Arctic Terns, or perhaps mixed flocks. In 1950 there were up to 500 on Staines Reservoir on 23 August, and in 1952 there were 304 at Staines Reservoir on 28 April. During the autumn of 1971 terns were first noted feeding over the power-station outflow at Dartford. In 1973 up to 330, most of which were probably Common Terns, were on the Thames at Thurrock in early September.

From the 1980s onwards flocks of terns were usually specifically identified and all the following flocks relate solely to Common Tern. In 1983 there were 150–200 birds at Woolwich Ferry on 5 September. In 1984 there were two on the Thames at Littlebrook Power Station on 8 December, one of which remained until the 26th. In 1988 about 200 were seen at West Thurrock on 11 August. In the autumn of 1989 an exceptional passage began with a flock of at least 150 over Beddington on 17 August; a total of 462 arrived on 20 August on six west London reservoirs, including 220 at Staines; 385 flew upriver past Battersea Park and 122 were at Staines Reservoir on the 26th; and there were 200 at West Thurrock on 16 September.

Large numbers continued to be seen on passage on the Lower Thames during the early 1990s, particularly at West Thurrock where terns were attracted by the power-station outflow. There was a very large passage in the autumn of 1992 in west London. Highlights were as follows: at Barn Elms Reservoirs 170 flew over on 20 August and 127 went over two days later; on 30 August there were 400 at King George V Reservoir and 155 at Barn Elms; and on 6 September 150 passed over Staines Reservoir in one hour and 123 flew over Wraysbury Reservoir. In 1994 the peak day for autumn passage was 15 September when at least 383 included 250 on the Thames at Littlebrook. In 1999 one flew up the Thames past Dartford Marsh on 1 January.

21st century

Although feeding flocks no longer linger on the Thames since the closure of West Thurrock Power Station, there are still occasional large movements along the river. In 2000 the highest count was 197 birds at Barking on 24 July, a relatively early date for the autumn peak. In 2007 a total of 840 passed Rainham, along the Thames, on 22 August. Elsewhere, the largest numbers are seen on the reservoirs but small groups may be seen anywhere on passage throughout the London Area.

Breeding

The first breeding took place in 1958 on the causeway at Queen Mary Reservoir; two eggs were laid but were thought to have been predated.

In 1962 Common Terns were present during the breeding season in the Lea Valley, at South Weald and Ockenden, although there were no signs of any nesting activity. The first successful breeding was on a gravel pit at Cheshunt in 1963. Several summered in the Lea Valley during the next two years but did not nest. In 1966 two pairs bred at Broxbourne GP and the following year this new colony increased to four pairs, while in west London two birds summered and may have attempted to breed at Shepperton. There was another increase in 1968 when up to 17 summered at Broxbourne; additionally, the first successful breeding occurred in west London when a pair bred at Colnbrook. In 1969 up to five pairs nested at Broxbourne; a pair bred in the Staines area; and up to three birds were present during the summer at South Ockenden.

By the beginning of the 1970s the breeding population had increased to 12 pairs and as the decade progressed the numbers continued to rise, reaching 23 pairs by 1974. The provision of nesting rafts at Rye Meads helped to establish a viable breeding population as 16 pairs bred and fledged 35 young. At the end of the decade there were at least 49 breeding pairs in the London Area and a minimum of 96 young were raised; the largest concentration was 38 pairs at Rye Meads.

The population continued to increase and by 1989 the breeding population exceeded 100 pairs for the first time. In 1994 there were at least 170 pairs, almost all of which bred on artificial rafts, the main exception being at Staines Reservoir where 31 pairs bred on islands on the partly drained southern basin. The provision of new rafts in the Docklands led to immediate success as two pairs bred. However, there were setbacks; virtually all the 22 chicks hatched in 1995 at Stocker's Lake were predated, probably by a Mink. The landmark of 200 breeding pairs was passed in 1996.

At the beginning of the 21st century the breeding population was still increasing and had reached 306 pairs, an increase of 25 per cent on 1999. The largest colonies were 68 pairs at Walthamstow and 58 pairs at Staines. The latter had just started to suffer as a result of pressure from breeding Black-headed Gulls. In 2001, 332 pairs bred, including 100 pairs at Rye Meads. The breeding population fell for the first time in 2002 and the colony at Rye Meads declined by over 50 per cent. In subsequent years the total number of breeding pairs varied from 200 to 255 but had decreased to about 162 pairs by 2009. The reason for the decline is unclear but it may be partly due to pressure from gulls. Black-headed Gulls have taken over the nesting rafts at Staines Reservoir and Lesser Black-backed Gulls have predated the tern chicks at Brent Reservoir; in 2009 they ate all but one chick but it was early enough in the season for most of the terns to re-lay.

Facts and Figures

The earliest migrants were nine on 9 March 1924 at Walthamstow Reservoirs.
The latest migrant was on 24 October 2004 at Walthamstow Reservoirs.
The largest flock was of up to 500 birds on Staines Reservoir on 23 August 1950.

Roseate Tern   *Sterna dougallii*

A scarce passage migrant (36)

Range: breeds coastal western Europe, Australasia and Americas, European population winters coastal West Africa; rare breeding bird in northern UK and a scarce passage migrant.

The first record for the London Area was of two seen at Brent Reservoir by Harting in 1866; he shot one, which was later identified by Bond as an immature. There were no more records for almost 90 years until one was seen at Perry Oaks SF in May 1953.

The 1960s saw a sharp increase in the number of records and the birds were seen on seven occasions. The first of these was on 12 May 1960 when two flew east over Staines Moor and King George VI Reservoir during a period of pronounced tern passage through the London Area. Apart from two at Sevenoaks in July 1964 all the others were in west London.

In the early 1970s they were seen at Sevenoaks Wildfowl Reserve in four consecutive years, including a group of six in the spring of 1972. After this succession of records only one was seen in the following 28 years, at West Thurrock in 1988 when one was present on the power-station outflow for four days in August. The decrease in numbers matched the large fall in the UK breeding population.

Since the beginning of the 21st century they have been seen more frequently and were recorded 10 times between 2000 and 2012. In 2002 a pair of adults was present at East India Dock Basin and was seen mating on 14 May; the birds were originally ringed at the breeding colony on Coquet Island, Northumberland. Out of the total of 36 birds, 22 were in May and 12 were in autumn between July and September.

1866: Brent Reservoir, two on 16 August
1953: Perry Oaks SF, on 16 May
1960: Staines Moor/King George VI Reservoir, two on 12 May
1961: Barn Elms Reservoirs, two on 18 July
1963: Barn Elms Reservoirs, on 24 July
1964: Sevenoaks WR, two on 11 July
1967: Queen Mary Reservoir, on 6 September
1968: Staines Reservoir, on 17 May
1969: Staines Reservoir, two on 6 May
1970: Staines Reservoir, on 11 May
1970: Sevenoaks WR, on 30 May
1971: Sevenoaks WR, six on 20 May
1972: Sevenoaks WR, on 7 July
1973: Sevenoaks WR, on 25–26 June
1988: West Thurrock, on 20–23 August
2001: Barking Bay, on 18 September

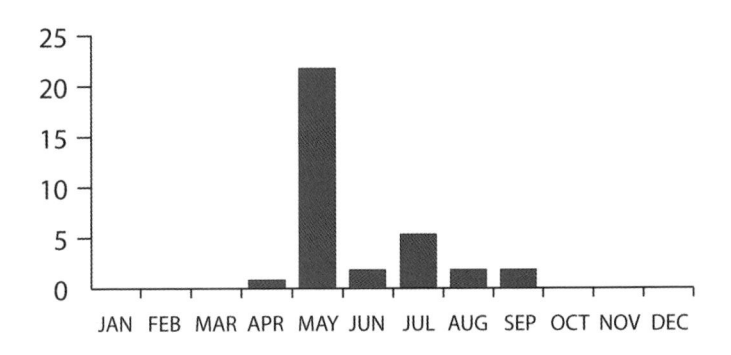

Seasonal occurrence of Roseate Tern: all records.

2002: East India Dock Basin, two on 14 May
2002: Staines Reservoir, on 3 June
2005: Queen Elizabeth II Reservoir, on 19–20 May
2005: Staines Reservoir, on 26 May
2006: East India Dock Basin, on 16 May
2007: Rainham Marshes, on 22 August
2011: Queen Mother Reservoir, on 9–13 May
2011: King George VI/Staines Reservoir, on 10 May
2012: Island Barn/Walton/Staines Reservoir, on 8 May

Unacceptable records

In 1921 four terns were seen at Barn Elms Reservoirs on 28 April, one of which had a rich pink flush on the breast. However, no other identification features were seen and it has not been counted by subsequent authors.

Arctic Tern *Sterna paradisaea*

An uncommon passage migrant

Range: breeds northern Europe, Asia and North America, winters Antarctic Ocean.

Historical

This species was first recorded in 1843 by Bond who included it on his list of birds observed at Brent Reservoir, and Bucknill who stated that it was seen on the Thames 'in some numbers'. The next record was of an immature shot on the Thames at Chertsey Weir on 6 October 1846. Harting stated that although they were less plentiful than Common Tern, 'small flocks visit us annually in spring and autumn, the two species frequently consorting together'. As he wrote this in 1866, there must have been more records than those listed above.

There are a further six documented records in the 19th century, all from Brent Reservoir: two on 19 May 1867; several on 9 October 1869; one shot on 17 October 1870; several on 15 May 1871; several on 8 May 1874; and two shot on 2 September 1881.

20th century

There were no further records for over 40 years. Although this can be partially explained by the cessation of shooting at Brent Reservoir, it seems odd that no birds were recorded from Queen Mary or Staines Reservoirs, which were receiving more observer coverage. It would seem that the difficulty in separating them from Common Tern was the most likely reason for the absence of records.

There were definite records of Arctic Tern in four springs and 10 autumns between 1926 and 1953. None of these sightings exceeded four birds apart from a flock of 15 at Walthamstow Reservoirs on 22 August 1931 and those recorded in the spring of 1947. During the latter spring there was a large passage of terns in the UK, many of which were Arctic; in the London Area, Arctic Terns were seen at several locations, including about 20 at Barn Elms Reservoirs on 26 April.

During the 1960s and 70s very few were identified apart from in 1978 when a flock of 103 was seen at Troy Mill GP on 6 May. By 1980 observers were able to identify more Arctic Terns than they had previously and a minimum of 185 were seen, including a flock of 116 at King George V Reservoir on 26 April. There was a large passage in 1981 with peaks of 162 on 29 April and 191 on 4 May, including 130 at Staines Reservoir on the latter date. There were several other large influxes during the 1980s, including 140 at Staines Reservoir on 29 April 1982; 60 at West Thurrock on 25 August 1983; and 96 at King George V Reservoir on 25 April 1988.

There was a similar pattern in the 1990s, with the maximum numbers occurring in 1998 when about 576 were seen. Most of these – 400 – were noted on 2 May, including 126 north over Banbury Reservoir. Unusually during this peak day only eight were seen on the west London reservoirs.

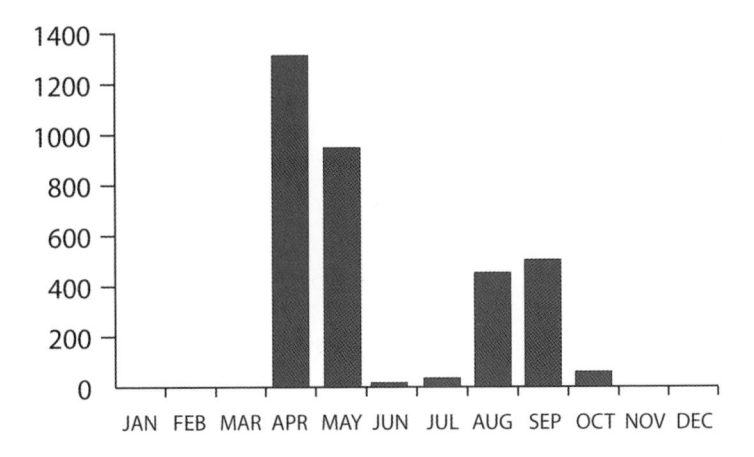

Seasonal occurrence of Arctic Tern: 2000–2009.

21st century

Since the turn of the century there has been a significant increase in the number seen in the London Area, with an average of more than 350 a year. Most of this is due to weather patterns in spring; the perfect conditions for a mass arrival are northerly winds, which halt the bird's overland passage and force them down onto the reservoirs.

The peak passage totals were almost 600 in 2006, including 550 in spring, and a record 931 in 2008. During the latter spring there were at least 195 at Staines Reservoir on 23 April but only two small flocks were seen elsewhere.

Facts and Figures

The largest flock was of 195 birds at Staines Reservoir on 23 April 2008.
The earliest migrant was on 5 April 2003 at Staines Reservoir.
The latest migrant was on 6 November 2005 at West Thurrock.

Guillemot *Uria aalge*

A scarce winter visitor

Range: breeds coastal Europe, North America and North Pacific islands, winters at sea.

Historical

This was a rare visitor in the 19th century with just four documented records: one flew over London Bridge and landed on the Thames by Cannon Street station following a stormy night in the second week of November 1879; one was shot at Aldenham Reservoir in November 1882; one was picked up alive in Russell Square after a gale in the spring of 1883; and five were on the Thames off Greenwich Pier on 20 November 1893. Additionally, Ruegg stated in 1847 that they were occasionally seen in the Woolwich area but gave no details; similarly Ticehurst stated that they could be found in small numbers every winter on the Thames as far upriver as Dartford.

20th century

In the first half of the century they remained a rare visitor and there were just four records: in 1911 an exhausted bird was picked up at Winchmore Hill on 22 November; two were on the Thames by London Bridge on

17 November 1930; one picked up alive at Bunhill Fields Burial Ground on 15 December 1933 subsequently died and was found to be of the northern race *aalge*; one was found in a garden at Limpsfield after a night of severe storms on 14 January 1934.

There were no further records until 1955 when one was seen on the Thames between Westminster Bridge and Lambeth Bridge on 8 October. The only two in the 1960s were both in central London: an oiled bird on the lake in St James's Park on 22 December 1965 and another oiled bird found dead in Regent's Park on 1 December 1968. There was a similar occurrence in 1972 when an oiled bird was found in Hyde Park on 3 January and taken into care; two more were seen in 1972 on the Lower Thames Marshes; and the following year there was one on Queen Elizabeth II Reservoir on 16 September.

The peak period was the 1980s when they were seen in seven years. Most records were on the Thames. In 1983 more than 10,000 were seen on the East Coast towards the end of winter and seven arrived in the London Area between 16 February and 10 March; this arrival followed a small influx of Razorbills. In 1985/86 one overwintered on the Thames around Barnes Bridge from mid-December to mid-February.

The best-ever year was 1986 when at least 90–125 were seen. In late 1985 unusually high numbers of auks, mostly Guillemots, were seen off the Kent coast, mainly in the Channel. In early 1986 they had started to occur in the outer Thames Estuary and several hundred were present at Southend. Easterly winds and cold weather began on 24 January and continued for most of February. The first London Area bird arrived on 26 January, on the Thames at Westminster. The main influx occurred from the middle of February onwards and the first ones were again seen in central London, with individuals picked up in St James's Park and Regent's Park. Several more arrived over the next few days and there was then a lull before the next arrival. On 26 February nine flew along the river past Mortlake and a few others were seen elsewhere. Two days later 28–42 were seen, mostly on the Thames, with flocks of 10 at Putney, eight at Rainham and seven at Chelsea Reach. The peak day was on 1 March, a day of light snow showers. In west London at least a dozen were seen between Teddington Lock and Grosvenor Bridge but the main movement occurred with the incoming tide when 27 passed London Bridge; most were carried along on the tide but a few actually flew west. Up to 14 were present on the river over the next few days and a few remained for most of the month. Although the vast majority were on the Thames, small groups were seen elsewhere with at least nine on the Grand Union Canal at Brentford from 27 February to 5 March when the lock gates were opened and they swam out onto the Thames. At least 20 were either found dead or picked up in a weakened state.

During the 1990s only eight more were seen. Four of these were on the Thames; three were picked up dead or taken into care; and one was seen in the Lea Valley.

21st century

Since 2000 they have become much scarcer in the London Area and only five were seen up to 2009. All of these were on the Thames apart from one in Surrey Docks on 22 May 2001. There were two on the Lower Thames, at Northfleet and Rainham; one at Limehouse Reach, which was also seen off Bishop's Park, Fulham, the next day; and one between Putney and Isleworth.

Razorbill *Alca torda*

A scarce winter visitor (53+)

Range: breeds coastal Europe and North America, winters at sea.

Historical

The earliest record was in 1831 when one was seen on a pond at Cobham. In 1847 Ruegg stated that they were occasionally seen at Woolwich. The only other reference in the 19th century was in 1894 when Cook wrote that he had received large numbers for preservation which had been taken on the Thames at Rotherhithe. He stated that it was unusual to find them so far up river and thought this was due to the severe weather.

20th century

They remained a rare visitor before 1930 with only two more records: a group of five was on the Thames by Blackfriars Bridge on 20 November 1911 after two weeks of stormy weather; and one bird was at Ruislip Lido on 31 December 1924.

In the 1930s they were recorded in five consecutive years between 1934 and 1938, including one that remained at Walthamstow Reservoirs for almost six months and one on the Thames at Mortlake in 1935, which was discovered during that year's boat race; it was seen swimming rapidly upstream between two motor-boats. There have been several influxes; the largest group was 16 at Queen Mary Reservoir on 22 October 1934. All of these birds were found dead four days later and upon examination were discovered to be infested by parasites. In 1948 a dozen arrived on 5–6 October, including seven on the Thames between Westminster and Lambeth and one on the Round Pond in Kensington Gardens which was present for 13 days.

After just one record in the 1950s there was a return to multiple records in the 1960s when four were seen, including two on the Thames. There were only three more in the 1970s: on the Thames at Swanscombe on 1 September 1972; at Queen Mary Reservoir on 22 October 1973; and on Ruislip Lido on 9 December 1978. In 1983 three were seen in a four-day period in early February, and in December 1988 one spent four days on Fairlop Lake.

Since the start of the 1990s there have been just two, both on the Thames: at Swanscombe on 30 December 1990; and at Rainham on 20 February 1996.

Black Guillemot *Cepphus grylle*

A very rare visitor (1)

Range: breeds coastal Europe, Asia and North America, winters at sea.

The only record was of one found dead on the Thames at Lambeth; the bird had died as a result of being jammed between two floating pieces of timber. Bucknill saw the specimen in the Charterhouse Collection although it is no longer there.

Pre-1900: Lambeth, found dead on the Thames

Unacceptable Records

This was included on a list of birds taken in Harrow published by Torre in 1838, but it was not accepted by Glegg and no other details are known about this record.

Little Auk *Alle alle*

A scarce winter visitor (71)

Range: breeds Iceland and Arctic Circle, winters at sea; uncommon winter visitor to UK.

Historical

This was a very rare visitor in the 19th century with a total of nine records. In the 1880s there were two on the Thames: one shot at Chertsey on 8 December 1883, and one taken alive on 12 February 1889 at Teddington. In 1890 an exhausted bird was picked up at Godstone on 2 December. Severe weather during the winter of 1894/95 brought an influx into the country and there were records from Beckenham, Bexley, Darenth, Dartford, Greenwich and Sutton Park.

20th century

In the first two decades of the 20th century it remained a very rare visitor with five records: in 1900 one was taken alive at Nunhead on 23 March; two were found at Cudham in 1905; there were two in November 1910; and in February 1912 one walked into a doctor's surgery in Finsbury Park but died a few days later. The records in 1910 and 1912 occurred during periods when there were 'wrecks' of this species in the UK.

From the late 1920s Little Auks started to become more regular and 10 were seen between 1927 and 1934; seven of these were at Staines Reservoir with the others at Dartford, Kensington Gardens and Walton Reservoirs. However, no more were seen until the 1950s when six were on Staines Reservoir or on the adjacent King George VI Reservoir; all were one-day birds apart from one from 23–29 October 1955, which was eventually found dead.

There were three in both the 1960s and 70s, all on the reservoirs in west and south-west London. As for many other seabirds in the London Area, the peak period was during the 1980s and 90s. There were 11 between 1983 and 1988, seven on reservoirs and four picked up. During the 1990s there were 13, five of which were taken into care. In the autumn of 1995 exceptionally large numbers were seen along the East Coast and a record five made it into the London Area, all arriving in a three-day period between 31 October and 2 November.

21st century

The first record since the turn of the century was of one flying along the Thames past Rainham on 25 November 2006. The following year there was a small influx in November, including eight on Queen Mother Reservoir on the 14th which later flew off; this appears to be the largest inland flock ever seen in the UK. Another two were seen the following day on the west London reservoirs.

Facts and Figures

The largest flock was of eight birds on Queen Mother Reservoir on 14 November 2007.

Puffin *Fratercula arctica*

A scarce winter visitor

Range: breeds coastal northern Europe and North America, winters at sea.

Historical

There were 10 records in the 19th century; additionally Ruegg stated in 1847 that they were occasionally seen in the Woolwich area. One taken alive on the Thames near Chelsea in May 1812 and kept for several days was the first auk to be seen in the London Area. The next two were both picked up, at Broxbourne in April 1882 and at Munden, near Watford, in 1884. In 1887 there was a rather bizarre record of one which flew through an open window into a house in Brook Street, Mayfair, on 16 May. In the winter of 1890/91 one was found at Totteridge; in 1891 there was one at Kew Gardens; in late 1893 one was shot at Roxford Farm, Bayfordbury; and in 1894 one was seen at Winchmore Hill on 17 January and another was found alive in St Albans. There was also one on Hackney Marshes prior to 1899.

20th century

Four were seen during the first two decades, two on the Thames in the Hammersmith area, one shot in Croydon and one picked up in a garden at Banstead. Between 1920 and 1939 there were 10, eight of which were picked up; these included one that spent five days on the lake in St James's Park in October 1927 before it was taken into care, and another that landed on the Strand in October 1935 and held up the traffic. There were six in the 1940s, several of which arrived after cold or windy weather.

There was a huge increase during the 1950s when 31 were seen; six of these arrived in the autumn of 1955, including one which remained on Staines Reservoir from 2 to 25 October before being taken into care. In 1957 there were an unprecedented 10 between 17 September and 9 December; of these, six were picked up and taken into care and one was found dead. 1958 was another good year with six records, including four found between 20 and 25 October. The following year one collided with a cyclist in St Paul's Cray on 20 October.

There was a return to more typical numbers in the 1960s when seven were seen. Three of these were found within a five-day period in early November 1961, including one found sheltering in a garden shed in Shirley. Another three were seen in the autumn of 1967. There were no further sightings until 1976 when one was on the Thames at Barnes on 14 December. Two years later an injured bird was found on Queen Mother Reservoir on 21 August.

In 1983 four were seen in a four-day period in February: two adults flew up the Thames past Swanscombe; one was picked up in Holborn and taken into care; and an immature was found on a doorstep in Shenley. Two more exhausted birds were taken into care the following year, at Harlow in February and at Sloane Square in September. Three more were seen between 1985 and 1989, two on the Thames and one on Wraysbury Reservoir. During the 1990s four were picked up and taken into care, at Richmond Lock, Thamesmead, Hounslow and Cuffley.

21st century

Only one has been seen since 2000, on King George VI Reservoir on 17 November 2007, a few days after an influx of Little Auks; it was later found dead.

Pallas's Sandgrouse *Syrrhaptes paradoxus*

A rare vagrant (*c.* 94)

Range: central Asia; vagrant to UK.

There have been several irruptions of this species into the UK. The first records for the London Area occurred in 1863 when there was a particularly large influx during the summer. For example, more than 60 were shot in Norfolk and Suffolk. During this influx some were shot at three sites in the London Area; the following year one shot on Wanstead Flats was presumably part of this influx.

A much larger irruption occurred in 1888 when thousands arrived in Britain and birds were seen at four locations in the London Area, including a flock of 40 at Jepps Farm, Hoddesdon, on 20 May (two of which were shot). The following year one was seen at Shirley and about a dozen were seen at Barkingside and Fairlop Plain.

In the third irruption in 1907–08 there were only two sightings in the London Area. Although the Hendon record was not included in *Birds of the London Area* due to the lack of detail, it was nevertheless accepted by Glegg.

1863: Forest Gate, one shot in June
1863: South Mimms, two males shot at Dugdale Hill in July
1863: near Neasden, one shot in August
1864: Wanstead Flats, one shot in July
1888: Hoddesdon, 40 on 20 May
1888: Stratford, about 12 on 10 June
1888: Wimbledon Common, 11 flew over on 16 June
1888: near Staines Moor, a flock of about 12 on 19 June
1889: Shirley, in February
1889: Barkingside and Fairlop Plain, about 12 present in July
1907: Hendon, flew over on 23 September
1908: South Ockendon, one shot on 1 September

Rock Dove (Feral Pigeon) *Columba livia*

A very common feral resident and widespread breeder

Range: Europe and Asia; cosmopolitan feral population.

Historical

During mediaeval times pigeons were kept as a source of food; the adults roamed freely to find food and returned to their dovecotes to breed and roost while the young birds were taken for human consumption. Some of these birds strayed from their dovecotes and bred elsewhere and these were the source of the 'London pigeon'.

They were known to be well established in London in the 14th century. The earliest reference dates from 1385 when the Bishop of London complained about people hurling stones and other implements at the pigeons and crows that were nesting or sitting on St Paul's Church, and breaking the windows or stone images in the process. Pepys wrote about the Great Fire in 1666, recording how 'the poor pigeons … were loath to leave their houses, but hovered about the windows and balconies, till they burned their wings, and fell down'.

In 1850 Feral Pigeons had recently started to nest on the tops of the pillars of the Bank of England and the Royal Exchange, and by 1865 they were nesting on London Bridge station. There were also colonies on St Paul's Cathedral and Westminster Abbey.

20th century

Fitter stated that 'the commonest ledge-nesting bird of London buildings is the semi-domestic London pigeon'. The species routinely nests on buildings, whether it be on ledges, in holes in the brickwork, behind signs and advertising hoardings, or in drainpipes. Favoured sites are bridges and ornate buildings but many of these now have wires and netting to keep pigeons away. Nest material could be any debris that they can pick up; in 1937 a nest at a cable works in Greenwich was constructed entirely from pieces of galvanised iron wire. During the Second World War pigeons often nested inside bomb-damaged buildings. When there were still active docks in east London, they nested on the cranes.

In the late 1930s the highest count in Trafalgar Square was 1,150 and this famous landmark continued to hold the largest number of pigeons in the London Area for much of the 20th century. Elsewhere, they could be found throughout central London during the 1940s and 50s and in many other parks and squares across the whole of the area. Large gatherings could also be found feeding on spilt grain in the docks. The population is regularly infused with lost racing pigeons which help to maintain a variety of colours.

Pigeons are quick to exploit many sources of food that become available, from the seeds of cut grass on reservoir banks to handouts from people. They can be found wandering around grassy spaces at lunchtimes, hoping to pick up scraps from office workers and tourists. The population depends on people for food handouts and there was evidence of a decline during and after the war years when these were not available. The recovery in numbers started when bread rationing was finally abolished in the early 1950s.

Pigeons have been recorded travelling on Tube trains since the 1970s, mainly in central London. One particular bird was regularly observed by commuters boarding trains at Paddington. They have not been seen to feed while on a train but appear to use this as a means of transport between stations. It is possible this habit was learnt accidently and they discovered a source of food that could not otherwise be reached by flying to it. Early observations were from Tube lines that are not fully underground, such as the Circle and District Lines, but they have since been noted on other lines where the stations are completely underground. Although most pigeons typically travel only one stop before alighting, they have been seen travelling into central London from as far out as West Ham. In November 1957 one was seen flying down the escalator at Trafalgar Square station (now Charing Cross).

The second breeding Atlas showed that the species bred in 503 tetrads, just under 60 per cent of the total. As might be expected it was present virtually throughout the urban areas but much scarcer in the rural areas.

21st century

Inevitably, when large numbers of pigeons congregate to be fed by the public, this can bring hygiene concerns. Ken Livingstone, as London Mayor, was keen to reduce numbers in Trafalgar Square, a traditional feeding place, to improve its condition for cultural events. Initially this met considerable opposition, with furious pigeon lovers gathering outside City Hall, chanting 'Blood Red Ken'. Harris's Hawks were brought in to scare the birds away (at substantial public expense) and bird feeding was banned in the square from 2001, and on the paving outside the National Gallery since 2007. Numbers have now substantially reduced from the 3,000 or so which were commonly seen at the start of the new millennium and the square is certainly cleaner.

Hawks are employed in several other central London locations, and netting and treatments to window ledges are extensively used to discourage the birds. How far any of these measures impact on the population, rather than simply dispersing the birds, is unclear. New natural predators have also become established in central London in the form of Sparrowhawk and Peregrine. Nonetheless, the pigeon remains ubiquitous across the capital, though BBS indicates a fall of 9 per cent in the Greater London area from 2004–2011.

Facts and Figures

The largest flock was of 4,000 birds in Trafalgar Square on an unrecorded date.

Stock Dove *Columba oenas*

A common partial migrant and widespread breeder

Range: Europe and Asia, northern population migrates to southern Europe and North Africa.

Historical

Stock Doves appear to have been relatively scarce in the London Area in the 19th century. The earliest record dates back to 1838 when the species bred in Richmond Park and two years later it was recorded breeding in Epping Forest.

Harting stated that he 'occasionally observed small flocks in the county in the autumn and winter, but have seldom seen more than ten or a dozen individuals in each'. They were described as 'formerly not uncommon in the Hampstead woods', suggesting a recent decline in the middle of the century and the odd pair was suspected of breeding not far away at Kingsbury. They only bred regularly as near to central London as Richmond, although the young were habitually removed from their nests by the keepers and eaten. Towards the end of the century they were first seen in the central London parks and were believed to be resident by 1898.

20th century

During the 1920s a decline was noted in the Harrow district and it was thought that this was due to old trees being cut down as well as to competition from Woodpigeons. The first confirmed breeding in Inner London was in 1927 in Holland Park.

In the 1930s they had a patchy distribution across the London Area. Single pairs bred in central London in Holland Park and probably in Kensington Gardens. They appeared to be generally absent in many suburbs with breeding only reported in Beddington Park, Hampstead Heath, Hendon, Mitcham, Richmond Park, Walthamstow Reservoirs, Wanstead and Wembley Park. However, they were widely distributed in rural areas, and occasionally could be found in flocks of up to 30; the largest flock comprised more than 100 birds at Aldenham Reservoir in March 1934.

In 1937 large numbers were seen at the construction site of King George VI Reservoir during autumn and there were estimated to be about 300 birds on 10 October; the following year a flock of 250–300 was seen feeding on farmland at Hayes in south-east London.

In the 1950s they frequently nested in parkland and open woodland, and less often in thicker woodland. The species was the most common breeding pigeon on the Thames marshes between Woolwich and Northfleet. In winter it was usually found in small flocks on arable land, often feeding with Woodpigeons, and flocks of up to 300 were sometimes noted. More unusually, large numbers were noted during the breeding season, such as 100–150 on a wheat field at Dartford in May 1951, and about 200 at Perry Oaks SF in late May to early June 1953; it was believed that these were birds that had flown in from a wide radius to exploit a particularly good source of food.

In the inner suburbs they were only reported breeding in the largest open spaces such as at Greenwich Park, where there were at least three pairs, and on Hampstead Heath. The population in central London slowly began to increase; in 1954 seven pairs were recorded within a radius of a few hundred metres of the Round Pond in Kensington Gardens.

From the mid-1950s through to the early 60s they suffered a large reduction in numbers due to the widespread use of agrochemicals on farmland. They disappeared from central London and there were population decreases elsewhere. However, there were still occasional counts of up to 250 outside the breeding season.

At the beginning of the 1970s they were a widespread but not particularly common breeding species in the outer areas of the London Area. The only ones in central London were two pairs in Hyde Park and Kensington Gardens. In 1980 the breeding population was at least 100 pairs and there was a gradual increase over the next few years; the largest breeding concentration was 20 pairs at Osterley Park. The highest count in winter was 320 at Erith Marshes in January 1989.

By the start of the 1990s the breeding population in Osterley Park had decreased by 50 per cent in 10 years. In central London there were single pairs at Buckingham Palace Garden, Kensington Gardens and Regent's Park. The largest count was 500 at Broadwater Lake in November and December 1991. During the 1990s the overall population increased and at least 160 breeding pairs were known by 1992; between the two breeding Atlases they almost doubled their breeding range in the London Area.

21st century

The breeding population has continued to increase since the beginning of the 21st century. Although most pairs are in the more rural areas of London – for example, there were 14 pairs in the Bricket Wood area – they can be found in open spaces within the suburbs and central London. Feeding flocks numbering several hundred are recorded in most years, however in the last few years some huge feeding flocks have been counted in late spring near Maple Cross and West Hyde: 1,251 in 2009 and 912 in 2010. These counts more than double the previous highest counts in the London Area and are possibly the highest gatherings of this species ever seen in the UK. They are clearly part of a substantial breeding population within flying distance of these sites.

Migration

They were first observed migrating over the London Area in the 1940s; several hundred at Sevenoaks in late March and April 1943 and over 300 flying north-east over Chingford on 27 November 1949 were considered to be birds on passage. They are only occasionally noted on migration, usually during the movement of Woodpigeons over the London Area in late autumn. They are typically seen in small numbers apart from in 1971 when 200 flew over Dartford Marsh with Woodpigeons on 31 October, and in 1995 when 350 flew south over Charlton on 27 November.

Facts and Figures

The largest flock was of 1,251 birds at Woodoaks Farm, Maple Cross, on 11 April 2009.

Woodpigeon *Columba palumbus*

A very common resident, passage migrant and widespread breeder

Range: Europe and Asia, north European population winters western and southern Europe and North Africa.

Historical

In the 19th century this was an uncommon breeding bird; Harting noted that 'a few pairs breed with us every year, but this species is most numerous in the winter and early spring, when large flocks pass over and occasionally drop down upon the cleared bean-fields and stubbles'.

In the built-up areas they were recorded breeding as far back as 1834 in parks such as Brockwell, Brondesbury, Clissold and Ravenscourt. In central London they were fairly scarce in the Royal Parks and by 1863 had yet to be recorded in Regent's Park. They had started to breed in Kensington Gardens where two or three pairs nested and the odd pair bred in Green Park and St James's Park, but these all disappeared during tree-felling in the 1870s. In 1883 a pair nested in Buckingham Palace Gardens and three or four pairs nested in St James's Park; they spread out from there over the next few years. By 1886 about a dozen pairs were breeding and shortly after this there came a dramatic increase in numbers. In 1893 about 200 were estimated to have been resident in Hyde Park and Kensington Gardens.

The first breeding record away from the parks occurred in 1890 when a pair bred in Great George Street in Westminster. Within three years they had spread to the City of London and by 1897 six pairs were breeding in Soho Square. During the 1890s they continued to prosper in central London and the north London suburbs and they became one of the most common breeding birds in the London Area.

About the same time their habits also changed and they became extremely confiding, having no fear of man, and even settling on people to take food. This tameness in the capital followed similar behaviour in Berlin and Paris. Away from central London, they continued to be naturally wary, apart from in some suburbs such as in Deptford, Greenwich and Lewisham where they fed in the streets with Feral Pigeons.

Pigott noted that they left central London in the winter to feed on acorns in the country and only returned if there was heavy snow or frost there. Winter breeding was noted first in December 1896 in St James's Park and this was followed by breeding in Green Park in January 1901 and Bayswater in January 1905.

20th century

Woodpigeons did not colonise the outer suburbs until the 20th century. In 1904 the population was increasing rapidly at Lee and by 1910 there had been a notable increase in Brixton, Dulwich and Herne Hill, although this was not a consistent increase as they were described as 'peculiarly rare' on Wimbledon Common in 1912. In the winter of 1907/08 large numbers were seen but many of these died from pigeon diphtheria.

Resident birds were joined by many others flying in to roost in central London. In 1902 about 200 were known to roost in Battersea Park. There was a roost of several thousand in Kensington Gardens in the mid-1920s, and more than 1,200 roosted in Regent's Park in 1930. All of these roosting sites were in trees adjacent to the lakes.

The breeding population in St James's Park had increased to 50 pairs by 1934. In the rural areas Glegg stated that 'it is a very common bird everywhere, but not so abundant as in some counties, as there is not much suitable woodland and very large flocks are unusual'. During the 1940s both the rural and central London populations were greatly reduced by shooting in an organised campaign. When the numbers in central London had significantly declined, the remaining birds became rather wary. After the Second World War, when the shooting campaign ceased, the numbers slowly increased and by 1950 had recovered to the previous high. In Regent's Park 75–100 pairs bred in 1959 and the maximum count was about 1,200 on 26 September, most of which were feeding on grassland.

There was an influx during the severe winter in early 1963; many were seen flying over the London Area on 6 January and two days later the numbers in Archbishop's Park in Lambeth increased from 10 to 153. Up to 5,000 fed on crops at Addlestone and large numbers fed along the tide-line of the Thames in the Barnes area. The prolonged cold winter had a significant effect on the breeding population, and the population in Dollis Hill, for example, fell by half.

Apart from the migrating flocks, large counts were either roosting or feeding flocks. Autumn feeding flocks were noted from Bushy Park, Hampstead Heath and Richmond Park, where they fed on acorns and grain. For example, an estimated 4,000 were in Richmond Park in December 1934 and thousands fed on the acorn crop at Kenwood in October 1937. In 1966 a feeding flock of 4,000–5,000 was seen at Clement Street on 5 December.

In 1978 at least 5,000 roosted at Perivale Wood on 4 November. In 1984, 8,000 were at Bulphan Fen on 29 January. In the winter of 1993/94 up to 5,000 were at Coursers Lane Fields in February and similar numbers were at Walsingham Wood in January. In December 1997 there was a roost of 10,000 at St Pancras Cemetery, Finchley.

The second breeding Atlas in 1988–94 showed that they were the most widespread member of the pigeon family, occupying 95 per cent of the area.

21st century

By the beginning of the new millennium they had become one of the most common breeding birds throughout the London Area. The largest flocks continue to be seen in the more rural areas, for example 3,000 feeding in a barley field at Woldingham in 2001 and 4,000 on stubble fields at Canon's Farm in 2007.

Artificial Nest Sites

Nesting on buildings in central London was first recorded in 1897 when pairs bred in St Martin's Church in Trafalgar Square and on India Office near St James's Park. Hudson also reported nesting on a mansion in Victoria Street. Other nest sites include a windowbox on a fifth floor in Chelsea in 1905, and a girder under the roof of St James's Park station that was used annually from 1913 to 1918.

This behaviour was noted again in the 1950s, particularly in the Westminster district where four pairs nested on buildings in 1951. A pair made a nest entirely from wire netting on a fire escape in Westminster in 1953; the same year a pair built a nest on a structure in the middle of the Thames adjoining Lambeth Bridge; and in 1954 a pair successfully bred on the inside window sill in an occupied flat in Kensington. In 1962, 11 nests were located in drainpipes in Belsize Park and a pair bred in an underground car park in Pimlico in 1965–66.

Movements

Occasionally, large-scale winter movements are noted; the first of these occurred during the morning of 31 December 1868 when 'millions' were said to have flown south over Barnet. Thousands were noted flying south over Watford on 9 December 1907. A large 'invasion' occurred in 1929; on 7 December, near Staines Moor, Glegg witnessed 'a flock containing some thousands, and with a front of three-quarters of a mile, swept due south against a south-westerly gale and disappeared over the Thames'; and it was reported that hordes of Woodpigeons covered 30 [metric] acres of farmland near St Albans on 28 December. In 1930 there was an 'exceptional and abnormal influx' near Leatherhead with an estimated 12,000 in mid-January. In 1960 at least 10,000 flew south-west over Brent Reservoir on 28 December.

By the 1950s studies had shown that there was a regular passage between late October and early December, with movement mainly between north and north-east and west and south-west. Most of these were believed to be birds from Scandinavia and Finland that were migrating to France or Spain for the winter. Occasional movements noted between late December and February were probably cold-weather related.

There was a good passage in the autumn of 1964, particularly on 27 November when an estimated 10,000 flew south over Broomfield Park and about 6,000 flew over Lambeth. In the autumn of 1971 many thousands flew south-west throughout November as the weather got very cold, but when mild weather returned about 10,000 flew north-east over New Malden in five minutes on 30 November, and a flock of 4,000 flew north-east over Hyde Park on 16 December.

In 1978 about 9,000 flew over Kensington Gardens on 27 November in one hour. In 1989 the peak counts were 13,000 over Hogsmill SF on 5 November and 10,000 over Romford the next day. In 2002 there were two very large counts over the London Wetland Centre: 14,000 on 28 October in three hours and 11,800 on 4 November. In 2006 a total of 36,065 was recorded over the London Area on 29 October, including 16,700 over the Wetland Centre.

Facts and Figures

There have been few long-distance ringing recoveries but a locally bred bird ringed at Elmers End in June 1961 was recovered in France that November.

Collared Dove *Streptopelia decaocto*

A common resident and widespread breeder

Range: Europe, Asia and North Africa.

20th century

Collared Doves were not recorded in the UK before the middle of the 20th century but they had been spreading west across Europe since the 1930s and were first seen in Norfolk in 1952. Five years later the first one was seen in the London Area, at Rye Meads on 23 July 1957. It remained there until 3 September, with two together there on 4 August. They continued to spread across England during the late 1950s and bred in neighbouring counties, in Hertfordshire in 1958 and in Essex in 1960.

No more were seen in the London Area until December 1961 when a flock of up to 15 took up residence at Carshalton for several months. The following year there was a small influx, culminating in the first breeding record at a churchyard in west London. Sightings continued to increase across the London Area and the first one in Inner London was seen in Regent's Park on 2 September 1963. Breeding occurred at Hounslow, Osterley and Shenfield, and probably near Banstead; additionally pairs were present at five other sites. By 1964 they had begun to concentrate in several areas and produce a stable breeding population. For example, in the Shenfield area three pairs bred and there was a total of 23 by the end of the year, and near Banstead at least 40 were seen in August. They continued to spread the following year, especially south of the river; in the Hook and Chessington areas new birds arrived almost daily during December and up to 50 were seen in Chessington Zoo.

The breeding population in 1966 was about 28–44 pairs. Flock continued to increase in size and about 150 were at Chessington Zoo; there were also 117 near Banstead. However, few were seen in the suburbs apart from at Southall where they were seen in increasing numbers. In 1967 the breeding range had extended in west London as far as Greenford.

Part of their success was the ability to breed virtually all year. For example, a recently fledged juvenile was seen in the snow at Weybridge in December. By the end of the 1960s they had not colonised the more heavily built-up areas, although a pair summered in Inner London at Dolphin Square in 1969.

The increase continued during the 1970s but they were slow to penetrate deep into the suburbs. In north-west London they were described as 'still a scarce visitor' around Kingsbury, Neasden and Willesden. Large increases, however, were noted in the Walton and Weybridge areas where about 60 breeding pairs were present, an increase of 50 per cent over the previous year; and about 150 displaying birds were counted in the Chessington area. During the first breeding Atlas, covering the years 1968–72, they bred in one-third of all tetrads, with the bulk of the population in the southern half of the London Area. In the mid-1970s the highest count was about 650 at Walton-on-Thames in October 1976. In 1977 they began to increase in the east of London but the large autumn flocks south of the river had dramatically reduced. In 1978 a pair bred in Inner London for the first time, at Regent's Park.

They failed to colonise central London; after 1978 the only breeding attempts were in Regent's Park in 1982 when a pair possibly bred, and in St James's Park in 1989 when a pair built a nest but did not raise any young. By the start of the 1990s the population in the London Area had virtually stabilised for the first time since birds had arrived in the UK. They had become a widespread breeding bird apart from in the most heavily built-up areas.

During the early 1990s they attempted to colonise Inner London; a pair was present during the breeding season at New Cross Gate in 1981; and three pairs were present in Kilburn in 1984, one of which bred. These may have come from the expanding population around nearby Hampstead Heath. By the second breeding Atlas in 1988–94 they had more than doubled their range. However, they were still absent from most of Inner London and were scarce in the inner suburbs. During the period 1994–2000 there was an increase of over 40 per cent in the Breeding Bird Survey in London. There was some evidence that they were moving into areas vacated by Turtle Doves in the more rural parts of east London.

21st century

By the beginning of the 21st century they had only been present in the London Area for 43 years yet they had become a widespread and common breeding bird. However, despite several attempts they have still not colonised Inner London and are seen there only infrequently.

Facts and Figures

The highest count was about 650 birds at Walton-on-Thames on 24 October 1976.

Turtle Dove *Streptopelia turtur*

An uncommon summer visitor and localised breeder

Range: breeds Europe, Asia and North Africa, winters Africa.

Historical

In the 19th century this was a common summer visitor and like many of the dove species it was taken for eating. Harting observed that during September 'these birds may be seen in small flocks upon the stubbles, gleaning the scattered grain'. In 1881–87 Booth wrote that when he was a boy Turtle Doves were abundant in Middlesex and were frequently found breeding around Harrow. They also bred on Wimbledon Common in 1881 and on Hampstead Heath in the 1890s.

20th century

In the 1930s Glegg described Turtle Doves as a fairly common summer resident in rural areas, being well distributed over suitable districts without being numerous. They were rarely seen in central London and were considered unusual in the suburbs. He believed that they had declined considerably, even without the habitat loss due to suburbanisation; however, they increased in the Harrow district between 1925 and 1930.

At the beginning of the 1940s the occasional pair could still be found at Hampstead but this was the only breeding location in the inner suburbs. Further out in the rural areas it was still much more common. A slow increase in numbers had been noted in the Edgware, Finchley and Mill Hill districts, and at Parndon, where it was previously considered local, it had increased enormously to become widely abundant by 1941. Turtle Doves were seen on passage in most years in central London, typically in Kensington Gardens or Regent's Park but also sometimes in squares and gardens. In 1949 a large flock of 150–200 was seen feeding among dredged material from the River Darent at Dartford Marshes on 14 August.

By the mid-1960s they had begun to disappear as a breeding bird from west and north London but were still widespread in the south and east. At the beginning of the 1970s the overall population was relatively stable but there had been a shift in the distribution with fewer in the western half of London and more in the south-east, where it was described as numerous at Dartford, Farningham, Shoreham and Swanscombe. By the start of the 1980s the closest breeding birds to central London were in Wimbledon Park but in 1981 pairs summered in Inner London at Holland Park and in the inner suburbs at New River Walk, Islington. In the mid-1980s the breeding population was at least 75 pairs.

At the beginning of the 1990s they had become a localised breeding bird in the rural areas of north and east London and were quite scarce elsewhere; the highest breeding concentration was five pairs at Rye Meads. Three wintering birds were seen in 1990/91; it was thought that the increase in winter records was due to odd birds associating with flocks of non-migratory Collared Doves.

Although most areas were declining, for example a 30 per cent decline was noted in Hainault Forest and they had disappeared from the Sewardstone area of Epping Forest, there were also localised increases such as at Berwick Ponds where up to six pairs bred, and at Rye Meads where there were eight pairs in 1995. The breeding Atlas for 1988–94 showed a decrease in the distribution of 44 per cent compared to the first Atlas in 1968–72. Breeding occurred in 202 tetrads, with most of the population situated from the Thames corridor through to the rural areas of east and north London but they had virtually disappeared from the North Downs.

21st century

By the start of the 21st century the Turtle Dove was in serious decline in the UK, and in the London Area there were only about 25 territories in 2000. Even as a passage migrant it had become scarce and the largest flock was just six at Beddington on 10 May 2000. It has continued to decrease and is rapidly heading towards extinction as a breeding species in the London Area. Although a few were heard calling in the summer of 2009 there were no reports of any breeding.

Facts and Figures

The earliest migrant was on 5 April 1960 at Sevenoaks and on the same date in 1982 at Regent's Park.
The latest migrant was on 7 November 2007 at Rainham Marshes.
The largest count was 240 birds at Horton Kirby on 13 and 26 August 1967.

Ring-necked Parakeet *Psittacula krameri*

A common resident and widespread breeder; naturalised population

Range: Asia and Africa; naturalised population in southern UK.

Historical

Escaped parakeets that were probably this species were first observed in Dulwich Park in 1893 and in Brixton the following year.

20th century

The first breeding record was in Epping Forest in 1930 (Morgan) but this appeared to be a one-off occurrence. There were no further sightings until the late 1960s when the ban on importing Ring-necked Parakeets was lifted and the population in the London Area derived from these escaped birds.

They were first seen in the Croydon and Shirley areas in 1969; they remained in the area and a pair bred there in 1971. In the early 1970s there was a small population in the Wraysbury area and up to 10 birds roosted just outside the London Area at Runnymede. These may have come from a number of birds that escaped from a pet shop at Sunbury-on-Thames in about 1970. In 1971 single pairs bred in Bromley and in the Esher area, and by the late 1970s a small population was present in the latter area; these birds went on to form the roost at Elmbridge. Also in the early 1970s a few escaped from an aviary in Syon Park after debris from an aircraft broke a glass pane.

The Atlas of Breeding Birds in Britain and Ireland (Sharrock 1976) revealed the presence of Ring-necked Parakeets in the London Area, Surrey, Essex and Kent. By December 1980 there was a roost of up to 110 at Beckenham. In 1982 they were present and probably breeding around Shepperton Studios; this may explain the false theory that the population was started after some were deliberately released after the shooting of a film. There were no reports of any parakeets being accidentally released during the 1950s when *The African Queen* was filmed.

By 1983 there had been a contraction of the population and some of the small breeding colonies of the previous decade had become extinct. The peak counts were 40–60 at Hither Green Cemetery and 51 at Wraysbury GP. The species was admitted to the British List in 1984 when the UK population was believed to be around 1,000 birds, all of which were in south-east England apart from a tiny population in Greater Manchester which soon died out. Probably half were in the London Area, mainly in the south-west but with a separate population in the south-east, around Hither Green and Foots Cray. The population at Shepperton had increased to 60 by 1985 and the first record in Inner London was of a bird seen the following year in Battersea Park. The largest counts were made at communal roosts, with the peak being 142 at Wraysbury GP in December 1989.

By the start of the 1990s the breeding population was still largely restricted to south London; the only breeding pairs north of the Thames were at Osterley Park and Staines Moor. In 1991 the population was estimated to be about 600 and birds were beginning to spread to east London; three were present all year in Havering. The breeding population was estimated to be at least 150 pairs in 1992.

In 1994 the total population was estimated to be 1,700; this was also the first year of the roost in the Elmbridge/Hersham area, which was used in subsequent years as a guide to the population growth; there were 697 birds in December, with some seen flying in from a considerable distance. At Hither Green the roost had increased to more than 200. By 1997 the population had doubled to at least 3,000 with up to 1,500 roosting at Hersham. Over the next two years the smaller populations remained relatively stable while the south-west London population continued to expand.

21st century

A lot had changed in just 30 years and by the beginning of the 21st century Ring-necked Parakeets had become a common and widespread resident throughout many parts of the London Area. They had adapted well to living in the suburbs and frequently visited gardens to take advantage of birdfeeders. The population was at least 4,000, of which almost 3,000 roosted at Hersham. They were beginning to spread clockwise around the London Area from the south-west but had not yet colonised north London; they were also virtually unknown in the rural areas and in central London.

Within a further three years the Hersham roost had more than doubled to almost 7,000. By 2005 they had started to colonise central London, where up to 21 were resident in Hyde Park and Kensington Gardens, including one seen evicting a Green Woodpecker from a nest-hole. The species was also recorded in St James's Park for the first time as well as in Battersea Park and Regent's Park.

The roost at Hersham was abandoned in 2006 after tree-felling and a new roost of 6,000 was formed in Stanwell although numbers rapidly declined there and a number of smaller roosts have since been established. It has become more difficult to calculate the total population since the loss of the main roost but the highest counts in 2009 suggested a total of about 10,000. The only large population elsewhere in the UK is a few hundred in Ramsgate, Kent. One estimate put the population in the London Area at about 30,000 in 2010, but this appears to be over-optimistic. However, the population continues to expand and the species is spreading east and north towards the boundary of the London Area.

Facts and Figures

The largest roost count was 6,918 birds at Hersham at dawn on 24 August 2003.

Cuckoo *Cuculus canorus*

An uncommon summer visitor and localised breeder

Range: breeds Europe and Asia, European population winters Africa.

Historical

Cuckoo was first recorded in London by Gilbert White on 8 May 1785 in South Lambeth. In the 19th century it was a regular and widespread summer visitor. A few were seen on passage in central London each year; breeding records come from Old Brompton in 1813 and Regent's Park in 1873 and 1884.

20th century

There were three confirmed breeding records in Inner London during the first decade of the century: a juvenile being fed by a Robin in a garden on the Marylebone Road in 1905; from Regent's Park in 1909, using the same host species; and in Battersea Park where an egg was discovered in a Common Whitethroat's nest. The only other later breeding record was in Holland Park in 1921.

In the 1930s it was still a common summer visitor but breeding was restricted to the rural areas. A Cuckoo was seen on three occasions in St James's Park in August 1932 feeding on the caterpillars of the Vapourer moth which were abundant at the time. After the first decade of the 20th century Cuckoos were generally found no further into the London Area than a radius of about 15 or 16 kilometres, the only regular exception being on Hampstead Heath where they continued to breed into the 1950s. By the early 1960s they had become scarcer in the London Area. A further decline was reported in 1963 but the following year was considered to be a good one for this species and 20 were calling in Broxbourne Woods on 14 May.

At the beginning of the 1980s the breeding population in the outlying areas of the London Area was fairly healthy and stable but by 1990 many sites had noticed a decline. The second breeding Atlas in 1988–94 found a range contraction of 10 per cent compared to the first Atlas. Much of the decrease occurred in south London, whereas in the outer areas of north and east London the population actually increased. The total population continued to decrease during the rest of the 1990s.

21st century

By the start of the 21st century Cuckoo could still be found breeding in the outer areas of the London Area but it was no longer common. The decline has continued and there were only about 50 territories recorded in 2009, almost all north of the Thames. The birds have also become much scarcer on migration in the suburbs.

Facts and Figures

The earliest migrant was on 25 March 1978, at Mill End.
The latest migrant was on 25 October 1994, at South Norwood Country Park.
Studies of the host species used by Cuckoos in the London Area showed that just over one-third were Dunnocks. The second most common was Robin with 13 per cent, followed by Reed Warbler, Pied Wagtail and Common Whitethroat. A total of 25 species has been parasitised, including Woodlark, Nightingale and Red-backed Shrike.

Yellow-billed Cuckoo *Coccyzus americanus*

A very rare vagrant (1)

Range: Americas; vagrant to UK.

The only record of this North American cuckoo occurred in the autumn of 1991 after very strong westerly winds. It was first found moribund at Oxted station and handed into the local pet shop from where it was taken to the home of a local RSPB member at Limpsfield. After being allowed to rest and recuperate for a couple of hours it showed enough signs of activity for it to be released into the garden where it immediately flew into a thick hedge and was not seen again. This was the fifty-fourth record in Britain and Ireland but only the third inland one.

1991: Oxted, picked up on 17 October and taken into care

Barn Owl *Tyto alba*

An uncommon resident and localised breeder

Range: cosmopolitan.

Historical

The earliest record dates back to 1738 when Albin stated that one had been shot in a field at Waltham Abbey. There were no other references until the 1830s when a pair nested at Tooting; however, according to Harting this was the most common owl in the 19th century. He knew of nests in hollow trees in Canons Park and in churches at Kingsbury, Northolt and Stanmore. When he examined some pellets found at his local church he discovered that the majority of prey items were small birds, invariably finches.

They were occasionally recorded in central London. In 1863 one was seen in Regent's Park; around the 1890s one was heard on several nights by Lambeth Palace and was thought to have made its home in the tower of Lambeth Church; and in 1898 a pair was probably resident in Kensington Gardens.

20th century

At the beginning of the century they were widely distributed in the rural areas where they were locally common. They could also be found in suitable urban areas; breeding was recorded at Hampstead in 1902; in Regent's Park prior to 1911; a pair bred in an old house on Clapton Common in 1923 and they probably bred annually in the area to about 1926.

They had declined by the 1930s, mainly due to persecution and loss of habitat, to such a point that Glegg described them as the least common of the breeding owls. Another contributing factor was the loss of traditional nesting sites, as church towers and belfries were secured to prevent Jackdaws from using them. They could still be widely found in the rural areas but were thinly distributed and only occasionally bred in the suburbs.

There were several sightings in Inner London between 1930 and 1960, including one flying over the Strand in 1936; one found in the City of London and taken into care in 1937; one in the West End along Berners Street at dusk on 7 October 1943; at Hyde Park Corner on 11 May 1950; and in the grounds of the Royal Hospital, Chelsea, on 24 May 1950. In the East End, one resided in a shed at Bromley-by-Bow for several weeks in the autumn of 1937.

In 1959 the breeding population was 17 pairs, including eight between Broxbourne and Brickendon; this was thought to have represented a genuine increase rather than just increased observation. In 1961 a pair bred in the suburbs at Southgate. A London-wide census carried out in 1967 found seven successful breeding pairs and a further 15 pairs that were present during the breeding season. In the outer suburbs pairs were found at Barking, Beddington, Kempton Park and Richmond Park; one or two were also heard in the Dulwich area until the land they were on was developed.

By the start of the 1970s the only known breeding pair in the suburbs was in Richmond Park, but they had ceased breeding there by 1978. The first breeding Atlas recorded Barn Owls breeding in 48 tetrads with a fairly even spread around the outer areas of London. The population contracted further during the 1980s and only a handful of breeding pairs were known, mainly in the area between the Lea Valley and the Lower Thames; the only breeding pair elsewhere was in Hatfield Park. In 1986 captive birds were being bred at one site and released in the Harefield area; however, this attempt at reintroduction appeared to be unsuccessful.

In the early 1990s the breeding population was at least 10 pairs, six of which were in the eastern part of London. In 1991 one was seen in Inner London for the first time in over 40 years, around Campden Hill on 13–14 March. Additional birds were released as part of a reintroduction programme. By the second breeding Atlas they had vacated their former breeding areas in south and west London and were found to be breeding in 22 tetrads, mostly in east London apart from a few pairs towards the northern boundary.

21st century

Since the beginning of the 21st century there have been up to 12 breeding pairs, mostly in the rural areas in the north and east, and on the Lower Thames Marshes. Elsewhere, they are occasionally seen outside the breeding season.

'Dark-breasted Barn Owl' *T. a. guttata*

The only known records of the continental race were both of birds shot in Epping Forest, in about 1843 and in December 1864; the first of these also appears to be the first British record.

Scops Owl *Otus scops*

A very rare vagrant (1)

Range: breeds Europe, Asia and Africa, most of European population winters Africa; vagrant to UK.

The only record was of one illustrated by Selby in 1833 that had been taken near London and was just the third record in Britain at the time. This record was accepted by Yarrell in his *British Birds* and later by both Harting and Glegg.

Pre-1833: 'near London', shot

Unacceptable Records

One was said to have been shot at Chelsfield in 1904. Although the skin was preserved, its identification was not confirmed by any competent ornithologist at the time and the possibility remains that it may have been a Little Owl as this species had been introduced into Kent in 1896. A pair that reportedly bred at Moor Park in 1902 is similarly presumed to relate to Little Owl.

Snowy Owl *Bubo scandiacus*

A very rare vagrant (1)

Range: circumpolar; vagrant to UK, bred on Scottish islands in the late 20th century.

The only record occurred during the severe winter in 1963 when an adult female was present at Sewardstone for most of March. This was the only sighting in the UK during that winter.

1963: Sewardstone, one on 2–28 March

Little Owl *Athene noctua*

An uncommon resident and localised breeder; naturalised population

Range: Europe, Asia and Africa; introduced population UK.

Historical

The earliest reference was in 1758 when Edwards described how one came down a chimney in St Catherine's parish, by the Tower of London; this was the first record of the species in the UK. Upon dissection it was found to be a pregnant female containing eggs. At first it was thought to have been a foreign bird which had escaped from a ship on the Thames but another Little Owl was later discovered after also falling down a chimney in Lambeth. It is not known where these birds came from as there is no record of any introductions in the 18th century. It is possible that they were genuine vagrants as in 1738 Albin recorded that one alighted on a ship off Ouessant, France.

The first introduction in Britain took place in 1843 in Yorkshire and the main introductions occurred near Edenbridge in Kent in 1874, 1896 and 1900, and also at Oundle in Northamptonshire prior to 1889. They soon colonised these areas and quickly spread into the London Area. The first one was found dead at Chigwell on 2 January 1865; further records came from Sunbury prior to 1866 and Highgate Woods on 29 October 1870. The population in Kent spread along the Darenth Valley to Dartford and became established at Westerham and Sevenoaks by 1900. One was caught near Lambeth Palace prior to 1884.

From the Midlands they made a rapid spread southwards and by 1897 they were breeding about 1 kilometre beyond the London boundary at Easneye, Hertfordshire. The following year they had crossed into London at Heronsgate, Moor Park and West Hyde.

20th century

By 1907 they were well established across Hertfordshire and a pair nested in the London Area between Hatfield and St Albans; they had also been breeding at Moor Park for several years. Little Owls continued to push further into north London and in 1909 one was shot in Hampstead. Breeding occurred near Enfield and they then began to become more widely spread.

In south London a pair was seen at a probable nest site at Coulsdon in 1900, and by 1910 they had spread to Bromley, Hayes and across the North Downs to Oxted, the Godstone Valley, Chipstead and Warlingham, where they nested in rabbit burrows. There were five pairs in Titsey Park in 1911. In east London 10 were released at Loughton in 1905 and these may have been responsible for sightings in Chingford in 1913 and Woodford Green in the following year; also, two pairs bred at Brentwood in 1914.

They continued to increase across the London Area during the 1920s. They were first discovered nesting in the Harrow district in 1920 when three pairs were found. They were occasionally recorded right in the heart of London; for example, there was one in Battersea Park in 1923 and one was seen perched on a chimney pot at New Cross in around 1929.

By the 1930s it had become a fairly common resident and Glegg considered it probably the most numerous breeding owl in Middlesex and as 'too common' in some rural localities. He stated that it had often been seen in central London and had occurred at a number of sites in the suburbs. However, in some areas the population had peaked and there were reports of decreases in many districts, which were probably due to persecution; for example, five were shot by one gamekeeper in the spring of 1936 and in 1944, 10 dead Little Owls were found on a keeper's gibbet at Bayfordbury Park. Little Owl habitat preferences were reported to be elms on farmland and oaks in parkland, although they had been recorded using a wide variety of trees in which to nest.

After the Second World War, the persecution lessened and the population began to increase, especially on farmland. During the 1950s they were widespread but never common, and birds were rarely noted from built-up areas; the nearest breeding pair to the centre of London was at Dulwich Wood where breeding was first recorded in 1947. They were still seen occasionally in central London and in 1960 one was heard calling in Regent's Park.

By 1967 the breeding population in south London was 50–60 pairs in about 50 locations; they were also widespread elsewhere with several population clusters, for example there were six pairs in the Hainault area and four in Black Park. During the first breeding Atlas in 1968–72 they were found breeding in 161 tetrads. Most of the population was situated beyond the outer suburbs apart from the large open spaces such as three pairs in Osterley Park and eight pairs in Richmond Park. The second breeding Atlas in 1988–94 showed there had been little change in the overall breeding population; they were mainly found in the outer areas of London with the majority in the north and east.

21st century

Since 2000 the population has been relatively stable with some signs of an expansion. In December 2006 two were seen in Regent's Park and they bred there the following year, the first-ever breeding record in Inner London. They have continued to breed in Regent's Park and in 2012 two pairs were also found in Kensington Gardens. However they are still mainly restricted to the outer suburbs and rural areas; the largest population is in Richmond Park where there are believed to be at least 30 pairs.

Facts and Figures

They can be vulnerable to attack from other predators and two were killed by a Peregrine at Queen Mother Reservoir in 2004.

Tawny Owl   *Strix aluco*

A common resident and widespread breeder

Range: Europe, Asia and North Africa; widespread in UK.

Historical

The earliest record was of one heard by Gilbert White at Vauxhall on 14 March 1778. Harting considered them to be rare in Middlesex in the 19th century, but elsewhere in the London Area they bred on Dulwich Common in 1811 and on Wimbledon Common in the 1830s. They were also considered abundant in the Ealing district and were present in Kensington Gardens towards the end of the century.

20th century

By the beginning of the century the population had increased although they were still fairly scarce across most of the London Area and were much less common than Barn Owl. In 1909 Dixon stated that their call was very familiar in the wooded districts of suburban London; they were numerous in Richmond Park and bred at Kew and Wimbledon. By 1913 they had become fairly common in Hampstead.

They continued to increase during the first half of the century and Fitter believed this was partly due to the maturing of trees. In the 1920s they were present in the central parks; in the 1930s they bred annually in Holland Park and a pair bred in Green Park in 1936. The population had significantly increased by the 1930s as Glegg described it as a common resident in the rural districts and widely distributed in the suburbs. An unfortunate pair was killed by a bomb which exploded in Kensington Gardens in February 1944.

In the 1950s they were often heard in many London squares and large gardens. They vacated Bloomsbury after their nesting tree in Queen Square was cut down in the late 1950s. In 1959 they were present in 14 locations in Inner London. In the exceptionally cold winter in early 1963 there were several reports of birds seen hunting during daylight hours. During an owl census in the spring of 1963, 28 were heard calling at Cuffley. In another census in 1967 about 91 breeding pairs were present across the London Area, including five in Inner London; however most were located in the suburbs and they were noticeably scarce east of the Lea Valley.

The second breeding Atlas in 1988–94 showed a range contraction of 17 per cent compared with the 1968-72 Atlas with most losses being in the suburbs. There were also noticeably fewer on the North Downs but this may have been due to under-recording as nocturnal birds are often missed on surveys such as this.

21st century

The Tawny Owl remains a reasonably common and widespread breeding resident throughout the London Area; there are at least three pairs in Inner London.

Prey

Although their typical prey is small birds and mammals, one was once seen pursuing a Grey Heron in Hyde Park. In 1964 Beven showed that Tawny Owls in urban areas were surviving on a more avian diet than those in rural areas where small mammals were the main prey. In Holland Park they were mainly taking House Sparrows, along with Feral Pigeons, Blackbirds and Starlings. One in Hyde Park was also preying on Feral Pigeons in 1995.

Long-eared Owl *Asio otus*

An uncommon winter visitor and scarce breeder

Range: Europe, Asia, Africa and North America.

Historical

The earliest reference was in 1734 when Albin stated that one had been shot in Enfield Chase. No more were recorded until the spring of 1815 when one was shot on Dulwich Common.

In the latter half of the 19th century it was a widespread but localised breeding bird. Harting knew of nests in Canon's Park (undated), Wembley Park (1861), Pinner (1865) and Kenwood (1871), and he believed it to be breeding in Ruislip Woods. They were also known to have bred in Langley Park prior to 1868. A serious decline must have taken place in west London not long after these breeding records as there were no further sightings of Long-eared Owl anywhere in Middlesex until the 1940s. In 1875 a pair bred in Balls Park near Hertford and it was also considered to be 'not rare' in Epping Forest.

Occasional birds were met with elsewhere, often in winter, such as at Kingsbury where two were seen together in December 1862; this appears to be the earliest example of a winter roost.

20th century

Between 1900 and 1930 there were isolated breeding records from Bushey Heath, Dartford Heath, Esher, Richmond Park, Totteridge and West Wickham; additionally four were shot near Watford in 1904. By the start of the 1930s they had became much scarcer and the only definite breeding record was at Coulsdon in 1934; there were also a couple of isolated winter sightings. During the 1940s they were known from two sites in the Darenth Valley.

After the Second World War there were no further breeding records although there was an increase in the number of sightings during winter and it would seem its status had changed from a breeding bird to a scarce winter visitor. The only possible signs of breeding were at Orpington in 1954 where they were heard calling often early in the year, and two were seen on 16 August; and at Well End near Shenley where a pair was present on 30 May 1956. In 1962 one was found dead at Elstree on 30 August, not far from Well End, and given the secretive nature of this species it is highly likely that there was a breeding pair in this area. Another bird was heard calling near Cuffley on 30 July 1963.

In 1968 a pair bred at Cockfosters, the first breeding record for over three decades. There was no real change in status by the start of the 1970s when a pair summered in east London and one was heard calling on several nights in spring at Cockfosters. In 1973 up to five wintered at Dartford and a pair bred there, although the young were taken from the nest; another pair bred at Joyden's Wood. They continued to overwinter at Dartford in subsequent years. In 1977 a pair bred successfully at Broxbourne Woods. There was a large influx in the first four months of 1979 when at least 38 were seen, including eight at two roosts on Bookham Common.

By the beginning of the 1980s they had become a well-established wintering bird, mainly on the eastern side of London, typically on the Thames marshes and in the Lea Valley. In the winter of 1986/87 at least 38 were present, including four roosts of at least six birds; and in 1989/90 up to 11 wintered at Dagenham Chase. There were very few signs of an established breeding population, with just isolated breeding records.

In 1991 one was recorded in Inner London for the first time in 20 years, at Tower Hill where a bird was picked up on 24 May and later released. In the winter of 1991/92 there was a large influx into the country and 76 were seen in the London Area; the peak counts were 17 near Ten Acre Wood and 15 at Dartford. The breeding population had increased by 1999 when a minimum of four pairs bred and there were possibly seven pairs.

21st century

At the start of the 21st century there were about 25 wintering at four sites; however, since then the wintering population has been much lower with as few as two birds. The breeding population has also remained low with up to three pairs recorded annually; in 2009 none were seen during the breeding season.

Short-eared Owl *Asio flammeus*

An uncommon winter visitor; rare breeder

Range: Europe, Asia and Americas.

Historical

This was a scarce winter visitor in the 19th century; the earliest record was of one shot near Wembley sometime between 1828 and 1832. Several were shot at Brent Reservoir prior to 1866 and elsewhere there were records from another nine localities.

In 1887 a nest belonging to this species was said to have been taken in the market gardens between Twickenham and Isleworth; unfortunately there is nothing to corroborate this record and according to Kerr they were 'exceedingly uncommon' in the Staines district so it must remain unproven.

20th century

There were influxes in 1903 and 1906; during the latter year, 13 were flushed from under a hedge on Woolwich Marshes. The only other record around this time was of one shot at Edmonton SF in about 1908.

Its status had changed little by the mid-1930s when Glegg described it as 'a very rare straggler'. Subsequently they became much more regular and were seen during most winters with occasional influxes, e.g. five wintered on Walton Heath in 1938/39. They used to hunt over fields and damp ground on Hampstead Heath in winter prior to 1949. During the influxes in the 1950s up to four wintered at Beddington or on the adjacent Mitcham Common, and in 1954/55 up to six were present at Dartford Marsh. In 1958 there was a large influx in autumn when about 24 were seen; this appeared to be connected to an abundance of Short-tailed Voles; some owls overwintered, with eight at Beddington and five at Staines Moor.

In the early 1960s they were relatively scarce with only a few seen each year. In 1964 one was seen hunting at Chertsey on the unseasonable date of 22 July. Wintering numbers continued to vary from year to year and there were further influxes in 1970/71 and 1978/79. In the latter winter season, 22 birds were seen in the Staines/Perry Oaks/Poyle area on 11 November 1978 and just over 100 were found across the London Area between January and April, including 15 at Perry Oaks SF, 14 at Rainham Marshes and 12 at Beddington.

They were fairly regular during the colder winters of the 1980s, especially on the Lower Thames Marshes. Another large influx occurred in late 1982 when about 70 were seen, including 15 at Rainham and eight at Walthamstow Reservoirs. During the relatively mild winters in the mid-1990s only a few overwintered.

21st century

In 2000 the wintering numbers were the lowest they had been for many years with just five at Beech Farm and none on the Lower Thames Marshes. The wintering population has subsequently remained low, but they have returned to Rainham and this is now the most regular wintering location, with up to five present in most winters. Elsewhere, they are occasionally recorded flying over on passage, even in central London.

Breeding

Breeding occurred for the first time during the 1980s at Rainham Marshes. A pair almost certainly bred in 1986; two years later they reared one juvenile and they may have bred again in 1989. In addition to the Rainham birds, a second pair summered in 1990. In 1993 a pair probably bred and in 1995 a pair summered at a third site in east London.

There were no further breeding attempts until 2008 when a pair bred again, although there were occasional sightings on the Lower Thames Marshes during the intervening summers.

Facts and Figures

The largest count was 17 birds at Rainham on 21 November 1985.

Tengmalm's Owl *Aegolius funereus*

A very rare vagrant (2)

Range: Europe, Asia and North America; vagrant to the UK, historically more common

In the 19th century this was a scarce winter visitor in the UK and there are two records in the London Area. The first of these was described by Christy, who commented that 'some boys noticed a bird in a tree near the iron bridge in the Barking Road, Poplar, which on being disturbed was soon killed'. Four years later one was shot near Dartford.

1877: River Lea, between Poplar and West Ham, killed in late January
1881: Near Dartford, shot on 18 November

Unacceptable Records

Harting refers to a recently shot bird found in a poulterer's shop in London in 1836 but he did not count it as he had frequently seen owls and hawks for sale in Leadenhall Market that had been shot in Holland.

Nightjar *Caprimulgus europaeus*

An uncommon summer visitor and scarce breeder

Range: breeds Europe, Asia and North Africa, winters Africa.

Historical

The earliest record was in 1731 when Albin stated that it was very common in Enfield Chase and that he had also seen it in Epping Forest. In 1813 Graves wrote that the birds could be found at Coome Wood, Enfield Chase and Hornsey. Harting described them as 'local rather than rare, evincing a partiality for ferny commons and tracts of low underwood surrounded by cover'; breeding occurred at Kenwood, Hampstead, Mill Hill, Scratchwood and Winchmore Hill. They were also said to be 'rather more than plentiful' in Elstree and 'not uncommon' around Ealing, and up to four were seen hawking over the river at Laleham.

South of the Thames they were said to be abundant on Sydenham Common in about 1821, and breeding was recorded from Ashtead, Barnes Common, Coombe Woods, Dulwich, Epsom, Richmond Park, Streatham Common, Tooting and Wimbledon Common. Ticehurst stated that they were plentiful at Blackheath and Shooters Hill in around the middle of the 19th century.

In 1892 a Nightjar was seen in one of the parks and apparently provided a rival attraction to a performance there by Buffalo Bill.

20th century

At the beginning of the century they were reasonably common in areas with suitable habitat and bred in the suburbs of Hampstead, Highgate, Wembley and Winchmore Hill to the north of the Thames, and at Bromley, Eltham, Kew Gardens, Richmond Park, Shirley and Wimbledon Common to the south. They were more widespread in the outer reaches of the London Area, such as in Epping Forest, Ongar Park Wood and around Oxhey. However, the population was in decline, mainly due to a loss of habitat and persecution from gamekeepers who considered them akin to hawks; for instance, one keeper killed around six annually because he thought they would damage his Pheasants.

They continued to nest regularly at Enfield up to 1912 and Hampstead Heath up to 1914 but most of the suburban sites where they bred had been abandoned by the end of the First World War, although they continued to nest in Richmond Park until at least 1930.

In 1929–30 at least 16–20 pairs nested in the more rural areas of north-west London, such as Harrow-on-the-Hill, Harrow Weald, Northwood, Ruislip and Stanmore Common. In the 1930s Glegg stated that they had declined because of suburbanisation, particularly with the loss of breeding grounds such as old parks and heaths.

By the end of the decade the only regular breeding sites were at Epping Forest, Northaw Great Wood and Ruislip in the north, and Epsom, Hayes, Shirley and Walton Heath in the south; they probably also bred on Hampstead Heath in most years as they were seen or heard annually during the summer. In the Ruislip district there were 13–14 pairs in 1937.

Disturbance at many sites during the Second World War only added to the general decline. In the late 1940s and early 50s they were found at more than 20 sites during the breeding season, possibly reflecting a recovery in numbers. Almost all of the sites were in the outer districts. At Darenth Woods, 10 breeding pairs were discovered in 1949; as this was an historic breeding site they were probably present throughout. Also in 1949, 10–15 were heard churring at Limpsfield during the breeding season. A pair nested in Richmond Park in 1950, the first to do so there for 20 years; unfortunately the nest was destroyed by a bulldozer.

In the early 1960s it was thought that 40–50 pairs were still breeding in the London Area. The largest populations then were 14 pairs in Broxbourne Woods and 13 pairs between Banstead and Walton Heath. In 1965 a pair summered at Harrow Weald, just on the edge of the suburbs. By the beginning of the 1970s there had been a large decrease in the breeding population; during the first Atlas in 1968–72 they were breeding in 23 tetrads. The heaths of south London provided the largest populations with five pairs at Headley and three pairs at Walton; at Broxbourne Woods they had decreased to just three pairs.

In 1981 there was a national survey and only seven males were located, including two in the Ongar area. However, they had abandoned the Ongar site within four years as it was no longer suitable. By the start of the 1990s the breeding population had been stable for seven years at an average of four known pairs, virtually all north of the Thames. However, by 1991 they failed to breed for the first time ever after finally abandoning their traditional site in Broxbourne Woods as it became too overgrown. In subsequent years at least one territory was occupied on the North Downs and in the mid-1990s a pair bred at a newly restored heathland site in the west of the London Area.

21st century

By the start of the 21st century there was still a small breeding population of about five pairs at three sites, one in the west of the London Area and two towards the southern boundary. However, the population has continued to dwindle and no territorial birds were located in 2009 or 2010. Migrants are still occasionally seen, even in central London. In 2006 a male held territory in the suburbs at Teddington and was regularly seen in the streets at dusk.

Facts and Figures

The earliest migrant was on 2 April 1926, at Gilwell Park.
The latest migrant was on 27 September 1986, at Orsett Fen.

Common Nighthawk *Chordeiles minor*

A very rare vagrant (1)

Range: Americas; vagrant to UK.

The only record occurred in 1984 when an adult male was found in poor condition on Barnes Common; it was taken into care but died five days later. The specimen was kept and taken to a taxidermist for mounting. This was just the twelfth record for Britain and only the third mainland one, following others in Nottinghamshire in 1971 and Dorset in 1983. There have been 22 accepted records in the UK up to 2010 all of which arrived in September or October.

1984: Barnes Common, one taken into care on 23 October

Common Swift │ *Apus apus*

A very common summer visitor and widespread breeder

Range: breeds Europe, Asia and North Africa, winters Africa.

Historical

The earliest record was in 1774 when Gilbert White saw a party of swifts around the Tower of London; they fed over the Thames but did not enter the crowded City.

In the 19th century this was a regular summer visitor although there is scant information about breeding. An early indication of a decline was noted in 1817 by Forster who wrote that the tower of Hackney Church used to abound with swifts but they had become less numerous. In 1842 Meyer stated that they were particularly abundant at Hampton Court Palace compared to other parts of the Thames. Harting observed them daily over reservoirs and about the Thames and thought that they 'must travel a considerable distance to breed, for there are but few favourable sites for nesting'. Breeding was recorded at Harrow, Isleworth, Kingsbury, Northolt and Stoke Newington. In July 1873 around 2,000 were seen passing over Hyde Park.

20th century

At the beginning of the century they bred in Enfield and at both Stonebridge and Tooting in good numbers. In 1909 Dixon stated that they bred regularly in what were then rural suburbs, such as at Enfield, Kilburn, Putney, Tooting, Willesden and Woolwich, adding that although he had seen them closer to the centre of London he knew of no actual nests.

Swifts usually shunned central London because of the smog but during the summer of 1921 there was a coal-miners' strike, which left the skies of London unusually clear and large numbers were regularly seen over the Serpentine in Hyde Park. In 1928 three pairs bred under the eaves of the Post Office in the City of London, the first confirmed breeding record from that district.

In the 1930s Glegg stated that they were seen only occasionally in central London but could be found in suitable locations in the suburbs and rural areas. Fitter estimated that they bred as close to central London as Notting Hill, St John's Wood, Islington, Norwood and Putney during the 1940s. In 1946 eight birds appeared to be nesting in Canning Town, the first there for some 40 years.

The LNHS conducted a survey in 1949 and divided the study area into three zones. In the inner zone, no nesting was found but swifts were seen regularly feeding over some park lakes. The highest density of nests occurred in the inner suburbs and was thought to be between 24 and 40 nests per 1,000 acres; the nests were concentrated into four main areas. In the outer suburbs the density was only about half this and they were mainly found in three large groups. Two of these groups were situated at Dawley and West Drayton and it is interesting to note that several of the surrounding villages were devoid of nests.

From the mid-1950s swifts started to return to central London. They may have bred continuously at Kilburn but they began to spread from there closer to the centre and had established colonies in Maida Vale and Regent's Park by the end of the decade. During the 1960s they also started moving into Hammersmith and Notting Hill but there was virtually no sign of any potential colonisation to the south and east. In 1971 they bred at Westbourne Terrace and by 1973 they were also breeding in Islington.

A few continued to breed in central London during the 1980s but they did not attempt to colonise the East End until 1994 when breeding was recorded in Bethnal Green and Whitechapel. There was very little change in the breeding distribution between the two breeding Atlases. The bulk of the breeding population was concentrated in the suburbs where there was a good supply of nest sites.

21st century

Since the beginning of the 21st century a few pairs have continued to breed in Inner London, mainly in the west and north-west fringes. Elsewhere, they breed in the suburbs and rural areas where they are regularly seen during the summer months. Across the UK as a whole there has been a 38 per cent decline from 1994-2010 according to BBS counts, with an even higher 43 per cent decline in London from 1995-2011. Modern style buildings and roof repairs sealing off entry are seen as key factors in the decline in urban areas.

Movements and Concentrations

Glegg noted that very large flocks would gather at the reservoirs prior to their return migration. Large feeding flocks also occur where there are concentrations of flying insects or to escape adverse weather. In 1960 an abnormally large number was seen in the country in early October, and in the London Area there were records of single birds from seven localities between 1 and 8 October.

In late spring and early summer large feeding concentrations involving flocks of over 1,000 birds are sometimes seen over some of the reservoirs. Large numbers often occur at Staines Reservoir, e.g. about 3,000 per hour flew over on 28 April 1962; more than 5,000 passed over there in three hours on 1 May 1968; and at least 5,000 have been recorded on two other occasions. Elsewhere, there were about 8,000 at Thorndon CP on 22 July 1978; 5,000 flew over Wraysbury GP in two hours on 10 May 1983; and 6,000 moved over King George VI Reservoir in one hour on 30 July 2009.

Facts and Figures

The earliest migrant was on 20 March 2010 at Wimbledon Common.
The latest migrant was on 11 December 1994 at Tottenham Cemetery.
A nest-hole in a house at Mill Hill was used for 43 consecutive years up to and including 1966.
At least 12 were killed by traffic while feeding low over a road in Danson Park in May 1965.
An adult ringed at Elmers End in 1962 was picked up dying at Bromley in July 1976, making it at least 16 years old (as adult plumage is not attained until the third year), one of the oldest recovered in the UK.

Alpine Swift *Apus melba*

A scarce passage migrant (33–34)

Range: breeds southern Europe, Asia and Africa, winters Africa; scarce passage migrant in UK.

The first record occurred in 1834 on the River Wandle at Garratt Copper Mills, near Wimbledon. A description of the bird was given to Blyth who took his gun with him. After several attempts to see the bird he eventually saw three or four although they were out of range for a shot. There were four other 19th-century records. Ticehurst also referred to one that was seen in a flock of Common Swifts over Stone Park.

There were no further sightings until 1965 when one was present at Troy Mill Lake in the Colne Valley; it was presumed to be the same bird seen four days later at St Albans. Following the next one two years later there was a gradual increase in the number of sightings with two in both 1975 and 1977, one of which, at Twickenham on 11 November, was a particularly late date for this species in Britain.

There were five in the 1980s, including one which remained in the Wraysbury area for 17 days. Although they continued to become less rare in the UK there was only one in the 1990s. However, since 2000 they have become an almost annual visitor. In 2006 one remained at Hampstead Heath for nine days, the first one to linger since 1983. Of the dated records, about 55 per cent were in spring with three in June or July and the rest in autumn.

1834: Wimbledon, three or four on 30 April
1841: Brent Reservoir, in August
1859: Lewisham, shot
1860: Finchley, shot in August
1895: Staines, on 19 May
Pre-1909: Stone
1965: Troy Mill Lake, on 26 September; also at St Albans, on 30 September
1967: Beddington SF, on 18 June
1975: Staines Reservoir, on 19 September
1975: Twickenham, on 11 November

1977: Hartsbourne GC, Bushey, on 14 May
1977: Staines Reservoir, on 25 August
1980: Walthamstow Reservoirs, on 21 June
1981: Orpington, on 26 July
1983: Barn Elms Reservoirs, on 24–26 April
1983: Wraysbury GP, on 9–25 May; also at Wraysbury Reservoir
1989: Croham Hurst GC, on 12 August
1990: Hilfield Park Reservoir, on 30 April; also at Watford
2000: Richmond Park, on 27 May
2001: Hooks Marsh, on 13 October
2003: Brent Reservoir, on 5 May
2004: London Wetland Centre, on 17 October
2005: Walton Reservoir, on 30 April
2006: Hampstead Heath, on 8–16 April
2006: Crossness, on 14 April
2009: Aldenham Reservoir, on 5 September
2009: Alexandra Park, on 6 September
2010: Hollow Pond, on 22 March
2010: Crossness, on 27 March
2010: Stoke Newington Reservoir, on 1 April
2010: Warren Gorge, on 3-5 April; also at Rainham Marsh on 5 April

Unacceptable Records

One at Fairlop Waters on 27 April 2003 was accepted by the BBRC but not considered acceptable by the LNHS Records Committee.

Kingfisher *Alcedo atthis*

A common resident and localised breeder

Range: Europe, Asia and North Africa.

Historical

The earliest reference was in 1833 when Blyth stated that they bred at Tooting. Harting knew this as a largely resident species, with some leaving the area in winter; adding that 'a few pairs breed on the banks of the Brent and Silk stream, and also along the Thames and Colne'. In 1893 Swann stated that they had declined in London and the only breeding pair was on the River Brent at Twyford Abbey. Ticehurst recorded that they bred once in Greenwich Park.

There was a record of one flying over the Serpentine on 25 December 1863 and another of one in Muswell Hill that visited a garden pond daily to take newly released carp, although the observer had never seen them before or since in the neighbourhood. In 1862 one was found dead in the New River at Enfield, its beak being firmly held by a freshwater mussel.

20th century

In the first decade of the century they were known to be breeding in Blackheath, Bromley, Lee and Lewisham, and probably also in the Hampstead and Highgate district. However, during the first half of the century the population in all of these areas declined. They were occasionally seen in central London; in 1902 Pigott noted that two had been caught in the grounds of the British Museum and a pair had taken up residence in Regent's Park. In 1920 six pairs bred along the Grand Union Canal at Southall but elsewhere in the suburbs they bred only occasionally with nesting known at Barn Elms in about 1924 and at Kew in 1936.

In the 1930s Glegg described the Kingfisher as a 'common but local resident in the rural parts of the county', adding that it was more widely distributed between August and March, when it could even be found in suitable locations in the suburbs and central London. One was seen flying along the Thames under London Bridge, and another took to visiting the lake in Regent's Park twice a day as it was being drained to take advantage of the fish that were stranded in the shallow pools.

After the breeding season some moved away from their territories and in August 1911, 22 were netted as they flew downstream along Ching Brook at Woodford. In the 1950s many could be found on the Thames-side marshes where they were common in autumn until the end of the year, before dispersing. Some also migrated further along the Thames, such as the one that was ringed west of London at Wokingham in July 1927 and recovered at Barking Marsh in August 1928.

Occasionally they breed some distance from water if there is no nearby suitable habitat, such as in the following examples: in 1873 a nest was found in a disused, dry gravel pit between Aldenham and Bushey, a mile away from the nearest river; and a pair nested in a sand pit in Kenwood on Hampstead Heath in at least 1932–33.

Many perished during the exceptionally cold winter of 1962/63 and only two were seen during the following breeding season, at Darenth Gravel Pit. In England and Wales the numbers crashed to just 5 per cent of the normal breeding population. There was a small recovery in 1964 when two pairs bred and three other pairs were present, and by 1966 they had recovered to the level before the severe winter, although this was still relatively small at about 10 pairs. The breeding population continued to expand and by the beginning of the 1970s 25 pairs were located. However, this was probably well short of the total as in the first Breeding Atlas breeding was reported in 83 tetrads.

There was a decrease of almost one-fifth in the summer of 1982 after severe weather at the start of the year. After a mild winter numbers recovered quickly and at least 30 pairs were present during each of the following two breeding seasons, including 13 pairs in the Lea Valley. Severe weather in early 1985 caused a decrease of some 40 per cent but a series of mild winters at the end of the decade helped the breeding population to increase to at least 74 pairs.

At the beginning of the 1990s the population continued to increase, but the geographical split showed that just over two-thirds bred north of the Thames. Following severe weather in February 1991 the breeding population fell again with only 36 pairs reported. The second breeding Atlas in 1988–94 recorded an increase of 88 per cent in the distribution of breeding Kingfishers; most of the increase was in the Thames corridor, the outer suburbs and the Colne Valley.

21st century

Since the beginning of the 21st century the breeding population has remained fairly stable due to mild winters, with an estimated 50–100 pairs. They can be seen on most rivers and other waterbodies, and occasionally even on the park lakes and the River Thames in central London.

Facts and Figures

At Rye Meads 38 different birds were ringed in 1970, one of which was recovered near Etal in France on 17 October of that year, the first foreign recovery of a British-ringed bird.

In 1990 a Kingfisher was recovered at Walthamstow Reservoirs on 26 October that had been ringed in Germany on 8 June in the same year, the first recovery of a German-ringed Kingfisher in the UK.

(European) Bee-eater *Merops apiaster*

A rare vagrant (22)

Range: breeds Europe, Asia and southern Africa, winters Africa; scarce summer visitor and occasional breeder in UK.

The first record was of a bird shot between Kingsbury and Hampstead 'many years' prior to 1866. Harting saw the specimen but was unable to obtain any further details about it due to the length of time since its capture.

Although Glegg considered this to be an unsatisfactory record, he gave no reason for that view, so it should still stand as its identity was confirmed by Harting. The only other 19th-century records were of one seen 'feeding on yew berries' in a large garden at Oatlands, Walton-on-Thames, in 1869 and one that Swann states had been seen at Wanstead prior to 1893.

There were no further sightings until 1955 when one flew low over Old Parkbury in July; that year the species bred in Sussex. Three years later, in August 1958, one flew north over Beddington; autumn records are rare in the UK and this is the London Area's only August record. The only one in the 1960s was at Staines Reservoir. In 1979 four Bee-eaters spent three days in Thorndon Park in April; this was the first sighting of more than one bird, and was also the first and only time the species has remained for more than a day.

There were three records in the 1990s, including one of a flock of four over Nore Hill, Chelsham, which occurred during a notable influx into the country in June 1997. The largest flock comprised six birds over Richmond Park in 1999. The only one since then was seen around the visitor centre at Rainham in 2007 before flying south over the Thames.

Of the eight dated records, six were in spring which is typical for this species with just the two records from the 1950s being in July and August.

Pre-1866: shot between Kingsbury and Hampstead
1869: Walton-on-Thames
1893: Wanstead
1955: Old Parkbury, on 12 July
1958: Beddington SF, on 5 August
1963: Staines Reservoir, on 30 April
1979: Thorndon Park, four on 24–26 April
1993: Streatham, on 22 June
1997: Nore Hill, Chelsham, four on 16 June
1999: Richmond Park, six on 9 June
2007: Rainham Marshes, on 11 May

(European) Roller *Coracias garrulus*

A very rare vagrant (4)

Range: breeds Europe and Asia, winters Africa; vagrant to UK.

The Roller has only occurred on four occasions, the first of which was in the summer of 1832 on Penge Common. At that time it was almost an annual visitor to the UK. There are three modern-day records, all within a 10-year period. Two occurred in 1959, including one of a bird which remained for a week at Oxshott; the other was seen on the causeway at Staines Reservoir. The only other record was in 1968 when an immature in Osterley Park was present for two days in August and was trapped and ringed on its second day.

1832: Penge Common, in summer
1959: Oxshott, one on 23–29 May
1959: Staines Reservoir, one on 26 September
1968: Osterley Park, one on 26–27 August

Unacceptable Records

One seen by the River Ravensbourne at Deptford on 31 January 1945 was considered to have been an escape because it was observed in the middle of winter; there have been no subsequent winter records in the UK.

Hoopoe   *Upupa epops*

A scarce passage migrant

Range: Europe, Asia and Africa, European population mostly winters in Africa; scarce visitor to UK.

Historical

The first appearance in the London Area was in the winter of 1666/67 when one was killed 16 kilometres from the City of London and was described and illustrated by Charlton. There were two in the 18th century, one illustrated by Albin (1731–38), which had been shot at Woodford, and one shot in Norwood near Osterley Park and illustrated by Edwards in 1743. During the 19th century there were more than 30 birds, including one in Inner London, in Kensington Gardens in 1855. Many of these were shot so were seen only on one day but one spent several days in a garden in Knotts Green, Epping, in 1886.

Ticehurst refers to two probable breeding records in south-east London. A female was shot at St Mary's Cray on 11 May 1863 and there was 'evidence of having a nest and eggs'; while at Green Street Green a juvenile shot in 1891 could hardly fly and was thought to have been reared in the neighbourhood.

20th century

Between 1900 and 1940 there were eight birds, including one which remained in a garden at Totteridge for about two weeks in August 1936 and three which arrived on 16 April 1934. The number of sightings increased towards the middle of the 20th century and between 1941 and 1959 there were a further 38. These include a territorial male in Whitewebbs Park from the first week of May through to 22 June 1941; one on a roof-top in Streatham in June 1945; four in various gardens in the Oxted area in 1949, at least two of which were present for a fortnight; and the only winter record was near Epping on 26 November 1954.

A further 23 occurred during the 1960s, including one which remained at Brentwood between 20 April and 8 May 1966 and one in Kensington Gardens on 7 November 1967, which was the first Inner London record for 112 years. There was a similar pattern in the 1970s when 26 were seen, with two in Inner London: at Hyde Park in November 1973 and at Surrey Docks in April 1975. There was then a significant decrease in records, with 20 during the 1980s and just 10 in the 1990s.

21st century

In the first years of the 21st century they were seen annually with 17 records up to 2010. All of these were in the outer suburbs and rural areas apart from one in Brixton in October 2005. Ten of these were in spring with April being the commonest month for arrivals.

Facts and Figures

The earliest date was 2 March 1969 at Woldingham.
The latest date was 26 November 1954 at Epping.

Wryneck *Jynx torquilla*

A scarce passage migrant; former breeder

Range: breeds Europe and Asia, winters Africa; scarce passage migrant in UK.

Historical

This was a regular breeding summer visitor in the 19th century; the earliest reference was at Epping where it was seen annually from 1828 to 1845. Christy stated that a decline had been noted as early as 1832 in Epping Forest and quoted Doubleday 'This bird which used to be heard a few years since in all directions, is now so scarce that I have not heard more than three or four in the neighbourhoods'. Around 1850 they could be found in central London in Kensington Gardens where they presumably bred.

Declining numbers were reported in several other areas during the second half of the century; in Harrow they were not uncommon in 1864 but a decline was noted by 1891. They had become scarce in the Norwood area of south London by 1885 and in the Lower Brent Valley by 1896. However, in some areas there was still a thriving population; Ruegg stated that they were 'tolerably abundant' in Woolwich in 1847; and around Hampstead they were still common in 1889.

20th century

At the beginning of the century Dixon stated that they nested closer to the centre of London than any other woodpecker. They bred to the south of the river at Battersea Park, Dulwich, Lewisham, Sydenham, Tooting and Wimbledon in the 1900s; and in the north London inner suburbs they bred regularly at Hampstead up to 1908 and around Willesden to 1910.

They continued to decline in the outer suburbs: they bred annually at Mitcham until 1905; at Winchmore Hill up to about 1906; at Harrow until 1909; at Wanstead until 1910; and around Enfield to 1912. They were not uncommon in the Staines district in 1906 and sporadic nesting was reported at Morden in 1910, at Upper Norwood and Croydon in 1913–14, in Orpington from 1913 to 1915 and at Stanmore in 1914 or 1915.

However, by 1920 they had declined to such a point that they had ceased breeding in the inner suburbs and were rarely seen even on passage. Further out, they were common at Mill Hill until 1926 but they disappeared after that; breeding occurred in Harrow and Loughton in 1927; and they had decreased around Epping Forest by 1929. Wrynecks continued to breed intermittently around Northwood and Ruislip until the mid-1930s.

The population held on for much longer in the far south of the area but local extinctions also occurred there; they last bred at Chelsham and Warlingham in 1932 and at Ashtead and Epsom in 1936. About 10 pairs once bred annually in the Godstone, Limpsfield and Oxted areas but by the late 1940s this number had declined to just one or two, and they last bred in 1949. Up to six pairs once bred in the Tadworth area but they, too, had disappeared by 1954. Migrants were occasionally seen in central London; one in St James's Park on 28 August 1955 was seen feeding on the ground with House Sparrows.

During the 1950s they were still breeding in very low numbers in south London, mainly on the North Downs, but it was thought that there were no more than half a dozen pairs annually. To the north of the Thames breeding was known from Park Street in 1950–51, at Osterley Park in 1952 and at Arkley from 1949 to 1951 and again in 1953. Elsewhere, they were occasionally seen on passage. By the start of the 1960s they were just about hanging on as a breeding species. In 1964–65 up to 10 pairs bred south of the Thames but the breeding population decreased rapidly after this and by 1969 was down to two pairs.

Wryneck finally lost its status as a breeding bird in the early 1970s; the last confirmed breeding was in 1972 when two pairs bred, in south-east London and in rural north London. The following year one pair probably bred but none summered in 1974 and they were virtually extinct as a breeding species in England. By the mid-1970s they had become a scarce passage migrant; in 1976 about 19 were seen, including a pair in a garden on 11–12 May where they had previously bred.

They continued to decline and by the beginning of the 1980s only a handful were seen on passage each year. There was an influx in autumn 1981 when 15 were seen. However, in 1985 none were seen for the first time ever. During the 1990s they were seen almost annually, mostly in autumn.

21st century

Since the start of the 21st century there has been an average of three to four migrants a year with a spring/autumn ratio of 1:7. There was a particularly strong showing in the autumn of 2008 when nine were seen between 28 August and 23 September.

Facts and Figures

The earliest migrant was on 19 March 1862 in Middlesex.
The latest migrant was on 7 November 1998 at Rye Meads.

Green Woodpecker *Picus viridis*

A common resident and widespread breeder

Range: Europe and western Asia.

Historical

The earliest sighting was by Gilbert White at Vauxhall on 14 March 1778. In the 19th century the species was common throughout England, although it was less common in the London Area than Lesser Spotted Woodpecker. Harting noted that 'it was nowhere plentiful, even in those woods which are strictly preserved' and listed it as being present at Alexandra Park, Brockley Hill, Canons Park, Fryern Barnet, Hampstead, Harrow, Kingsbury, Mill Hill, Muswell Hill, Stanmore, Trent Park and Wembley. It was common in Langley Park in the 1860s, and in south London it was frequently seen on Wimbledon Common in 1872 and was also recorded at Dulwich and Richmond Park. The only documented record in Inner London was in Hyde Park in November 1885.

20th century

At the beginning of the century Green Woodpeckers were reasonably common in the rural areas but fairly scarce in the suburbs and built-up areas. In the 1930s Glegg stated that it was 'probably the least common of the three woodpeckers'; it was rare in Inner London, where he knew of only four records, and largely absent in the northern suburbs apart from at Hampstead.

By the middle of the century the population had increased in the inner suburbs and the parks in central London. It was probably breeding at Dulwich Woods, Greenwich Park, Hampstead Heath, Kew Gardens and Wandsworth Common. Breeding took place in Holland Park in 1952–53. However, after this it became very rare again in central London. In the outer suburbs and rural areas it was still relatively common despite some local population fluctuations after 1900.

After the severe winter in early 1963 reductions were reported across the London Area. The population was very slow to recover and at the start of the 1970s there were just 34 breeding pairs reported and the species was virtually absent in central London. During the 1990s the population began to increase and birds were present at more than 70 sites during the breeding season. However, the more heavily built-up areas proved to be a barrier and there were none in central London. The highest concentrations were 12 pairs at Hainault Forest and up to nine in Richmond Park.

The population continued to expand through the 1990s: at Bricketwood it increased to at least 12 pairs in 1993 from seven the previous year; and in Tottenham Cemetery no birds had been recorded until 1990 but then it was breeding within three years. Green Woodpeckers had also become sporadic visitors to central London. By the time of the second breeding Atlas in 1988–94 there had been an increase of over 50 per cent in the distribution throughout the London Area. Much of this increase was in the suburbs but the species had yet to colonise Inner London and the more heavily built-up areas in east London.

21st century

By the beginning of the 21st century the population was already at an all-time high. Nationally, the population has more than doubled since 1970 as measured by the BBS; one significant factor is thought to be a trend towards milder winters as this species suffers when the ground is frozen for long periods and it cannot reach its favoured prey. Green Woodpecker has expanded its range into more urban areas, occurring in the central parks; a pair bred in Victoria Park in 2000, the first confirmed breeding in Inner London. Since then the increase has continued and four pairs bred in Regent's Park in 2009.

Great Spotted Woodpecker *Dendrocopos major*

A very common resident and widespread breeder

Range: Europe, Asia and North Africa.

Historical

Great Spotted Woodpecker appeared to be genuinely scarce for much of the 19th century, though it is known to have bred in Kensington Gardens at the beginning of the century. In 1832 Doubleday stated that it was scarce around Epping and he had never seen a nest. Harting noted a decline, stating that it was formerly not uncommon around Kingsbury and Harrow but was now 'one of our rarest birds'.

During the 1880s they had begun to make a comeback and the population increased. Bucknill speculated that a possible reason for the increase was the end of the fashion for having cases of stuffed woodpeckers in houses and inns. However, they were still very rare at Wimbledon in about 1895.

20th century

During the early 1900s they were locally common in parts of east London such as Epping Forest and Upminster, as well as in various woods in south-east London, particularly around Bromley where Ticehurst stated that they were more common than Green Woodpecker. However, they were relatively scarce elsewhere at this time. There were none in Harrow in 1903 but by 1918 they had recolonised the area.

In the 1920s the population had increased to such an extent that they had become widespread throughout the London Area and were the most common woodpecker in the suburbs. They were increasingly seen in gardens and started to take advantage of food on bird tables. They had also colonised Inner London, breeding in Holland Park from 1922 onwards and in Regent's Park in the 1930s. One of the largest populations was in Richmond Park where there were up to seven pairs in the 1930s.

By the 1940s they had become the most common woodpecker across the London Area, and had returned to nest in Kensington Gardens. One factor in their return to central London was considered to be the maturing of the large number of trees planted in parks and gardens in the 19th century. During the 1950s the central London population expanded and they spread to Battersea Park, Campden Hill, Hyde Park and St John's Wood. However, by the start of the 1970s there had been a decline and they had ceased breeding in central London.

During the first Atlas in 1968–72 the species was located breeding widely across the London Area apart from in the Thames valley from central London eastwards. A pair bred in Kensington Gardens in 1976 but there was no further confirmed breeding in central London until 1981, in Regent's Park. In 1986 they were seen at more than 130 localities during the breeding season; some of this increase was due to better coverage as a Woodland Survey was carried out but it was also apparent that they had become more widespread during the 1980s.

During the second breeding Atlas they were found breeding in 500 tetrads, an increase of 75 per cent since the first Atlas. There was a large expansion of the population during the 1990s, notably along the Lower Thames and into the inner suburbs. They were still quite scarce in the more heavily built-up areas of east London but by the end of the decade they were beginning to colonise even these. The largest populations were 20–25 pairs on Bookham Common in 1999 and 18 pairs in Hainault Forest in 1995.

21st century

At the start of the 21st century the population was increasing; there were over 300 pairs, including at least seven in Inner London. The increase has continued since then throughout the London Area which may be due to a number of factors including increased habitat as more trees mature in the capital; increased young survival rate due to reduced competition with Starlings linked to the starling decline; and an increase in garden bird feeders, particularly in the suburbs as this species seems to be very partial to peanuts.

This species' fortunes in the London Area mirrors the national pattern. It has a huge range right across the Palearctic and well into Indochina. It would appear to have been widespread across Great Britain in the 18th century, but then disappeared almost completely from Scotland and northern England in the late 18th and the first half of the 19th century, as well as declining elsewhere, before recovering slowly after this. The UK population has increased by a factor of 368 per cent between 1970 and 2010, of which 141 per cent took place between 1995 and 2010, and it would seem that this dramatic increase has been reflected in the capital.

Lesser Spotted Woodpecker *Dendrocopos minor*

An uncommon resident and localised breeder

Range: Europe and Asia.

Historical

This was the most common woodpecker throughout the London Area in the 19th century. In central London it was frequently seen in Kensington Gardens.

20th century

At the beginning of the century it was still regarded as the most common woodpecker, although it appeared to be absent from central London. A decline had set in by the 1930s when Glegg described it as 'a local but not uncommon resident', adding that 'it is generally distributed over the rural parts of the county where there are suitable belts of trees or copses'.

By the 1950s it had become the least common of the three resident woodpeckers with its population still decreasing while the other two greatly increased. The decline in breeding numbers continued and by 1964 breeding was recorded at just 15 locations, including four pairs in Epping Forest. A recovery started during the late 1960s and in Richmond Park they were more numerous than Great Spotted Woodpecker and possibly also Green Woodpecker in 1967.

At the beginning of the 1970s the breeding population was at least 35 pairs. In 1977 exceptionally high numbers were reported in the breeding season and it was thought that this was due to the effects of Dutch Elm Disease which temporarily led to an increase in food supply of beetle larvae. They started to be seen more often during the breeding season in the central London parks; a pair probably bred in Regent's Park in 1979. The Woodland Survey in 1986 found the species in 90 localities during the breeding season; the highest concentration was six territories on Wimbledon Common.

At the start of the 1990s the population was relatively stable with some signs of an increase. For example, a pair bred in Alexandra Park for the first time; and the number of breeding pairs on Hampstead Heath increased to six by 1994. By the 1988-94 Atlas survey, an increase of 40 per cent was recorded in the number of tetrads with evidence of breeding compared with the 1968-72 Atlas, although it is difficult to know how far this may have reflected more thorough fieldwork, given that the national Breeding Atlases showed a decrease of 11 per cent between the 1968-72 and 1988-91 Atlas surveys. However, towards the end of the decade the population started to decline.

21st century

In 2002 the species was placed on the UK 'red list' on detecting a decline of 71 per cent from 1970-2010, based on a combination of CBC and BBS data. In the London Area the breeding population also appeared to be falling sharply. In 2005 about 50 territories were reported but three years later only 20 territories were located. Somewhat higher numbers were reported in 2010 with birds found in 69 locations, and possible evidence of breeding of 23-27 pairs as work on a new Breeding Atlas got underway. They are more widespread outside the breeding season but are rarely seen in the more urban areas, although territorial birds were still present in Wanstead Park and on Wanstead Flats in 2012.

Golden Oriole *Oriolus oriolus*

A scarce passage migrant; former breeder

Range: breeds Europe, Asia and North Africa, winters Africa; rare breeder in southern UK.

Historical

The first accepted record was in 1850 when a pair spent several days in a garden in Leyton; the female was shot on 25 May and upon dissection was discovered to have contained two fully developed eggs. Subsequently the species was seen infrequently across the London Area and a nest with three eggs was found in the parish of Amwell in 1881.

All 19th-century records are listed below.

1850: Leyton, pair in late May
1851: Harlesden Green, male shot on 11 May
1857: Well Springs, Wembley Park, male shot
1862: Hampstead, male seen in a garden in Frognal
1863: near Edgware, on 21 April
1870: High Trees, Redhill, male shot on 22 April
1870: Gatton Park, pair
1881: Amwell, pair nested
1887: Leatherhead, in April
1888: Harrow, shot on 21 May
1888: Broxbourne, pair on 20 July
1890: near Redhill, shot
1899: Addington, on 9 June
Pre-1899: Hackney Marshes

20th century

In the first half of the 20th century there were about 20 records; of these 15 were in spring or early summer, including a pair that was shot at Kenwood prior to 1912. Additionally, Ticehurst stated that they had occurred at Dartford and Eltham prior to 1909 but gave no further details. In 1930 a pair bred just inside the London Area boundary, near Godstone, and successfully reared young. The birds returned to the same site for at least the next two years although it was not known if they bred successfully again. In 1940 a pair was seen in the grounds of Chiswick House and a half-completed nest was found on 4 May but the birds later deserted the site. At some time prior to 1943 a pair was shot near Bayford. Five records between 1900 and 1950 come from the Watford area, three of of which were in June, which suggests they may have been breeding there.

From 1950 to 1979 there were just over 20 records, mostly of singing males on one day only although a pair was present at Keston from 1 to 3 June 1953, and two were in Epping Forest on 3 October 1976.

The peak period was the 1980s and 90s when a total of 45 was seen, including a run of at least one every year between 1980 and 1996. They were seen for three consecutive years at Northaw Great Wood from 1980 to 1982 and on Hampstead Heath from 1993 to 1995. Long-staying males were on territory for six days in Richmond Park in 1992 and on Horsenden Hill for five days; a pair was also present in Bushy Park from 13 to 22 May 1984.

21st century

The UK has had a very tiny breeding population, with its main focus in the fens of East Anglia. Numbers were thought to be increasing up to the early 1990s, but have since dwindled, although conservation efforts are underway. Birds seen in the London Area may be en route towards UK breeding sites, but are more likely to be heading towards breeding areas in mainland Europe, or are just young wandering adults. Between 2000 and 2010 there were 13 records in the London Area; all of these were of males in May apart from a female on 30 April 2010 at Foots Cray Meadows. The only one to be present on more than one day was in Bushy Park from 5 to 8 May 2004.

Unacceptable Records

One at Little Chelsea in the autumn of 1813 was later rejected by Glegg along with a number of other sight-only records. There was also a claim that they had bred at Warley Place on more than one occasion but there is no evidence for this.

Facts and Figures

The earliest date was 11 April 1935 at Wimbledon.
The latest date was 3 October 1976 in Epping Forest.

Brown Shrike *Lanius cristatus*

A very rare vagrant (1)

Range: Asia; vagrant to UK.

The only record was of a first-winter bird at Staines Moor from October 2009 to January 2010 although it was not seen from 18 to 31 December. Its identification was not established until the second day of its stay due to the difficulties of distinguishing young birds from the similar looking Red-backed Shrike. As well as being the first overwintering record and the longest-staying bird in the UK, it was only the seventh-ever British record.

2009: Staines Moor, on 11 October–2 January 2010

Isabelline Shrike *Lanius isabellinus*

A very rare vagrant (1)

Range: breeds Asia, winters Asia and Africa; vagrant to UK.

The only record occurred in 1994 when a cat brought a long-dead bird into a house on Lambert Avenue, Richmond, on 21 March. It was later identified as belonging to the race *isabellinus*, known as Daurian Shrike. Most UK records are in autumn and this was only the second one in March.

1994: Richmond, found dead on 21 March

Red-backed Shrike *Lanius collurio*

A scarce passage migrant; former breeder

Range: breeds Europe and Asia, winters Africa; former breeder in the UK, now scarce passage migrant.

Historical

The earliest reference was on 23 May 1787 when Gilbert White saw a pair nesting in South Lambeth; another was seen there in June 1788.

In the 19th century Harting described this as a common summer visitor although he commented that it 'has not been numerous here of late years', putting the decline down to the prevailing habit of 'plashing' hedges. It was often referred to as the 'butcher-bird' since prey items were frequently stored in a 'larder' on thorn bushes. In Harrow, Blue Tit and Common Whitethroat were found to be the most commonly stored prey items. Ticehurst referred to an exceptional year for this species in 1875 when 20 nests were taken on Dartford Heath.

20th century

In the first decade of the century it was a relatively common breeding bird outside the built-up areas. Its range extended into the suburbs at Barnes Common, the Brent Valley, Dulwich, Hampstead Heath and Kingsbury. Even in these areas the situation did not last long and a pair that nested on Hampstead Heath in 1908 was the last to have bred in north London's inner suburbs. Away from these districts the birds bred around Beddington, Croydon, Epsom, Leatherhead and Reigate in the south; they were common in the Dartford area and across all the outer districts north of the Thames. The first record from Inner London was on 29 August 1904, although the exact location of the bird is not known.

By 1914 they had become scarce in parts of east London and this was considered to be a result of the drastic trimming of hedges along lanes and on railway embankments. However, they were still common around Epping Forest. During the 1920s they ceased breeding at Kenton, Kingsbury and Preston but they did breed at Southall and Wembley in 1928.

During the 1930s the species retreated away from the suburbs but was still relatively common in favoured haunts such as in the Harrow district where seven pairs bred in 1930. It had declined further by 1935 when Glegg described it as 'a not uncommon summer resident in the rural districts', adding that 'it cannot be described as a suburban bird'. To the south of the Thames it bred regularly on Wimbledon Common until 1936, the same year that five pairs bred on Mitcham Common; the following year it was reported to be present on practically all of the commons in south London and along the North Downs. In west London 11 pairs were located in an area of approximately 26 square kilometres between Harefield, Northwood, Ruislip and Denham.

In the mid-1940s, 12 pairs bred in a 4-square-kilometre area around Connaught Water with an additional five pairs at Sewardstone; but by 1949 there had been a substantial reduction as a result of the cutting down of bushes. At Ashtead Common the large breeding population was reduced in 1936 due to building, and again in 1944 by ploughing up of part of the area; eight pairs continued to breed there until 1949, when only two pairs remained after further habitat loss. Odd pairs nested in the suburbs, at Greenwich Park in 1944, Putney Vale in 1947–48 and Dulwich in 1949.

At the start of the 1950s it was still a regular but local breeding species; in 1952 an estimated 61 pairs were known. During the early 1950s breeding took place in the suburbs at Beddington, Brent Reservoir, Edgware, Finchley, Mitcham Common and Southgate; and on Wandsworth Common in 1950, only 5 miles from the centre of London. Most pairs were found in the south, especially on the outer heaths and commons where a total of 27 pairs was located in 1952, along with a dozen pairs in south-east London.

The UK breeding population begun to decline during the 1960s, partly due to a loss of habitat. In the London Area, breeding birds were virtually restricted to the outer rural areas, having abandoned the outer suburbs. In 1962 there were at least 23 pairs reported; they were all south of the Thames apart from four between Epping Forest and Brentwood. The following year there had been a further reduction and just 11 breeding pairs were located; none were found in Epping Forest and only one pair bred north of the Thames, at East Barnet. The rapid decline continued and none bred in 1967 for the first time ever, although at least one pair bred in subsequent years.

At the beginning of the 1970s the species was just about holding on as a breeding species, albeit just a single pair. In 1973 two pairs bred successfully in south London and single pairs bred in 1974–75. By 1976 it was all over for breeding Red-backed Shrikes in the London Area and only one bird was seen all year; two years later not a single bird was seen. During the 1980s and 90s occasional birds were seen on passage in most years.

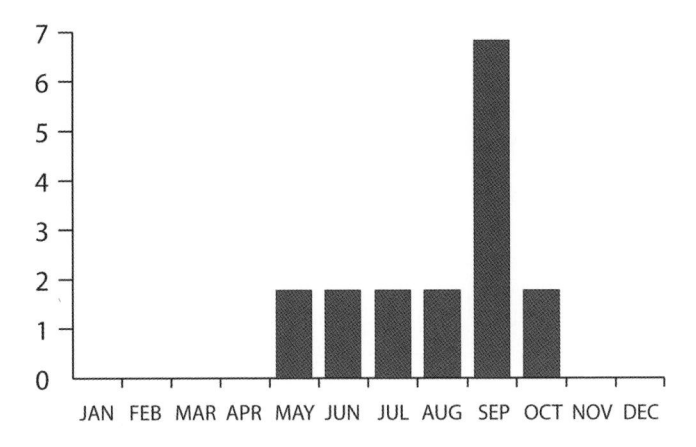

Seasonal occurrence of Red-backed Shrike: 2000–2009.

21st century

By the start of the 21st century this former breeder had become a scarce passage migrant. Between 2000 and 2009 there were three in spring and 11 in autumn. A male was present in summer at Lake Farm Country Park, Hayes, from 11 to 22 July 2012.

Facts and Figures

The earliest date was 16 April 1863 in Middlesex.
The latest date was 1 November 1980 at Wanstead SF.
A male was seen in winter at Harrow-on-the-Hill from 2 to 5 February 1934.

Lesser Grey Shrike *Lanius minor*

A very rare vagrant (3)

Range: breeds Europe and Asia, winters Africa; vagrant to UK.

There have been three records, one in the late 19th century and two in consecutive years in the 1950s. The second one at Perry Oaks was probably also seen in the same area on 4 and 7 October.

1897: Bromley, on 15 May
1956: Banstead Downs, on 21 May
1957: Perry Oaks, on 6 Oct

Great Grey Shrike *Lanius excubitor*

A scarce winter visitor

Range: Europe, Asia and North America; winter visitor to UK.

Historical

The earliest reference was recorded by Hayes (1808–1816) when he noted that one had been shot in Southall. This was a scarce winter visitor during the 19th century with about 37 records; all were between autumn and spring apart from one at Tyttenhanger Green in July 1879. The closest ones to the centre of London were on Hampstead Heath in April 1840 and in Kilburn in 1850.

20th century

In 1909 Ticehurst stated that they were seen almost every winter in the Dartford district but this may have referred to the late 19th century. In the first three decades of the century they remained a scarce visitor to the London Area with an average of one sighting every three years although none were seen between 1928 and 1935. In 1920 one was present on Ham Common between 15 May and 13 June. From 1936 to the early 1950s there was a notable increase in the numbers seen and they were recorded annually, including in three consecutive years at both Brent Reservoir and Mitcham Common. During this period there were usually only one or two each year but there were four in 1946. In central London one was seen in Camden Town on 23 May 1949.

Sightings continued to increase during the 1960s when there were typically two or three records each year. They were more frequent in the 1970s and 29 were seen in 1975, the most ever in a single year; some of these overwintered but the majority were passage birds. Up to two overwintered in most years in the Lea Valley during the 1980s but far fewer were seen elsewhere compared to previous years and none were seen in 1989–1990. The species remained scarce during the 1990s although individuals overwintered at Mitcham Common and South Norwood CP early in the decade.

21st century

Since the start of the 21st century they have ceased wintering in the London Area and have become scarce passage migrants. Just six were seen between 2000 and 2009 with none from 2001–04. Four of these were on passage, two in spring and two in autumn, while the other two were short-staying birds in January. Migrants are occasionally recorded in green spaces deep in the suburbs with recent records from Brent Reservoir, Kew Gardens and Wormwood Scrubs.

Facts and Figures

The earliest migrant was on 10 September 1938 in Bushy Park.
The latest migrant was on 30 May 1950 in the grounds of Hampton Court Palace.

Woodchat Shrike *Lanius senator*

A rare vagrant (17)

Range: breeds southern Europe and Middle East; scarce visitor to UK.

There were four in the 19th century, the first of which was a female shot at Winterdown, Esher, in 1853. The record of a male seen in the spring of 1882 by the Vale of Health on Hampstead Heath was square-bracketed by Glegg without giving a reason; however, as it matches the pattern of other records it should stand.

The first 20th-century record was at Longfield where a male was seen on 24 April 1934 before being found dead the following month. There were two in the 1950s, including a first-summer bird which was present for three weeks in Richmond Park. Five more were seen over the next 20 years. The only record in the 1980s was of an adult hit and killed by a car at Moorhouse, near Limpsfield.

In 1992 a male was present in Bushy Park on 25 May; this was the first to be seen north of the Thames for over a century; it was closely followed by another three years later when a moulting female spent three days on Barking Marsh. Since the beginning of the 21st century there have been two more. Just over half of the dated records arrived in May which is typical for this species when it overshoots its breeding range in spring.

1853: Esher, shot on 7 May
1873: Near Hertford, in May
1882: Hampstead Heath, in spring
Pre-1888: Ham
1934: Longfield, on 24 April
1951: Bookham Common, on 26–27 May

1953: Richmond Park, on 13 April–5 May
1960: Addington, on 13 May
1970: Oxshott, on 8 June
1971: Littlebrook, on 5 and 7 May
1971: Fetcham Downs, on 10 July
1973: East Ewell, found dead on 11 May
1986: Moorhouse, on 8 October
1992: Bushy Park, on 25 May
1995: Barking Bay, on 5–7 July
2003: Langley Park, on 1–3 May
2005: Richmond Park, on 12 June

Magpie *Pica pica*

A very common resident and widespread breeder

Range: Europe, Asia and North Africa.

Historical

The earliest record is of a pair that nested in St James's Park in 1638. In the early 1700s a pair nested at Temple but their eggs were removed and substituted with those of a Rook. They also nested by St Giles Church (near Covent Garden) in 1734.

During the 19th century the species was sparsely distributed across the London Area. Harting described it as scarce in the inner districts but more frequently observed in the more wooded and rural districts. Hamilton, writing in 1879, stated that it was formerly common in the parks of central London and nested in Kensington Gardens up to 1856. There was some evidence of local declines, for example in the Harrow district where it was formerly common and bred regularly from 1831 to 1838 but had become only an occasional visitor by 1891.

It appears to have been relatively scarce in the inner suburbs during the late 19th century and only a few pairs were known to breed around Hampstead before they were all shot. In the 1890s breeding took place in Regent's Park and there were several other birds in central London, all of which were thought to be escapes, as at least one Magpie was known to have escaped from cages in St James's Park. In 1899 the known escape had paired up with a Jackdaw and they occupied an old Magpie's nest.

20th century

At the start of the 20th century it had become extremely scarce due to persecution. In west London it was only sparsely distributed as far out as Staines and was just about surviving around Watford. In the east it was uncommon around Epping and only single pairs were known from Brentwood and Ongar. In the south-east it was 'all but exterminated' and at Oxshott it was said to be 'one of our rarest breeding birds'. On the North Downs only a single pair at Upper Warlingham in 1900 was known until 1909 when a pair was seen near Oxted. From this date there was a slow spread into this area from beyond the London Area border in Surrey where it had begun to recover after being extirpated from the entire county. A pair that nested in Green Park from 1909 onwards was considered to have escaped from captivity, as was a pair which frequented Kenwood on Hampstead Heath some years later.

There was a slow increase in numbers during the next decade and by 1918 there were six pairs near Godstone, and the gradual spread of birds from Surrey continued. During the 1920s they became fairly widespread in the outer districts of the London Area. In the Harrow district they bred in only one location where up four pairs nested annually between 1924 and 1930. In 1926–27 a flock of 10 at Harefield was the largest number known in the London Area at the time.

By the mid-1930s Glegg observed that it had an uneven distribution in Middlesex, being uncommon and local in the northern area, and common in the southern half; ascribing this as being due to there being a considerable area of orchards in the southern half where at least 30 pairs nested. Elsewhere, they were increasing rapidly although in some areas they were still being persecuted by gamekeepers. They were virtually unknown in urban areas; apart from a resident escaped pair at Kenwood and another pair that attempted to nest at Dulwich in 1938–39. Occasional birds were seen nearer to central London outside the breeding season.

After the Second World War the population increased greatly, although there was still the odd area where they were persecuted, such as at Addington where 57 were killed in the autumn of 1946. Several large gatherings were reported including about 80 on a 1-acre plot at Farleigh, about 70 at West Longfield and 60–70 at Reigate. Magpies had still not colonised the more heavily built-up areas by the beginning of the 1950s but by 1954 one pair bred at Finchley and there were nesting attempts on Dulwich and Mitcham Commons.

On 11 October 1960 two were seen in Regent's Park, the first to be recorded in Inner London since the start of the century. One or two remained in the area until the end of the year; they were thought to have come from Hampstead Heath which had recently been colonised. The following year there were reports of increases from all around the London Area and they had started to spread further into the built-up areas. As the decade ended a pair built a nest in Hyde Park/Kensington Gardens but later deserted after harassment from Carrion Crows, and a pair was present in Regent's Park from May to the end of the year.

At the beginning of the 1970s Magpies were still attempting to colonise central London; a pair finally bred successfully in Hyde Park/Kensington Gardens in 1971. During the time of the first breeding Atlas in 1968–72 they were breeding in over half of the London Area but were still absent from much of the built-up areas. By 1975 the breeding population in central London had increased to eight pairs; they continued to prosper and at least 20 pairs were breeding in central London in 1987.

By the start of the 1990s the breeding population in central London was at least 29 pairs. During the second breeding Atlas in 1988–94 Magpies increased their breeding distribution by over 60 per cent and were seen in all but four tetrads. The Breeding Bird Survey showed an increase in the population in the London Area of a quarter between 1994 and 2000. The highest counts in the 1990s were roosts of 200 in Devilsden Woods, Coulsdon, on 9 January 1993 and 153 at Stanstead Abbotts GP on 11 January 1992.

21st century

By the start of the 21st century Magpie had become one of the London Area's most common and widespread breeding birds. It has colonised all areas of the capital and even breeds in the streets of Westminster. However, its predations on nests of some song birds mean that it is less than popular with many bird lovers.

Facts and Figures

The largest roost was of about 250 birds at Farthing Downs on 27 December 2000 and 13 January 2002.

Jay *Garrulus glandarius*

A common resident and widespread breeder

Range: Europe, Asia and North Africa.

Historical

The remains of one found at Oatlands Palace, Walton-on-Thames, date to no later than 1650.

In the 19th century Harting refers to Jays as being 'resident throughout the year, but nowhere numerous'. Although he knew of them in many districts from Wembley and Kingsbury outwards to Elstree, they were absent from Hampstead and most other areas within the metropolis. The only record from Inner London was at Kensington Gardens in 1855. They were regular in the Norwood district and on Streatham Common at the end of the 19th century.

Jays suffered a great deal of persecution which kept the population down. For example, they were scarce on Wimbledon Common where they were shot, and in Epping Forest an edict was issued to order the destruction of Jays due to their habit of taking other birds' eggs. On one day in April 1888 a total of 106 was killed and this persecution continued at Epping for at least the next 70 years. However, they were so abundant in Oxhey Woods in the 1880s that their blue wing feathers were collected annually and sent to Scotland, where they were used to create artificial flies for salmon fishing.

20th century

Prior to the First World War they were regarded as declining across the London Area as a result of persecution, apart from in the Bromley district and at Wimbledon where they were increasing. From then on they started to increase slowly across the London Area, although they remained only occasional visitors to central London until the autumn of 1923 when a flock of 15 arrived in Holland Park; two remained in the area for some time and were thought to have bred.

In the 1930s they bred in the inner suburbs at Chiswick, Gunnersbury Park, Highgate Woods and Kenwood, and they had also become abundant at Wimbledon. During the 1930s they began to breed in Kensington and slowly colonised central London during the 1940s; breeding was first recorded from Battersea Park in 1946 and St James's Park in 1947. Jays also spread southwards from Hampstead and a pair nested on Primrose Hill in 1938. Four fledglings from one of the parks were examined in 1950 and were all found to be suffering from the results of a poor diet, which presumably consisted of a lot of bread. By 1951 three pairs were nesting in Kensington Gardens and a fourth pair bred in South Kensington.

Following the end of the Second World War, there was a notable increase in the population of Jays as persecution diminished. At the beginning of the 1960s at least five pairs bred in central London; by 1963 the population had increased to nine or 10 pairs. Persecution continued in some areas in the 1970s; for example, 70 were shot at Badgers Mount on a single day in April 1971.

In the first breeding survey in 1968–72 they were found breeding in almost half of the area but there were very few records from the built-up areas in east London although some of this may be due to under-recording. The population continued to increase and in the second breeding Atlas in 1988–94 Jays were found breeding in 552 tetrads, an increase of almost one-third since the first survey. In 1992 the highest breeding population was 20 pairs at Hainault Forest.

21st century

By the beginning of the 21st century the Jay had become a common and widespread breeding bird throughout most of the London Area. Nationally, the population has increased by 15 per cent from 1995–2010 and there has been a slightly smaller increase in the London Area. Although regularly seen in the Royal Parks where they even take food from the hand, Jays have not colonised central London in the same way as Magpies and Crows.

Movements

Occasionally large flocks have been recorded, typically in autumn when they are searching for acorns.

In 1959 there was a considerable influx of Jays; it began on 20 September when 19 flew over Richmond Park. Within a week flocks of up to 19 were being recorded at other sites and on the 28th two flocks, each 28-strong, flew over Dulwich Woods; two days later another 50 flew over the same site. On 2 October 30 flew over Greenwich Park with another 20 the next day. Just outside the London Area, 94 were counted moving north-west or north into the London Area at Wisley on 11 October. More sites continued to report Jays flying over, and on 24 November a flock of 63 was seen in Dulwich Woods.

In 1962 two groups totalling 57 flew over Broxbourne on 16 November. In the autumn of 1963 they were particularly noticeable south of the Thames, and flocks of over 40 were seen at Foots Cray and Sundridge Park in October. In 1992 a large influx was noted on Hampstead Heath when up to 100 were seen in the air together on 19 September.

Nutcracker *Nucifraga caryocatactes*

A rare vagrant (13 records/14 birds)

Range: Europe and Asia; irruptive visitor to UK.

The first record was of one taken at Darenth and documented by W.E. Leach in 1816; the specimen is in the Natural History Museum at Tring. The only other record from the 19th century was of one shot at Dulwich Meadows in around 1865.

There were two shot early in the 20th century, at Epping and Croydon; when the latter specimen was examined it was found to be of the slender-billed form *macrorhynchos*. The skin is now in the Manchester Museum. Fifty years later one was trapped at Springhead, Northfleet. In 1968 there was a large irruption into the country involving a total of 315 birds from 5 August onwards and within a few days the first one reached the London Area. By the end of the year 10 had occurred; most moved on within a day but two remained at Brentwood for about a month and the first one that was seen also stayed for a month in South Croydon.

Pre-1816: Darenth
c. 1865: Dulwich, shot
1900: Epping Forest, shot on 5 November
1913: Addington Park, Croydon, one shot on 13 October
1963: Northfleet, trapped on 26 August
1968: South Croydon, one on 8 August–7 September
1968: Headley, one on 10, 22 and 26 September
1968: Brentwood, two first seen on 14 September remained for about a month
1968: Thrushesbush, Harlow, one on 3–4 October
1968: Hendon Park Cemetery, one on 2 November
1968: Brookman's Park, one on 10 November
1968: Mill Hill School, one on 20 November
1968: Watford, one in November
1968: Shenfield, one on 8 December

Unacceptable Records

In 1905 Power claimed to have seen one flying over the golf course at Lordship Lane in Dulwich on 14 April and later wrote that he was 'satisfied to my correct identification'. Unfortunately he provided no details about the sighting and it was treated as doubtful by the LNHS. One seen at Kew Green on 6 July 1936 was considered to have been an escape.

Jackdaw *Corvus monedula*

A common partial migrant and widespread breeder

Range: Europe, Asia and North Africa.

Historical

Remains of Jackdaw dating back to Roman Britain have been found at Ewell. The earliest reference to Jackdaw comes from 31 May 1604 when a young bird, which had presumably come from a nest somewhere on the Palace of Westminster, entered the House of Commons; it was regarded as a bad omen for the bill which was being discussed at the time.

In the 19th century Harting described it as 'thinly distributed in the county'. In the 1870s a pair bred under the wings of a stone angel on a house near Bond Street and several pairs bred in the steeple of Grosvenor Chapel in Mayfair; six or seven pairs bred at Hyde Park and Kensington Gardens in 1879; and breeding occurred at St Paul's Cathedral up to 1889 and at Piccadilly until 1893. By the end of the century the only large colony left in Inner London was centred on Kensington Palace, and they bred in Holland Park and Kensington Gardens. In 1898 Hudson stated that Jackdaws were probably only breeding on three of the 1,600 churches in the inner suburbs but were resident in a few places in central London.

20th century

Jackdaws survived in the City until 1905 when they last bred on the Tower of London. A pair bred in a church in Brixton Hill regularly up to 1910, and on a church in Clapton up to 1911. They continued to be seen in Kensington Gardens throughout the 1920s and 30s although no breeding was known from 1921 to 1935; in 1936 three or four pairs were present and this had increased to six by 1941; and there was a maximum count of 24 birds on 1 November 1946.

By the 1930s Glegg considered that they were 'well distributed over the rural parts of the county, in some localities being abundant'. In 1930 there were 12–14 pairs within a 1.5-kilometre radius of Harrow-on-the-Hill church. However, by the beginning of the 1940s they had disappeared from almost all of the inner suburbs and had ceased breeding in Hampstead and Wimbledon.

During the early 1950s there was a considerable increase in the number of Jackdaws in rural areas. The small colony of up to four pairs in Kensington Gardens was still present and accounted for most of the sightings in central London squares and gardens. There was a large-scale clearance of diseased elm trees in Kensington Gardens in 1954 and even though nest-boxes were erected, the last breeding took place in 1957. Jackdaws remained in the area for several years afterwards and in 1963 a pair attempted to breed in Kensington Gardens but they were harassed by Carrion Crows; they bred again from 1965 to 1969.

By the beginning of the 1970s they finally ceased breeding in Kensington Gardens and only two birds were seen up to 4 February. After this date only migrants were recorded in Inner London and no birds were seen in 1972 for the first time ever. At the time of the first breeding Atlas in 1968–72 they were found to be breeding in 30 per cent of all the tetrads in the London Area; most of the population was located in the rural areas. From the late 1970s onwards large flocks began to be noted in the more rural areas; for example, in 1979 there were 300 at Wraysbury GP. In 1983–84 the peak count had risen to 2,000 on the tip at Wraysbury, and by 1985 the peak count had further increased to a record 5,000 roosting birds on 12 January.

The first signs of an increase in the breeding population were seen in 1991 in the rural fringe of east London. In 1993 a pair bred on Hampstead Heath for the first time in 50 years. The 1988-94 breeding atlas showed a 7 per cent decline in breeding distribution compared with the first atlas (although there was an increase in the number of tetrads where the species was simply recorded).

21st century

At the start of the 21st century the population was slowly increasing although Jackdaws were mainly restricted to the more rural areas and outer suburbs. Nationally there has been a 15 per cent increase in the breeding population from 1995–2010. Very few pairs breed in the inner suburbs apart from on Hampstead Heath where at least five pairs breed. The largest roosts are at Beech Farm, Broadwater Lake, Holyfield Farm and Sevenoaks Reserve.

Jackdaw Movements

The passage of Jackdaws over the London Area was first noted over a century ago. Towards the end of the 19th century, when there were few breeding Jackdaws left in the inner suburbs, a flock of 15 descended into Clissold Park where they spent a couple of hours before continuing their migration.

Although known to be largely resident, there were occasional large movements noted in the first half of the 20th century, in particular a flock of 100–200 flying over Regent's Park on 14 February 1923 in fog. A hint of their origin came when one was recovered at Radlett on 10 March 1940; it had been ringed in Sweden in the previous summer.

There were no further records of migration until the 1950s when about 40 flew south-west on 1 November 1951; and about 250 flew over Bloomsbury on 12 October. Since then they have been regularly recorded in autumn and, to a much lesser extent, in spring. There was a very large movement in the autumn of 2010 when they were seen across the London Area, including 355 that flew west over Brent Reservoir on 17 October and 310 over Regent's Park on 11 October.

Facts and Figures

The largest count was 5,000 birds at Wraysbury GP on 12 January 1985.
Harting once noted 'a curious variety of this bird, in the collection of a friend, was killed some years ago at Hammersmith. The scapulars and part of the back are of a beautiful bronze colour'.
In 1976 a bird of the northern race *monedula* with a clearly defined 'collar' was seen with four Hooded Crows at Sevenoaks on 24 November; a similar bird was seen there again on 10 January 1977 and during December of that year. In 2000 a bird resembling the eastern race *soemmerrungii* with a very pale nape was seen several times in April at Beddington.

Rook   *Corvus frugilegus*

A common partial migrant and widespread breeder

Range: Europe and Asia.

Historical

The earliest record was of a rookery that was found in Inner Temple in 1666; it remained in use until at least 1774 but had disappeared by 1831. The only other 18th-century rookeries were at St Giles-in-the-Fields in 1734 and at Gray's Inn in 1777.

In the 19th century Rooks were widespread breeding birds throughout the London Area, even in the City of London. Indeed, one or two pairs nested at Wood Street, just around the corner from St Paul's Cathedral, between 1836 and at least 1845. Due to the shortage of trees in the City, they sometimes nested in unorthodox sites; a pair nested on St Olave's Church in Hart Street in 1838, and some years earlier a pair nested on Bow Church, between the wings of the dragon on the vane. After being disturbed from their previous location of the burial grounds of St Dunstan's in 1817, Rooks established a rookery on the Tower of London in 1817–18, with some birds nesting on the weather vanes of each turret on the White Tower. Although they later returned to St Dunstan's, they began to nest on the Tower of London again in the 1850s.

In Inner London the largest rookery was in Kensington Gardens; it had 100 nests in 1836. Elsewhere, there were about 50 nests in Chesterfield Gardens, Mayfair, in 1846, 45 in Holland Park in 1881, and 38 at Gray's Inn in 1872. Away from central London the largest rookery was at Hampton Court Park, where the average number of nests in the four years up to 1832 was 750.

Rookeries began to disappear in the latter part of the 19th century. A large one was active in Greenwich Park until the 1860s when all the nesting trees were felled. A rookery at Stamford Hill ceased to exist after 1876; and in 1880 the large rookery in Kensington Gardens disappeared when the 700 elm trees in which the birds were nesting were cut down. The rookery at Gray's Inn disappeared for three years in the 1870s after some of the branches were lopped off the nesting trees. When the Rooks returned their numbers remained reasonably constant in the 20 years up to 1898, fluctuating only between 17 and 30 nests; the birds had no need to fly out to rural areas for food as they were well fed by local people.

Away from central London rookeries could be found in many places, including Camden Town, Ealing and Kilburn. In 1890 the biggest rookery in the inner suburbs was at Brockwell Park, near Herne Hill, but when the park was opened to the public the following year the majority of the Rooks abandoned the park and by 1898 no more than 10 pairs nested. The largest rookery in east London was at Wanstead Park, where 430 nests were counted on Rook Island in 1898.

Both Rooks and Jackdaws used to pick up the scraps discarded by horses from their nosebags, and Rooks regularly fed in the open grassy areas in Hyde Park. In 1864 there was a very dry spring and Rooks could not find enough food to feed their young; some landowners in and around Kingsbury 'abstained from the annual shooting in May, in consequence of the scarcity of birds'. Between 1874 and 1909 Power regularly recorded flocks of Rooks migrating over his home in Brixton in October, often with other species.

20th century

They continued to nest in central London right up to the early years of the century, although by this time they had quite a long flight each day to the nearest fields in which they could feed. In 1900 there were 12 nests at Connaught Square, and on the southern edge of Hyde Park there was one nest by the Prince of Wales Gate and seven in a garden off the Kensington Road; the last two sites were abandoned the following year. The Connaught Square rookery remained active and 10 pairs nested in 1908; after a three-year absence, three pairs nested in 1911. The Gray's Inn rookery lasted until 1915 and the last breeding in Inner London was in 1916 when four pairs nested at Temple. The combination of nesting trees being felled and increasingly long flights to feeding areas seem to have been the main causes for the loss of breeding Rooks in central London.

Further out, urbanisation was the main factor in the decline of Rooks; they disappeared as outlying villages became part of suburbia. Rookeries ceased to exist at Hampstead in 1903; Herne Hill in 1919; Wandsworth in about 1924; Harringay and Ealing in 1928; Greenford and West Ealing in 1929; Barn Hill in 1930; and Dulwich at around the same time.

By the 1930s there were still rookeries in many of the outer districts of west and north-west London, such as at Bushy Park, Harrow Weald, Mill Hill, Pinner, Ruislip, Southall, Staines and Uxbridge. The largest ones were at Kempton Park where there were 150 nests in 1930, and at Harefield where there were 128 nests in 1930. The same year Harrisson surveyed the Harrow district and counted 361 pairs within a 5-mile radius of St Mary's Church. Most of the rookeries in this area had grown over the previous couple of years but four had ceased to exist. A follow-up survey the following year found only 278 pairs, a decline of about 25 per cent; some of this was due to tree-felling at two rookeries. In 1936 the closest rookery to central London was at Ham Green where there were 14 nests. Rooks were rarely seen in central London but in 1939 a flock of 15–16 flew north over London Bridge on 29 October, and the following year three were seen in a tree in Regent's Park on 12 November.

In 1941 the rookeries in Bushy Park and Hampton Court Park were destroyed by enemy action but the following year a new colony of eight nests was created near Hampton Wick station. A complete census of Rooks in the London Area was made in 1945–46 and it produced some interesting results. There were 9,971 nests in 355 rookeries, but within a 16-kilometre radius of the centre of London there were only three rookeries: at Lee Green, Woodford and Bromley. The rookery at Lee Green was abandoned in 1947. More than one-third of rookeries were located in the eastern part of London. In four localities, around Eynsford, Godstone, North Weald and Watford, there was a high density of breeding Rooks with about 10 nests per square kilometre. In Eynsford there was a huge pre-roost gathering of about 10,000 Rooks and Jackdaws in Lullingstone Park in 1947, the former species being the majority.

Flocks were still occasionally seen flying over central London, for example about 50 were seen with a flock of about 250 Jackdaws over Bloomsbury on 12 October 1952, and about 100 flew over Battersea on 28 December 1954. In the autumn of 1959 there was a notable passage over central London and the Lea Valley between 26 October and 5 November, with a peak of 165 over King George V Reservoir on 3 November. In 1961 an abnormally large number of sightings in central London included one flock that lingered in Regent's Park throughout March and much of April, another in Bloomsbury Square from 28 March to 17 April and a flock of 11 over Waterloo on 27 November. Also that autumn about 500 birds flew south along the Lea Valley at Rye Meads on 18–19 November.

In 1967 the rookery at Hainault Lodge, which had about 128 pairs in the early 1950s, became extinct. Eight more rookeries between the River Roding and Rainham went between 1948 and 1968. The demise of four was due to tree-felling but the reasons for the others' extinction are not known. In 1969 a roost of some 8,000 corvids was seen at Cole Green, Hertford, on 5 January, of which 90 per cent were Rooks.

An attempt was made to reintroduce Rooks into central London in 1972; four birds were released in Hyde Park, two of which survived into the following year. Two more were released in October 1973. The following year a pair built a nest but deserted due to disturbance by Carrion Crows; by 1976 all the introduced birds had disappeared. In 1975 a flock of 2,000 fed on a ploughed field at Symondshyde on 16 January.

A national survey of rookeries was carried out in 1975; in the London Area 135 rookeries were located with a total of 3,712 nests. This showed a reduction in the number of rookeries of 60 per cent (from 355) since the previous survey in 1945–46. There were seven rookeries with more than 100 nests, the largest being 242 at Bayford. Further analysis (Sage and Cornelius) showed that the decline had not been consistent since there had been large increases in some areas, mostly up to the 1960s, before a general decline set in. The biggest losses were in the north-west which was affected by suburbanisation more than any other part of the London Area; in 1945 there were 190 nests at six rookeries but by 1975 only eight nests were located at four sites: two in Greenford and at Monken Hadley and Southall.

In 1979 a new rookery of at least 36 nests was started at Wraysbury GP; by the following year there had been a huge increase to over 120 nests. Also in 1980 a new rookery of four or five nests was started at Kempton Park, but this and the Wraysbury rookery were the only two in west London.

At the start of the 1990s rookeries were continuing to decrease. For example, the rookeries at North Ockendon and Passingford Bridge had declined by at least half in four years. Several new rookeries were found in eastern London in 1991, including 30–40 at Aveley landfill site with the birds clearly taking advantage of the rubbish tip at Rainham, where up to 500 birds were seen in February of that year. The second breeding Atlas (1988–94) highlighted a 28 per cent reduction in the breeding distribution of Rooks since the previous (1968–72) Atlas. It showed that almost all the surviving rookeries were beyond the suburbs, mostly in the far north-east and north-west. Since the first Atlas, the Great Storm of 1987 had brought down a huge number of trees and caused a further decline. The closest rookery to central London had been at Northwick Park Hospital in Harrow, where there were three nests, but this was deserted after 1988. A survey of rookeries in 1993–96 found a total of 102 rookeries with 2,549 nests. This equates to a reduction of 22–30 per cent since the previous survey in 1975, below the national decline of 41 per cent. Several large rookeries with more than 100 nests had disappeared since 1975. The largest one found in the 1993–96 survey, with 114 nests, was at North Weald; indeed, it was the only one with 100 or more nests, compared with seven such in the 1975 survey.

21st century

Since the beginning of the 21st century Rooks have remained reasonably common in the rural areas north of the Thames but there are only a very few rookeries south of the river. A new rookery of 12 nests was formed at Ten Acre Wood in 2000, the only one in the old county of Middlesex. There were 55 rookeries with about 1,500 nests in 2002, mostly in the eastern part of the London Area.

Wandering birds are still occasionally seen flying over central London but rarely land; one feeding in St James's Park with Carrion Crows on 4 December 2006 was the first one on the deck for several years. The largest count was 2,000 birds at Holyfield Hall Farm on 30 January 2000.

Carrion Crow *Corvus corone*

A very common resident and widespread breeder

Range: Europe and Asia.

Historical

In the middle of the 19th century Carrion Crow was widespread in the London Area, although not numerous, due to persecution. Harting noted that 'owing to its destructive habits, the Carrion Crow has but few friends', adding that 'amongst the long list of crimes attributed to this bird, not the least heinous are those of carrying off game-eggs and killing weak and sickly lambs'.

Large numbers were present in Richmond Park and birds from there and nearby Syon Park used to fly daily down the Thames and out into the estuary to search for food at low tide. Harting observed them feeding on the muddy shores of reservoirs at low water in search of mussels; when they found a shell that was hard to open they would fly up and drop it from a height to try and break it. In the period 1866–77 they were common at Grays, Orsett and Upminster, and fairly common at Thorndon and Warley. One nest in Hampstead was used for at least 60 consecutive years.

Towards the end of the 19th century Carrion Crows were scarce across much of the London Area. Occasional pairs were known in central London, such as the one that bred by the Serpentine between 1887 and 1894 until its habit of predating ducklings proved to be too much for the authorities; the latter pulled down their nest and shot the adults. Despite the loss of this breeding pair, Carrion Crows were seen daily in Hyde Park as they moved around the London Area seeking out food. In two or three other parks Carrion Crows were protected by the keepers and allowed to breed; by 1898 these were the only ones known to nest in the metropolis.

20th century

By the early part of the century, the species was still rare in many rural areas but had become numerous throughout much of the London Area, 'breeding in every part except the most densely populated districts of Inner London'. An indication of its status at the time comes from several local studies around north and west London: a dozen nests were counted around Hanwell on one afternoon in 1902, and about 20 nests were counted in a day around Kenton in 1908; however, in the Staines district they were rare in 1906.

By 1912 the suburbs had become the stronghold of the Carrion Crow and a few pairs bred in the large green spaces in central London such as Kensington Gardens and Buckingham Palace Gardens. After the First

World War it continued to increase in most areas of London; much of this increase was as a result of less persecution, especially in the rural districts. The last area to witness this increase was in the outer and rural districts south of the river where it was still relatively scarce in the 1920s. In the spring of 1922 some 19 nests were destroyed in the Lea Valley 'in the interests of the many wildfowl' which nested on the reservoirs. An example of their level of abundance could be seen at Harrow-on-the-Hill where there were 24 pairs within a radius of 2 kilometres in 1929.

Although typically sightings were of one or two birds around this time, there were occasional records of flocks, with the largest being 50-strong at Hampton Court Park in 1928; at Staines Reservoir 12 large parties flew north in one hour on 20 April 1929.

The Second World War further reduced the number of gamekeepers and during the 1940s the number of Carrion Crows continued to increase. By the end of that decade they had become common throughout the London Area apart from in a few heavily built-up areas with little open space, such as the East End and just south of the Thames in central London. They were still controlled in the Royal Parks in the late 1940s because they predated ducklings. In 1949 a flock of 120 fed on farmland at Stanmore on 14 April and roosting figures at Woodmansterne peaked with at least 209 on 19 February.

The large feeding groups continued to increase and in 1950 flocks of 150–160 were noted at Hilfield Park Reservoir and Richmond Park. In 1951 there were 21 pairs nesting in central London and by 1966 the known figure had increased to 26 pairs although the real population was estimated to be 30–50 pairs.

In the first breeding Atlas in 1968–72 they were found breeding in 72 per cent of the area; they appeared to be mainly absent in parts of east London although some of this area was under-recorded. In 1979 about 1,500 were on Wimbledon Common on 1 January, and there were 1,000 on a rubbish tip at Holmethorpe in 1981.

By the beginning of the 1990s the breeding population in central London was at least 35 pairs. Large numbers had started to congregate on the rubbish tip at Rainham Marshes and reached 550 in 1991. During the second breeding Atlas in 1988–94 Carrion Crows had increased their range by a quarter and were found breeding in over 90 per cent of the area.

21st century

By the start of the 21st century the Carrion Crow had become an extremely common bird throughout the London Area, even in the most built-up areas reflecting their ability to adapt to a wide range of habitats. They have increased in London at a greater rate than the UK as a whole, with the BBS showing an increase of 65 per cent from 1995–2010 in London compared with a 10 per cent increase for the UK as a whole.

Crows can be found congregating at waste tips, scavenging along the Thames foreshore, searching the rubbish bins and litter in the parks, occasionally predating young waterfowl. They are often fed by the public; many birds display white wing feathers, showing that the poor-quality food they are given hinders the production of sufficient melanin. In some of the Royal Parks numbers are controlled; for example, more than 200 were culled in St James's Park in 2003–04.

Roosts

There has been a roost at Kenwood, on Hampstead Heath, since at least 1926. Another old roost was known at Walthamstow Reservoirs in 1928 when 76 were counted; by 1934 this had increased to 100. The Kenwood roost had increased to about 300 by 1967 and had reached about 420 by 1976. A new roost located at Bayhurst Wood in 1978 peaked at 1,000 on 12 November. The roost at Perivale Wood built to a peak of at least 750 on 4 November 1978. In 1980 the pre-roost gathering on Hampstead Heath was 600-strong and by 1988 had reached 700. In 1993 the roost at Kenwood peaked at 900 in winter; many of the birds flew in from central London. A second roost was found in the Lea Valley in 1995, when 1,000 roosted at Fishers Green on 13 January.

The largest roosts since the start of the 21st century were 1,567 in Perivale Wood on 28 January 2012 and 1,000 on Wimbledon Common in 2001.

Facts and Figures

Although Carrion Crows typically build nests in trees, they have been known to use artificial sites; they often nested on the gasholders at Bromley-by-Bow and a pair nested on the Houses of Parliament in 1934.

Hooded Crow *Corvus cornix*

A scarce winter visitor

Range: Europe, Asia and North Africa; breeds Scotland.

Historical

In the 18th century this was a fairly common bird in winter. Although it was known to breed on the Essex coast, Glegg could not find any evidence that it had ever bred in the London Area. The earliest reference is from 1734 when Albin wrote that he had seen many of them around Hackney. Graves, writing in 1811–21, noted that they were formerly very common in the London Area, particularly in the Hoxton and Hackney districts but were now rare. In the severe winter of 1813/14, when the Thames froze, they took to feeding in gardens.

Harting described them as 'an occasional winter visitant from the north, appearing about the end of October, and leaving in April' and that 'on their arrival in this country, frequent marshes near the sea, and the banks and shores of tidal rivers, inhabiting both sides of the Thames as high up as within a few miles of London'. They were also seen in Regent's Park, robbing the ducks of their food. Between 1850 and 1866, Harting was only aware of a few individuals that had been killed away from the Thames; these were at Bushey, Cricklewood, Hendon and Kingsbury, and he also saw one flying over Regent's Park one April. Some years earlier a pair was seen in several successive springs between Kilburn and Kensal Green before the area was developed. Most records away from the Thames in north and west London were of single birds, but at Ruislip there were seven on 30 October 1887 and 13 on 5 November 1893. They were occasionally seen at Epping Forest and there were 50–60 in a ploughed field near Ilford on 19 November 1889.

Between 1874 and 1909 Power recorded the late autumn passage of birds which flew over his home in Brixton and noted four Hooded Crows passing over. One was seen feeding on the lawn of the gardens at Temple in November 1874 despite the adjacent Embankment heaving with crowds of people waiting to see the Lord Mayor's Show.

20th century

By 1900 the numbers visiting the London Area started to reduce. In the first decade of the century they were still being seen regularly in the Lea Valley around Haileybury and Broxbourne, and also south of the Thames at Blackheath, Dartford and Greenwich. On the North Downs in the Limpsfield, Titsey and Woldingham districts flocks of up to ten birds occurred regularly. They continued to be seen in the North Downs into the 1920s, sometimes in large numbers; for example, there was a group of 60 at Beddlestead Farm on 22 February 1920, and in the following year there were 20 at Warlingham on 20 February.

Until the late 1920s up to three were often seen feeding at low tide along the Thames at Chelsea Reach. Two on the Thames between Chiswick and Barnes from 3 March to 14 April 1929 were described by Glegg as being unusually long stayers. Hooded Crow remained an occasional winter visitor to the London Area until the 1930s when it became a lot rarer. The only regular sites were at Beddington, where up to three were seen in most winters until 1935/36, and at the Lea Valley reservoirs where single birds were seen in six consecutive winters between 1933/34 and 1938/39. No more were seen until December 1945 when there was one at Beddington; another overwintered there in 1947/48.

During the 1950s only occasional birds were seen apart from in 1952/53 when there was an influx and four wintered at Horsenden Hill, along with three at Ruislip and up to six at Aveley. At the start of the 1960s a few were still overwintering and there were also a few other records between November and April; most were of single birds but in early 1969 four were present at Rainham Marshes.

By the start of the 1970s they had virtually ceased to overwinter and had become scarce passage migrants. In 1975 they were more numerous than they had been for many years with 13 seen at 10 sites, including up to four at Fairlop Waters between 23 February and 9 March. Four overwintered at Sevenoaks in 1976/77.

Overwintering birds were last seen during the 1980s when one remained at Barnet between January and 7 April 1984. Over the next couple of winters several were present for short periods and by 1989 the only record was at Blackheath on 27 April. By the start of the 1990s they had become scarce winter visitors and the only records during the decade were at Girling Reservoir on 4 April 1990; at Rainham Marshes from 18 February to 9 March 1991; and over Dagenham Chase on 29 October 1997.

21st century

The UK population is largely sedentary and restricted to western Scotland where they replace the Carrion Crow. The Scandinavian birds that used to winter in England now rarely venture south due to climate change resulting in very few Hooded Crows now being recorded in southern England. Since 2000 the only record in the London Area was at Leyton on 8 April 2010.

Raven *Corvus corax*

A scarce visitor; historic breeder

Range: northern hemisphere.

Historical

Remains dating back to Roman times have been found at Southwark. They were formerly a widespread breeding bird in the London Area; the earliest reference was in 1317 when Stow referred to a cattle plague in London in which some died after feeding on the poisoned carcases. *A Relation of the Island of England*, written in about 1500 by the Camden Society, stated that there was a penalty for destroying Ravens, which helped to keep the streets clean.

However, by 1768, when Robert Smith published his book *The Universal Directory for Destroying Rats and other Kinds of Four-footed and Winged Vermin*, he wrote that he was allowed as much per head for killing Ravens as for kites and hawks. His remarks indicate that the Raven was still a common bird in the London Area as he relates how as soon as he had trapped one 'great numbers will keep around him', adding that he had caught large numbers in a single day. He referred to the 'London Ravens', which frequented the outer districts of the city and whose plumage was stained a dusky brown colour because they raided warrens for young rabbits, whereas 'country Ravens' were as black as jet.

Ravens still bred in the London Area in the early 19th century. A pair bred regularly in Hyde Park until about 1826 when one of the park-keepers pulled down the nest with young still in it and the adults left the park. One of the young was hand reared and was released back into the park. Its tameness made it well known to the park visitors and its favourite trick was to rap people's ankles with its bill; after one of these attacks a startled lady dropped a valuable gold bracelet and the Raven flew off with it to Kensington Gardens where it was believed that the bird hid the jewellery in a hollow tree.

They continued to decline as a result of persecution. Among the last breeding records were a pair that nested in Harrow in 1828 or 1829 and another in Bush Hill Park in the 1830s. In 1833–34 they bred in the heronry at Wanstead, and in 1846 a pair bred at Copt Hall, Epping. In 1847 Ruegg stated that the Woolwich area was 'seldom without a pair'. The last breeding pair was at Enfield where a pair bred regularly in a tree known locally as 'The Raven's Tree' until it was cut down; they then bred in a stand of seven elm trees known as the 'Seven Sisters' until about 1850.

The number of sightings declined dramatically in the second half of the 19th century, with only a few records documented by Harting and Glegg. In Regent's Park two were seen fighting in May 1850, ending with one of them killing the other; and another was taken there prior to 1866 and was in Bond's collection. A pair was seen over Hampstead in the spring of 1859 and 1860. Away from north-west London, a pair was seen at Mimms Wood on 25 February 1880; one flew over Balham on 9 January 1894; a pair was seen daily during the winter of 1895/96 at Penge; and there was one at Stoke D'Abernon on 23 May 1896.

One was seen over Kensington Palace in October 1889 and another was seen nearby in Kensington Gardens for some weeks in March and April 1890 before being captured and taken away. It seems likely that the last two records relate to one of the Ravens that had been introduced to the Tower of London and was capable of flight; it was often seen soaring over the Tower and regularly perched on St Paul's Cathedral before it eventually disappeared.

20th century

In the first decade of the 20th century there were three records, all of which were of two birds together that were presumably pairs. After this succession of records, no more were seen in the London Area for 50 years and they were generally absent from south-east England.

Ravens remained rare throughout the second half of the 20th century and just 13 were seen. There were three records in the autumn of 1973, although conceivably they may have related to one wandering bird as they were all within a 40-kilometre stretch along the North Downs. The two records in 1990, both fly-overs, could relate to the same individual as the locations are quite close, although they were a month apart.

All 20th-century records are listed below:

1902: Eltham, two on an unspecified date in the autumn
1905: Aldenham Reservoir, two on 27 July
1907: Bromley, two on 19 September
1957: Orpington, one 6 and 19 January
1957: Orpington, two on 22 November, with one on the 27th
1960: Sevenoaks, two on 2 July
1962: Brent Reservoir, on 19 August
1975: Sevenoaks, on 12 April
1978: South Harlow, two on 28 October
1983: Wraysbury GP, on 13 March
1984: Bermondsey, on 28 October
1990: Wallington, flew south on 20 July
1990: Beddington/Mitcham Common, flew north on 20 August

21st century

Since the beginning of the 21st century the number of sightings has significantly increased, coinciding with more sightings in the Home Counties, especially in Hertfordshire where they are now breeding. They are currently making a comeback in the London Area and it may not be too long before they are breeding again. A clear indication of their increased frequency came in 2008 when about 22 were seen, including several records of two to three birds. Although most birds have only been seen flying over, a pair was present for several weeks in the Copped Hall area during February and March 2011.

Escapes

There were several records during the late 20th century which were believed to be of escapes as wild birds were rare then. These include a tame bird present at Headley Heath from mid-August to 18 September 1973 that may also have been responsible for sightings at Sevenoaks on 25 September and at Westerham on 30 October; one at Orpington from 17 to 21 March 1984 that had a well-worn metal ring on its left leg and was barely able to feed itself; a tame bird present at Sutton on 6–7 July 1995 and presumably the same bird later seen at Wallington; one was present on Wanstead Flats for most of 1996; and singles over Ashtead Common on 29 October 1997 and Leatherhead station on 20 December 1997.

Goldcrest *Regulus regulus*

A common partial migrant and widespread breeder

Range: Europe and Asia.

Historical

The earliest reference is prior to 1738 where it was known from Upminster. Harting described the Goldcrest as appearing to be unable to bear much cold and that it 'moves southwards at the approach of winter, and is more numerous here during that season than at any other time of year'. He added that it was exceptional if any stayed to breed although he knew of nests at Elstree, Hendon, Muswell Hill and Pinner; they also nested occasionally in Hampstead. Elsewhere in the London Area, a pair bred at Kings Langley in 1877 and they were winter visitors to Dulwich, Wandsworth and Wimbledon.

20th century

There was little change at the beginning of the 20th century; for example, in the Staines district they were found during winter and early spring but did not breed. By the 1930s it had become a scarce breeding resident, mainly restricted to pines in the rural areas of London. One obtained in the Sevenoaks district on 7 November 1937 was identified as being of the continental race *regulus*.

In the 1950s it was considered to be a localised breeding species across the London Area, most common on the North Downs and surrounding heaths but more widespread in autumn and winter. From late September through to early November there was a regular influx into the London Area, mostly from the Continent, with the majority of birds arriving in the second half of October. Although many of these just passed through, some remained throughout the winter. The numbers varied every year and sometimes exceptionally large numbers were seen.

The resident population took a severe battering during the winter of 1962/63 and Goldcrest were absent from many former breeding areas during the following summer; in all, only 10 adults at seven sites were located during the breeding season. There was also a large reduction in the number of autumn migrants in 1963 with just a handful seen at three localities. The breeding population was very slow to recover; the following year only one pair was recorded breeding and by 1966 there were about eight pairs. There was a very large influx in the autumn of 1965 and about 200 were present in the Hosey Common/Brasted area. In 1972 a pair bred in Holland Park, the first confirmed breeding in Inner London.

The breeding population suffered large losses in 1986 after the coldest February since 1947; for example, in Hainault Forest there were five pairs the previous year but none in 1986. The population recovered fairly quickly and within two years there were four pairs in Hainault. By the start of the 1990s the breeding population had increased after a run of mild winters to about 80 pairs across the London Area. However, very cold weather in February 1991 was probably responsible for another decrease in the breeding population of about a quarter. The second breeding Atlas in 1988–94 found an increase in the breeding distribution of one-third since the first survey. They were found breeding in just under half of all tetrads in the London Area, being absent from much of the built-up areas of London as well as the Thames corridor. By the late 1990s the breeding population had increased to at least 160 pairs, including about 30 pairs in Broxbourne Woods.

21st century

At the beginning of the 21st century the breeding population was relatively high at around 150–200 pairs located, including four in Inner London. Mild winters helped the population to increase to more than 200 pairs by 2003. The largest overwintering population was of 50 birds in Hainault Forest in 2000/01. Although the breeding population has increased, the species has become scarcer on passage, particularly in autumn.

Firecrest *Regulus ignicapilla*

An uncommon winter visitor and rare breeder

Range: Europe and North Africa.

Historical

There were only two 19th-century records, in a garden at Blackheath prior to 1859 and taken by a bird-catcher at Epping on 26 November 1878.

20th century

Between 1900 and the mid-1930s there was a gradual increase in the number of sightings and six birds were seen, all in south London between October and March. In 1938/39 a pair wintered on Ruislip Common and in subsequent winters they were regularly seen there up to 1944/45. Most sightings were of one or two birds but on 23 January 1944 at least six were present. Winter records also started to increase elsewhere and up to 1954 there were about 28 records, all of which were between September and April, with the majority being in December and January.

By the 1960s they had started to become regular in autumn and were also being recorded during the spring migration. In 1961 the first one was seen in Inner London, at Regent's Park on 1 September. After the severe winter of 1962/63 none were seen in the following two years and they remained rare until the spring of 1967 when five were seen. At the start of the 1970s they were continuing to increase and were mainly seen on passage with the occasional winter record. There was a good spring passage in 1975 with 13 seen, including three or four in Regent's Park on 21 April; this corresponded with the first records during the breeding season.

An analysis of all records between 1950 and 1987 showed that the main passage was in spring with a peak during the first week of April. During the early 1990s there were far fewer seen on passage and they had become scarcer in winter, but towards the end of the decade small flocks had started overwintering, including six at Beaulieu Heights Wood and up to five at Maryon Wilson Park.

21st century

Since the beginning of the 21st century there have been around 10–15 wintering with a similar number also seen on passage. In 2004 around 70 were seen during the year, including up to four at Dagenham Chase and Warley Place Nature Reserve.

Breeding

The first signs of potential breeding occurred in 1975 when singing males held territory at two sites in east London and one summered at Harrow School; additionally, there were birds in the breeding season at three other sites. The following year two males were again singing in east London and a female with a brood patch was trapped and ringed at Ruislip on 5 June. The first confirmed breeding took place in 1980 at Hainault Forest. They continued to be seen during the summer in subsequent years but the next confirmed breeding was not until 1987 when a pair bred in Hainault Forest.

Breeding was fairly sporadic in the 1990s; a pair bred on Hampstead Heath in 1992 and three pairs bred in 1999. In 2000 the breeding population was seven pairs, at least two of which raised young. The population has fluctuated since then from none to a potential nine pairs.

Facts and Figures

The largest concentration was of six birds at Beaulieu Heights Wood in November 1996.

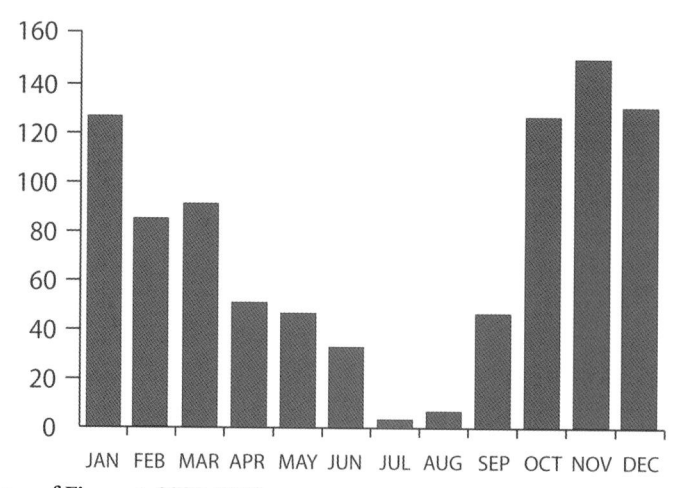

Seasonal occurrence of Firecrest: 2000–2009.

Penduline Tit *Remiz pendulinus*

A rare winter visitor (15)

Range: Europe and Asia; rare visitor to UK, sometimes overwinters.

The first record was at Ockendon GP in 1982; this was just the seventh British record at the time and the first one in June. There were no further sightings until the late 1990s when they were seen in two consecutive years at Brent Reservoir; a juvenile was seen on 16–17 November and again on the 22nd. Almost exactly a year later an adult and two juveniles were present for two hours.

No more were seen until 2004, since when they have been recorded annually on the Lower Thames Marshes, usually at Rainham. This is the only place in the country where they overwinter on a regular basis. In 2005 there was one in October, then no further sightings until December when three arrived on the 18th. The following day six were found, including three males, and at least five of these overwintered. It is virtually impossible to tell how many birds have been at Rainham as the number of returning birds varies each winter, but as none were seen in 2007/08 subsequent birds are treated as new ones.

1982: Ockendon GP, on an unspecified date in June
1996: Brent Reservoir, on 16–22 November
1997: Brent Reservoir, three on 2 November
2004: Rainham Marshes, three adults on 29 December; one on 9 October 2005, then up to six on 18 December 2005–30 March 2006; two on 22 December 2006–26 March 2007
2007: Swanscombe Marshes, on 29–30 January
2008: Rainham Marshes, two on 6 December 2008–29 March 2009; one on 30 January–14 March 2010; two on 1–23 October 2010

Blue Tit *Cyanistes caeruleus*

A very common resident and widespread breeder

Range: Europe and Middle East.

Historical

Like many very common species, the Blue Tit is sparsely written about in historical publications, Harting referring to it only as 'resident throughout the year, and in its habits one of the most interesting and amusing birds to observe'.

Among the nest-sites recorded for this species are lamp-posts, a stone vase, a letterbox and an old Kingfisher's nest-hole in a disused gravel pit. Harting noted that they were nesting in lamp-posts in Hampstead in 1889; within 20 years this had become a regular habit in the district.

20th century

In the first half of the century there was no change to the status of Blue Tit in the London Area; it was common and widespread throughout the London Area, even in the more heavily built-up areas where it was able to take advantage of a wider range of habitats than Great Tit. In 1909 Dixon stated that it was probably breeding all the way 'from St James's Park outwards to the limits of our radius' but he described it as being 'so grimy in exploring the soot-covered trees that one is often puzzled … to recognise it'.

Glegg confirmed that in the 1930s it was still a very common resident and 'freely frequents the gardens and other open spaces of the suburbs, and is abundant in the rural parts of the county', adding that it was a breeding resident in central London and most numerous there in autumn.

In winter Blue Tits could often be found in weed-grown bombed sites after the Second World War. The habit of feeding from milk bottles was known from many areas of London in the 20th century, even from central London. One was even watched trying to remove a cardboard cap from a bottle on a window sill in Lambeth.

During the autumn of 1949 there was an exceptionally large influx of Blue Tits into central London and they were occasionally seen on tall buildings and scaffolding where there was unlikely to be any food. The peak of this influx was during the first half of October but the source was never traced.

In 1957 there was an even larger influx of tits during autumn. It was first noted in north-west Europe and an increase in the numbers in the London Area was noted before coastal movements were recorded. The first increase in Blue Tit numbers in the London Area was in August but larger numbers were reported from the middle of September onwards, particularly in central London where the influx was more obvious. Numbers continued to increase throughout October and into November. At Dulwich Woods flocks of up to 50 were seen flying over, mainly in a north-easterly direction, between 23 September and 3 November; at Romford SF the main influx took place in October when up to 60 were seen. By December the numbers had dropped significantly in most areas of central London but higher numbers were seen again at the start of 1958, possibly as a result of birds beginning to return. As a result of this influx there was a corresponding increase in reports of milk bottles being attacked, especially in areas where this habit had not been previously reported. In one particular incident 57 bottles out of 300 were attacked at a school in south London. In the vast majority of cases Blue Tits were the culprits.

In 1970 there were 58 breeding territories in Hyde Park and Kensington Gardens but by 1972 this had increased to 70 territories, probably the highest ever. The exceptionally hot and dry summer of 1976 caused a drastic drop in the number of juvenile Blue Tits in south-east London due to the drought conditions.

During the second breeding Atlas in 1988–94 Blue Tits were seen in every tetrad in the London Area with breeding recorded in more than 98 per cent of all tetrads. A 13 per cent increase in distribution since the 1968–72 Atlas was recorded, although doubtless some of the gaps in the first atlas were due to under-recording. On Hampstead Heath there were estimated to be some 150 territories in 1994, while in Hainault Forest the number of pairs increased from 80 to 110 between 1993 and 1995.

21st century

Blue Tit remains a very common and widespread breeding bird throughout much of the London Area with the Breeding Bird Survey recording a 40 per cent increase from 1995–2009. They were the third commonest bird in RSPB's Big Garden BirdWatch in London in 2013. In central London Blue Tits are common in the smaller parks and garden squares as well as the Royal Parks. They were recorded in 55 per cent of 293 small parks and squares in central London visited once only in a survey in 2004; they were more common in those with plenty of trees and shrubbery, and where nest boxes and bird feeders were provided. They can also be found foraging on street trees where they are able to feed on the hanging fruits of the ubiquitous London Plane and will take food from an outstretched hand in Kensington Gardens.

Facts and Figures

1,011 were trapped and ringed in a garden in Potters Bar during 1977.

Great Tit *Parus major*

A very common resident and widespread breeder

Range: Europe, Asia and North Africa.

Historical

The status of Great Tit appears to have changed very little since the mid-19th century when Harting described it as a common resident, the only exception being in the central parks where Hudson stated in 1898 that they were no longer to be found.

20th century

Throughout the 20th century Great Tits were common across the London Area, nesting in tree holes, rabbit burrows, nest-boxes and a variety of non-natural holes such as in drainpipes, vents, hollow iron railings, letterboxes and stone vases. In 1932 one was watched flying through an open window of the Natural History Museum, then taking fur from a mammal skin that was lying on a table in a storeroom and flying off with it to line its nest in the museum gardens.

Great Tit is a highly adaptable species and it was first noted taking food from the hand in Kensington Gardens in 1927. This behaviour was later seen in Kew Gardens. In several parks Great Tits have also been watched foraging for scraps in litter baskets. They were first recorded taking the cream from milk bottles at Dartford in 1924, two years after the behaviour was first observed in the London Area, at Croydon (by an unspecified tit). In 1925 the behaviour had spread to Chalfont St Giles and in 1929 it was recorded in Richmond. By 1946 milk bottles were being raided across a wide area of London, from Harold Wood in the east to Sipson in the west, and from Aldenham in the north to Oxshott in the south.

In 1935 Glegg recorded that it was still a fairly common breeding resident in central London, with influxes in autumn. In autumn 1959 there was a notable influx of several species including Great Tits; an adult ringed at Esher in October was recovered in Germany in April 1960, an indication of the distance some of these birds had travelled.

Melanistic birds with completely black heads have been seen since 1945 when one was first recorded at Oxshott on 24 December. Since then, many others have been seen in this area, a genetic anomaly appearing to be present in the population.

In 1966 there were about a dozen pairs breeding in both Regent's Park and Hyde Park/Kensington Gardens. By 1972 the breeding population at the latter site had increased to 29 territories, the highest since the Common Bird Census started there eight years earlier. Breeding numbers had fallen in the central London Royal Parks by the early 1990s due to blocking up nest holes in old lamp-posts but after nest boxes were erected the number of territories increased.

The second breeding Atlas in 1988–94 showed an increase in the breeding distribution of 16 per cent since the first survey in 1968–72. The species was found to be breeding in 96 per cent of the London Area's tetrads, marginally fewer than Blue Tit. The population increased during the 1990s; at Hainault Forest about 80 pairs bred in 1995, an increase of one-third over the previous year. The Breeding Bird Survey showed a massive rise in the London population of 96 per cent between 1994 and 2000.

21st century

Since the start of the 21st century Great Tit has remained a very common breeding bird, resident throughout the London Area with the Breeding Bird Survey recording an increase of a third from 1995-2010. However it is less widespread in central London than Blue Tit – it was found in just 16 per cent of 293 small parks and squares compared with 56 per cent for Blue Tit in a 2004 survey (Sibley et al); it was commoner in sites with plenty of trees and tall shrubbery and seemed to have more need for under-storey trees and shrubs than Blue Tit.

Facts and Figures

A bird of the continental race *major* was ringed at Esher on 22 October 1959.

Crested Tit *Lophophanes cristatus*

A rare vagrant (4)

Range: Europe and western Asia; breeds Scotland, vagrant to England.

There are four acceptable records, at least two of which occurred in spring. They are assumed to be of wanderers from the Continent as the Scottish population is not prone to wander far; indeed, there have been no proven records of this race (*scoticus*) in southern Scotland, let alone in England.

The first was seen in the autumn of 1839 by Meyer in a fir wood near Claremont House during a rough north-westerly gale. Harting found one in the collection of Mr Warner, a Kingsbury resident who assured him he had shot it in a small spinney in Cool Oak Lane in Kingsbury (this road separates the two areas of Brent Reservoir). Harting also knew of one shot 'a few years ago' in Blackheath by Mr Engleheart in his own garden.

There were about 30 records of Crested Tit in England in the 19th century, but maybe fewer than 10 of these could now be considered acceptable, along with a similar number from the 20th century. Although it can be difficult to review very old records since many are poorly described or the shot specimens have no documented provenance, some are clearly mistaken identifications or outright fraud. It is interesting that several of the older records come from the East Coast where continental birds would be most likely to be found. However, despite the spread of Crested Tits along the French coast in the 1980s, very few have since been claimed in England; the only acceptable recent record is of one in Northumberland, which was present for five weeks in the autumn of 1984.

The 1945 record in Godstone was described by the observer thus 'The colouring was rather different from the rest of the tit family; the underparts were yellowish buff and the back was darkish brown. The crest was very prominent, as was the little black collar. All this was observed within 12 yards of the actual bird'. The record was later accepted by the LNHS. In the same year there were two records in Devon. Just two years later in 1947 two more were reported in England – on Scilly and near Dawlish, Devon – so there is some evidence of immigration into the UK in the 1940s coincident with a population expansion on the near Continent at the time.

1839: Claremont, in autumn
1860: Brent Reservoir, shot in the spring
Pre-1866: Blackheath, shot in a garden
1945: Godstone, on 10 April

Unacceptable Records

On 24 April 1904 a tit with a marked crest was seen flying about the tops of some tall trees in Croydon. The observer heard a note which he described as *tur, tee-er, teere* and considered to be that of Crested Tit but was unable to get a closer view of the bird. He reported this sighting to *British Birds* in 1922 and it was published as a probable. In view of the brief details and length of time between sighting and submission this record has since been deemed unacceptable by a number of reviewers. A record of one in Kensington Gardens on 25 April 1968 and published in *British Birds* was not accepted by the LNHS.

Coal Tit *Periparus ater*

A very common resident and widespread breeder

Range: Europe, Asia and North Africa.

Historical

In the 19th century Harting noted that the 'Cole Tit', as it was known at the time, 'appears most numerous in autumn and winter, but resides with us all the year round'. Hibberd stated that it was the most common tit in the suburbs, but in central London Power considered the Blue Tit the most common in Bloomsbury and seldom saw Coal Tit there.

20th century

Although Coal Tits almost exclusively breed in conifers, a pair once bred in a hole on one of the platforms of Mill Hill station in 1924. The nest was within half a metre of a busy track and the birds had to wait for trains to depart before they could enter the nest. In October 1930 unusually high numbers were seen passing through central London.

Glegg referred to it as 'a thinly but well-distributed resident', adding that it was occasionally seen in central London and was not seen as frequently in the suburbs as Blue and Great Tits. Coal Tits were sometimes seen in flocks of mixed tits in winter and there are also two December records of flocks of about 30 birds, all consisting of Coal Tits.

By the 1940s there had been a decrease across much of the London Area but they had begun breeding in Kensington Gardens. In the early 1950s a pair bred successfully in Regent's Park and they probably bred in St

John's Wood but there were only three other breeding sites within an 8-kilometre radius of St Paul's: at Dulwich, Greenwich Park and Hampstead.

In 1954 a flock of 50–60 was seen in Battersea Park on 30 October. In the autumn of 1957 there was a considerable influx of Coal, Blue and Great Tits into north-west Europe, and there was an increase in numbers in the London Area, which was particularly evident in central London. On 20 October there was a party of 20 Coal Tits in Battersea Park. In 1959 there was a similar influx in autumn noted in central London and some of these were believed to be of the continental race although this was never confirmed.

From the late 1950s they began to increase in central London reaching a maximum of 22 territories reported in 1976, after which fewer were recorded. Elsewhere, the largest breeding concentration was 36 pairs on Wimbledon Common in 1983. The second breeding Atlas in 1988–94 showed a relatively stable population since the first survey with an increase in the distribution of 5 per cent. They were found breeding in just over half of the tetrads in the London Area, with the highest concentration located in the outer suburbs and rural fringes, and substantial gaps in the inner eastern suburbs.

21st century

Since the beginning of the 21st century this has remained a relatively common breeding bird in most parts of the London Area, however it is fairly scarce in built-up areas due to a lack of its preferred habitat of coniferous trees. There is a small population in the parks of central London and a few have been recorded in some of the better wooded garden squares.

Facts and Figures

At least one bird of the nominate continental race *ater* was at Tower Hamlets Cemetery on 22–23 October 2008.

Willow Tit *Poecile montana*

A scarce visitor; former breeder

Range: Europe and Asia.

Historical

For most of the 19th century Willow Tit was not known in the UK as it had not been separated from the similar Marsh Tit. It was first recognised as a British species in 1897 when two specimens from Hampstead were discovered in the British Museum; the same year the museum received a further two Willow Tit specimens from Coldfall Wood in Finchley. One of the latter became the type specimen for the British race *kleinschmidti*.

20th century

As news of this 'new' species began to circulate, more were found in the London Area. In 1905 it was realised that some of the 'Marsh Tits' breeding in the Bromley district were in fact Willow Tits. In 1919 the first nest in the London Area was found, at Godstone.

By the end of the 1930s Willow Tits were being seen at many more locations all around the London Area, including the lower Colne Valley, Epping Forest and other nearby woods, Ruislip Lido, Stanmore Common, and the heaths and commons south of the Thames. Willow Tits were well established in many of these areas and it is likely that this was a result of observers being aware of this species rather than any apparent increase.

In the 1940s and 50s the largest population was found on damp wooded commons in the far south of the London Area, and there were several records from Wimbledon Common although they were not thought to breed there. In the south-east of London they were also present in many areas. Their favoured areas north of the Thames were in the Colne Valley between Uxbridge and Rickmansworth; Ruislip Common; Stanmore; Mimmshall Wood; Cuffley; and Epping Forest. Although they were not usually recorded from the inner suburbs, one was heard in Parliament Hill Fields in January 1944.

By the start of the 1960s the population was considered to be stable with around 21 known breeding pairs at 12 locations. The population began to expand and in 1965 a pair bred at Brent Reservoir, within 13 kilometres of the centre of London. During the first breeding Atlas in 1968–72 they were found breeding in 190 tetrads, 22 per cent of the London Area, considerably more widespread than had previously been thought, and slightly more common than Marsh Tit. They began to decline towards the end of the 1980s and only three were seen south of the river in 1989; they had also begun to disappear from locations in the eastern part of the London Area.

At the start of the 1990s only 24 pairs were reported breeding; the largest breeding concentration was at Cobbins Brook where there were seven pairs. By 1992 the population was in free-fall in the London Area and they had slumped with only 11 pairs reported; the decrease was particularly evident south of the Thames with only one breeding pair located, at High Elms. By the time of the second breeding Atlas in 1988–94 the breeding distribution had decreased by almost half since the first survey, in contrast with a national decrease of 10 per cent. In 1999 only three pairs were recorded during the breeding season, including a pair that bred at Fir and Ponds Woods near Potters Bar.

21st century

Willow Tits require damp woodland with damp timber to nest in so climatic factors such as dry summers may be involved in their virtual extinction from southern England. Predation by Great Spotted Woodpeckers has also been suggested as a contributing factor. Nationally, the BBS shows a relentless decline since 1994 with an 85 per cent decrease recorded from 1995–2011 and it is now 'red listed' as a species of conservation concern.

By the beginning of the 21st century the population in the London Area had decreased so much that extinction was inevitable and the only probable breeding pair was at Fir and Pond Woods in June 2000. On Bookham Common two were present in April 2001 and a male was on territory in 2003, but there were no other signs of breeding. Also in 2003 there was one at Moorhouse on 16 September. In 2004 none were seen for the first time since they were discovered to be widespread in the London Area just 70 years earlier. There have been no further records.

Marsh Tit *Poecile palustris*

An uncommon resident and localised breeder

Range: Europe.

Historical

In the 19th century this was described by Harting as the scarcest of the resident species of tit, adding that 'it can hardly be said to be rare'. It bred as close to central London as Highgate and was not uncommon near Dulwich; one was seen in Kensington Gardens in November 1898. However, it must be borne in mind that these comments could refer to either Marsh or Willow Tit, as the two species had not been separated at the time.

20th century

Not all documented records at the beginning of the century can be relied upon to have eliminated Willow Tit, so the first definitive status was not written until the 1930s when Glegg considered that Marsh Tit was scarce in the suburbs though one or two pairs could be found in most woods in the rural areas. In 1937 an adult was seen feeding four young in Holland Park and presumably bred there or nearby, the only breeding record in Inner London. It bred irregularly on Wimbledon Common during the 1940s and in Greenwich Park until 1945; in Dulwich it was resident up to at least 1946.

In the mid-1950s Marsh Tits were widespread throughout the rural areas and in wooded suburbs. They had ceased breeding in the inner suburbs of north-west London although they were still resident at Kenwood. Outside the breeding season they could be seen almost anywhere and would often join mixed flocks feeding on bird tables; at Kenwood and Whitewebbs Park they were fed by the public and they were still seen in most years in central London. In 1976, 30 pairs were located in Epping Forest, the highest number ever there.

At the beginning of the 1990s the largest breeding concentrations were 15 pairs in Wintry Wood and 14 at Hainault Forest. During the second breeding Atlas in 1988–94 the species was found breeding in 126 tetrads, a decrease of 29 per cent, well above the national decrease of 17 per cent. Almost all of the population had retreated beyond the suburbs and the biggest losses were suffered south of the Thames. A survey in south-east London in 1993 located about 20 pairs. The population continued to decrease and by the end of the 1990s the population in Hainault Forest had crashed to just three pairs.

21st century

At the start of the 21st century it appeared that Marsh Tits were heading for extinction in the London Area since the breeding population had severely contracted. This decrease mirrored the national decline which showed a drop of 24 per cent from 1995–2010 and has led to it being red listed as a species of conservation concern.

In 2003 there were at least 25 territories but 16 of these were concentrated on Bookham Common. The situated had deteriorated further by 2009 when there was a maximum of only five breeding pairs reported, located at the northern and southern extremities of the area. Outside the breeding season they are seen a little more widely, although almost always beyond the suburbs.

Facts and Figures

The largest count was 30 birds at Walton Heath on 11 October 1983 and on Headley Heath on 2 February 1985.

Long-tailed Tit *Aegithalos caudatus*

A very common resident and widespread breeder

Range: Europe and Asia.

Historical

This was a common resident species in the 19th century; the earliest reference was in 1832 when Jesse described a nest in Bushy Park. Harting noted that 'in the autumn and winter it may be observed in small parties, varying from five or six to ten or more'. It would seem that mixed tit flocks were unknown as Harting described 'an unusual sight, namely, four species of *Parus* within a few yards of each other. A little family of Long-tailed Tits, and a pair of Blue Tits, were climbing about the low branches of an oak, while a Cole Tit and a pair of Great Tits were chattering and hunting in the hedge immediately below them.'

20th century

In 1935 Glegg stated that it was 'a fairly common resident in the rural areas' and was more widespread in winter. He added that it was 'the least suburban of the breeding titmice' and was a scarce visitor to central London. In 1938 there was an influx into the built-up areas of London, noticeably in central London where a flock was seen for two months in October.

There was evidence of a decline as they had ceased breeding at Bostall Woods, Chingford, Gilwell Park, Perivale, Shooter's Hill and Wimbledon Common by the 1940s. Numbers were also greatly reduced after the severe winter in 1947. Another influx in central London occurred during the autumn of 1949, and a flock of 14 was tracked from near St Paul's Cathedral to Bloomsbury on 3 October. During the 1950s they were reasonably widespread in scrubby areas and commons in the rural areas, and in some of the larger open spaces in the suburbs, but they were still declining in some areas and on Hampstead Heath, where they had previously been common, they last bred in 1951.

In the autumn of 1960 exceptional numbers were seen throughout the London Area; the movement was first noted in suburban gardens on 9 September and by 2 October they had started to appear in central London. Throughout October small flocks were seen passing through the London Area and a flock of about 20 was present in Holland Park on 21 October. A fresh wave of arrivals was recorded on 5 November and more continued to move through during the month, the highest count being a flock of 50 flying north-west near Walton on 26 November. By December most had left central London and the only high count elsewhere was about 30 at Walthamstow.

The breeding population was decimated during the severe winter of 1962/63 and the species was absent from many sites where it had bred the previous year; for example, none were seen on Bookham Common compared to up to 10 pairs the previous year. They started to recover slowly in 1964 and two or three pairs were present on Bookham Common. By 1965 there was a widespread recovery and they were present during the breeding season at more than 30 locations with at least a dozen pairs successfully breeding.

At the start of the 1970s they were still scarce in central London but in 1972 three pairs bred in the parks; within two years the population had increased to eight pairs. There were three counts of more than 100 in the mid-1970s: at Walton Oaks, a roost on the banks of Queen Mary Reservoir and at Headley Heath. Numbers reported in the LBR fluctuated during the 1980s, with no records from central London in some years.

During the early 1990s the population at Hainault Forest doubled to 30 pairs within three years. Further evidence of the population increase came during the second breeding Atlas in 1988–94. There was an increase of more than 100 per cent in the breeding distribution in the London Area since the first survey in 1968–72. They were still largely absent in the most heavily built-up areas on the eastern side of London, although almost every tetrad in the western half of Inner London was occupied.

21st century

Since the start of the 21st century they have become a widespread and common breeding bird throughout the London Area with a minimum of 11 breeding pairs in central London. Nationally the BBS recorded an increase of almost a third from 1995–2011; a GLA commissioned study showed a 23 per cent increase in London from 1994–2011. Milder winters are thought to be a factor, and the national graph shows sharp falls following more severe winters. A roost of 100 was found in Welwyn Garden City in 2007.

Facts and Figures

The largest flock was of about 200 birds on Alderstead Common on 8 February 1939.
Although considered to be fairly sedentary there is some evidence of long-distance movement; one ringed at Walberswick, Suffolk, in October 1978 was trapped at Wanstead in the following February.

'Northern Long-tailed Tit' *A. c. caudatus*

The only record of this white-headed race was at Hampden Wood, Warley, on 23–29 February 2004; this coincided with a small arrival into southern England. A bird resembling this subspecies at Waltham Abbey Arboretum on 17–23 April 2004 was paired with a UK bird and may have been an intergrade. One at Chelsham in April 1954 was thought to probably be a pale-headed bird of the British race.

Bearded Tit *Panurus biarmicus*

An uncommon winter visitor and rare breeder

Range: Europe and Asia.

Historical

In 1743 the Countess of Albemarle brought a large cage full of Bearded Tits with her from Denmark when she returned to Britain. It was alleged that some escaped and 'a colony was formed in England', but they were already known in the country as Edwards stated that he saw the birds brought over by the Countess and wrote that 'I have seen some others of the same kind... shot among the reeds in marshes near London'.

Another early record came from Coombe Wood near Edgware, where a male was shot on 15 December 1794, in the company of a second male. Meyer stated in 1800 that they bred all along the Thames between Oxford and the Essex marshes; while on the south side of the river Montagu stated in 1802 that they could be found between

Erith and London. They were also known from Sydenham Common and in the area between Bermondsey and Deptford in the 1810s. Later in the 19th century there were also records on the River Brent near Stonebridge Park and on Stratford Marshes.

20th century

By 1900 they had disappeared from the Thames-side marshes and the only record during the first half of the century was of a pair at Enfield Chase on 14 May 1913. No more were recorded until the winter of 1959–60 when there was a large irruption from the breeding population in East Anglia and a single male was seen at Northfleet on 13 December. The following year there was another irruption and about 30 reached the London Area, including 15 at Walthamstow Reservoirs on 16 October, six at Stanborough and five at Swanscombe; the flocks at Swanscombe and Walthamstow remained for four months. Further irruptions occurred during subsequent winters with peak counts of 20 at Swanscombe and 17 at Rye Meads.

By 1970 they had become scarce annual visitors, mainly to the Lower Thames Marshes and at Rye Meads. There was a larger than usual irruption in the autumn of 1971, when there were 35 at Swanscombe, 20 at Rainham Marshes and 17 at Brent Reservoir. Over the next few winters even larger numbers were seen, with up to 50 at Rainham Marshes. From the mid-1970s through to the early 80s fewer were seen although they were slightly more widespread.

Up to 50 wintered at Rainham from 1984 to 1987 but only three were seen in 1988; smaller numbers were seen at other sites. By the start of the 1990s there had been a large reduction in wintering numbers and none were seen at the former stronghold of Rainham.

21st century

By the beginning of the 21st century they had become rare winter visitors again with typically only a few birds seen each year, mostly on the Lower Thames Marshes. On 14 October 2006 there was a bizarre record of seven flying over the shopping centre at Kingston-upon-Thames.

Breeding

In 1966 two pairs bred at Stanborough Reedmarsh, rearing at least seven young, the first breeding record for the London Area. None were reported there the following year but two pairs bred again in 1968 and 1971. Unfortunately there is no record of whether these birds were looked for in the intervening years. Two pairs started nesting in 1972 but the reed bed was destroyed by fire; the following year one pair started nesting before the reed bed was again burnt down.

There were no further breeding attempts until the 1990s when they began to breed in east London. In 1991 two pairs almost certainly bred and the following year the breeding population increased to three pairs. None were seen during the breeding season in 1993 but a pair probably bred in 1994–95. One or two pairs were present at Rainham early in the breeding season in 1998 but did not breed due to disturbance. In the first few years of the 21st century there were several more breeding attempts at Rainham Marshes, and in 2003 one or two pairs bred in the Lea Valley. However, since then there have been no further breeding records.

Facts and figures

The largest count was 50 birds at Rainham Marshes on several occasions.

Short-toed Lark *Calandrella brachydactyla*

A rare vagrant (5)

Range: breeds Europe, Asia and North Africa, winters Africa and Middle East; scarce passage migrant in UK.

The first record for the London Area was taken at Orpington in 1883; it was originally dismissed as the specimen was believed to have been imported but Ticehurst investigated the circumstances and accepted the record. The specimen was later given to the Natural History Museum. The next record, at Staines Moor in June 1960, occurred at an unusual time of year – they are most often seen in spring and autumn. The other four records were between late April and early May. Although Short-toed Lark is an annual visitor to the UK it is rare in the London Area and relatively scarce in the south-east; for example, there have only been three in Essex up to 2007. The bird at Wraysbury Reservoir showed characteristics of one of the southern reddish races, probably *brachydactyla*, which are more often seen in Britain in spring.

1883: Orpington, killed in June
1960: Staines Moor, one on 8–16 June
1966: Beddington SF, one on 24 April
1985: Wraysbury Reservoir, one on 29 April–1 May
1991: Tyttenhanger GP, one on 7–10 May
1999: Parkside Farm, Enfield, on 30 April

Crested Lark *Galerida cristata*

A very rare vagrant (2)

Range: Europe, Asia and Africa; rare vagrant to UK.

The only record for the London Area was in 1947 at the end of a long spell of very cold weather when two were seen together on the foreshore of the River Thames between Chiswick Eyot and Hammersmith Bridge. They were seen on the mud and exposed gravel at low tide for about 10 minutes at close range, the description stating 'the conspicuous crests and flesh-coloured legs were clearly seen and the outer tail feathers were not white'. The last part is a key identification feature to eliminate Skylark. This was the first inland record and only the tenth in Britain; it was also the first sighting in the 20th century.

1947: Hammersmith, two on the Thames foreshore on 8 March

Woodlark *Lullula arborea*

An uncommon passage migrant and scarce breeder

Range: Europe and North Africa.

Historical

During the 19th century Woodlarks were relatively abundant on the heaths and commons south of the Thames, but they declined so much that by the end of the century they were considered to be rare and extremely localised. They were seen on Wimbledon Common in summer and a pair nested near Kingston in 1884.

North of the river they were much scarcer. Between 1831 and 1838 they were a common winter visitor to Harrow. Later, they had become an uncommon resident in other parts of north London, for example at Barn Hill, near Wembley, and at Hampstead. However, this new status appears to have been short-lived as Harting wrote that 'bird-catchers have much contributed to exterminate the species'. The last one was taken on Hampstead Heath on 27 October 1869. In Epping Forest the population fluctuated and most disappeared after the severe winter in 1859.

20th century

At the beginning of the century a pair bred in Selsdon Park and they were occasionally seen in Richmond Park and on Wimbledon Common; in Epping Forest one was singing in 1905. The only regular breeding sites were in the Orpington, Sevenoaks and Westerham districts. A pair bred at Symondshyde Great Wood in 1915.

During the 1920s there was a sudden increase in the population; from 1923 onwards they started breeding on the North Downs and at one site eight pairs bred successfully in 1925. Breeding occurred on Wimbledon Common in 1925–26, and in Richmond Park up to three pairs bred annually from 1926 to 1935. In north-west

London they could be found at Harrow, Mill Hill and at Stanmore Common from 1924 onwards. The only localised decline was on Hampstead Heath where they had probably ceased breeding by the mid-1920s.

Towards the end of the 1930s the population suddenly declined, especially after the severe winter of 1938/39; the following summer only one was heard singing on the North Downs and they failed to breed in Richmond Park. None were seen in the breeding season during the early 1940s. In 1943 a flock of about 20 was seen at Selsdon on 16 January. They began to recover during the mid-1940s and a pair bred at Shoreham in 1944; by 1949 there were about 27 pairs across the London Area.

The population continued to increase across southern England and there were 45 pairs in the London Area in 1950. These included four to five pairs in Richmond Park, five on Wimbledon Common and one on Putney Heath, the last being less than 12 kilometres from the centre of London; there was also a thriving colony at Stanmore. Most of the population at this time was resident all year although some congregated in flocks of up to 20 to feed on stubble after the breeding season. Towards the end of the 1950s a rapid decline took place and by 1956 they had decreased to about 16 pairs; also that year a flock of eight was present during severe weather on the football pitches in Battersea Park from 19 to 27 February. Research published by Harrison in 1959 showed a correlation between climatic conditions and the rise and fall of Woodlark populations.

By 1961 there were only nine breeding pairs. At the start of the cold winter of 1962/63 a flock of 12 arrived at Brent Reservoir on 28 December. By 1964 the breeding population had retreated from the London Area and no pairs were present for the first time since 1943. A pair bred in south-east London in 1965–66 and they slowly began to recover. During the first breeding Atlas in 1968–72 they bred or probably bred at Headley Heath, Hunton Bridge, Joydens Wood, Mickleham, Northaw Great Wood, Sundridge and Walton Heath. However, after this they yet again ceased breeding. During the early 1970s a few were seen on passage and up to five wintered at Foots Cray Meadows in 1972. Breeding occurred again in 1975 at Esher but the species remained scarce after this with just a few seen in winter in Alexandra Park and a few on passage.

Woodlark became scarcer during the 1980s and by the start of the 1990s had become a rare visitor, not even recorded annually. Breeding may have occurred in 1990 as a group of five at Limpsfield Chart in May was probably a family party. Breeding was confirmed again in 1994 when one pair raised two broods in Black Park, where management work had cleared an area of woodland. The birds probably bred there in subsequent years and breeding also occurred on the North Downs. By 1999 the breeding population was about 10 pairs at seven sites.

21st century

Since the start of the 21st century there has been a small breeding population, mainly confined to a few sites on the North Downs. Elsewhere, it is a scarce passage migrant, and sightings are mostly of fly-overs in October. They are typically recorded at the best-watched sites, even in central London. A flock of up to nine was present in a field at Sidcup from 3 December 2008 to 13 January 2009.

Skylark *Alauda arvensis*

A common partial migrant and widespread breeder

Range: Europe, Asia and North Africa.

Historical

Skylark was first documented in London by Gilbert White in 1780. In Harting's time Skylarks were a 'common resident, and one of the earliest birds to commence singing in the spring'. A pair even nested on Primrose Hill despite the crowds of people that congregated there on a Sunday. Skylarks bred on Highbury Fields until they were opened to the public in 1886. As early as the 1880s they were becoming scarcer in some areas as the spread of London continued. A Skylark that summered in the vicinity of Victoria Park in 1895 was the target of many people who tried to shoot it, including 'sportsmen' from Hackney Marshes who went there every Sunday to try and bag it; it managed to evade them and remained on territory until the winter. In the really severe winter of 1890/91, several were seen daily feeding on the Thames Embankment. A few wintered in the grounds of Lambeth Palace in the comparatively mild winters of 1896/97 and 1897/98.

20th century

At the beginning of the century the closest that they bred to the centre of London was at Hackney Marshes. In the 1920s they were still resident in Clapham, Tooting, Wandsworth Common and Peckham Rye. In Regent's Park a pair appeared to be prospecting for a nest site in 1932. At Bromley-by-Bow they were considered to be common in the meadows near the gasworks in 1937. By the 1930s the species was still a very common and widespread resident, but wintering numbers had apparently increased.

During the 1950s Skylarks could still be found breeding well into the suburbs at Brent Reservoir, Hampstead Heath, Wimbledon Common and Wormwood Scrubs. In the rural districts they were widespread and relatively common.

Between 1971 and 1993 up to 10 pairs bred in the disused Surrey Docks; by 1994 they had disappeared as the redevelopment work made the area unsuitable. In 1980 the largest flocks in winter were of 250 birds at both Dartford and Rainham Marshes.

The second breeding Atlas which covered the years 1988–94 found Skylarks breeding in 539 tetrads, about 63 per cent of the London Area; there had been a decrease of 10 per cent since the first Atlas in 1968–72. A census of Rainham Marshes in 1990 found a total of 103 territories. A small population in the suburbs at Brent Reservoir had disappeared by the start of the 1990s. In 1997, 10 pairs bred on Wanstead Flats and seven or eight pairs bred in the Royal Docks. In 1998 about 20 pairs bred near Barnet in set-aside fields.

21st century

At the beginning of the 21st century there was still a reasonable population of breeding Skylarks, mostly in the rural areas apart from one to two pairs on Hounslow Heath and 16 pairs on Wanstead Flats; a pair also attempted to breed on Wormwood Scrubs in 2009. The largest population is of 80–90 pairs at Rainham Marshes. Nationally, there has been a 58 per cent decline from 1970–2010 leading to it being red listed as a species of conservation concern, but efforts are underway to encourage farmers to improve nesting opportunities in arable land.

In southwest London great efforts are being made to protect the small remaining population at Richmond Park of 13–14 pairs. Inevitably there is a tension between public recreation, dogs and ground-nesting birds; volunteers have mapped out their territories and dog owners are requested to keep dogs on their leads in the most sensitive areas. Nearby, another seven territories were recorded at Bushy Park in 2008. A few large flocks continue to overwinter.

Movements

Harting stated that after the breeding season 'they begin to flock, and by the end of the autumn may be seen congregated in great numbers'. An example of this is related by a newspaper correspondent, who when walking in Regent's Park at one o'clock on 31 January 1857, observed 'an immense flock of these birds coming over the Zoological Gardens; their numbers were countless, and they literally darkened the air; they were flying so low that the flock had to divide in order to pass the observer. They took two or three minutes in going over, and shortly after this flock had passed, there came another almost as numerous; and so it continued for the space of an hour, flock succeeding flock, at short intervals, like divisions of a great army, all coming from, and going in, the same direction; sometimes it was a detachment of a few hundreds, and then as many thousands. One of these flocks settled on the ground not far from the observer, covering half an acre, and standing within a few inches of each other. The park-keeper and others witnessed this extraordinary sight. How long the flight had been going on before it was observed, and how long it continued after he left, cannot be told. Here was evidently a migration on a large scale, from one part of the country to another, most likely in search of food.'

In the 1930s the large autumn immigration from central Europe was documented; this movement peaked in October and stretched from the Humber all the way down to Kent, with the Thames Estuary being at the centre. In addition to this regular autumn passage, occasional large-scale movements were seen in hard weather. After a very cold spell in January 1936 numbers at Beddington rose from about 80 to 900–1,000.

Another big movement occurred on 22 December 1938 following a day of almost continual snowfall. Thousands were seen flying west over the London Area throughout the day, mostly in flocks of about 50. Over the next few days there were many reports of Skylarks looking for food in the squares and streets of central London, and on bird tables in the suburbs. They were also seen feeding on the tide-line of the Thames at Hammersmith, which was one of the few places not covered by snow. Thousands were seen at Beddington and many hundreds died of cold and starvation. A group of about 250 was seen at allotments in Loughton eating cabbage leaves and roosting under the cabbages at night. They were still present in some central London locations in early January 1939, with 30 counted on Primrose Hill and 15 in Regent's Park on the 5th.

In 1959 there was an exceptional late autumn passage and it was estimated that around 360,000 flew over London in just four days, between 30 October and 2 November; this movement was witnessed throughout the London Area, including in the very centre, where about 500 flew over Regent's Park on 1 November and about 1,325 flew over Kensington Gardens the following day. On 12 November a total of 2,200 flew north-west over the Lea Valley.

On 31 December 1961 there was a large movement of Skylarks during a day of almost continuous snow; up to 1,000 birds an hour were noted flying over and a total of 17,647 were counted across London, although it was estimated that about 200,000 may have flown over the London Area. The movement continued on New Year's Day with, for example, 4,900 flying over Dollis Hill.

At the start of the extremely cold winter of 1962/63, almost 1,000 flew west over Brent Reservoir during a four-hour period on 28 December. A week later nomadic flocks containing hundreds of birds appeared, desperately searching for food. At Beddington the numbers increased dramatically from about 600 on 6 January to about 2,000 a week later, and peaked at 3,000 on the 23rd and 25th; as the cold spell continued they began to move on. In central London up to 200 were seen in Regent's Park in early January.

In 1971 severe weather at the start of the year brought an influx into the area with 2,000 at Beddington being the maximum count. Snow on 31 December 1978 saw a large and widespread movement with 3,000 over Wraysbury and 1,000 over two other sites. In early 1979 there were several large cold-weather movements beginning on New Year's Day with 5,000 at Epsom; on 24 January five sites witnessed counts of 1,000 or more, including 6,335 over Northfleet and 5,700 over Thornton Heath in just 80 minutes. Cold-weather movements in the 1980s included 6,810 over Erith Marsh on 22 December 1981 and 1,400 at Osterley Park in January 1987.

During snowy weather in early March 2000 there was an influx across the London Area, including about 200 over Hyde Park; over the next two days up to 130 were seen feeding on the ground there. On 5 March about 10,000 were seen around Heathrow Airport, and at Beddington they peaked at 5,000 on 8 March.

Facts and Figures

Skylarks were taken for human consumption as late as 1939 when they were included in the steak and kidney pie that was served at the Cheshire Cheese pub in Fleet Street.

Although considered to be a diurnal migrant, groups were heard calling after dark over Croydon on 20 January 1960.

Shore Lark *Eremophila alpestris*

A rare winter visitor (21+)

Range: Europe, Asia, North Africa and Americas; scarce winter visitor to UK coasts.

This was a rare visitor to the UK in the 19th century and there were four records in the London Area, all of which were of captured birds. The first was of a bird taken alive by a bird-catcher on Hackney Marshes in March 1865; five years later, several were caught at the same site in November. In the 1880s two more were captured at Stamford Hill and Staines.

There were no further sightings until 1957 when one was present on ploughed land on Hampstead Heath. This was followed by three more records in the 1960s, including those of three birds at the construction site of Queen Elizabeth II Reservoir and two at Swanscombe, one of which was seen several times during the severe winter in early 1963. One overwintered at Wraysbury Reservoir in 1971/72 and possibly returned there the following winter. In 1979 one flew over Barn Elms Reservoirs on 1 January, on the same day as a considerable cold-weather movement of Skylarks took place.

There were three in the 1980s, including two at Barn Elms Reservoirs; one arrived on 29 September, a particularly early date in southern England. There were two in late autumn 1996, one of which lingered for an hour in Kensington Gardens and one present for just three minutes at Queen Mary Reservoir. There was an exceptionally large influx on the coast of south-east England in 1998/99 and one was seen at Rainham Marshes on 7 November, having probably been heard there the previous day.

The only record since then was in the autumn of 2011 when one spent a week at Queen Elizabeth II Reservoir. In the UK Shore Larks typically winter on the coast and despite the Lower Thames Marshes being the most obvious location, only three have occurred there.

1865: Hackney Marshes, one captured in March
1870: Hackney Marshes, several captured in November
1881: Stamford Hill, caught at the end of October
1883: Staines, captured about in January
1957: Hampstead Heath, on 24–25 March
1961: Queen Elizabeth II Reservoir, three on 28 October
1963: Swanscombe Marsh, on 24 January–9 March
1963: Swanscombe Marsh, on 8 December
1971: Wraysbury Reservoir, on 22 December–1 April 1972
1972: Wraysbury Reservoir, on 3 December
1979: Barn Elms Reservoirs, on 1 January
1980: Queen Mary Reservoir, on 25 October
1983: Barn Elms Reservoirs, on 12 November
1985: Barn Elms Reservoirs, on 29 September–8 October
1996: Kensington Gardens, on 21 October
1996: Queen Mary Reservoir, on 16 November
1998: Rainham Marshes, on 7 November
2011: Queen Elizabeth II Reservoir, on 25–31 October

Sand Martin *Riparia riparia*

A common summer migrant and localised breeder

Range: breeds Europe, Asia, North Africa and North America, Western Palearctic population winters Africa.

Historical

The earliest reference was in 1774 when Gilbert White stated that a few 'haunt the skirts of London, frequenting the dirty pools in Saint George-Fields and about White-Chapel'.

Harting noted that 'in some seasons they are numerous, but although they may be seen almost every day during summer at our reservoirs, they do not breed there' adding that 'this species breeds in some chalk-pits at Pinner, and at Hampstead…where an old sand-bank is completely riddled with their holes.' They also bred along the banks of the Thames; in the early 19th century there was a colony of over 100 pairs in a chalk pit by Hanging Wood, Charlton, and another colony was found at Dartford. At Blackheath they were breeding in sand banks in 1827, and in the 1890s they bred at Hanwell, Kew Bridge and Southall.

20th century

During the early part of the century they were widespread in the London Area with breeding reported in chalk pits at Beddington in 1909; in military trenches at Gidea Park in 1917; and in the Bexley, Bromley and Dartford districts, as well as in the Hornchurch, Ilford and Romford districts, prior to 1920. In the Lea Valley there were many colonies in gravel and ballast pits but these were later abandoned.

The first use of an artificial nest site came in 1902 when they bred in drainpipes in the railway embankment at Clapham Junction. In 1918 they started to nest in drainpipes in the aqueduct south of King George V Reservoir. As the century progressed they started to make further use of artificial nest sites, such as drainpipes along the canal at Harefield, on an island in the Thames near Teddington Lock and along Beverley Brook. They bred in holes in the brickwork of Rye House station in 1923 and seven colonies used railway embankments, including one by Earlsfield station that was active from 1924 to 1935. A pair bred in a drainpipe by Pymme's Brook in Palmers Green from 1926 to 1928. At Putney Bridge, a pair attempted to nest in a hole in a wharf beside the Thames in 1929.

Although the overall population was increasing there were also local losses. They ceased breeding at Barn Elms in 1924 when the sand extraction began in their chosen pit, and at Hampstead in 1926; a colony of at least a dozen pairs at Hurlingham was abandoned after 1927 when the gravel pit was filled in; a colony at Beddington was deserted in 1933 due to land development; and a nearby colony at Mitcham Junction was deserted after 1937 when the gravel pit began to be filled in.

The only completely natural nest site was in a sandy bank on the Thames at Shepperton, where they continued to breed into the 1940s. They were also quick to colonise new sites with suitable nesting habitat; for example, 65 nests were found in 1946 at Old Parkbury in a new extraction site which was still being dug. Half of all breeding sites up to this time were in gravel pits and one-third were in sand pits.

A census of colonies in 1960 found a total of 50 occupied colonies with 3,859 burrows. The largest colonies were at Swanscombe with 535 burrows, and Caterham and and Westerham, which both had 300 burrows. During the wet summer of 1958 the colony beside King George V Reservoir at Sewardstone was flooded and many birds drowned. In 1964 the colony at Holmethorpe was estimated to contain 450–500 pairs, and in 1966 an enormous colony containing 1,230 pairs was discovered at Sundridge; there were also 400 pairs at Chipstead and two colonies totalling 595 pairs at Sevenoaks. In late summer 1967, 2,000–3,000 birds were at the three main breeding colonies at Holmethorpe. During the 1970s there was a large decline in the population; a colony of about 100 pairs at Iver GP was deserted in 1971 due to disturbance, and by 1976 the colony at Sevenoaks had decreased to just 24 pairs.

In the early 1980s the total breeding population had declined to 734 known pairs with the largest concentration being 131 pairs in the Lea Valley. In 1984 spring migrants were very late to arrive and many did not return to their breeding colonies, some of which had up to 60 per cent fewer nesting birds than in the previous year. However, by 1986 there had been a good recovery and at least 743 pairs bred. The increase continued and by the following year there were at least 948 pairs; the largest colony was of 525 at Fishers Green.

In 1990 the largest breeding colony was at Apps Court GP where 150–200 pairs bred; however, in 1993 the colony was reduced to just five nests after new excavations on the site. In 1995 at least 200 pairs bred again at Apps Court but that site was also later destroyed. The second breeding Atlas recorded a contraction of 30 per cent in their distribution across the London Area. The main loss was in west London, particularly in the Colne Valley. Sand Martins were recorded breeding in very few suburban areas, and the main breeding concentration was in the Lower Lea Valley.

21st century

At the start of the 21st century the breeding population was about 500 pairs; the largest colony was of 80–100 at Walton Reservoirs but in 2004 Foxes destroyed all the nesting burrows. The breeding site at Swanscombe was partly destroyed in 2000/01 when spoil from the high-speed rail link was dumped on it. In 2000 one pair bred at or near the London Wetland Centre and an artificial bank was built there; by 2004, 10 pairs bred there and this new colony expanded quickly, reaching 66 pairs in 2009. Another artificial bank was built at Beddington; birds were slow to utilise it but 12 pairs were nesting in it by 2008. Populations at individual sites vary year on year, e.g. numbers at Mill Wood, Chafford Hundred dropped from 26 pairs down to zero then back up to 17 pairs within four years.

New sites are occasionally exploited by pioneering birds; for example, two pairs have been breeding in the old Surrey Docks area in Inner London since 2009, and two pairs bred in drainage holes in the dam wall at Brent Reservoir for the first time in 2011.

Movements

In August 1863 Harting saw about 600 on the ground near Brent Reservoir and presumed that they had congregated there as a prelude to migrating. He also noted a similarly sized flock there the following year. Hamilton stated he once saw hundreds flying over the Serpentine one April.

During the autumn passage, large numbers congregate along rivers and wetland sites, especially reservoirs. In 1953 about 2,500–3,000 flew over King George V Reservoir in one hour on 30 August. At Perry Oaks SF about

4,000 flew south-west on 6 August 1958, and in 1966 a huge group of about 8,000 was at Staines Reservoir on 11 August. In 1967 there were about 7,000 on 12 August at Queen Mary Reservoir. In 1968 about 5,000 flew over Staines Reservoir on 20 August, and in 1969, 3,000–4,000 were present at Staines Reservoir on 25–26 August.

A flock of 1,300 was seen resting on a quiet concrete taxiway at Heathrow on 28 July 1982. In autumn 1990 there were more than 3,500 in the Ingrebourne Valley on 28 August and over 2,000 at Rainham Marshes on 1 September; large numbers remained at both of these sites for over two weeks. The largest counts in spring were 1,000 passing through Amwell on 12 and 28 April 2006.

Facts and Figures

The earliest migrants were on 29 February 2008 at Amwell GP, Holmethorpe SP and Island Barn Reservoir.
The latest migrant was on 15 December 2000 at Walton Reservoir.
The largest count was about 10,000 birds roosting at Ashford GP on 6 September 1963.

Swallow *Hirundo rustica*

A very common summer migrant and localised breeder

Range: breeds Europe, Asia, North Africa and North America, Western Palearctic population mainly winters Africa.

Historical

In 1774 Gilbert White wrote that Swallows bred in Aldersgate which, being outside of the City of London, was mainly open fields. They also bred in Lambeth in 1780.

This was a regular breeding summer visitor in the 19th century. Harting did not specify any breeding sites, which implies that at least in the outer districts of London, they were widespread. In 1865 Hibbert wrote that Swallows bred on the fringes of London at Kensington, Lambeth, Kentish Town and Stoke Newington although his description of the nests in Stoke Newington could relate to House Martins. In the 1860s a pair nested in Regent's Park; also in central London they bred on a church in Battersea in 1884 and in Battersea Park in around 1896.

20th century

By the start of the century they had ceased to breed in central London, probably as a result of a lack of food, although in 1907 and 1908 a pair bred in one of the deer sheds in the grounds of London Zoo. They continued to breed in many suburban districts; in the 1900s they were common in Enfield and Willesden, and bred in Acton, Barnes, Chiswick, Clapham Park, Dulwich, Finchley, Greenwich, Hampstead, Highgate, Hornsey, Neasden, Stratford, Streatham, Sydenham, Tooting, Wanstead, Wembley and Wood Green.

They last bred in Greenwich Park and at Walthamstow Reservoirs in 1928. In 1935 Glegg wrote that it was 'a common summer resident, well distributed in the rural districts', adding that it was now only a passage migrant in central London. They bred on Putney Heath in a boathouse annually until 1932 and at Beulah Hill, Streatham, until 1933.

A pair bred again in central London in 1941, at Eccleston Square, probably taking advantage of a bombed building. Apart from this isolated record, the closest that they bred to the centre was in Camberwell in 1945. Occasional pairs bred in the south London suburbs and they bred regularly on Wimbledon Common until the late 1940s. By the mid-1940s they had stopped breeding in Richmond Park, and by the 1950s they had ceased breeding in the inner suburbs and were declining in the outer suburbs. In north-west London they bred at Harrow, Harrow Weald and Mill Hill, and in 1952 a pair bred in Golders Hill Park; in the same year 39 pairs bred at Heathrow Airport. A few pairs continued to breed along the Thames-side marshes from Plumstead downstream. In 1968 a pair bred in Regent's Park, the first breeding in central London since the Second World War. The following year a pair bred on Hampstead Heath for the first time since 1952.

In 1982 a pair frequented Bankside Power Station (now Tate Modern) during early June but did not stay to breed. In 1990 up to 10 pairs bred at Rainham Marshes. The second Atlas in 1988–94 showed a slight reduction in the breeding distribution. Beyond the suburbs they were found breeding in almost all areas.

21st century

By the start of the 21st century there appeared to have been a considerable reduction in the breeding population with less than 100 pairs reported; however more breeding pairs may go unrecorded in farmland. This appears to be against the national trend which showed an increase of 35 per cent from 1995–2010 following a dip in the 1980s. There is insufficient data in the London Breeding Bird Survey to monitor this species. The only suburban breeding sites are at Dagenham Chase and Stanmore.

Movements

In the 18th and 19th centuries, large numbers of Swallows used the Thames as a migration route and stop-off. In 1767 Gilbert White wrote about 'the myriads of the Swallow kind seen annually on the Thames at Sunbury in the autumn'. In about 1834 Jesse referred to 'the vast collection of emigrating Swallows seen every year on the banks of Thames at Hampton', adding that thousands would assemble on a small eyot, and in 1831 large flocks left the eyots in the first week of September, some a week later and the last on the 16th of the month.

Cornish stated that there was a large passage of Swallows, House Martins and Sand Martins along the Thames every year but on 16 September 1896, after a period of wet and windy weather, he saw 'a great migration of swallows down the Thames…between five and six o'clock immense flights of swallows and martins suddenly appeared above the [Chiswick] eyot arriving, not in hundreds, but in thousands and tens of thousands. The air was thick with them, and their numbers increased from minute to minute…thousands kept sweeping just over the tops of the willows, skimming so thickly that the sky-line was almost blotted out from the height of from three to four feet.' He went on to describe how they all roosted overnight in the osiers on the island and were all gone by 6:30 hrs the following morning.

Glegg described a passage of Swallows on 12 September 1931 in the Thames Valley between Sunbury and Chertsey 'which were migrating steadily to the west, along a front of at least a mile, perhaps much wider, during the four hours I was present. The birds were all flying low and just topping the hedges'.

In 1963 an estimated 20,000 roosted in a reed bed at Northfleet on 22 September; they continued to roost there over the next few days but there was a gradual reduction in numbers.

There was a large passage in the autumn of 1999, mainly from 22 to 27 September; on the 26th a total of 31,500 flew over the London Area and the following day about 10,000 flew over Hampstead Heath.

Facts and Figures

The earliest migrant was on 28 February 1970 at Merchant Taylor's School.

Up to 1972 there had been about seven occurrences in December, the latest being on 28 December 1934 when a party of six was seen in the Colnbrook Valley; there is also one extreme record, on 29 January 1809 at an unknown location in Middlesex.

The largest count was about 20,000 roosting at Northfleet on 22 September 1963.

Single hybrid Swallows x House Martins were seen at Hainault on 6 September 1980, and at Fairlop on 16 May and 1 June 1991.

House Martin *Delichon urbicum*

A very common summer migrant and widespread breeder

Range: breeds Europe, Asia and North Africa, Western Palearctic population winters Africa.

Historical

In the 18th and 19th centuries House Martins bred at various sites in central London. One of Gilbert White's correspondents wrote in 1767 that they were breeding in Borough. White had also seen them nesting in the Strand and Fleet Street prior to 1773.

They bred in Hyde Park and Kensington Gardens in the 1870s and in St James's Street, Porchester Place, Upper Seymour Street and Lancaster Gate in the 1880s. Also, in the late 19th century they bred in the cloisters in Westminster Abbey.

Christy stated that a correspondent of his from Stubbers near North Ockendon stated that they 'are being steadily exterminated by the injurious and obnoxious Sparrows' which drove them out of their nests. After all the House Sparrows were shot the number of House Martin nests increased from seven in 1870 to 237 by 1885.

20th century

By 1900 they had ceased breeding in central London and this was thought to be due to the loss of flying insects on which they fed. However, they did breed in the inner suburbs at Barnes, Brixton, Dulwich, Hammersmith, Peckham Rye, Putney and Willesden. In 1909 they were nesting in many streets in Woolwich but numbers had fallen considerably in many other suburbs in the south-east of London. In some areas a terminal decline had set in and they had ceased breeding on Wimbledon Common and in Roehampton by 1923, and in Highgate by 1926.

In the 1930s Glegg stated that they were still a common summer visitor, breeding in the suburbs and rural areas. The continued suburbanisation had led to some declines but also provided opportunities for new breeding sites; a colony was started in Ruislip in 1937 on houses that had been built less than a year earlier. In the 1940s they were still common in parts of south-west London, particularly around Putney and Wimbledon, and they had returned to breed in Highgate after a 15-year absence. However, there had been a considerable decrease in east London over the previous 30 years; for example, in the Epping area they had declined more than Swallows.

By the middle of the 20th century there was a gradual decline in the population in the suburbs; Fitter suggested that a lack of muddy pools from which to collect nest-building material may have been a determining factor in this decline. In the 1950s they were still breeding in small numbers in many north-western suburbs. South of the Thames House Martins were still common in Bexley, the Darenth Valley, along the Thames between Greenwich and Swanscombe, and between Kew and Hampton. Three pairs bred in Tooting, the last remnants of a previously healthy population in the district. Further out, scattered colonies could be found in many rural districts.

A survey conducted in 1966 located six breeding pairs in Harley Road, St John's Wood, the first breeding occurrence in Inner London since 1889. The survey also found increases in the inner suburbs in west and north-west London. For example, there were 45–60 pairs in the Hammersmith and Fulham areas, a considerable increase on the five known pairs in 1949; also, colonies had returned to Hampstead and Shepherds Bush. It was thought that this was attributable to reduced air pollution which dropped by about two-thirds between 1952 and 1965 after the introduction of the Clean Air Act of 1956 (Cramp and Gooders). By 1969 the Harley Road colony had risen to 15–19 nests and one pair bred by Victoria Park.

A survey in 1974 of representative areas of London found increases in the inner suburban area of from 45 nests in 1949 to 272; in the outer suburban area the increase was from five nests to 71; and on the outskirts from three nests to 75. In central London two new colonies were established in 1975, and by 1982 the population had increased to over 120 pairs.

As in the case of Sand Martin, 1984 saw a large reduction in the breeding population. For example, in the Tadworth area there was a decrease of 80 per cent and only 41 nests were located in central London. The following year further decreases were reported in some areas; on one road in Staines there was a reduction of 50 per cent. By 1986 there had been a partial recovery and at least 69 pairs were nesting in central London.

At the start of the 1990s House Martins were breeding at more than 20 locations in central London, mainly in the western part. In 1991 reductions were reported in breeding numbers and many birds were late to arrive; there was also a very late arrival in 1993–95 and a further decrease in breeding numbers which continued throughout the decade. There was virtually no change in the breeding distribution of House Martins in the London Area between the two breeding Atlases; during the second survey in 1988–94 they bred in almost three-quarters of all the London Area's tetrads. The decline in the population began almost as soon as the survey was concluded and by the end of the decade large decreases were being reported. In the Hornchurch and Romford area there was an 80 per cent decline between 1997 and 2000.

21st century

By the beginning of the 21st century House Martin was still a fairly common and widespread breeding bird throughout the London Area despite declines in the 1990s. However, it has continued to decrease. It bred in at least six sites in central London in 2000 although a traditional site in Bermondsey was abandoned. A survey carried out in three north London boroughs in 2003 failed to find any nests at 47 sites where the birds had bred in 1985. In 2009 about 350–400 breeding pairs were located. There are very few colonies in Inner London, the most notable being the long-established site in the French Embassy on Knightsbridge which continues to be occupied by birds which feed over the Serpentine in Hyde Park.

Movements

The autumn passage can start as early as July and in some years many birds will remain until the first half of October with a few late migrants seen in November and even into December. There have also been two midwinter sightings of single birds: at Hampton Court on 10 January 1912 and at Sevenoaks on 1 January 1975.

Counts of up to 2,000 in autumn are recorded in most years; occasionally much higher numbers are seen. In 1971 more than 6,000 flew over Ongar on 29 September during a day of considerable passage throughout Essex. In 2001 some 6,000 flew over Dartford in one hour on 30 September.

Facts and Figures

The earliest arrival date was 9 February 2004 at Dagenham Chase.
The latest-ever record was 12 December 1974 at New Addington.

Red-rumped Swallow *Cecropsis daurica*

A rare vagrant (14)

Range: breeds southern Europe, Asia and North Africa; Western Palearctic population winters Africa.

The first record occurred in 1964 at Ruxley GP, where there was also a Hoopoe on the same day. Two years later there was one at Hilfield Park Reservoir, still the only autumn record. The species remained rare over the next three decades with two in both the 1970s and 80s. There were national influxes in 1992 and 1996 when two more occurred, and another was seen in 1999.

Since the start of the 21st century they have been seen more frequently in the country and this increase has also been noted in the London Area. In 2000 one was present for three days at Hilfield Park Reservoir, the first time one had stayed for more than a day. There was another national influx in the spring of 2003 and two arrived in the London Area on the same day.

1964: Ruxley GP. on 25 April
1966: Hilfield Park Reservoir, on 1 October
1973: Staines Reservoir, on 17 May
1975: Wanstead Park, on 4 June
1982: Hilfield Park Reservoir, on 18 May
1987: Amwell GP, on 15 April
1992: Hainault Forest, on 23 May
1996: Staines Reservoir, on 6 May
1999: Beddington SF, on 16 May
2000: Hilfield Park Reservoir, on 27–29 May
2003: Beddington SF, on 28 April
2003: Staines Reservoir, on 28 April
2007: Stocker's Farm, on 11 May
2009: Stoke Newington Reservoir, on 9 May

Cetti's Warbler *Cettia cetti*

An uncommon partial migrant and localised breeder

Range: Europe, North Africa and Middle East.

20th century

Cetti's Warbler was not recorded in the UK until 1961. It first bred in 1972, in east Kent, and its expanding population was probably responsible for the push into the London Area. The first London Area records occurred in 1975: a male at a site in south-east London from 20 to 25 May; a pair at Wraysbury on 18 July; and one ringed in the Colne Valley on 14 December, which remained until 9 January 1976. Over the next few years it continued to increase and several took up territory.

In 1980 they bred for the first time in the London Area, in the Lea Valley. The following year there had been a further expansion with two pairs breeding in the Lea Valley and a pair breeding in the Colne Valley; there were also up to seven other singing males during the breeding season. However, the birds failed to breed in 1982 and despite another mild winter the attempted colonisation of the London Area had failed – none were present during the 1983 breeding season. Two consecutive severe winters decimated the national population in the mid-1980s. The rest of the decade witnessed only one or two wintering in the Lea Valley with the occasional record elsewhere until the autumn of 1989 when there was a sudden increase in the Lea Valley, with up to six at Amwell.

In 1990 two males held territory in the breeding season, at Brent Reservoir and at Nazeing GP. There was an increase the following year to five singing males, and in 1992 two or three pairs probably bred in the Lea Valley. However, this attempted colonisation was short-lived and in 1994 none were seen at all for the first time since they had arrived two decades earlier. One or two were seen during the next two summers and they probably bred in 1997 and 1999.

21st century

Numbers continued to fluctuate during the first few years of the 21st century; most sightings were again in the Lea Valley and a pair possibly bred at Cheshunt in 2001. Since then the national population has increased dramatically and there has been a corresponding increase in the London Area, mainly in the east of the area.

In 2005 there were 11 territories in the Ingrebourne Valley and up to two resident birds at Amwell. At this time larger numbers were being seen outside of the breeding season across the London Area. The following year the numbers in the Ingrebourne Valley doubled to 22 and another nine were on territory elsewhere in the London Area, including a pair that bred at Amwell. By 2009 the population had increased to more than 70 territories, almost all of which were in the area between the Lower Thames Marshes and the Lea Valley. In the western half of the London Area Cetti's Warbler is mainly a winter visitor in small numbers, particularly in the Colne Valley. It has recently colonised the London Wetland Centre with breeding first recorded in 2010.

Eastern Crowned Warbler *Phylloscopus coronatus*

A very rare vagrant (1)

Range: east Asia; vagrant to UK.

There have only been two sightings of this rare vagrant in the UK. The first was in Durham in October 2009; two years later one was found in a mist net at Hilfield Park Reservoir. After release it was not positively seen again.

2011: Hilfield Park Reservoir, ringed on 30 October

Pallas's Warbler *Phylloscopus proregulus*

A very rare vagrant (1)

Range: Asia; vagrant to UK.

There has only been one record in the London Area, on Wandsworth Common in 1985. The bird was seen for only 15 minutes in the morning and was even heard singing but could not be re-found later in the day.

1985: Wandsworth Common, one on 29 October

Yellow-browed Warbler

Phylloscopus inornatus

A rare vagrant (25)

Range: Asia; vagrant to UK.

The first one was seen by Dr Butler in his garden at Beckenham on 15 April 1899. The record was originally square-bracketed by Ticehurst only because no specimens had been obtained in Kent; he had no reason to doubt Butler.

No more were seen until 1930 when one was seen and heard singing in a garden in Sutton. Thirty years later one was seen in a garden full of other migrants at Reigate; this coincided with a fall of Yellow-browed Warblers on the East Coast of England and in Ireland.

There was one in 1978 and 10 years later there was a large influx into Britain and Ireland with almost 400 seen in the autumn; two arrived in the London Area on consecutive days. The 1990s saw a big increase in the numbers arriving in the country and this was reflected in the London Area. There were four in 1990–91, including two at Buckhurst Hill; one was also seen in winter in a garden at Stoneleigh, Epsom. After a two-year absence three were seen in 1994. The first of these was at Brent Reservoir in mid-September; this was found on a day when there were virtually no other migrants present although there had been a large arrival in the UK a few days earlier. Two more were seen in October after another large arrival in the country. In 1996 there was one in Hainault Forest, which arrived at the start of a large influx into the country.

Since the start of the 21st century the number of sightings has continued to increase and it has become almost annual. In 2007 one remained at the Wetland Centre between 13 October and 29 November, the longest-staying individual so far. The following year two were present at Rainham Marshes on 8 October, which represents the first multiple arrival. London Area records reflect the national trend with 83 per cent occurring in September and October.

1899: Beckenham, on 15 April
1930: Sutton, on 10 October
1960: Reigate, on 28 September
1978: Sevenoaks WR, on 8 October
1988: Staines, on 24 October
1988: Wimbledon Common, on 25 October
1990: Ladywell Fields, Catford, on 14 September
1990: Buckhurst Hill, on 27–28 October
1991: Stoneleigh, Epsom, on 17 February–2 March
1991: Buckhurst Hill, on 19–23 September
1994: Brent Reservoir, on 18–20 September
1994: Dagnam Park, on 9 October
1994: Kenton, on 11 October
1996: Hainault Forest, on 21 September
2001: Waterworks NR, on 28 October
2003: Brent Reservoir, on 31 October–2 November
2005: Alexandra Park, on 8 October
2007: London Wetland Centre, on 13 October–29 November
2008: Rainham Marshes, two on 8 October, one on 10 October
2008: Fairlop Waters, on 18 October
2009: London Wetland Centre, on 20–25 September
2009: Ridlands, on 8 November
2010: Bromley-by-Bow, on 27 September
2010: Alexandra Park, on 1 November

Hume's Warbler *Phylloscopus humei*

A very rare vagrant (2)

Range: Asia; vagrant to UK.

The first record was of an overwintering bird at Fairlop Waters in 2004. This was quickly followed by a one-day bird at Brent Reservoir. In the BBRC report these two records were presumed to relate to the same bird but they are listed here as separate records. The Fairlop bird was well watched and searched for but was not seen after the last sighting on 25 April when it was heard singing. The Brent bird arrived six days later, probably during a late-morning shower that downed several other warblers. Its calls had not been heard before the shower but it was very vocal throughout the afternoon and into the evening and was heard singing on a number of occasions. There had been a large influx during the previous autumn and it is likely that it was one of these birds heading north rather than the Fairlop bird relocating 24 kilometres to the west.

2004: Fairlop Waters, on 11 January–25 April
2004: Brent Reservoir, on 1 May

Radde's Warbler *Phylloscopus schwarzi*

A very rare vagrant (1)

Range: Asia; vagrant to UK.

The only record was in 2000 when one was found at Fairlop Waters. That autumn was a record year for Radde's; up to 30 were found across the country and there were seven at Spurn, East Yorkshire, on the same day. The Fairlop bird showed only briefly on the first day and was extremely elusive on the following morning.

2000: Fairlop Waters, on 1–2 October

Dusky Warbler *Phylloscopus fuscatus*

A very rare vagrant (1)

Range: Asia; vagrant to UK.

The only record occurred in February 2010 when one remained for eight days at Walthamstow Reservoirs. Although this species is predominantly a late-autumn migrant there have been several records of overwintering in the UK.

2010: Walthamstow Reservoirs, on 14–21 February

Wood Warbler *Phylloscopus sibilatrix*

An uncommon passage migrant; former breeder

Range: breeds Europe and Asia, winters Africa.

Historical

In 1794 this was described as a new species by Lamb, after having first been observed in 1792; he also documented that he had heard it near Uxbridge. In the 19th century it was evidently a widespread summer visitor. In 1823 Sweet stated that he had frequently heard it in Kensington Gardens, while in the 1830s it was not uncommon around Tooting.

In north-west London they bred at Edgware, Hampstead and Kingsbury, and were described by Harting as being 'more plentiful...in the more wooded neighbourhoods of Stanmore, Bushey and Pinner'; by the late 19th century they were also discovered breeding near Perivale and in Highgate Woods. Elsewhere, they bred in Epping Forest and at many localities in south London, including right into the built-up area at Dulwich.

20th century

Early in the century they were still a common breeding bird in many areas, for example Ticehurst stated that they were plentiful around Orpington, Bromley, Keston and Hayes; and at least eight pairs were located in one wood in the Watford district in 1911. They were still breeding at Highgate and Hampstead in 1920 but by the mid-1930s they had ceased breeding in both areas; the reason for the extinction in Highgate Wood was the loss of the understorey.

By the 1930s they had declined further in number and range, Glegg describing the breeding range in north and west London as 'very local, being confined to a few districts', namely Ruislip Woods, Stanmore Common and Trent Park. In the early 1930s they ceased breeding in Broxbourne Woods. However, they had started breeding in a number of small copses around Harrow. By the start of 1940s they had disappeared from the inner suburbs apart from on Wimbledon Common where around six pairs nested annually, and on Hampstead Heath where breeding occurred from 1945 to 1949. The London Area's breeding population was around 50 pairs at this time, with about half being in their stronghold of Epping Forest; another 12 pairs were located along the North Downs.

In the 1950s it was considered that the overall population in the London Area had remained largely unchanged since the start of the 20th century despite some local losses. South of the Thames they were common in woods along the North Downs, and breeding also occurred in the suburbs at Putney Heath, Richmond Park, Shirley Hills, Shooters Hill and Wimbledon Common. North of the river they were locally common in Epping Forest and bred irregularly on Hampstead Heath. In 1953 four pairs were found in Mimmshall Woods. In central London they were considered to be an uncommon passage migrant by the middle of the century, usually confined to the parks.

The population began to decline. During the early 1960s up to 14 males held territory during the breeding season but only a few pairs actually bred. The largest population was in Northaw Great Wood where at least four pairs bred and another four or five pairs were present.

Apart from the occasional bumper year, the general trend was downwards and by the start of the 1990s the breeding numbers reported were down to just two pairs, at Hainault Forest and Ruislip Woods; additionally,

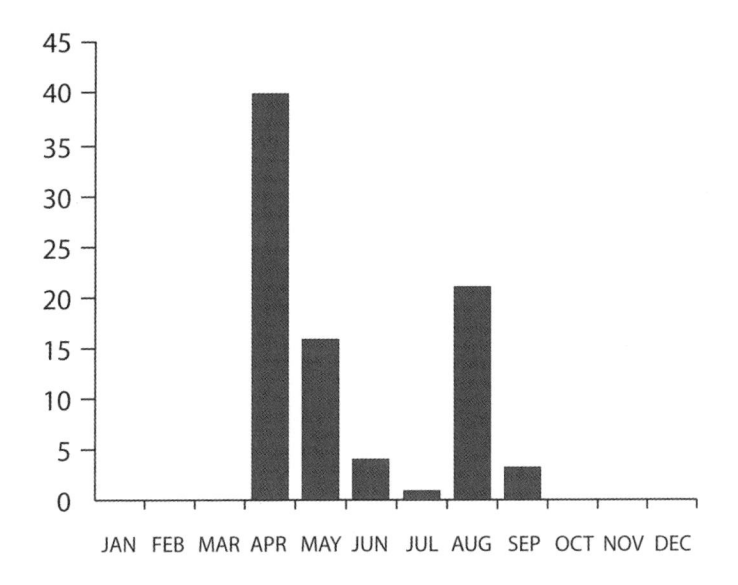

Seasonal occurrence of Wood Warbler: 2000–2009.

singing males were recorded at 12 localities, most of which were passage birds. There was a 40 per cent decline in the breeding distribution between the two breeding Atlases. During the second survey in 1988–94 they were found breeding in 34 tetrads. Much of the loss was on the North Downs and it was thought that the Great Storm of 1987 may have been partly responsible as the loss of many trees opened up the canopy in exposed woodlands and made them less suitable. The last pair to breed was on Bookham Common in 1996.

21st century

By the start of the 21st century they had ceased breeding and had become much scarcer on passage. Only 12 migrants were seen in 2000, half of them in Regent's Park, including four singing males on 26 April. Over the next few years even fewer were seen and in 2004 only one was seen all year. A male held territory in Epping Forest in 2006 and there were several on territory in south London in 2008 but there were no further signs of breeding. Nationally there has been a 65 per cent decrease in the breeding population based on the Breeding Bird Survey, resulting in it being red listed as a species of conservation concern. The first migrants are usually seen in late April or early May; they are comparatively scarce in autumn, usually in August.

Facts and Figures
The earliest date was 9 April 1949 at Ruislip.
The latest date was 5 October 1947 at Headley.

Chiffchaff *Phylloscopus collybita*

A very common summer visitor, uncommon winter visitor and widespread breeder

Range: breeds Europe and Asia, winters western and southern Europe and Africa.

Historical

The Chiffchaff's breeding status has changed very little since the mid-19th century when Harting wrote that it was a common bird and 'one of the first of our summer migratory birds to arrive'. What has changed since those times is its presence in winter – Harting knew of none wintering although south of the Thames one overwintered at Wimbledon in 1833/34 and another was shot at East Molesey on 23 December 1890. Another one seen on 6 February 1898 was considered to be an exceptionally early migrant at the time but was more likely to have been an overwintering individual.

Comparing the relative abundance of the Chiffchaff, Willow Warbler and Wood Warbler, Harting stated that the Chiffchaff was 'the commonest of the three Willow Wrens'.

20th century

Glegg stated in the 1930s that it was 'a fairly common summer resident in the rural parts of the county', adding that it was a passage migrant in central London but may have bred. He also stated that it had become less common than Willow Warbler. In the 1940s and 50s they were widespread breeders in the southern suburbs and were common in the large open spaces at Kew Gardens, Richmond Park and Wimbledon Common. In the more rural districts it was relatively common where there was sufficient habitat. In central London migrants were often heard singing in spring but the only known nest was found in 1937 in Holland Park. Further breeding occurred in Regent's Park in 1972 and 1981.

There was only one winter record in the first half of the 20th century, at Walton-on-Thames on 1 February 1948; during the early 1950s two more were seen. The number seen in winter started to increase during the late 1950s. In early January 1959 there were three at Beddington SF, including one of the race *abietinus* which was trapped on 3 January, and re-trapped at Elmers End SF on 15 April. In 1961 they were recorded at 11 localities, including five at Perry Oaks. During the exceptionally cold winter of 1962/63 there was a complete absence in January and February and the first migrant was seen on 10 March. Up until the beginning of the 1970s there were only one or two seen each winter. From 1971 they began to increase and in 1975 at least

23 were seen in the first two months of the year. This increase in the wintering population continued and in 1984 at least 78 were present in the last two months of the year, including at least 13 in the Wraysbury/Horton area and 11 at Rye Meads; one trapped at Rye Meads on 21 December showed characteristics of 'Siberian Chiffchaff' *tristis*. Fewer were seen in subsequent years as the winters were much colder but they increased again during the 1990s. In January 1994 more than 80 were noted. Unlike Blackcaps, which were mostly found in suburban gardens, Chiffchaffs usually wintered in scrub and woodlands by water, particularly in north and west London.

During the second breeding Atlas in 1988–94 there was an increase in the breeding distribution of 10 per cent compared with the 1968-72 Atlas. They were found breeding in 70 per cent of all the tetrads in the London Area; almost all of the vacant areas were situated in the more heavily built-up areas. In the mid-1990s the population continued to increase with large rises noted in Hainault Forest, on Hampstead Heath and on Wimbledon Common, for example.

21st century

This is a common and widespread breeding bird throughout much of the London Area and the population has increased in some areas since the beginning of the 21st century. It is reasonably common on passage in central London but rarely remains to breed there.

Since 2000 birds resembling 'Siberian Chiffchaff' have been recorded at Alexandra Park, Amwell GP, Arthur Jacob NR, Brent Reservoir, Dagenham Chase, Horton GP, Mudchute City Farm, Rainham Marshes, Stoke Newington Reservoirs, Walton Reservoirs and Wraysbury GP; most of these occurred in December and January.

Mixed singers

There have been some records of 'mixed singers', i.e. individuals that use elements of the songs of both Chiffchaff and Willow Warbler. The first one that was described was reported to be a Chiffchaff in Hampstead in 1908. This bird 'invariably concluded its song with an exact reproduction of that of the Willow-Wren, with no interval between the two'. One in Epping Forest on 3 May 2003 had a Willow Warbler's song with sections of Chiffchaff song interspersed. Similar individuals have also been recorded at Alexandra Park and Brent Reservoir.

Facts and Figures

The largest count was 400 birds at Wraysbury GP on 9 September 2000.

Iberian Chiffchaff *Phylloscopus ibericus*

A very rare vagrant (1)

Range: breeds Iberia and North Africa, winters Africa; vagrant to UK.

In June 1972 an unusual song was heard at Brent Reservoir and tracked down to an odd looking *Phylloscopus* warbler singing from some willows adjacent to the reservoir. A full description of the bird was taken along with a sound recording, which later helped to identify the bird as an Iberian Chiffchaff, at the time considered to be just a subspecies. It was the first record in the UK and was accepted by the LNHS.

Almost 25 years later the taxonomy of this form changed and it became a separate species. A formal submission was made and subsequently accepted as the first British record. There were no further sightings in the UK for 20 years after the first; since then at least 14 more have been seen, several of which have remained on territory during the summer.

1972: Brent Reservoir, singing male on 3 June

Willow Warbler *Phylloscopus trochilus*

A common summer visitor and widespread breeder

Range: breeds Europe and Asia, winters Africa.

Historical

Harting described this as a 'common summer visitant' whose 'nest and eggs [were] found every year'. Towards the end of the 19th century they bred as close to central London as Ravenscourt Park and Dulwich Wood.

20th century

In the late 1920s there was a significant increase in the numbers breeding in the Harrow district; in 1925–26 there were none but by 1930 there were eight pairs. During the 1930s Glegg described it as 'one of the commonest summer residents', adding that it was 'probably an annual breeder in Inner London and considerable numbers pass through, mainly at the end of July and in August'. He also observed that 'in the rural districts it is probably the most numerous Warbler'. In central London they were more frequent breeders than Chiffchaff during the first half of the 20th century with breeding recorded from Hyde Park and Kensington Gardens, Holland Park and Regent's Park.

In the 1950s it was more common and widespread across the London Area than Chiffchaff; breeding occurred occasionally in the parks in central London. In 1969 a pair bred in Hyde Park/Kensington Gardens, the first breeding there since 1923.

In 1972 there was a large influx on 9 April at Cheshunt GP when over 100 were seen; they were reportedly singing from every piece of cover available. During the 1970s the population began to increase and the largest concentration was 60 pairs in Hainault Forest in 1979. The peak count on Wimbledon Common was 119 breeding pairs in 1982. During the second breeding Atlas in 1988–94 they were found breeding in 72 per cent of the tetrads in the London Area, marginally more than Chiffchaff and slightly higher than during the first survey in 1968–72.

By the beginning of the 1990s a decline had set in with 337 pairs/singing males reported compared to 456 in 1989; the only known breeding birds in Inner London were at Surrey Docks where there were up to 10 pairs. In Epping Forest there had been a decline of 25–30 per cent over the 10 years up to 1992, and at Dagnam Park the population had almost halved in eight years. In 1995 the population at Hainault had decreased from 65 pairs to 48 in five years. The most dramatic reduction was on Wimbledon Common where there were just five territories in 1999, a loss of over 100 pairs in 17 years. However, some sites did manage to buck the trend. For example, at Hornchurch Country Park there was an increase to 35–40 pairs in 1999 from 18 three years earlier.

21st century

At the start of the 21st century the breeding population was in a steep decline in the London Area. The Breeding Bird Survey combined with earlier CBC data indicated a decline of 34 per cent from 1970 to 2010 for the UK as a whole, but this masks the fact that for England only the decline was 28 per cent whilst in Scotland there had been an increase of 33 per cent. This apparent shift in population may be linked to climate change. They have continued to decrease since then and have ceased breeding at many localities. In Inner London they are regular on passage but do not breed.

Singing migrants are widely recorded in spring, from the end of March to early May; in autumn most pass through in August and September with the last ones usually seen in the middle third of September; they are rarely recorded in October.

Facts and Figures

The earliest date recorded was 4 March 1957 at Barnet.
The latest date was 25 November 2008 at Rainham Marshes.
There are winter records in St James's Park on 13 February 1935 and at Rye Meads on 14–15 January 1978.
Birds of the northern race, *acredula*, which occurs across Scandinavia, have been recorded twice: one was found freshly dead at Aldenham on 25 April 1949 and one was seen at Shoreham on 13 April 1952.

Blackcap — *Sylvia atricapilla*

A common summer visitor, uncommon winter visitor and widespread breeder

Range: breeds Europe and Asia, winters western Europe and Africa.

Historical

Harting described the Blackcap as 'an annual summer visitant, but not common, arriving early in April and leaving about the last week of August'. He also knew of some instances where it had occurred during winter.

It was first recorded breeding in Inner London in Battersea Park in 1869 and in London Zoo in 1879. In the 1870s it bred regularly in one area of Kensington Gardens but deserted when the shrubbery was 'beautified' by the authorities.

20th century

In the 1930s Glegg thought that it probably bred annually and he described it as a common summer visitor in suitable localities. He added that it was generally more common than Garden Warbler, although some recorders held the opposite view and thought that the relative status of the two species varied periodically. The average arrival date prior to 1935 was 17 April with March dates being very rare; by 1954 the first arrivals were often recorded in the last 10 days of March with most birds arriving from mid-April onwards.

In the 1940s they did not breed in central London but by the early 1950s a pair was breeding in Holland Park and two or three pairs bred in St John's Wood in 1953. Around the same time they bred in the suburbs at Barnes, Blackheath, Camberwell and Charlton, as well as in Greenwich Park, Streatham Common and Hampstead Heath where there were around 12 pairs. When the bombed sites in St John's Wood were redeveloped in the late 1950s, they ceased breeding there but by the start of the 1960s they had begun to nest in nearby Regent's Park. By 1968 there were up to 16 breeding pairs in central London.

During the 1980s and 90s the population increased across the London Area; for example, in Hainault Forest Blackcaps increased from 30 pairs in 1980 to 55 pairs by 1993. By the time of the second breeding Atlas in 1988–94 they had expanded their breeding range by 27 per cent compared to the first survey. Much of this increase was in the suburbs with almost all of the outer areas hosting breeding birds. They were found breeding in 82 per cent of all the London Area tetrads.

Blackcaps have been wintering in Britain since at least the beginning of the 20th century but in fairly small numbers up to 1950. During this period there were only four reported instances in the London Area. By the early 1960s, however, there had been an increase in the numbers seen during the winter months though this dropped off by the middle of the decade and in some winters only one was seen. At the start of the 1970s wintering birds had begun to increase again with up to seven seen; by the early 1980s more than 40 wintered.

21st century

At the start of the 21st century the breeding population was still increasing. For example, at Hainault Forest the number of pairs increased from 30 to more than 100 in 20 years. Apart from Hainault, the only other site with more than 100 pairs is Wimbledon Common. Blackcaps breed widely throughout the area apart from in central London where they are scarcer, tending to be found where shrubbery has a more natural quality often with some bramble or other native species, although they are more common on passage.

The wintering numbers have also increased; in January 2004 a record 95 were seen. Despite the slightly higher temperatures very few overwinter in central London and most prefer to winter in large gardens in south London.

Garden Warbler *Sylvia borin*

A common summer visitor and widespread breeder

Range: breeds Europe and Asia, winters Africa.

Historical

In the 19th century they were common in the Hampstead woods and were also known from Dulwich Woods, Richmond, Tooting and Wimbledon. Swann stated that they were perhaps more common than Blackcap though Ticehurst maintained that 'In any given locality the Blackcap will be plentiful and the Garden Warbler scarce, while the next year the positions may be reversed', adding that Walpole-Bond had also noticed this in the Bromley district.

20th century

At the beginning of the century Garden Warblers were widespread across the London Area, but as suburbanisation took hold they were lost from some areas such as at Lee and Tooting Common. In the 1930s Glegg described them as common in the rural districts but absent as a breeding species in suburban areas, apart from Hounslow Heath.

By the 1950s they had all but disappeared from the inner suburbs. The closest they were known was from Bostall Woods, Shooters Hill, Dulwich Woods, Wimbledon Common, Richmond Park, Ham Common and Epping Forest. Up to five pairs bred on Hampstead Heath but they ceased breeding there after 1950. In 1961 at least 20 males were heard singing in the Ruislip area in late May. They continued to increase during the 1980s and by 1985 there were around 190 territories, almost as many as those of Common Whitethroat.

During the second breeding Atlas in 1988–94 they were found breeding in 42 per cent of all the tetrads in the London Area, an increase of 13 per cent over the first survey. They were mostly located beyond the suburbs but they bred on many of the large open spaces within the urban areas apart from in central London where they were only occasionally seen on passage. There were 24 territories on Bookham Common in 1999, an increase of 10 in two years.

21st century

At the start of the 21st century the population was relatively stable and there was a minimum of 167 territories at 55 sites. The largest breeding concentrations were 22 territories at Broadwater Lake and 17 in Hainault Forest. In some areas the population has increased since then, probably due to maturing woodland, for example on Wimbledon Common where the number of territories increased from nine in 2000 to 25 by 2009, and at Brent Reservoir where they increased from one to five over the same period. Usually only small numbers are seen on passage but in 2000 there were 25 at Wraysbury GP on two dates in autumn.

Facts and Figures

The largest count was about 50 birds at Thorpe on 21–22 August 1970.
The earliest migrant was on 27 March 1954 at Epsom Common.
The latest migrant was on 5 November 1988 at Queen Mary GP.
In the cold winter of 1945, one wintered in Hampstead Garden Suburb from January to at least 10 March; this was the first recorded overwintering in the UK.

Barred Warbler *Sylvia nisoria*

A rare vagrant (9)

Range: breeds eastern Europe and Asia, winters Africa; scarce passage migrant to UK.

There have been nine records since the first one in 1972. Eight have occurred in the area between the Lea Valley and the Thames, the exception being in Richmond Park. The first two were both trapped and ringed at Rye Meads. Two thirds arrived in the period from mid-August to mid-September which is the national peak period.

1972: Rye Meads SF, on 19 August
1975: Rye Meads SF, on 16 August
1985: Richmond Park, on 21 September
1986: Buckhurst Hill, on 31 August
1996: Barking Marsh, on 11–13 September
1998: Tottenham Marsh, on 29 August
2003: Sewardstone Marsh, on 1 October
2006: Sewardstone Marsh, on 9 September
2006: East India Dock Basin, on 25–26 September

Lesser Whitethroat *Sylvia curruca*

A common summer visitor and localised breeder

Range: breeds Europe and Asia, winters Africa and Middle East.

Historical

In the early 19th century Montagu stated that they were 'a frequent inhabitant of market gardens near London'. Harting referred to this and Common Whitethroat as 'two of the commonest Warblers we have'. In south London they were variously described by Bucknill as 'abundant, especially on migration' in Dulwich, 'very plentiful' at Richmond and 'not common' at Wimbledon. They bred in large gardens in Brixton and Camberwell up to 1894 and a pair bred in Battersea Park in 1898.

20th century

In the first half of the 20th century there were very few confirmed instances of breeding in central London, the only recorded ones being in Holland Park in 1915 and 1920–21, and in Kensington Gardens in 1921. However, they regularly sang at Holland Park between 1910 and 1940 and were presumed to have bred there in other years; they were also present in Kensington Gardens in subsequent years. In the 1920s Harrisson recorded that in Harrow they tended to arrive a day or two earlier than Common Whitethroat.

Between 1900 and 1950 there was no detectable change to their status apart from in central London where they had ceased breeding by 1940. In the 1940s and 50s there were few breeding records in the suburbs; breeding was recorded on Wandsworth Common in 1942 and at Finchley in 1951. They bred regularly on Hampstead Heath until about 1946 and were present subsequently during the breeding season to at least the mid-1950s. In 1961 one was seen at Ruxley GP on 9–10 December; uniquely, a Common Whitethroat was also present on the latter date.

Since the 1960s the breeding population has fluctuated and this appears to be due more to a natural variance rather than to observer bias. In the 1970s they bred again in central London: at least one pair bred in Regent's Park in 1977–78 and a pair also bred in Kensington Gardens in 1978, the first breeding there since 1921. There were clear signs of an increase in 1981 with records from 71 localities compared to 45 three years earlier; additionally, 106 birds were ringed at Queen Mary Reservoir during the year compared to 27 in 1978.

The second breeding Atlas in 1988–94 found breeding in 391 tetrads in the London Area, an increase of 77 per cent over the first survey in 1968–72. The main increase was in the northern half of London, with an additional spread into the suburbs; however, there was a reduction in the population along the North Downs. Two birds seen during December at Dagenham Chase in the 1990s may have been attempting to overwinter.

21st century

Lesser Whitethroat remains a widespread but sparsely distributed breeding bird where its favoured habitat of light scrub and dense hedgerows can be found. It is difficult to calculate a breeding population as many of the singing males recorded each spring are migrants and breeding birds can be quite elusive, but there are likely to be fewer than 100 breeding pairs. They can be common on passage in some years and overwintering birds were seen at Chafford Hundred in 2004/05 and at Seventy Acres Lake in 2008/09.

Facts and Figures

The earliest migrant was on 4 April 1994 at Bushy Park.
The largest count was 40 birds at Wraysbury GP on 19 August 1995.

Common Whitethroat *Sylvia communis*

A common summer visitor and widespread breeder

Range: breeds Europe, Asia and North Africa, winters Africa.

Historical

This species was so common in the rural districts of the London Area during the 19th century that Harting wrote 'no one, while taking a country walk in June, can have failed to notice this noisy little bird, sometimes on a high spray, sometimes dancing and jerking in the air, pouring forth its garrulous song'. A pair was recorded nesting in a hanging flower-basket in St John's Wood in 1862.

20th century

Although they were abundantly common in the early years of the century there was the occasional dip in their numbers. In 1909 it was noted that they were less numerous in the Sevenoaks district than 10 years previously. In 1935 Glegg stated that it was 'one of the commonest of the summer residents', adding that it was only seen on passage in central London and did not breed in the suburbs. In the 1940s the only regular breeding site in the inner suburbs was Wimbledon Common where it was quite abundant. There were two records of birds being seen in winter in the UK prior to 1950, one of which was in the London Area, at Poyle on 17 January 1948.

By the 1950s they bred no closer to central London than Greenwich, Shooters Hill, Eltham Park, Dulwich Woods, Streatham Common, Mitcham Common, Wimbledon Common, Richmond Park, Hanger Hill, Horsendon Hill, Hampstead Heath and Mayesbrook Park. Inside this area they were just a passage migrant apart from in 1953 when a pair bred successfully in Regent's Park, the first confirmed breeding record in Inner London. In 1960 two pairs bred in the inner suburbs at Old Ford in Bethnal Green. A bird with a damaged wing was seen from 23 November to 28 December 1958 at Ruxley GP.

There was a major crash in the population during the winter of 1968/69 due to a drought in the species wintering area in the Sahel and very bad weather around the Mediterranean during the return spring migration. Examples of the declines in the breeding population came from 10 Common Bird Census plots where there was a reduction from 89 territories to just 18 in 1969, a decline of 79 per cent. This matched the national decline of 80 per cent. At Brent Reservoir there were 24 pairs in a study area in 1968 but only three the following summer.

At the start of the 1970s the population had already started to recover and by 1979 the species had returned to its previous status as the more common of the two whitethroats; the population was at its highest since the crash. During the 1980s the population continued to increase and went from at least 132 reported territories in 1980 to almost 300 by 1989.

In 1990 the largest breeding concentration was at Rainham Marshes where there were 40 territories; three pairs bred in Inner London in Surrey Docks. By 1993 some areas of east London recorded that the populations had recovered to 'pre-crash' levels although this was not noted throughout the London Area. The second breeding Atlas showed a modest increase of 5 per cent compared to the first Atlas in 1968–72. However, the first year of that survey was before the big crash so the population was already at a high level. During the second survey they were breeding in 65 per cent of all the tetrads in the London Area with most of the gaps being in the urban areas. In 1996 a pair bred in Regent's Park, the first breeding there since 1953; three pairs bred in central London the following year. The population continued to expand and there were, for example, 164 territories at Rainham in 1999, a 400 per cent rise in 20 years.

21st century

By the beginning of the 21 century this warbler fully justified its 'common' prefix; in 2000 more than 1,000 singing males were located, including 193 at Rainham Marshes and 140 at Ongar Park Wood. Since then they have increased at some locations; for example at Tottenham Marshes the number of territories increased from 23 in 2008 to 36 in 2009. Since 2000 they have been largely absent from central London with only a few migrants seen on passage, mostly in the large parks.

Facts and Figures

The earliest arrival was on 23 March 1957 at Brent Reservoir.
The latest migrant was on 17 December 2001 at Island Barn Reservoir.
One overwintered at Walthamstow Reservoirs in 2011/12.

Dartford Warbler *Sylvia undata*

An uncommon partial migrant and scarce breeder

Range: western Europe and North Africa; localised in southern UK.

Historical

The first specimens found in this country were a pair shot at Bexley Heath, near Dartford, on 10 April 1773, from where this species received its English name. After this they were found on gorse-covered commons in many southern counties. The only other 18th-century record in the London Area was in 1783 when several were shot on Wandsworth Common in the winter.

It was much more widespread in the London Area in the 19th century although, as it is now, it was still restricted to areas where there was sufficient gorse. In north-west London it was recorded breeding at Finchley, Hampstead, Highgate, Harrow Weald Common and Stanmore Common, and away from these areas it was seen or obtained on Hounslow Heath, Old Oak Common and Wormwood Scrubs. In the 1840s and 50s it was relatively common at Hampstead and Highgate but the last pair was seen there in 1872, the last record from north London in the 19th century.

In south-east London they could be found on Blackheath, Bromley, around Dartford, Hayes, Keston and Shooters Hill but had decreased due to the spread of urban London, increased disturbance, gorse fires and severe winters. By 1830 they had disappeared from the Dartford district but one or two pairs continued to breed on Hayes Common until at least 1886. In south-west London they were resident on most commons and heaths, although by 1836 they had become extremely scarce in Tooting as a result of collecting.

By the 1880s a terminal decline had set in and they were last reported from Kew Gardens in about 1880, at Wandsworth Common in the autumn of 1881 and on Hayes Common in 1891. However, they still frequented Wimbledon Common in the 1890s.

20th century

By the beginning of the century Dartford Warbler was virtually extinct in the London Area. In 1908 two or three pairs were discovered on Putney Heath and a pair bred there in 1910; they remained there until at least 1913. Between 1912 and 1915 one or two were found on Banstead and Walton Heaths in winter.

There were no further records until 1927 when a pair was seen again on Walton Heath and the birds probably bred there the following year. Breeding occurred several more times in the 1930s and the population on the North Downs had reached eight pairs by 1938 but after a very cold winter the numbers fell to three pairs. A pair also bred on Wimbledon Common in 1936. With a succession of cold winters in the early 1940s and the ploughing up of heathland as part of the Second World War effort, they became extinct again in the London Area until the 1970s.

After a 33-year absence, one wintered at Rainham Marshes in 1972/73; another four were seen during the 1970s, including a juvenile ringed in the New Forest and killed by a cat in Barnes on 1 November 1975. For much of the following decade they remained equally rare but in 1989 a pair bred in south London, the first breeding record for over half a century. They also started to occur during winter with one present for much of February 1989, and one overwintered at Thamesmead in 1989/90. Although the breeding pair remained until the following summer it did not breed and there were no further instances of breeding during the 1990s. However, they did establish themselves as scarce winter visitors and were also seen on passage; in 1994 a singing male was seen at Hampstead Heath on 1 May, the first record there for over 120 years. On several occasions they were seen with Stonechats.

21st century

At the beginning of the 21st century Dartford Warbler was attempting to recolonise south London. One or two pairs bred annually on the North Downs up to 2003 but subsequently disappeared. A pair bred at a site in west London from 2003 to 2005 but since then none have bred anywhere in the London Area. Regular overwintering sites include Bushy Park, Rainham Marshes and Richmond Park with a few others elsewhere; they are occasionally seen on migration, often accompanying Stonechats.

Subalpine Warbler *Sylvia cantillans*

A very rare vagrant (4)

Range: breeds southern Europe and North Africa, winters Africa; scarce visitor to UK.

There have been four records in the London Area, three of which were of singing males in spring. The first occurred at Orpington in 1976 and this was followed by one on Walthamstow Marsh in 1994, part of a national influx with 25 others seen in May, all on the coast. In 1996 a male was present in a garden at Hampton on 15 August, an unusual date for this species in the UK and certainly not in typical habitat. A first-summer male at Lonsdale Road Reservoir in 2003 was probably of the eastern race *albistriata*.

1976: Darrick Wood, Orpington, on 9 May
1994: Walthamstow Marsh, on 15 May
1996: Hampton, on 15 August
2003: Lonsdale Road Reservoir, on 21 April

Sardinian Warbler *Sylvia melanocephala*

A very rare vagrant (1)

Range: southern Europe, North Africa and Middle East; vagrant to UK.

The only record occurred in 1992 when a female was trapped and ringed at Hogsmill SF; there had been a small influx into the UK that spring with four others seen.

1992: Hogsmill SF, adult female trapped on 2 June

Grasshopper Warbler *Locustella naevia*

An uncommon summer visitor and localised breeder

Range: breeds Europe and Asia, winters Africa.

Historical

The earliest record was in August 1823 when one was caught in a Nightingale trap near Grosvenor Place; this was the only record in Inner London for almost 150 years.

During the 19th century it was a widespread breeder on marshes and commons across the London Area but had started to decline as these areas were 'improved'. It last bred on Streatham Common in the 1830s and at Tooting in 1880 but was considered to be relatively abundant elsewhere south of the Thames. In 1832 Doubleday noted just one pair in Epping Forest but by 1885 it was apparently common in the area.

Harting frequently observed it in north-west London; he knew of nests that had been taken at Harrow, Bishop's Wood, Hampstead, Kingsbury and Stanmore Common. Swann knew of a nest in Highgate Woods in 1890 and they were still breeding on Hampstead Heath in 1898.

20th century

Throughout the 20th century there was a pattern of decline; on Wimbledon Common it was thought to be breeding up until 1909; it had disappeared from Bexley, Blackheath and Shooters Hill by 1909; Fitter considered that it was lost as a breeding bird in the inner suburbs by the end of the First World War; and at Enfield Chase it bred regularly until 1924.

During the 1930s it was still thought to be breeding in parts of north-west London and was recorded annually at Ruislip until its nesting area was bulldozed in 1942. Large parts of Epsom Common were also destroyed during the war and Grasshopper Warblers declined there from 15–20 pairs in 1938 to just four pairs in 1948.

The number of migrants heard singing in spring had declined rapidly and very few were reported between the end of the war and 1954. In the 1950s there were only three areas where the species regularly bred: on Ashtead Common and Bookham Common, and in the Colne Valley between Uxbridge and Rickmansworth. However, there was a bumper year for the species in 1959 and 50 singing birds were recorded. During the 1960s and 70s the population fluctuated across the London Area, even at the two main sites, Ashtead Common and Broxbourne Woods.

By the beginning of the 1980s the breeding population was about 27 pairs with 10 at Ashtead Common and six in Broxbourne Woods, but by the end of the decade there were territories at neither site, and the main population was in the Lea Valley. Between the two breeding Atlases there was a decrease in the distribution in the London Area of 65 per cent; during the second survey in 1988–94 almost all of the breeding population was situated beyond the suburbs. In 1995–96 there were up to 12 territories at Wraysbury GP.

21st century

Since the beginning of the 21st century the breeding population has remained at a very low level with about 15–20 territories each year; most of these are in the area between the Lea Valley and the Lower Thames Marshes. Some previous breeding sites are no longer suitable, particularly around gravel pits where low bushes have given way to dense scrub. They typically begin to arrive in the last 10 days of April and, because of their skulking nature, few are seen in autumn.

Facts and Figures

The earliest arrival was on 3 April 1999 at Walthamstow Marsh.
The latest date was 22 October 2006 at Hilfield Park Reservoir.

Savi's Warbler *Locustella luscinioides*

A very rare summer visitor (4)

Range: breeds Europe, Asia and North Africa, winters Africa; rare breeder in UK.

There have been only four records in the London Area, all of singing males in the Lea Valley. The first was present for six days at Stanstead Abbotts in 1979. At the time a few pairs were breeding in Kent and occasional birds were seen elsewhere in southern England. Two years later there was one at Cheshunt for two days and in 1989 another spent four days at Rye Meads, the third sighting in the Lea Valley in 11 years. Before the next sighting 20 years later the small breeding population in the UK became extinct and since 2000 they have become nationally rare.

1979:   Stanstead Abbotts GP, on 22–27 April

1981: Cheshunt GP, on 20–21 May
1989: Rye Meads, on 2–5 May
2009: Seventy Acres Lake, on 30 April–30 May

Unacceptable Records

A nest was allegedly found at Dagenham in May 1850 but this was dismissed by Christy who knew that the finder was a dealer in rare bird specimens. A nest with eggs was allegedly collected at Erith Marshes in May 1853 but again this was by a disreputable dealer and Ticehurst did not accept this record.

Icterine Warbler *Hippolais icterina*

A very rare vagrant (2)

Range: breeds Europe and Asia, winters Africa; rare breeder and scarce passage migrant in UK.

Although this is a scarce, rather than rare, migrant in Britain, most are found on the coast and they are rare inland; there have only been two occurrences in the London Area. The first bird was found by the causeway at Staines Reservoir in autumn 1965 and the only other record was of a singing male in Cheam in 1983.

1965: Staines Reservoir, on 9 August
1983: Seear's Park, Cheam, on 10 June

Melodious Warbler *Hippolais polyglotta*

A rare vagrant (10)

Range: breeds Europe and North Africa, winters Africa; scarce visitor to UK.

There have been 10 records in the London Area, a relatively high total for an inland county (even in Essex there have only been 10 records). They have been split equally between spring and autumn.

The first bird was seen at two sites in the Lea Valley in 1961. Three years later one was trapped and ringed at Navestock. They continued to occur irregularly after this. There were three in the 1980s, all in spring: in 1981 one was trapped and ringed at West Thurrock; in 1983 a singing male at Ruxley GP arrived a day after a singing Icterine Warbler elsewhere in the London Area; and in 1987 a singing male was present for three days at Croham Hurst Golf Course. There was only one in the 1990s, but three have occurred since the beginning of the 21st century, including a singing bird in a tiny patch of scrub in Leyton, surrounded by an industrial estate, housing and Leyton Football Ground.

1961: Rye Meads, on 12–13 August; also seen at Nazeing GP
1964: Navestock, on 10 August
1973: Romford, on 20 September
1981: West Thurrock, on 26 May
1983: Ruxley GP, on 11 June
1987: Croham Hurst, on 30 April–2 May
1990: Barn Elms Reservoirs, on 5 May
2000: Regent's Park, on 23 August
2004: Fairlop Waters, on 1–2 October
2012: Leyton, on 16–17 May

Icerine/Melodious Warbler *Hippolais icterina/polyglotta*

(3)

There have been three records of these similar species which were not specifically identified: singing near Connaught Water on 25 April 1948; on the causeway of Queen Mary Reservoir on 18 September 1984; and in Richmond Park on 10 October 1991.

Aquatic Warbler *Acrocephalus paludicola*

A rare passage migrant (13)

Range: breeds eastern Europe and Asia, winters Africa; scarce autumn visitor to UK.

The first record was in August 1924 when one was seen on the causeway of Staines Reservoir. No more were seen until the 1950s when there was a succession of records, three at sewage farms and one at Brent Reservoir.

The 1960s was the peak with five recorded, all of which were in August. The only record in the 1970s was also the first record in Inner London where one was present at Surrey Docks for two days in September. The last two records occurred at Rainham Marshes in 1981 and 2008.

1924: Staines Reservoir, on 6 August
1951: Perry Oaks SF, on 25–26 August
1955: Perry Oaks SF, on 29 July
1955: Brent Reservoir, on 28 August
1959: Beddington SF, trapped on 20 September
1960: Hilfield Park Reservoir, on 14–15 August
1964: Sevenoaks, on 21 August
1965: Beddington SF, on 17 August
1965: Perry Oaks SF, on 30 August
1966: Queen Mary Reservoir, on 20–21 August
1977: Surrey Docks, on 23–24 September
1981: Rainham Marshes, on 12 August
2008: Rainham Marshes, on 3–8 September

Sedge Warbler *Acrocephalus schoenobaenus*

A common summer visitor and widespread breeder

Range: breeds Europe and Asia, winters Africa.

Historical

Harting described this as 'a noisy little bird, singing all day and throughout half the night'. It was common in his time as he related how 'the marshy banks of the Thames, on either side of the river, where beds of willows or reeds abound, are well stocked with this bird'. Towards the end of the 19th century they were breeding on Hampstead Heath.

20th century

At the beginning of the century it was considered to be a common and widespread breeder in suitable habitat although it rarely penetrated into the inner suburbs; however, in 1902 Pigott noted that pairs had been seen in Hyde Park and Regent's Park. In south-east London they bred as close to the city as Deptford and Bermondsey.

In the 1930s Glegg recorded that it had recently colonised the banks of Staines Reservoir. Although the population was generally stable there were some local losses. After 1930 the birds ceased breeding at Twyford Abbey on the River Brent when the river was both diverted and polluted; at Chiswick Eyot, where they were previously common, they were not seen after 1934; and after 1938 they no longer bred at Epsom SF. On the other hand, breeding was first recorded at Brent Reservoir and Hampton Court in 1934.

By the 1950s there had been little change and the species was still widespread, breeding on the marshes at Abbey Wood and Plumstead although it was less common there than Reed Warblers as most of the area was reed beds. Sedge Warblers were absent from most of the suburban area, central London and across the North Downs where there was no suitable habitat. In the mid-1960s Brent Reservoir was the only breeding site in the inner suburbs. In 1976 there were 72 territories at Rye Meads and a pair bred at Surrey Docks, the first breeding record in Inner London.

In 1980 there were at least 367 territories, including more than 260 in the Lea Valley. The second breeding Atlas in 1988–94 showed that the breeding distribution had changed very little since the first survey and that the species occupied 22 per cent of all tetrads in the London Area. The majority of the population was situated in the Thames corridor and in the Colne and Lea Valleys. At the start of the 1990s a minimum of 382 territories were recorded, showing just a small increase in 10 years; the highest concentrations were 95 at Rye Meads and 74 at Rainham Marshes. Intensive survey work at Rainham from 1995 to 1999 showed an increase in territories of from 105 to 156.

21st century

Since the start of the 21st century Sedge Warbler has remained a reasonably widespread breeding bird. The largest population is at Rainham Marshes, where it had increased to 179 territories by 2000. Within the suburban area the stronghold is on Tottenham Marshes with small numbers breeding at other sites. It is occasionally recorded on migration away from breeding areas, sometimes even in the central London parks. Most birds arrive in April and the majority have usually left by the end of August; although they are not unusual in September they are rare in October.

Facts and Figures

The earliest migrant was on 15 March 2007 at the London Wetland Centre.
The latest migrants were on 20 November 2004 at Rye Meads and 20 November 2011 at Rainham Marshes.

Paddyfield Warbler *Acrocephalus agricola*

A very rare vagrant (1)

Range: breeds eastern Europe and Asia, winters Indian subcontinent; vagrant to UK.

The only record occurred in 1999 when one remained for three days in the Lea Valley, at the Cheshunt Bittern Watchpoint. Typically it left just before the weekend although during its stay it was still seen by around 800 people. There have been very few inland sightings of this vagrant.

1999: Seventy Acres Lake, on 26–28 October

Blyth's Reed Warbler *Acrocephalus dumetorum*

A very rare vagrant (2)

Range: breeds eastern Europe and Asia, winters Asia; vagrant to UK.

There have only been two accepted records in the London Area. The first of these was in 2001 for three weeks in October at the unlikely setting of Canary Wharf; the tight security restrictions in place at the time meant that news of this bird could not be released. The second record occurred just two years later and involved a singing male at Fishers Green. Although it was suspected to be this species when first found, its identity was not confirmed until very late in the day.

2001: Canary Wharf, on 6–28 October
2003: Fishers Green, on 16 June

Marsh Warbler *Acrocephalus palustris*

A scarce summer visitor and rare breeder

Range: breeds Europe and Asia, winters Africa; rare breeder in UK.

Historical

In the 19th century there was a series of odd breeding records of Reed Warbler in the London Area that may all have been Marsh Warblers. The first occurred in a poplar in Fulham and was documented by Sweet in 1823. In Hampstead, 'Reed Warblers' bred in large gardens far from water from 1861 to 1865. The first year one pair bred in a *Corchorus*, and in the following years they bred in lilac bushes, with at least four pairs in 1863. One of these birds was shot and when compared with Reed Warbler specimens was found to differ in several respects, notably in the longer wings and tail and shorter tarsus. In 1862 at least two pairs of 'Reed Warblers' bred in lilacs in London Zoo. The following year a pair bred in some lilac bushes in Ealing at least 1 kilometre from the nearest stream. Prior to 1866 a pair was recorded breeding annually in lilac bushes in a garden in Little Ealing. In 1867 a nest was found in a lilac in a garden in Teddington, and a pair bred in a lilac at Hampton Court Park for successive years in around 1879.

Harting thought that these birds were interesting and warranted further investigation; he speculated that they might be a distinct species. These records were reviewed by Fitter who concluded that they were more likely to be Marsh Warblers than Reed Warblers. It seems likely that all of these breeding records in atypical Reed Warbler habitat relate to Marsh Warbler, which was unknown in the UK at the time. Further evidence of this is the specimen taken in Hampstead which was directly compared with a Reed Warbler at the time and described by Harting. There were no further records of 'Reed Warblers' breeding in gardens before the end of the century.

20th century

The first time Marsh Warbler was properly recognised in the London Area was in 1903 when a nest was found in an alder by the River Colne at Harefield. There was another breeding record in 1905 in Chalfont Park and two years later two nests were located in a bed of willows, nettles and other rank vegetation near Thorpe. These breeding records add further weight to the 19th-century records and suggest that the species was a scarce but widespread breeder.

Apart from one singing in Kensington Gardens on 5 June 1924 there were no further records until 1931 when a pair bred again at Chalfont Park. The next year a male was on territory in north-west London but no more were seen until 1949 when a male was present on Plumstead Marshes on 6 June 1949. During the 1950s they were extremely local in south-east England and were only known to be breeding regularly from one site in Kent and a few localities in Sussex. In the London Area, breeding occurred on Swanscombe Marshes in 1952 and in Gatton Park in 1958.

In the 1960s there was a series of records from Rye Meads. One was trapped in 1960 and the biometrics matched Marsh Warbler. In 1961–62 at least four pairs bred in overgrown ditches and long grass, and their habits and songs were reported to be different from those of Reed Warbler. At the time they were considered to be Marsh Warblers but following a review of all of these records they are no longer accepted as that species (Gladwin and Sage, 1986) and at best could be described as being inconclusive. Elsewhere, there were six records, three of which were at Sevenoaks. During the 1970s it was thought that there was a colony at Sevenoaks; breeding occurred from 1971 to 1973 with up to 10 breeding pairs. In 1974–75 only one singing bird returned. There has been some speculation about the identification of these birds, despite some being trapped and ringed, and it was believed that they were in fact Reed Warblers with unusually mimetic songs.

Marsh Warblers were exceptionally rare in the 1970s and 80s and the only genuine sightings were at Stocker's Lake in 1976 and 1978, and at a site in south London in 1980. In 1990 there was an influx into the country and two singing males were seen in south-east London, the first for 10 years. This was the start of an increase of sightings in the London Area with several territorial birds present over the next few summers, leading to successful breeding at Barking Bay in 1996. The following year a pair summered and may have bred but only single males were seen in 1998–99.

21st century

In the first few years of the 21st century there were occasional breeding records. In 2001 three pairs bred: at Dagenham Chase, Rainham Marshes and the London Wetland Centre. Two males held territory in 2002 and may have bred at Rainham.

Since then they have become scarce summer visitors with the only records being singing males at Barking Bay in late May 2004; at East India Dock on 15 May 2004; and at Rainham Marshes from 2 to 4 June 2007. An autumn migrant was also seen at Rainham on 14 September 2008.

Reed Warbler *Acrocephalus scirpaceus*

A common summer visitor and widespread breeder

Range: breeds Europe, Asia and North Africa, winters Africa.

Historical

This was first described as a British bird in 1783 by Latham from specimens taken at Dartford, and also by Lightfoot who received the nest and eggs of this species that had been taken along the River Colne near Uxbridge; he then saw a nest there in July. Latham later noted that they were not uncommon on Erith Marshes.

Compared to Sedge Warbler, this species was more localised in the 19th century and, according to Harting, was 'seldom seen in the north and north-west portions of the county, although common along the Thames and the Colne'. In the 1830s it bred annually in reed beds in Regent's Park, along the Thames at Battersea and in the extensive marshes between Hammersmith and Shepherd's Bush. In the 1840s it was known from Rotherhithe, although it had decreased there since the 1830s when a railway was built on much of the suitable breeding habitat. Around the same time it also bred in reed beds along the Thames between Greenwich and Erith, and on the north side at Orsett. In 1844 Meyer stated that it was abundant on both shores of the Thames and on some of the islets.

There were breeding records from central London in Battersea Park, Regent's Park and St James's Park; however, as there was no record of the habitat they bred in, they could easily have been Marsh Warblers. In the 19th century there were several documented records of 'Reed Warblers' breeding in gardens across parts of the London Area, especially in lilac bushes, that were in fact Marsh Warblers (a species unknown in the UK at the time). These have been omitted from this account where it is apparent that the identification was incorrect.

20th century

Reed Warblers bred on Hampstead Heath until the early years of the century when Bog Pond was cleared of vegetation. Elsewhere, they bred along the Thames and started to expand along the Colne. Breeding was first recorded in the Lea Valley at Walthamstow Reservoirs in 1912; there was a colony of up to 11 pairs which bred in privet and elder but all the bushes were cut down in 1934. However, given the nesting habitat it is a possibility that these were Marsh Warblers.

In the 1930s they were widely distributed but fairly localised, breeding mainly in the more rural parts of the county, along the rivers, at sewage farms and at larger ponds. The largest colony was at Erith Marshes where 20 males were singing in 1938. Glegg recorded that they were occasionally seen on passage in spring in central London, especially in Kensington Gardens.

In the 1940s and 50s they were abundant in some areas, such as the Thames-side marshes from Dagenham and Plumstead eastwards; on one 3.5-hectare reed bed in Abbey Wood Marshes there were estimated to be 130–150 pairs in 1949. They were common in the Lea Valley where there were about 100 pairs in 1954, and they bred throughout the Colne Valley. The only breeding pairs in the suburbs were at Barn Elms Reservoirs, Beddington SF, Brent Reservoir, Hampstead Heath and Richmond Park.

The first breeding in Inner London in the 20th century was in Surrey Docks in 1977. During the 1980s the population expanded considerably; in 1980 it was at least 450 pairs and by 1986 it had increased to 706 territories. At the start of the 1990s the breeding population continued to increase. This was at least partially due to increased habitat, for example reed planting at Brent Reservoir where the numbers increased to 43 pairs in 1990 from 28 the previous year. One or two pairs still bred at Surrey Docks but these disappeared as the site was redeveloped, and a pair bred at Camley Street Natural Park.

The second breeding Atlas in 1988–94 showed an increase of almost 60 per cent in the breeding distribution across the London Area. The main population was located on the Lower Thames Marshes, along the rivers of east London and in the Colne and Lea Valleys. By the end of the decade there were more than 1,000 territories, including 503 at Rainham Marshes.

21st century

There were more than 1,200 territories in 2000 and the breeding population has continued to increase. The increase is mostly due to habitat creation, particularly as reed beds have developed in disused gravel pits, but there has also been a general increase. For example, at Rainham Marshes the population increased from 212 territories to 517 in the period 1986–2000. They have also begun to colonise central London following the creation of several small reed beds; there were two pairs in Victoria Park in 2000; in Regent's Park the breeding population had increased from one pair in 2001 to four pairs by 2009; and they have begun to breed at Canada Water, one of the redeveloped Surrey Docks.

Apart from breeding birds they are commonly seen on migration. The first arrivals are usually in the middle of April with the majority arriving in May; in autumn most have departed by the end of September with a few being seen during October.

Facts and Figures

The earliest migrant was on 1 April 2008 at Richmond Park.
The latest migrant was on 24 November 2003 at Rye Meads.

Waxwing *Bombycilla garrulus*

A scarce winter visitor

Range: breeds northern Europe, Asia and North America, European population winters central and western Europe.

Historical

The first record was of a bird killed at Eltham in the winter of 1781; this was followed by a pair shot at Hanwell in December 1783. The female was killed but the male was only wounded and lived for some time in a menagerie in Osterley Park.

Only a few were seen in the first half of the 19th century; some were shot around Camberwell in late December 1803; several were shot at Tooting in 1831; and there was one near Chertsey on 17 January 1847. There was a particularly large influx in Britain in 1849/50 and during this invasion a considerable number were obtained in the southern and south-eastern counties of England; in January no less than 586 are recorded to have been killed, including birds shot at Eltham, Kilburn, Harrow, Rainham and Wimbledon.

Further national influxes occurred in 1866/67 and 1872/73, although there were no London Area records from the latter period. In November 1866 a dozen were seen on Hampstead Heath. The last one of the century occurred at Epsom in November 1892.

20th century

Early in the century more than a dozen were seen at Hampstead Heath on 3 January 1902. Up to the end of the Second World War there were a number of irruptions into Britain that produced the London Area sightings. Small numbers were seen in the winters of 1903/04, 1921/22 and 1936/37; and larger numbers were recorded in 1913/14 and 1943/44; during the latter the maximum count was 26 at Hillingdon on 7 April. The only irruption that failed to produce a London Area record was in 1931/32.

A large invasion occurred in 1946/47 and more than 12,000 were recorded in the UK. About 500 were seen in the London Area, mainly in small parties, with the highest count being 60 at Watford on 19 February. There were further irruptions in 1948/49 and 1949/50. Waxwings were also recorded in non-invasion years; between 1900 and 1954 they were seen in 19 of these years.

After 1949/50 there were only a few small winter influxes until October 1965 when there was a large irruption in Britain and many were seen in the London Area. They were fairly scarce during the 1970s and 80s apart from during the winter of 1970/71 when they were recorded at more than 30 London Area localities, including a flock of 60 at Westerham. Waxwings remained scarce until early 1996 when about 1,000–1,500 were seen in the London Area, mostly between February and April; the largest flock was of 215 birds at Gallows Corner/Harold Hill on 20 March with no other flocks exceeding 90.

21st century

Since 2000 they have become far more regular in the UK and this has been reflected in the London Area. The first influx was in early 2001 when about 292 were seen, and there was an exceptionally large influx in the first four months of 2005 with sightings at 103 locations; the largest flocks were of 160 birds in central London at Warren Street; 160 between South Ockendon and Grays; and 150 at Greenhithe. The London Area's largest-ever influx took place in 2010/11. Large flocks arrived in the London Area directly from the Continent rather than filtering down the country as in previous winters, and at the start of 2011 the London Area held more Waxwings than most other counties. They remained for several months, with some still being seen into May.

Facts and Figures

The latest one recorded was on 4 May 2011 at Vicarage Farm.
The largest flock was of 367 birds at Lakeside on 23 January 2011.

Nuthatch *Sitta europaea*

An uncommon resident and widespread breeder

Range: Europe, Asia and North Africa.

Historical

This species was described by Harting as 'so thoroughly a wood bird, that in this county it is local rather than rare'. Nuthatches were present in many of the outer districts of the London Area where there was sufficient woodland. They have bred on Hampstead Heath ever since they were first recorded there in the 19th century.

There was some evidence of a decline after the middle of the 19th century, possibly due to urbanisation, although Fitter speculated that it may have been as a result of soot deposits on the trees. Harting recorded how the species was once comparatively common around Ealing but had recently become much rarer. Yarrell reported it as being common in Kensington Gardens in 1843 but it had become only a casual visitor there by 1879 and had disappeared from central London by the end of the 19th century. It was also plentiful in Hanwell Park but had become scarce after the area had been developed. Hamilton was one of the first people to attract Nuthatches into his garden in Putney by wedging nuts into a tree in 1881.

20th century

Early in the century it was still present in the suburbs but had become rare in the Willesden district by 1907. In Regent's Park it reportedly bred annually until at least 1911.

By the 1930s its status was unchanged in the outer districts but it was no longer present in some of the inner suburbs where it had previously occurred. Nuthatches had disappeared from central London with only three sightings during the 1940s, including a rather lost individual on the fire escape of a building near Cannon Street in August 1946.

In the early 1950s they were still fairly common throughout most of the outer suburbs although they were scarce in some areas even where there was suitable habitat, such as to the south of Dartford, in Hainault Forest, and around Romford and Brentwood. They became extinct in Greenwich Park in 1953 and it was thought this was due to the pollution caused by a nearby power station. By the end of the 1950s they bred in Battersea Park, Holland Park and Kensington Gardens but this was short-lived as they had ceased breeding in central London by 1970. A pair bred in Kensington Gardens in 1976, the first breeding record there since 1964.

During the 1980s some local populations began to increase and there were rises of 50 per cent at Hainault Forest and Osterley Park. In Epping Forest the population was in the order of 150–200 pairs. Increases were later seen at other sites; for example, on Hampstead Heath the population increased from at least 12 pairs in 1990 to 25–30 pairs in 1994. The second breeding Atlas in 1988–94 found evidence of breeding in almost half of the London Area tetrads and breeding distribution had increased by 25 per cent since the first survey in 1968–72. They were found breeding in more suburban areas of south London than of north London.

21st century

Since the start of the 21st century up to 70 breeding pairs have been reported each year although the actual population is considerably higher than this. The largest monitored population is on Wimbledon Common where there are at least 20 pairs. In inner London, one or two pairs are recorded in Kensington Gardens most years. They are more widespread outside the breeding season but remain relatively scarce in the suburbs away from breeding sites.

Treecreeper *Certhia familiaris*

An uncommon resident and widespread breeder

Range: Europe and Asia.

Historical

In the mid-19th century Harting stated that Treecreepers were 'resident throughout the year, although not very abundant'. However, in the more wooded south London, Bucknill considered them to be fairly abundant.

There was some evidence of a decline later as Hamilton wrote in 1879 that it was formerly resident in Kensington Gardens.

20th century

In the 1930s Glegg described it as 'a common resident, well distributed over the rural parts of the county which are suitable to its nesting habits'. He added that it was not regularly found in suburban areas and was only an occasional visitor to central London outside the breeding season. It started to return to central London in the 1940s and a pair bred in or near Kensington Gardens in 1945 and 1947.

In the 1950s it was considered to be thinly and unevenly distributed across the London Area and only common in the rural wooded areas. In the inner suburbs it was present in the breeding season at North Ealing, Streatham, Tooting Bec and Wimbledon Common but was more widespread at other times of the year. In central London it was resident in Kensington Gardens but rarely seen anywhere else, but by 1960 it was no longer resident in central London and remained rare until 1968.

During the 1970s and 80s the population increased in central London and by 1990 there were around seven pairs.

In Epping Forest the population was estimated to be in excess of 200 birds in 1988. In 1994 there were about 25 territories on Hampstead Heath and 18 at Hainault Forest; populations at both of these sites had increased significantly since the start of the decade. The second breeding Atlas in 1988–94 showed a 35 per cent increase in the breeding distribution since the first survey in 1968–72, but they appeared to be absent from much of the suburban area.

21st century

There has been little change in the overall population since the beginning of the 21st century and they remain relatively uncommon, but in central London they appear to have decreased as only one pair was recorded in 2010. Away from breeding sites they are only occasional visitors despite plenty of apparently suitable habitat.

Facts and Figures

Eight or nine were seen on one tree in Rickmansworth on 4 April 1940.

Short-toed Treecreeper *Certhia brachydactyla*

A very rare vagrant (1)

Range: Europe and North Africa; vagrant to UK.

The sole record occurred in 1975 when one was located in Epping Forest by its distinctive song. It was seen by one observer and is the only UK record not trapped or photographed as well as being the only one recorded inland. It does not fall into the typical pattern of late autumn migrants and occurred during a period when several birds were reported in Epping Forest but only this record was accepted. This was just the sixth British record at the time; most birds are trapped at coastal observatories.

1975: Epping Forest, on 26 May

Wren *Troglodytes troglodytes*

A very common resident and widespread breeder

Range: Europe, Asia, North Africa and North America.

Historical

Harting knew this as a common and widespread breeding bird despite it suffering from having its nests robbed.

20th century

In the 1930s it occurred throughout the suburbs and was found everywhere in the rural areas. Glegg stated that a few bred in central London where they were resident, although there appeared to be a temporary increase in numbers in autumn; this movement into central London was also noted in later years. After the severe winter of 1939/40 some observers noted a reduction in the breeding population. In 1949 Fitter noted that it had ceased breeding in the parks in central London.

In the 1950s it was considered to be common and widespread, and even bred in areas with little cover such as the Thames-side marshes. In central London there had been a recovery by 1954 and Wrens were breeding regularly in Battersea Park, Kensington Gardens, Regent's Park and probably St John's Wood. There was little proof of occupancy in the more built-up areas apart from at Hampstead and Peckham Park but they were known to be breeding in many large suburban gardens and open spaces. In the autumn of 1959 there was an influx into the City of London with a peak count of eight in Cripplegate on 16 October and up to four at various other sites; elsewhere in the London Area this influx went unnoticed apart from at Sewardstone where numbers increased from about 50 to about 200 in the third week of November. A similar influx was noted in 1961.

The population was severely reduced during the extremely cold winter in early 1963. Wrens were completely eliminated from some areas; for example, the regular 10 pairs at Downe were wiped out, as were the seven pairs in Dollis Hill; at Bookham Common there had been 12–17 territories in a 13-hectare woodland but none were located after the winter until a solitary male was heard in late May. In the inner suburbs and more heavily built-up areas the breeding population was much less affected. Within two years the population had virtually returned to the level before the big freeze and continued to increase until cold weather in the winters of 1976 and 1979 caused another crash. Reductions in many local breeding populations occurred in 1986 after a severe winter; for example, on one census plot in Osterley Park there was a 75 per cent crash.

At the start of the 1990s the breeding population was continuing to increase at most sites where comparative numbers were available. Following a prolonged spell of cold weather in February 1991 there was a reduction in the breeding population; the largest decline reported was at Cheshunt GP where only 35 singing males were present compared to 84 in the previous year. The second breeding Atlas in 1988–94 showed breeding in 98 per cent of the London Area tetrads. There were increases in central London and in east London, though the latter area was probably under-recorded during the first survey in 1968–72. The cold weather of 1995/96 caused reductions in local populations of up to 50 per cent, but over the next three years there were large increases in the population.

21st century

The Wren has remained a widespread breeding bird throughout the London Area during the first decade of the 21st century due to its adaptability and the range of habitats it can be found in – its loud song can be heard in woodlands and large parks as well as suburban gardens, garden squares, churchyards and scrub, wherever suitable low cover is available. The Breeding Bird Survey has shown a significant increase in the London population of 41 per cent from 1995–2010 compared to the national population which recorded a small decrease of two per cent over the same period.

Unusual Breeding Records

Breeding in winter is almost unheard of but a nest with two eggs was taken on 25 December 1873, at Kingsbury. Wrens are known for nesting in strange places and the following have been recorded in the London Area: an old shirt on a thorn-bush; the body of a Carrion Crow which was being used as a scarecrow; and old cans and kettles put up in the Brent Valley Bird Sanctuary. A pair also bred on a second-floor balcony in Paddington Green in 2001.

Starling *Sturnus vulgaris*

A very common resident and widespread breeder

Range: Europe and Asia, some winter North Africa.

Historical

In the 19th century Starling was widespread and common across the London Area. Surprisingly, there are no records prior to the 1820s when it was recorded breeding at Canonbury Tower and Sussex Place in central London, but it was undoubtedly present well before then.

In about 1850 it was recorded breeding in Gray's Inn Gardens and Temple Gardens, as well as being numerous in Hyde Park and Kensington Gardens; about 10 pairs were noted breeding in each of the last two parks in 1879. In Richmond Park it was observed that a pair had bred in 1875, suggesting that they had not done so before. Price, writing between 1899 and 1903, stated that in Harrow 'nests swarmed under the eaves of the houses of the town and that at Druries every gutterpipe, chink and crevice had a nest'. Since then there have been many instances of Starlings nesting in and on buildings and other artificial structures such as the bridge across the lake in St James's Park.

20th century

By the start of the 20th century the population had increased. In 1900 Bucknill stated that the species was more common than it had been not many years previously.

In the 1930s Glegg considered that Starling numbers had not been affected by urbanisation and that the species was a common breeder 'in the suburbs, outer suburbs and through the rural parts of the county'. He added that it was a common bird, even in central London, as far back as he could trace it, yet it continued to increase in numbers.

During the 1950s it was considered to be the most ubiquitous bird in the London Area, breeding throughout the area, right into the heart of the city where it nested in parks, squares and bombed sites. In 1966 it was estimated that the breeding population in Greenwich Park was 200–250 pairs.

During the second breeding atlas there was evidence of breeding in 99 per cent of tetrads and birds were present in every tetrad, a distinction shared only with Blue Tit. This represented a 7 per cent increase in range compared with the first atlas, although some of the gaps in the first atlas, particularly in east London, were undoubtedly due to under-recording. This was probably the peak of the species' population as the Breeding Bird Survey showed a decline of over a quarter in the London Area population between 1994 and 2000.

21st century

Although it is a still common and widespread throughout the London Area, the breeding population has continued to decrease since the start of the 21st century. Research by the BTO has shown that the national population has declined by 80 per cent since the 1970s and it is known that far fewer juveniles survive their first year; it is thought that they are unable to find sufficient food during drier summers. The Breeding Bird survey in London has shown a 40 per cent decline from 1995–2010 compared to the national figures of 50 per cent. The highest counts since 2000 have been on the landfill site at Rainham where 8,000 were present in 2008.

Roosts

Harting observed that Starlings formed large flocks after breeding and remained together until the following spring. He described an increase in numbers as winter approached and stated that many moved south before the cold weather set in. Harting also wrote that Starlings prefer to roost in reed beds but 'frequently roost in shrubberies among the evergreens'.

The roosting habits of Starlings were picked up by many other authors. Leach visited a known roost in Park Wood, Ruislip, in 1893 and observed that they came in from all directions in thousands. Although he had seen them congregate there before, it was never in such great numbers. In 1896 Read stated that he had observed large numbers roosting among the reeds of a pond in winter. In 1904 Kerr wrote that one autumn evening he had seen a flock of countless thousands but made no reference to a roost. Harrison, writing about the district of Harrow between 1925 and 1930, noted that Starlings in the western half of the district went to roost at Harefield but those in the eastern side flew towards central London.

The first record of Starlings roosting in central London was in 1894 when more than 1,000 roosted in St James's Park. In 1898 roosts were first observed in Battersea Park, Regent's Park and Buckingham Palace Gardens. By the end of the 19th century they had begun to roost on buildings in central London. Roosting was first observed on Nelson's Column in 1914, and three years later they were seen roosting on the House of Lords and on the church in Smith Square.

During the First World War they were first seen roosting on St Paul's Cathedral and this roost soon came to the notice of poets and writers like H.J. Messingham, who wrote 'The birds fly to perch gustily, in a flurry of forms. They whirl about the dome and settle inconsequently on the ledges, on the burly backs and swollen heads of the statues, and on the plane-trees in front of the Cathedral. Then, when the stack on the right, which looks unaccountably they prefer to its neighbour on the left, looks like a gigantic cake teeming with currants, they start whistling and cheering like mad.'

By 1922 the roosting numbers had increased, although there was considerable variation from year to year. Observations of these roosts showed that flocks arrived from various directions but the majority came from the west. In December 1923 it was noted that several thousand congregated at a pre-roost on an island in the Serpentine in Hyde Park before flying off, probably to join the thousands roosting on buildings in the centre of London; these included the British Museum, St Paul's Cathedral, Royal Exchange, National Gallery, Nelson's Column and buildings in St Martin's-in-the-Fields. There was a large roost in Victoria Park in the winter of 1926–27.

The first attempt to control Starlings took place in 1930 when thousands were roosting all year round on Duck Island in St James's Park; as they were causing damage to the trees, efforts were made to drive them off. However, they continued to roost there up until the mid-1940s; in 1944 they roosted between April and September despite efforts to drive them away. In 1945 they returned to roost in April, but in May floodlighting for the VE day celebrations caused them to abandon the roost and they did not return.

In 1941 it was found that the birds roosted in two main areas: in the West End, centred on Charing Cross; and in the heart of the City. There were also smaller roost centres around Marylebone and at Marble Arch. The catchment area for these roosts was between 10 and 20 kilometres from the centre of London, but many resident Starlings continued to roost in their own breeding territories or small local roosts. Roost counts include 2,410 in Trafalgar Square in November 1938 and 2,350 on the Royal Exchange in March 1931. On 12 August 1949 a pre-roost flock settled on the hands on one of the clock faces on the Houses of Parliament causing the clock to stop.

By 1950 around 100,000 Starlings were roosting in Trafalgar Square during the peak months of June and July. During the 1950s there was a pre-roost gathering in Regent's Park and up to 15,000 gathered there in September and October. Many continental birds remained in the area and roosted in central London. Ringing recoveries of these birds came from Denmark, Germany, Holland, Latvia, Lithuania, Poland and Russia.

There were also large roosts in the rural areas. In 1962 'uncountable numbers, estimated at hundreds of thousands' flew in to roost at Westerham from a northerly direction on 25 February, and at Cuffley up to 25,000 roosted in December 1962. The following February about 50,000 roosted on Ruislip Common. In 1968 two flocks, each containing about 100,000 birds, were seen flying south over Rye Meads to a nearby roost at dusk on 24 November. Near Cudham, on 9 December the same year, a vast flock flew over which was considered to 'have been close on a million, and stretched from one horizon to the other and came over as a continuous band for an appreciable time'.

In the suburbs there was a post-breeding roost of several thousand in willow carr at Brent Reservoir. The majority of these birds were juveniles which fed at a nearby rubbish dump and the roost was largely abandoned by September. Birds ringed at this roost were later recovered nearby at Kensal Rise, New Southgate and Paddington, and further afield at Catford, Elmers End, Northfleet, Romford and Fakenham in Norfolk.

During the 1970s large roosts were present at various locations around the London Area, including of 120,000 birds at Banstead Downs; 50,000 on Dartford Marsh; 38,000 at Staines Moor; 30,000 north of Watford; 20,000 at Great Westwood; and 10,000 at Ruislip. In the 1980s the largest roost counts were 60,000 at Dartford Marsh; 50,000 at Darenth; and 20,000 at Walton-on-Thames.

In the mid-1980s action was taken to drive the Starlings away from the roosts on buildings in central London because of the mess caused by their droppings. Initial attempts at preventing them from roosting were largely unsuccessful (banging dustbin lids together), and even playing distress calls over loudspeakers failed to stop them, more recently treatments to window ledges seem to have been more effective.

Although smaller numbers can still be found, the really large roosts are no longer a feature of the bird life of central London. Elsewhere, the large roosts began to decline and during the 1990s the maximum count was about 20,000 at Ponders End in 1992. Since the beginning of the 21st century the largest roost count has not exceeded 5,000 birds and the spectacle of watching flight lines heading into town has ended.

Movements

On 21 December 1938 cold weather spread into the UK from the Continent and up to 1,000 Starlings took refuge on a small steamer crossing from Brussels to London. When the ship docked the following day many birds were still on board and were later seen searching for food on nearby buildings.

A large ringing programme took place in Trafalgar Square from 1949 to 1952 and there were 234 recoveries from dead birds. In half of all the cases where the cause of death was known it was due to cats. Studies of these birds and others ringed in the suburbs show that the majority of adult Starlings were sedentary as 95 per cent of those ringed in summer were recovered where they were originally ringed. Young birds were found to move around a lot more; about one-third of London Area-raised Starlings move away in their first winter and juveniles from elsewhere in southern England move into, or migrate through, the London Area.

During the 1950s, from October to mid-November, there was a large passage of Starlings from the Continent through the London Area, and large flocks were seen flying in a westerly direction in the early hours of the morning. On 27 October 1956 there was a notable movement, including over 3,000 flying over Sewardstone and many hundreds elsewhere, including one flock over Harold Wood that was at least 400 metres wide. On 19 October 1958, on a day of considerable visible migration, about 2,500 were counted from a train between Dartford and London Bridge and at least 1,500 flew over Dulwich. In 1959 large numbers were again noted in late autumn and more than 1,000 birds an hour were noted flying over Honor Oak in the early morning of 25 October; on 1 November about 2,000 flew over King George V Reservoir; and the following day about 3,330 flew west over Kensington Gardens in 65 minutes in the early morning. In the autumn of 1964, 4,000 flew west in just 25 minutes over North Cray on 19 October and the following day 2,000 flew west over Weybridge. In 1966 about 3,350 flew west at Tooting Bec Common during a cold spell in 40 minutes on 19 January. In 1969 about 3,500 flew west over Worcester Park on 5 November. Starlings continue to migrate in a westerly direction every autumn, but in smaller numbers.

Facts and Figures

Starlings occasionally breed very early in the season; a nest with half-grown young was found at Staines in January 1874, and a nest with two eggs was found at Palmers Green on 5 December 1913.

Rose-coloured Starling *Sturnus roseus*

A rare vagrant (11)

Range: eastern Europe and Asia; scarce visitor to UK.

The first record for the London Area was of a bird killed at Norwood; it was also the first occurrence in the UK. The specimen was stuffed and displayed in a coffee house in Chelsea and was drawn by Edwards for his *Natural History of Uncommon Birds* (1743–51). Unfortunately, it is not clear where this bird was taken as there are two Norwoods in London, near Osterley Park and in south London, and two counties have laid claim to this individual. Harting published the record in *The Birds of Middlesex* as did Bucknill in *The Birds of Surrey*.

The second record was of a bird shot near Cranford in 1802, where it had been for some days in a meadow consorting with Blackbirds; this bird was illustrated by Hayes (1808–16) and described as being the third English occurrence. In 1845 a female was shot at Ditton Marsh in May and in 1861 another was shot nearby at Thames Ditton in July. Other historical records were also taken in the home counties of Essex, Hertfordshire and Buckinghamshire.

There were no more records for another hundred years. There was an influx into the UK in 1961 and one was present at Sunbury for four days in August. There were two adults in the 1970s, in Inner London at Surrey Docks in autumn and at Cheshunt in winter. Two more adults were seen in the 1990s, one briefly in Rensbury Road, Walthamstow, in June when there was a small influx into the country, and one spent three days in a Bookham garden where it fed with Starlings.

Since 2000 there have been two further records: a juvenile seen for just 10 minutes in a garden in New Malden and an adult in a garden at Custom House, near Canning Town.

Pre-1743: Norwood, one shot
1802: near Cranford, shot in winter
1845: Ditton Marsh, shot in May
1861: Thames Ditton, shot in July
1961: Sunbury, on 9–12 August
1971: Surrey Docks, adult on 11 August
1978: Cheshunt, adult on 4 December
1994: Walthamstow, adult on 22 June
1996: Bookham, adult on 23–25 September
2001: New Malden, juvenile on 6 October
2003: Custom House, adult on 6 July

Unacceptable Records

In the early 1900s a dozen were released in St James's Park. One at New Covent Garden Market, from 27 July to 12 August 1997 and also seen at Battersea Power Station on 15 August, was accepted by the BBRC but not by the LNHS rarities committee.

Dipper *Cinclus cinclus*

A rare winter visitor (20)

Range: Europe, Asia and North Africa; breeding resident in west and north of UK. The continental race *C. c. cinclus* ('Black-bellied Dipper') is a rare visitor to the UK and accounts for some of the London Area records.

There are five 19th-century records, three of which are undated. The first of these was documented by Meyer in 1842 when he wrote that he had seen one by the River Mole at Claremont. This was followed by one on the River Colne at Wraysbury, just below Bell Weir, some time prior to 1843 and one shot in Wembley Park prior to 1866. The only dated records are in the spring of 1862 when one was seen flying along the Silk Stream by Colindeep Lane in Hendon, and in the autumn of 1895 when one was shot on the River Mole at Mickleham. In 1900 Chipperfield referred to one that had been seen on Hackney Marshes.

The first record in the 20th century was during very cold weather in 1915 on the River Mole at Leatherhead; another was seen in the same area in 1926. During the intervening years one was on a flooded field near Hatfield in 1916 and there was also a record of one in Kelsey Park, Beckenham, some years prior to 1926.

There were no further records until the 1960s when one was trapped and ringed at Gibbs Brook, Godstone, in 1965; it was found to be of the British subspecies. The first one of the Black-bellied subspecies was present at Foots Cray Meadows in 1969. One seen flying into a pump tunnel at one of the empty reservoirs at Walthamstow in 1976 was described as being of the British subspecies.

In 1981 one was present between Lemsford Springs and Stanborough from January to March; a second bird was also present in the area from 8 to 28 March and occasionally crossed the 'boundary' into the London Area. Both birds were of the Black-bellied subspecies. In the winter of 1987/88 another Black-bellied Dipper was present on the River Ver, in the Sopwell area south of St Albans; it had been seen just outside the London Area at Lemsford Springs NR on 28 November.

There were two more in the mid-1990s: a Black-bellied Dipper on the River Gade in Cassiobury Park, and one on the River Lea. The latter was first seen just outside the London Area at Kimpton Hill on 11 September; it was then trapped and ringed near Lemsford Springs on 29 September where it was present intermittently until January 1996. In the hand it showed some characteristics of the central European race *C.c. aquaticus*.

Four of the Dipper records were of birds seen at Lemsford Spring NR. Most of the reserve is just outside the London Area but the southern part, notably the stretch of the River Lea between Lemsford and Stanborough, which these birds often favoured, is within the London Area.

Pre-1842: Claremont, on the River Mole
Pre-1843: Wraysbury, on the River Colne
1862: Hendon, on the Silk Stream in spring
Pre-1866: Wembley Park, one shot
1895: Mickleham, shot on the River Mole in autumn
Pre-1900: Hackney Marsh
1915: Leatherhead, on the River Mole on 28 March
1916: near Hatfield, on 5 April
Pre-1926: Kelsey Park, Beckenham
1926: Leatherhead, on the River Mole on 3 May
1965: Godstone, on Gibbs Brook on 13 April
1969: Foots Cray Meadows, on 11 January–10 March
1975: Lemsford Springs, on the River Lea near Lemsford and Stanborough on 19 January
1976: Walthamstow Reservoirs, on 1 January
1981 Lemsford Springs, on the River Lea between Lemsford and Stanborough on 25 January–7 March with two on 8–28 March
1982: Sevenoaks, on the River Darent on 11–20 January and again on 17 February
1987: Sopwell, on the River Ver on 27 December–25 February 1988
1994: Cassiobury Park, on 9 March
1995: Lemsford Springs, on 29 September–11 November; also seen on 1 and 21 January 1996; also seen at Stanborough Lake

Unacceptable Records

There is one possible nesting record from Pinner Brook in May 1876. The eggs were taken and later identified as belonging to Dipper by Harting and the description of the nest also fitted. However, Glegg rejected the record 'on the grounds that the eggs were bought from a dealer, that no mention is made of anyone having seen the birds, and that six years had elapsed before the publication of the record'; there were also no sightings of Dipper from Pinner, before or after this record. There are two unsubstantiated records mentioned in the Epsom College Natural History Society Reports from the Staines area in 1890 and on 5 May 1907.

Grey-cheeked Thrush *Catharus minimus*

A very rare vagrant (1)

Range: Americas; vagrant to UK.

The only record for the London Area occurred in 2005 when one was found in Northaw Great Wood. It remained there for almost two weeks and was seen by hundreds of birders. The chances of bumping into an American thrush in an inland wood in the UK are extremely remote. This was the forty-sixth record in the UK, most of which were seen on the west coast.

2005: Northaw Great Wood, on 13–25 November

Ring Ouzel   *Turdus torquatus*

An uncommon passage migrant

Range: breeds Europe and Middle East, winters southern Europe and North Africa.

Historical

The earliest reference was in 1831 when Doubleday noted that it was 'only seen now and again in spring and autumn' at Epping. Several were seen at Tooting in the autumn of 1833; prior to 1843 one was trapped in South Lambeth and a small flock was seen on Wimbledon Common in October. In the second half of the 19th century it was recorded at various locations in south London and occasionally seen in Inner London with records from Lambeth and Regent's Park.

Harting knew this as a passage migrant and recorded several instances of birds being taken. He shot a pair in Kingsbury on 25 April which had been present for more than a fortnight and upon dissection discovered that the female contained rudimentary eggs and that both birds appeared to have been sitting as 'the breast of each was destitute of the soft down which always covers it before incubation has commenced'. Three other birds that were with this pair remained until 1 May.

20th century

By the start of the century a decline in the numbers using North Downs stop-over sites had been noted and attributed to increasing development. Glegg described it as an irregular passage migrant, most often recorded in April and September. A pair wintered at Hampstead in 1909/10 and another pair was seen in Richmond Park in December 1909.

In the first half of the century there were 44 records; a decrease was already apparent as 27 were in the first quarter and only 17 in the second when there were more observers. Their favoured areas were still on the heights to the south of the Thames, with almost three-quarters of the records being in spring. An analysis of all records between 1947 and 1984 showed that the spring migration occurred between 19 March and 28 May with the peak from 16–22 April; the autumn passage lasted from 1 September to 17 November with the peak period being the first two weeks of October.

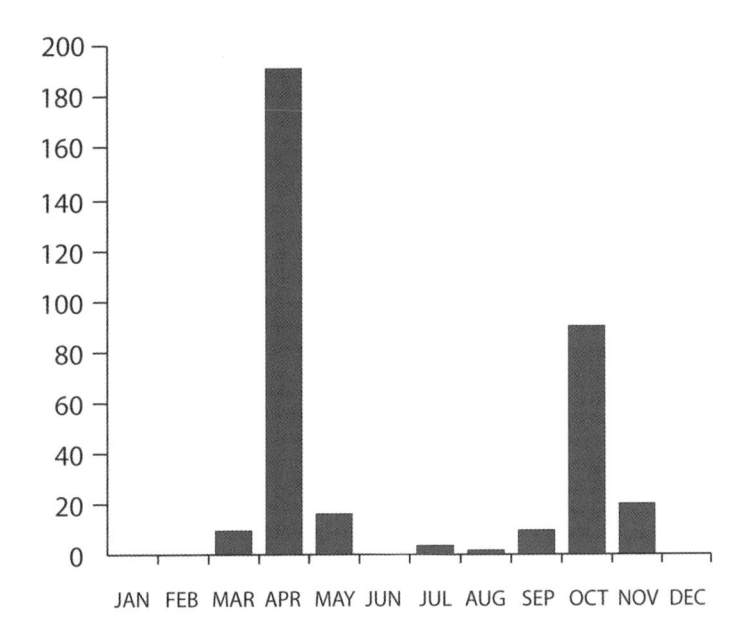

Seasonal occurrence of Ring Ouzel: 2000–2009.

In the late 1950s there were two overwintering records in the Staines area. There were several more winter records, including one of a bird at the Tower of London where it fed with Blackbirds in the moat from 8 to 12 December 1972; at Epsom from early January to mid-February 1975; at Sidcup from early January to mid-March 1986; at the Chingford Reservoirs from 17 November to 1 January 1988.

21st century

Since the beginning of the 21st century the total annual passage through the London Area has varied from 12 to 53 birds. The peak times for migrants are the middle of April and the middle of October. They can be seen throughout the London Area, sometimes even in the central parks; sites such as Wanstead Flats and Wormwood Scrubs regularly attract birds in both spring and autumn. One overwintered in a garden in Kingston-upon-Thames from January to March 2001.

Facts and Figures

The earliest migrant was on 6 March 1977 on Epsom Common.
The latest migrant was on 9 December 2000 at Staines Reservoir.
The largest flock was of 20 birds on Hampstead Heath on 14 October 1971.
The largest annual total was about 55 in 1998.
One at Epsom between January and 17 February 1975 was of the race *alpestris*.

Blackbird *Turdus merula*

A very common resident and widespread breeder

Range: Europe, Asia and North Africa.

Historical

Remains dating back to Roman Britain have been found at Southwark. Very little information on the status of Blackbird in the London Area is known prior to 1900. Cornish noted that he used to see a regular passage along the Thames at Hammersmith during the 19th century.

20th century

It seems likely that the population increased during the first half of the 20th century. Some of this is probably due to suburbanisation, which has benefitted Blackbirds by providing new habitats in the form of parks and gardens. By the 1930s the Blackbird had become a common breeding resident both in the central London parks and squares and suburban gardens. A study in the winter of 1948/49 showed that there had been a five-fold increase in the numbers in Kensington Gardens since the winter of 1925–26.

In the 1950s Blackbird was the second most common passerine in the central parks after House Sparrow; at least 200 singing males were counted in Inner London in 1959, more than half of which were in Regent's Park. They were also able to exploit new areas; for example, 11 pairs were found to be breeding in the bombed areas of Cripplegate in 1952. The exceptionally cold winter of 1962/63 reduced the breeding population in some suburban areas, so at Dollis Hill, for example, there were 86–87 pairs in the summer of 1963 compared to 185–195 pairs in the previous year. In 1969 there were at least 250 territories in the central London parks, but by 1974 there had been a 50 per cent reduction in Regent's Park, although they had increased in Hyde Park and Kensington Gardens. By the end of the 1970s there were 175 territories in Regent's Park and 150 territories in Hyde Park and Kensington Gardens.

The second breeding Atlas in 1988–94 found breeding Blackbirds in 99 per cent of all the tetrads in the London Area, more than any other species. Although they could be found virtually throughout the London Area, the population in central London declined during the 1990s, particularly in the Royal Parks where they are well monitored. This was also borne out by the Breeding Bird Survey, which showed a 21 per cent decline across London between 1994 and 2000.

21st century

At the beginning of the 21st century Blackbird was one of the most common and widespread birds throughout the London Area. Although this is still true, the population has continued to decline; between 1994 and 2008 the Breeding Bird Survey registered a decrease of 26 per cent compared to a 24 per cent increase nationally. In Kensington Gardens just 28 were found during the annual autumn count in 2010, a drop of almost 90 per cent since 1966. Such a sharp decline in one of our commonest species must be a cause for concern. Possible factors include climate change e.g. drier summers, or changes in parks management, both of which could affect the availability of soil invertebrates for ground-feeding birds, increasing predation now that the Sparrowhawk has become established in central London, and increasing public pressure on green spaces.

Movements

Glegg knew Blackbirds to be partially migratory, which was borne out by ringing recoveries showing continental birds wintering in the London Area. There was also one instance of the reverse happening, when a juvenile ringed in Wimbledon on 13 May 1945 was recovered at Brest, France, on 25 January 1949.

On Hampstead Heath there has often been an increase in numbers during the second half of October and into November, and similar influxes have been noted from other areas. In 5 November 1961, a day of large thrush movement throughout the London Area, there was a peak of about 150 birds at Rye Meads.

In the autumn of 1971 there was a notable influx, particularly on 5 October when hundreds were seen in Langley Park. In 1978 there was a large influx in the Dartford area with a peak count of more than 350 on 17 December. In 1979 cold weather produced counts of more than 300 at Wilmington on 18 February and 200 at Stonehill Farm/Joyden's Wood in January. In 1980 there were more than 300 at Joyden's Wood on 27 January and 200 at Rye Meads in the first week of November.

Studies

Breeding Blackbirds have been studied at a number of sites. In 1890, at Lordship Park, Stoke Newington, a female was seen incubating her fourth clutch in the same nest, having already raised three broods of four. In 1897, at Stamford Hill, three broods were raised in the same nest in 14 weeks between 10 April (when incubation first began) and 17 July when the third brood fledged. On 2 January 1991 one was seen sitting on eggs at a nest in Camden.

Communal roosting has been recorded from several areas in the London Area. The largest concentrations were at Brent Reservoir in the 1960s and 70s when birds flew in from the surrounding districts to roost in the scrub on the wall of the dam and in the adjacent woods. Although the roosts were in use all year round, the peak numbers were in late autumn with up to 2,000 roosting there. By 1972 more than 6,000 roosting Blackbirds had been caught and ringed; of the 235 recoveries only four were outside the London Area, and fewer than 6 per cent had moved more than 5 kilometres, suggesting that these birds were predominantly resident. The only foreign recovery was of a bird ringed on 10 February 1968 and found dead in Sweden in October 1969.

Ringing studies at Brent Reservoir showed that urban Blackbirds are more likely to survive prolonged snowy weather as they were found to weigh more during these times than rural birds, presumably because of the abundance of food put out for them. The studies also showed that Blackbirds are found at a higher density in suburban areas than in woodland (Batten).

Naumann's Thrush *Turdus naumanni*

A very rare vagrant (2)

Range: Asia, vagrant to UK.

There have only been two records of this species in the UK, both in the London Area. The first record in 1990 was of a bird discovered at Chingford on 3 February in a birder's garden, which was later realised to have been present in a neighbour's garden since 19 January. The news sparked a major national twitch and the bird was seen by many during its seven-week stay; it was probably killed by a Sparrowhawk.

Seven years later a second bird was found, only 3 kilometres from the first. This individual was only seen by the two finders despite other observers searching for it on the final date of its stay.

1990: Woodford Green, male on 19 January–9 March
1997: South Woodford, first-winter on 6–11 January

Fieldfare *Turdus pilaris*

A common winter visitor; former breeder

Range: Europe and Asia; winter visitor throughout UK and very rare breeder.

Historical

In the 19th century it was known as a winter visitor throughout most of the London Area, occasionally being seen in central London.

20th century

In the 1930s Glegg stated that its distribution was dependent on the weather conditions; typically it could be found anywhere in the rural districts but at the onset of severe weather it became more widespread. No large flocks were known at the time; the largest was a mixed flock of Fieldfares and Redwings that numbered 250.

The first large counts were made in the 1940s; in early 1945 up to 1,000 were seen at Borehamwood. During the cold winter of 1946/47, there was a large influx with peak counts of 2,000–3,000 in yew trees in Norbury Park on 15 December and about 2,500 at Beddington on 16 February. The following winter about 2,000 were at Berwick Ponds on 26 December 1947. On 5 November 1961 an estimated 90,000 flew over the London Area in a movement that lasted most of the day. In 1968 about 3,000 flew south-west over Swanley in three hours on 28 December at the onset of severe weather.

Since then the highest counts have been 2,000–3,000 at Dartford and Stone Marshes on 28 November 1970; a roost numbering thousands at Stocker's Lake in December 1974; about 9,000 roosting at Cheshunt GP in January 1979; about 3,000 flew over Wilmington on 18 February 1979; a roost of 2,400 in the Colne Valley in February 1980; an overnight fall of 2,000 at Stanstead Abbotts GP on 1 December 1985; and 3,000 over Hampstead Heath in two hours on 10 November 1999.

21st century

Due to the relatively mild weather during most winters since 2000 Fieldfares have not been particularly numerous; the highest count was 1,750 over Regent's Park on 2 November 2006. In most years there is a passage through the London Area in late autumn and relatively few are then seen until the New Year when feeding flocks move into the area and remain until early spring. Apart from overflying migrants, they are fairly scarce in central London and the inner suburbs.

Breeding

In the first half of the 19th century a pair once summered at Brent Reservoir. In 1975 a juvenile was seen at

Primrose Hill on 24 July; it was presumed to be from one of the British breeding pairs as there was no indication that Fieldfares had bred locally. A pair built a nest and laid eggs in Barking Park in 1980 but the nest was predated by Carrion Crows. The pair returned the following year and built a nest in the same tree but did not breed. The only other breeding record in the London Area took place in 1991 when an adult and a recently fledged juvenile were found in the Cray Valley on 30 May. There have been several other sightings in June and July.

Facts and Figures

The earliest migrant was on 19 August 1984 at Brent Reservoir.
The largest count was of about 9,000 birds roosting at Cheshunt GP in January 1979.

Song Thrush *Turdus philomelos*

A common resident and widespread breeder

Range: Europe and Asia; UK population joined in winter by continental birds.

Historical

During the 19th century this was a very common bird; for example at Blackheath, Witherby found three Song Thrush nests to one Blackbird nest.

20th century

At the beginning of the century they were very numerous throughout the London Area. In Harrow and around Woodford Song Thrushes and Blackbirds had been roughly equally numerous but by the late-1920s Blackbirds had become about five times as numerous; this was thought to be the result of a decline in Song Thrush numbers in the area. In some parks Song Thrushes were almost as tame as House Sparrows and would approach people for food.

In the 1930s Glegg described it as 'one of the commonest residents, although not quite so abundant as the Blackbird'. He wrote that it was well distributed in central London and common in the suburbs 'where it frequents even small gardens'. He knew it to be partially migratory and rarely seen in any number together. Studies in the Tadworth and Walton districts in the 1940s showed that the population of Blackbirds outnumbered that of Song Thrushes by four to one. After the hard winter of 1946/47 there was a decline in the population.

By the 1950s it was clear that there had been a marked decrease since the beginning of the 20th century, particularly in built-up areas. In central London they were still breeding in all the larger parks and in many large gardens but they had become rare in the squares. The breeding population away from the more heavily built-up areas suffered a serious reduction following the severe winter of 1962/63.

The population continued to decline across the London Area; by 1980 there were only 24 territories in Hyde Park and Kensington Gardens, a decrease of over 60 per cent since 1968. In 1993 the breeding population in Hainault Forest was 30 pairs, a decline of 25 per cent from the previous year; and in Battersea Park the number of territories had declined from an average of 30 in 1974–80 to seven by 1994. However, at several monitored sites the breeding population increased during the 1990s. During the second breeding Atlas in 1988–94 they were found breeding in 814 tetrads, 95 per cent of the London Area's total. The Breeding Bird Survey showed a small decline in London between 1994 and 2000 compared to a 12 per cent national increase.

21st century

The Breeding Bird Survey shows a decrease in the London Area population of 35 per cent from 1995-2010, however some sites have recorded an increase since the beginning of the 21st century. In central London, the species has continued to decline; this may be partly due to a loss of habitat and change of management. Dry summers are likely to be a significant factor for a bird which forages mainly on soil invertebrates and sympathetic soil management may help to encourage them.

Movements

Continental immigrants are regularly recorded in late autumn, occasionally in large numbers. In 1961 migrant Song Thrushes were particularly noticeable in late autumn when the numbers at Rye Meads SF increased from 10 to more than 100 on 21 October and a flock of about 50, containing some obvious greyish birds, was seen at Hersham SF. During the very cold winter of 1962/63 there was a huge concentration of about 200 at Beddington on 6 January. In 1978 more than 100 were seen in the Dartford/Littlebrook area on 2 November and 90 per cent of the birds ringed at Dartford Marsh in early December were of continental origin. In 1981 more than 100 continental birds were seen at Littlebrook in December. In 2005 a total of 255 flew over the Wetland Centre between 4 September and 13 November.

Some of the London Area's breeding birds make the opposite journey. The earliest ringing recovery of a Song Thrush in the London Area was of a nestling ringed in Stanmore on 20 May 1920 and recovered at Vendes, Calvados, in France, on 20 December the same year. In 1950 another young bird was ringed in the London Area in May and recovered at Caen, France, in November 1953. A nestling ringed at Shenfield was recovered in Spain in the following spring.

Facts and Figures

The largest number seen was about 200 birds at Beddington SF on 6 January 1963.
A pair was incubating eggs at Little Stanmore on 3 December 1888.

Redwing *Turdus iliacus*

A very common winter visitor

Range: breeds northern Europe and Asia, winters Europe and North Africa.

Historical

During the 19th century this was a widespread winter visitor across the London Area; the largest flock recorded was 300–400 newly arrived birds in a meadow at Denmark Hill in November 1880.

Harting gave two instances of breeding, at Barnet and Harrow, but these were dismissed by Glegg who wrote that these statements do not bear investigation.

20th century

In hard weather Redwings were sometimes seen foraging in the built-up areas of London. In February 1917 some were seen feeding on the mud by the Thames at Chiswick and then running along the pavement at high tide; several were seen on the iced-over lake in St James's Park, competing with ducks and gulls for scraps of food in 1929; and in December 1935 one fed with House Sparrows and Starlings on the pavement in Albany Street.

In 1935 Glegg described them as a common winter resident that could be found anywhere in the rural areas and even in central London during severe weather. The largest feeding flocks usually occured early in the year when up to several hundred could be seen; occasionally groups numbered into the thousands. In the hard winter of 1946/47 Redwings were particularly numerous with up to 2,000 at Beddington on 8 February; and in 1951 around 3,000 were seen on hawthorn bushes bordering Stone Marshes on 4 February.

In 1962 about 4,000 were located roosting at Caterham-on-the-Hill on 19 December and over 2,000 were at Addington the following day. As the cold winter progressed large numbers were seen flying over the London Area. There was a massive increase in the numbers present at Beddington; on 5 January there were about 200 but three days later this had increased to about 5,000. Many moved on within a week as the snow remained on the ground, and there were widespread reports of Redwings feeding in gardens. In 1980 there was a large roost at Headley Heath, which peaked at 3,500 on 1 November.

Summering birds were seen in 1971 at Old Slade on 21 May and 17 July, and in 1976 in Richmond Park between 10 June and 10 July at least.

21st century

Since the beginning of the 21st century the largest numbers of Redwings have been seen on passage and there have been fewer feeding in the outer suburbs and rural areas due to a series of mild winters. In the more urban areas they are much scarcer and are mostly recorded passing overhead although they are sometimes seen feeding in parks and gardens during colder weather, for example during a cold snap in early 2010 flocks of 50 or more turned up in several central London parks.

Movements

The largest movements usually occur in autumn or during winter when cold weather sets in. The first of these that was recorded was in 1938 when 2,000–3,000 flew over Ladbroke Square in just over an hour on 22 December. In 1952 about 5,000 flew south-west over Bookham Common on 9 November. In 1958 there was a large-scale movement after dark on 17 October when more than 2,000 calls were heard at Dollis Hill between 18:30 and 23:10 hrs. In 1969 there was a large movement on 12 October when 3,500–4,500 were seen flying over Thorpe and more than 2,000 birds passed over Dartford Marsh; and on 27 November about 3,000 flew over Burgh Heath.

In 1996 about 4,000 flew over Beddington during cold weather on 27 January. There was an exceptionally large passage in the autumn of 1997 with a total of some 41,000 on 12 October, including 12,000 over Hampstead Heath and 10,000 over Beddington. The first large movement of the 21st century occurred in 2004 when more than 16,000 flew over the London Area on 9 October.

Facts and Figures

The earliest migrants were on 8 September at Regent's Park in 1993 and Waltham Abbey in 2001.
The latest migrant was on 12 May 1996 at Grovelands Park.
Redwings ringed in the London Area during winter have been recovered in Germany, Norway and Sweden, and one was even recovered in Italy the following winter. One trapped at Hersham SF on 5 March 1983 was of the Icelandic race *coburni*.
An injured bird was picked up on Wimbledon Common on 19 August 1950; it was presumed to have summered in the area as it was not capable of flight.

Mistle Thrush *Turdus viscivorus*

A common resident and widespread breeder

Range: Europe, Asia and North Africa.

Historical

In the mid-19th century it was described as a common but sparsely distributed resident which gathered in small post-breeding flocks. Although the national population increased across the UK, it had declined in some areas by the end of the century; in south London it had decreased rapidly in the suburbs, and it had ceased breeding in the parks in central London.

20th century

At the beginning of the century the population began to increase, particularly in the more rural parts. By 1929 the population in central London was still only a few pairs and remained at this level throughout the 1930s.

Glegg described it as a common resident, adding that it was known from suitable locations in the suburbs and was widely distributed in rural areas. In the 1940s it bred in central London in Kensington Gardens, Regent's Park and St James's Park, as well as on Brompton Road and Park Lane; it was thought that the population increase in the London Area was due to the extermination of the Grey Squirrel as the latter was known to predate Mistle Thrush nests.

The population in central London continued to increase and by 1967 there were 24–26 breeding pairs, including 11 in Regent's Park. During the first breeding Atlas in 1968–72 it was found breeding in 71 per cent of all the tetrads in the London Area. During the second breeding Atlas in 1988–94 there was an increase of 17 per cent in the breeding distribution in the London Area compared to the first survey. However, the Breeding Bird Survey showed a 37 per cent decline in the population between 1994 and 2000. The largest breeding concentration was 15 pairs in Hainault Forest in 1990.

21st century

Since the start of the 21st century Mistle Thrush has remained a reasonably common bird throughout the London Area although there has been further evidence of a decline in the breeding population with the BBS showing a decrease of 51 per cent from 1995–2010 and this species is now an Amber listed Species of conservation concern. The London Biodiversity Partnership's survey of small open spaces in central London found this species was reasonably widespread in larger squares and similar open spaces which offered both a good number of mature trees and wide expanse of open grassland. It can sometimes be found feeding in small grassy areas right in the heart of the city, for example on the lawn around Westminster Abbey.

Movements

At some sites large post-breeding gatherings are often recorded, such as in 1919 when about 150 were at Caterham on 3 August. In late autumn and winter even larger concentrations have been recorded; birds regularly congregated to feed on yew berries on the North Downs and about 250 were seen on Mickleham Downs on 10 October 1937. At Selsdon at least 1,000 were present on 6 December 1942; some of these may have been immigrants from the Continent given the time of year. The same probably applies to a flock of 400 at South Croydon on 22 January 1946.

In 1959 there was a large influx of many passerines into the London Area and large numbers of Mistle Thrushes were recorded; 200 flew south-east over King George V Reservoir on 31 October and 285 flew south-west at Sewardstone on 7 November. In 1962 the first evidence of nocturnal migration was obtained in the London Area on 8 November when several were heard calling at 23:00 hrs, along with other thrushes.

Since 2000 the largest post-breeding gathering was 65 at Parkside Farm in late July 2000.

Nesting Sites

Although typically nesting in a fork of a tree, Mistle Thrushes have been recorded nesting in various artificial sites in the London Area, such as in scaffolding at Temple; in the iron canopy of a statue at Hammersmith; on a window ledge in Richmond; and on temporary stands used for the Coronation in Hyde Park and Green Park.

American Robin *Turdus migratorius*

A very rare vagrant (1)

Range: Americas; vagrant to UK.

The only record occurred in 2006 when one wintered in Peckham. It was seen regularly in residents' gardens from early January into March but the news was not released until late into the bird's stay. It was seen by four birders early on the morning of 28 March and then promptly disappeared. This was the twenty-first record in Britain.

2006: Peckham, in early January to 28 March

Unacceptable Records

One in Richmond Park in May 1912 was assumed at the time to be an escape.

Spotted Flycatcher *Muscicapa striata*

A common passage migrant and scarce breeder

Range: breeds Europe, Asia and North Africa, winters Africa.

Historical

The earliest known record was in 1786 when Gilbert White saw a pair with young in the London Area on 20 June.

During the 19th century it was common throughout the London Area, in some years even numerous, breeding in enclosures and gardens, and squares in central London. Although typically nesting in trees or on buildings, one was noted nesting in the ornamental crown on top of a lamp-post near Portland Place, and another among the ornaments on an iron gate in Hampton Court Gardens. There was evidence of a decline towards the end of the century as Hudson could only find one pair breeding in the central parks in 1897, in Kensington Gardens.

20th century

For the first half of the 20th century it was a relatively common breeder throughout the London Area, even in central London where some of the Royal Parks had two or three pairs. Towards the end of the 1940s it was the only summer migrant which still bred regularly in central London, breeding in most of the parks and some of the squares in Bloomsbury and Mayfair, at Lincoln's Inn Fields and in gardens in Chelsea and Lambeth.

In the late 1960s there were about 20 breeding pairs in Holland Park, Hyde Park, Kensington Gardens and Regent's Park. During the first breeding Atlas in 1968–72 they were found breeding in 55 per cent of all the London Area tetrads. The first signs of a decline had been noted by the start of the 1970s and they had begun to disappear from some sites. In 1974 further declines were noted in several areas of south London and in the central parks. By the late 1980s there were large decreases in the breeding population across the London Area; there was a 33 per cent reduction in Alexandra Park and a 50 per cent decline in Osterley Park.

By the time of the second breeding Atlas in 1988–94 there had been a decrease in the breeding distribution of 12 per cent since the first survey; most of the losses were in the south-west area. This was above the national decrease of just over 2 per cent. There was a large reduction in the breeding population at the start of the 1990s from 131 reported pairs the previous year to just 77; the population continued to plummet during the decade. Nationally, there was an 85 per cent decrease in the population from 1966 to 1999, the largest decrease of any breeding woodland bird.

21st century

At the start of the 21st century the breeding population was continuing to decrease rapidly and Spotted Flycatcher had finally ceased breeding in central London. By 2009 the breeding population had decreased with only six pairs located, all in the rural north and east, apart from a pair in Greenwich Park. They are still reasonably common during the autumn migration and can often be seen in the central London parks.

Facts and Figures

The earliest migrant was on 22 April 2000 at Fairlop Waters.
The latest migrant was in the first week of November 1828 in Fulham.
The highest count was at least 50 birds on Hampstead Heath on 20 September 1992.

Robin
Erithacus rubecula

A very common resident and widespread breeder

Range: Europe, Asia and North Africa.

Historical

In 1865 Hibberd recorded that Robin was very common in central London and could be found in Cumberland, Smithfield and Whitechapel Markets, the squares at Gray's Inn and Lincoln's Inn, on Blackfriars Road, Farringdon Street, Ludgate Hill and the Strand, and in St Paul's Churchyard.

Harting surmised that this species was so common due to 'a current superstition that it is unlucky to kill a Robin, or to take its eggs'. Even in the 19th century individual birds could become very tame and Harting records a brood of young Robins that came 'through the window to their breakfast table for crumbs, and would follow my father in the garden and light upon his head'.

Hudson, writing in 1898, stated that Robins were the fifth most common passerine to be found in London, behind House Sparrow, Starling, Blackbird and Song Thrush, even though they had declined in the more built-up areas. They had disappeared from all the urban areas where Hibberd had seen them 30 years earlier and were becoming scarce in the parks.

20th century

During the 1930s Glegg stated that it was one of the most common residents and could be found throughout central London and the suburbs. However, it decreased considerably in central London between the middle of the 19th century and the middle of the 20th century to such an extent that by the 1940s it did not breed in all of the central parks. This localised decline was attributed to predation by cats, which was also a factor in poor breeding success in the suburbs. Robins prospered after the war, breeding in bombed-out areas and in many squares and open spaces.

In the 1950s they were common and widespread across most of the London Area apart from in the most heavily built-up areas and on the Thames marshes. After the severe winter of 1962/63 large reductions in the breeding population were noted across the London Area; for example, in a 13-hectare survey plot on Bookham Common the numbers fell from 32 territories to 21–22 but by 1964 they had fully recovered. In the 1970s the population in central London was estimated to be 50–60 pairs.

In 1991 there was a reduction in the breeding population in some areas following a long spell of very cold weather in February. The population quickly recovered, though; for example, there were 60–70 pairs at Hainault Forest in 1992, an increase of 20 per cent on the previous year. The second breeding Atlas in 1988–94 found breeding Robins in 98 per cent of the London Area tetrads.

21st century

Robin remains a very common and widespread breeding bird throughout the London Area, even in the most urban areas; the population is actually increasing in some areas. The Breeding Bird survey indicates that the population in London increased at a greater rate than the population for UK as a whole from 1995 to 2010, with an increase of 75 per cent during this period. The largest breeding concentration recorded was 125 territories in Bury Wood, Epping Forest, in 2001.

Movements

Kerr stated in 1906 that it was subject to considerable local movements. Later evidence of this include: a large influx of birds in St James's Park on 20 September 1920 which had all gone within a week; a sudden increase at Beddington in September 1932; and a small but definite passage at Barnes along the Thames towpath on 11 October 1943.

Most of the ringing recoveries were close to where the birds were originally ringed, and up to the middle of the 20th century there was only one foreign recovery, of a bird ringed at Woodford Green on 10 May 1938 and recovered in the French Pyrenees in September of that year.

Nest Sites

Robins often use artificial sites for nesting. In 1915 a pair nested between two flowerpots in a hothouse in Kensington Gardens, and they have also been recorded breeding in the dressing room of a football club and in a public lavatory. While the Crystal Palace was being built at Sydenham several Robins nested very close to the workmen; 'the din around them was perfectly deafening, and the bustle incessant, but there they sat and hatched and reared their young, without displaying any signs of fear'. In December 2006 a pair built a nest in a Christmas wreath on the door of a house in Roehampton; the eggs hatched in late January.

Nightingale *Luscinia megarhynchos*

An uncommon summer visitor and localised breeder

Range: breeds Europe, Asia and North Africa, winters Africa.

Historical

The earliest record was in the 11th century; in 1768 Morant stated that in the time of Edward the Confessor (1003–1066) the Havering area was said to abound with warbling Nightingales which, according to a legend 'disturbed him at his prayers; and he earnestly desired God of their absence. Since which time, as the credulous neighbouring swains believe, never nightingale was heard to sing in the park'.

In the 17th century Pepys wrote in his diaries that he heard Nightingales singing at Vauxhall when he attended a concert there. In 1703, at the site where Buckingham Palace is now located, the Earl of Buckingham found 'under the windows...a little wilderness full of Blackbirds and Nightingales'. In 1730 they were considered to be common in Fulham. In the 18th century they were regularly heard in gardens in Marylebone.

Early in the 19th century they could still be heard singing in Hyde Park. John Keats wrote 'Ode to a Nightingale' after he had heard one from a tavern garden on Hampstead Heath in 1819. In 1830 they frequented the area between Hyde Park Corner and Kensington gravel pits. In the spring of 1858 a bird-catcher caught 34 at Leytonstone. In 1865 Hibberd stated that they could be heard anywhere beyond a 5-kilometre radius from St Paul's Cathedral, except eastwards.

Harting wrote that 'there are still many copses within a short distance of London where the wonderful song of the Nightingale may be heard'. By the late 1870s they had ceased breeding in many central London locations, such as Cadogan Gardens in Sloane Street and in Kensington Gardens, but they were occasionally still heard in Regent's Park. In 1873 they were apparently so plentiful in Enfield Chase that a Member of Parliament relinquished his lease on a lodge after a few weeks because 'his family could get no sleep for the singing of the Nightingales'.

Harting recorded that it 'is now by no means so common in the county as formerly'. This decline may have been due to capture for the cage-bird trade as he related how Harrow Weald was a favourite locality until its discovery by bird-catchers. This was a lucrative trade at the time; an acquaintance of Harting rented a cottage for £10 a year and in 'a good Nightingale season' made more than enough to pay his rent by the capture and sale of these birds; in one season alone 'he caught fifteen dozen, receiving eighteen shillings a dozen for them in London'. However, by 1891 this persecution had begun to take its toll and Nightingale was a much-decreased species around Harrow.

The species bred annually in Gunnersbury Park until at least 1889. In 1892 it was stated that Highgate was one of the best places for Nightingale as five pairs were in Bishop's Wood and several pairs were in other woods. A pair nested in Battersea Park in the 1890s. At the end of the 19th century it still bred in many parts of the London Area but the new century saw a rapid decline in numbers. It last nested on Hampstead Heath in 1899.

20th century

At the start of the century it bred around Hanwell and Winchmore Hill but fewer bred on Wimbledon Common due to the cutting of cover. In 1906 Kerr recorded that it still occurred in large numbers in the Staines district but was not as numerous as it had been; it had also ceased breeding at Lee in the Lea Valley. Ticehurst stated that it had increased in south-east London since the bird-catchers had stopped their trade and was quite numerous between Dartford and Blackheath.

There were no records from Inner London until 1928 when one was heard singing in Regent's Park. Migrants were occasionally heard elsewhere in central London in subsequent years but there have never been any records in Berkeley Square despite the reference in a popular song. In the 1930s Glegg stated that they no longer occurred in the suburbs.

By the 1950s they could be found at only a few places in urban areas south of the Thames; the closest breeding area was on Wimbledon Common where one or two pairs bred. Towards the outer edge of the area in places such as Beckenham, Epsom, Purley, Walton and along the North Downs they were still reasonably common. They were more widespread north of the Thames and could be found much closer to the centre of London. At Northaw Great Wood up to 12 pairs were known in just one small area and in 1959, 83 males were counted to the west of the Lea Valley between Broxbourne and Hertford Heath.

In 1961 a census of Broxbourne Woods found more than 100 singing males and in the southern part of Epping Forest a further 29 were located. The following year at least 20 singing males were located south of the Thames. By 1963 there had been a large reduction at the main strongholds: at Northaw Great Wood the species had declined from at least 20 pairs a few years prior to just four, and in Epping Forest only 16 singing males were located; at Broxbourne Woods 12 were heard singing in a 0.7-hectare plot but no direct comparison could be made with earlier years.

In the first breeding Atlas (1968–72) they were found to be breeding in 70 tetrads in the London Area, mostly beyond the suburbs. In 1973 none were heard singing in Epping Forest during the summer for the first time ever; they had also declined at other former strongholds with just three at Broxbourne Woods and five at Northaw. In 1974, only one singing male was located at Bookham Common. By 1982 the breeding population had decreased further and only about nine pairs were reported. Over the next few years there was a partial recovery and by 1985 there were 32 territories located, including nine or 10 in Epping Forest and eight in Hainault Forest.

At the start of the 1990s the population had begun to decrease again and there were only about a dozen pairs; half of these were in Epping Forest, mainly around Connaught Water. During the second breeding Atlas in 1988–94 they were found breeding in 34 tetrads, a decline of 50 per cent from the first survey. The largest decrease was along the North Downs where they had almost disappeared. By 1998 they had recovered to about 36 breeding pairs, including seven at Bookham Common but just one at Connaught Water.

21st century

Since the start of the 21st century around 25–35 pairs have been recorded breeding annually. In 2000 the highest concentrations were of eight birds at Bookham Common and five at Fishers Green; subsequently the population at both of these sites has varied, with a peak of 10–12 pairs. Apart from those at Bookham, almost all of the other breeding Nightingales are north of the Thames. They rarely breed in the suburbs but a pair bred at Brent Reservoir in 2008.

Away from breeding areas Nightingales are occasionally recorded on migration; the main passage occurs in early May and they are infrequently seen in autumn.

Facts and Figures

The earliest arrival date was 22 March 1901 at Winchmore Hill.
The latest date was 24 September 1960 at Addington.

Bluethroat *Luscinia svecica*

A rare vagrant (20)

Range: breeds Europe and Asia, European population winters Africa; scarce passage migrant to UK.

The first record for the London Area was at Castle Bottom, Banstead, in September 1862 and was said by Bucknill to be of the 'red-spotted' subspecies (*svecica*). Another red-spotted was shot at Dartford Marshes in 1881. No more were seen until the autumn of 1936 when an immature was present for two days at Walthamstow Reservoirs. The 1960s saw the start of an increase in the number of records and four were seen, all in autumn.

In 1997 a first-winter male was trapped and ringed at Beddington on 3 August; it was re-trapped two years later on 13 June and it remained until 10 July.

Out of those that were identified to subspecies level there have been six red-spotted and five white-spotted.

1862: Banstead, 'red-spotted' in September
1881: Dartford Marshes, 'red-spotted' shot on 3 September
1936: Walthamstow Reservoirs, on 19–20 September
1942: Aldenham Reservoir, 'red-spotted' on 4 March
1955: Dagenham GP, on 3 September
1960: Northfleet, 'red-spotted' on 22 September
1963: Beddington, first-winter female on 22 October
1965: South Ockendon, first-winter male on 4 September
1967: Romford SF, on 23–24 September
1970: Berwick Ponds, 'white-spotted' on 30 August
1976: Beddington, immature trapped on 10 October
1977: Perry Oaks SF, 'red-spotted' on 10–12 September
1982: Tadworth, 'red-spotted' on 20 May
1983: Rye Meads, 'white-spotted' on 4 April
1994: Rainham Marshes, 'white-spotted' on 26 April
1996: Cornmill Meadows, 'white-spotted' on 3–5 April
1997: Beddington, first-year 'white-spotted' male trapped and ringed on 3 August; also on 13 June, and 4 and 10 July 1999
2004: London Wetland Centre, on 29 April–3 May, with two on 30 April
2006: Staines Moor, first-summer male on 22 April

Red-breasted Flycatcher *Ficedula parva*

A rare vagrant (10)

Range: breeds eastern Europe and Asia, winters southern Asia; scarce visitor to UK.

The first record was in Ladbroke Square, Notting Hill, in August 1939; it still remains the only sighting in Inner London.

The 1950s was the peak period with four recorded. The first of these was an adult male on a fence by the mouth of the River Darent at Dartford Marsh in September 1954. Remarkably, the following year two were seen in August, at Chiswick and at Beddington SF, where one remained for a week. The final record of the decade was in a garden at Banstead.

There were no further records for 10 years when one was seen in Wanstead Park. Two occurred in successive years, 1973 and 1974, the second of which was at French Street, Westerham, on the exceptionally early date of 18 July. The next two both arrived on 5 November: one was present for less than 10 minutes in Kew Gardens while nine years later one was found in Beaulieu Heights Wood, where it remained for three days.

1939: Ladbroke Square, on 8 August
1954: Dartford Marsh, on the River Darent on 12 September
1955: Chiswick, on 13 August
1955: Beddington SF, on 31 August–6 September
1958: Banstead, on 18 August
1968: Wanstead Park, on 4 September
1973: Upminster, on 30 August
1974: Westerham, on 18 July
1989: Kew Gardens, on 5 November
1998: Beaulieu Heights Wood, on 5–7 November

Pied Flycatcher *Ficedula hypoleuca*

An uncommon passage migrant; historic breeder

Range: breeds Europe and Asia, winters Africa.

Historical

This was a fairly scarce species in the London Area in the 19th century, mainly seen on migration with sporadic breeding records. There were at least seven documented breeding records in the 19th century although the species probably used to breed on a regular basis, particularly in the southern half of the area. The earliest reference was in 1812 when a pair bred in Peckham; a nest with three eggs was taken at The Grove, Harrow, in around 1836; a nest with three young was taken near Hampstead Heath in 1866; two nests were taken near Tooting prior to 1873; a pair nested at Mottingham in 1889; a nest was found in the early 1890s at Harrow-on-the-Hill; and there is an undated record from Tooting Common where a pair lost its eggs to a snake.

Hampstead Heath was a particularly good site for this species on passage as five were caught there in May 1859 and four were taken in August 1868. The only known sighting in Inner London was in Kensington Gardens on 29 April 1890.

20th century

In the first quarter of the century, it had ceased breeding in the London Area and was rarely seen on migration; it was only recorded in six of these years.

By the 1950s it had become 'an irregular and scarce double passage migrant' and was being seen more frequently. It was seen most often in large open spaces surrounded by suburbia, in particular on Hampstead Heath. It was also being seen in central London, mainly in the Royal Parks in autumn. In 1951 exceptional numbers were recorded in autumn, peaking at 15 on 3 September in Regent's Park. In spring they were usually seen in April and May, while in autumn the main months were August and September. Between 1929 and 1954 there were five occurrences in June and three in October.

In 1961 a pair was present in Dulwich Woods from 12 April to early June but there was no evidence of any breeding activity. During the 1960s more were seen at well-watched sites such as Regent's Park, when there were

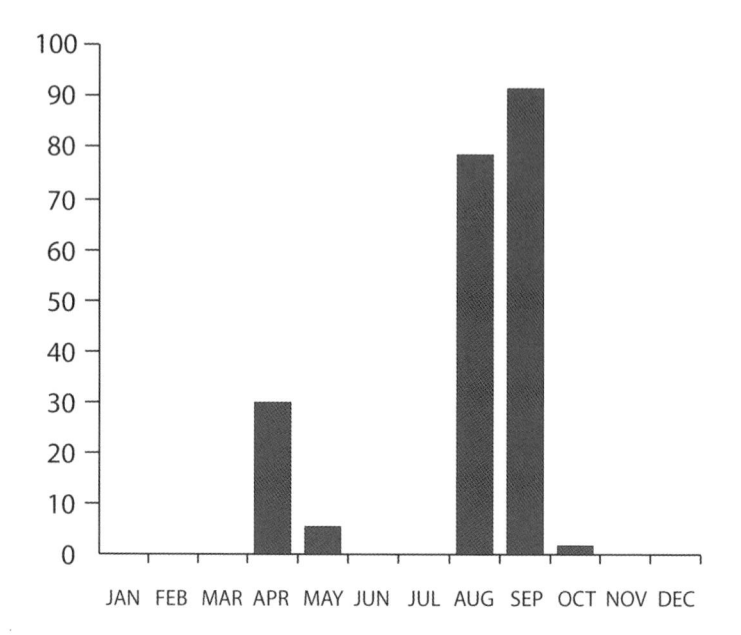

Seasonal occurrence of Pied Flycatcher: 2000–2009.

increased numbers at coastal observatories and it was thought that many of those seen in the London Area were continental migrants. An analysis of all migrants between 1974 and 1986 showed a 3:1 ratio of autumn to spring migrants, with the peak being the middle of August. In 1987 a male held territory in Osterley Park between 25 April and 2 June, and built a nest in a nest-box although no female was ever seen.

In the 1990s exceptional numbers of migrants were recorded in two autumns: 83 in 1995 and 92 in 1997. In the latter year the peak was on 24–25 August when 21 were seen.

21st century

Since the start of the 21st century it has become much scarcer on migration with an average annual total of 22. Migrants are considerably scarcer south of the Thames; for example, in 2008–09 eight times as many were seen north of the river. This may be due to their migration strategy as most birds in the London Area are seen in autumn when they undertake shorter movements, whereas in spring most fly direct to their breeding grounds, bypassing southern England. They are sometimes seen in the central London parks; for example there were four in 2008.

Facts and Figures

The earliest migrant was on 5 April 1995 at Bletchingley.
The latest migrant was on 6 December 1970 by Barn Elms Reservoirs.
The largest count was 17 birds in Hyde Park and Kensington Gardens on 16 August 1975.

Black Redstart *Phoenicurus ochruros*

An uncommon partial migrant and localised breeder

Range: Europe, Asia and North Africa; scarce breeder in southern UK and uncommon passage migrant with some overwintering.

Historical

In the 19th century this was a rare winter visitor to the country; the first British record was of a bird shot by Bond in a brickfield near Kilburn on 25 October 1829. This was followed by singles at Regent's Park and Shepherd's Bush, the latter also being shot in a brickfield in October 1849; and in a garden at Stone in the autumn of 1855. There were three records from Hampstead, one of a bird caught on 14 April 1868; one of a pair captured in autumn 1868; and two in spring 1869. One overwintered in the grounds of the Natural History Museum in South Kensington from the end of November 1885 to 8 January 1886, and in 1893 one was seen near Rickmansworth on 28 March.

20th century

Black Redstart had slowly started to colonise England in the early years of the 20th century. Following an isolated instance of breeding in Durham in 1845 a pair bred in Sussex in 1909 and two pairs bred there in 1923. This followed a spread across north-west Europe over the previous century and a half. In the first quarter of the 20th century it was a scarce migrant in the London Area; of the 16 records, 10 were between February and April, and only three were in October and November, whereas in other inland areas most records were in early November.

In 1926 a pair bred at Wembley among the buildings where the British Empire Exhibition had been housed the previous year. Three pairs then bred there annually up to 1941, apart from in 1937 when there were four pairs. They bred again in 1944 after a three-year absence due to disturbance during the Second World War. Elsewhere, a male and female were seen separately in the grounds of the Natural History Museum in South Kensington during the summer of 1927 and a pair bred at Woolwich Arsenal in 1933.

It began to summer in central London but breeding was not confirmed there until 1940 when a pair raised two broods in Westminster School; a further five males held territory. From then on at least two pairs bred annually in the London Area. In 1942 Black Redstarts began to take advantage of buildings destroyed by the Blitz to expand their population, and single pairs bred in the City for the first time as well as in Notting Hill and Wandsworth; a further 19–24 males held territory in the built-up areas of London. By 1948 eight or nine pairs were breeding in the City and Westminster, and more than 20 other singing males were located across the London Area. The following year 11 pairs bred in the City and fledged 76 young. From 1945 through to the 1950s the majority of breeding birds were found around Cripplegate (now rebuilt as the Barbican).

During this period of colonisation they were also increasingly seen in winter. From 1926 to 1950 there were 28 records away from known breeding territories between December and February. From the late 1940s onwards occasional birds started to overwinter in the Cripplegate area.

The population was presumed to have remained stable at around 15 pairs during the 1950s but actual breeding figures were harder to come by as observer effort diminished. Studies showed that the males arrived at the end of March and the females in April, and that they departed between late September and late October. The birds bred almost exclusively on bombed buildings. Away from central London breeding also took place at Beddington cement works and in Wandsworth, with up to four pairs breeding at Croydon Power Station in the mid-1950s.

During the late 1950s and into the 60s Black Redstarts' favoured bombed sites were increasingly developed and the birds moved to other sites, such as Dartford, Paddington and Ponders End. In 1964 they were found breeding in the Lea Valley for the first time; four pairs bred between Ponders End and Stratford. The following year they failed to breed in the City of London for the first time since 1942. The breeding population increased during the 1970s as they spread into new areas with five pairs at Beckton Gas Works and four pairs in Surrey Docks. By 1978 there were 23 pairs and another 23 singing males. The population peaked in 1985 when at least 26 pairs bred.

The second breeding Atlas in 1988–94 found breeding birds in 48 tetrads, an increase of 45 per cent on the first survey in 1968–72. At the start of the 1990s there was a breeding population of at least 12 pairs with another 10 singing males recorded. The largest concentration was now at Rainham Marshes where three pairs bred; there was also a small overwintering population. They continued to decline in some areas, such as Kings Cross, due to redevelopment. Increased fieldwork in 1998 led to the discovery of 23 pairs and 11 other singing males.

21st century

In 2000, 31 pairs were located throughout the London Area; these were found during an intensive survey rather than as casual submissions and show how difficult this species is to track by conventional means. A similar study the following year managed to locate only 20 pairs. The bulk of the population is situated along the Thames and in central London. There was a notable influx into central London in 2010 with up to 20 pairs/territorial males present; most of these were in the West End and the City.

A few remain in their breeding areas all year and there is a small wintering population on the large reservoirs. Elsewhere, a few migrants are seen on passage.

Common Redstart *Phoenicurus phoenicurus*

An uncommon passage migrant; former breeder

Range: breeds Europe, Asia and North Africa, Western Palearctic population winters Africa.

Historical

The earliest documented record was on 26 April 1785 when Gilbert White heard one singing in South Lambeth.

In the 19th century it was a fairly common breeding bird across the London Area. In 1832 Doubleday, writing about Epping Forest, stated that 'the Redstart has arrived this year in immense numbers. I have never seen so many before'. Subsequently he stated that they had become even more abundant and 'the forest literally swarms with them'. A decline was noted in some areas before the effects of suburbanisation took over; for example, in Harrow it was plentiful between 1830 and 1840 but by 1891 was seldom seen there. At Hampstead 16 were taken between April 1870 and October 1871, which may have led to a local decline.

Two pairs bred in Kensington Gardens in 1876 and it also bred near Regent's Park. In the inner suburbs breeding occurred at Dulwich Woods, Hampstead Heath, Highgate Woods and Streatham Common.

20th century

Around the turn of the century it was still common in many areas; for example, it bred commonly in gardens in the Bromley district, was plentiful between Gravesend and Dartford, and 12 pairs bred in Richmond Park. However, there was a notable decline in the population in Epping Forest between 1890 and 1910. Kerr stated in 1906 that it had never been very plentiful around Staines but was becoming rarer.

Losses were recorded with increasing regularity in the 1920s. Breeding ceased at Colindale and Edgware after 1925 due to building development; it last bred at Finchley and in Osterley Park in 1927; and in the Hampstead and Highgate district it last bred in 1928. In 1935 Glegg described it as 'a scarce and local summer resident in the rural areas' that could sometimes be seen on passage in the suburbs; he added that it 'was formerly very regular on the spring passage in Inner London, but in recent years it has become scarce, and it is rarely seen in autumn'. Richmond Park remained one of its strongholds and 20–25 pairs bred there in 1935–36.

By the 1940s it had ceased breeding in the inner suburbs. In 1944 Alexander and Lack stated that it had suffered a 'very marked decrease throughout southern, eastern and central England' and this was clearly reflected in the London Area. Although the cause of the national decline was unknown suburbanisation was a major factor in the London Area. Dent considered that the decline in Epping Forest was due to the growing up of old pollards, which closed the canopy and resulted in a reduced supply of food. By 1953 the number of pairs breeding in Richmond Park had dropped to nine and the decline was attributed to the loss of nest sites when a considerable amount of old timber was removed.

In the 1950s they were a very local breeder, confined to the well-wooded areas of the London Area. They continued to breed regularly at Bushy Park, Epping Forest, Hainault Forest, Northaw Great Wood, Petts Wood, Richmond Park, St Paul's Cray Common, Shirley Hills, South Weald Park, Stanmore Common and near Whitewebbs Park. Occasional pairs bred elsewhere, such as at Hampstead in 1954.

At the start of the 1960s the breeding population was mostly north of the Thames and centred on Broxbourne, Northaw Great Wood and Epping Forest. In 1961 there were 24 territories in Broxbourne Woods and a pair bred again on Hampstead Heath. In 1963 concerted fieldwork found a total of 90 territories across the London Area with at least 59 breeding pairs; the largest concentration was in Epping Forest where 56 males were located. However, only 11 singing males could be found in Epping Forest just two years later despite good coverage. The population scarcely recovered from this sudden decline and by 1970 there were still only 17 territories in Epping Forest. At Northaw Great Wood there had been a similar crash with just one pair in 1970 compared with 15–20 pairs four years earlier.

The population continued to decline and by 1976 breeding had become restricted to just four sites: two localities in Epping Forest, Limpsfield Chart and Northaw Great Wood. In 1980 there was only one breeding pair left in the London Area, in Epping Forest, although another three singing males were also present. Over the next few years singing males were present in Epping Forest and at Northaw Great Wood but no breeding occurred until 1985 when two juveniles were seen in Epping Forest, the first confirmed breeding since 1977. Although none bred the following year, the breeding population in Epping Forest rose during the late 1980s and in 1988 there were six pairs, at least five of which bred. However, in the early 1990s the breeding population went into terminal decline; from 1991 to 1994 just a single pair bred in Epping Forest and by 1995 the species had ceased to breed in the London Area.

21st century

Since the start of the 21st century they have remained a fairly widespread but uncommon passage migrant with annual totals varying from 45 to 90. In spring most are seen in April while in autumn they typically occur from late July onwards; most have departed by the middle of September with a few being seen up to the middle of October. On 12 December 2004 one was found at Wormwood Scrubs and remained until 3 January 2005, the first case of attempted overwintering by this species in the UK; there have also been two other records in December. A male held territory at a former breeding site in Epping Forest in 2005.

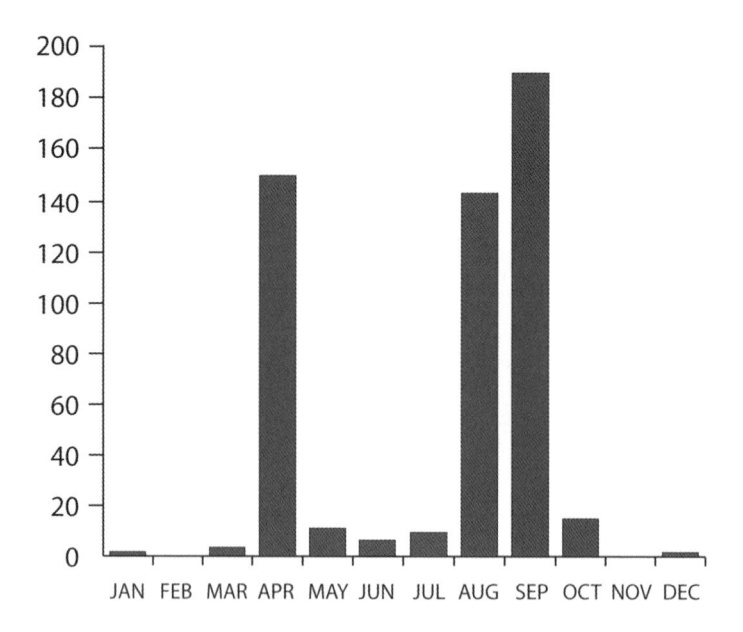

Seasonal occurrence of Common Redstart: 2000–2009.

Facts and Figures

The earliest arrival was on 17 March 1928 at Walthamstow Reservoirs.
The latest migrant was on 19 December 1970 at Perry Oaks SF.

Whinchat *Saxicola rubetra*

An uncommon passage migrant; former breeder

Range: breeds Europe and Asia, winters Africa.

Historical

In the 19th century Whinchat was a fairly common breeding bird in the London Area, though subject to periodic fluctuations. In 1831 Doubleday noted that it was numerous in Epping Forest but by 1839 he remarked that it had become scarce. In some years it was particularly numerous, especially so in 1861 when Harting noted that there was 'an extraordinary visitation of Whinchats'.

20th century

At the beginning of the century Whinchat was a locally common breeding species, being more populous north of the Thames. In the Enfield area it was said to be exceedingly abundant with one or two pairs in every field and meadow. In the inner suburbs, breeding was recorded on commons and other open sites in south London at places such as Dulwich, Norwood and Streatham between 1890 and 1910. Subsequently the birds began to nest in Richmond Park and up to four pairs bred regularly. They were relatively scarce on the North Downs, probably due to the pastures being closely cropped.

In the 1930s Glegg described it as 'a well-represented summer resident in the rural areas'. It was considered to be a comparatively abundant breeding species around South Harrow on rough land and particularly along the banks of the Metropolitan Railway. Locally common breeding populations included about 12 pairs breeding within a 1-kilometre radius of Aldenham Reservoir; at least six pairs on Chingford Marsh; six pairs at Edmonton and three pairs at Barn Hill, Wembley. The colony in Richmond Park ceased to exist in 1940 because of military manoeuvres; it had also disappeared throughout south London after 1938, except on the Thames marshes. North of the river it had declined to about 18 pairs.

The decline continued during the 1940s when there was a significant decrease in numbers. The reason for the decline appears to have been loss of habitat due to building, uprooting of scrub, earlier cutting and trimming on railway embankments, and the ploughing of pastures for crops. During the early 1950s one or two pairs returned to south London where they bred at Walton GP. Along the Thames, 10 pairs bred on the marshes between Abbey Wood and Swanscombe; on the opposite shore they bred on the marshes and in the Ingrebourne Valley and Mar Dyke. At least 13 pairs bred in the Lea Valley. In north and west London there were only a few scattered pairs apart from around Staines and Perry Oaks where they were well established. The breeding population in 1955 was at least 40 pairs.

By the start of the 1960s there had been a serious range contraction and Whinchats were very scarce away from the Lea Valley and Staines area; the only breeding record south of the Thames was at Swanscombe Marsh. The population rapidly decreased and in 1964 the only confirmed breeding pairs were at Elstree Aerodrome and Rainham Marshes. Although they bred at five sites the following year the population remained at a very low level. During the early 1970s the only breeding pairs were all in east London, mainly on the Lower Thames Marshes and in the Ingrebourne Valley, but after 1974 they ceased breeding although a pair summered at Rainham Marshes in 1979.

At the beginning of the 1980s two pairs were present during the breeding season at Rainham Marshes; elsewhere, they were a reasonably widespread passage migrant. During the decade there were sporadic breeding attempts, mainly at Rainham Marshes; for example, in 1989 four pairs summered, one of which bred at Rainham. By 1991 they had ceased breeding in the London Area although one or two occasionally summered at Rainham in subsequent years.

21st century

Although Whinchat is still a regular and widespread passage migrant, the annual totals are lower than those in the 1990s. Typically 250–350 are recorded in most years. It is scarce in most urban areas apart from at a few traditional stopping-off sites such as Wormwood Scrubs and Wanstead Flats; even at the former site the peak counts have declined since 2000. The main passage occurs between mid-August and mid-September.

Facts and figures

The earliest migrant was on 8 March 1853 in Middlesex.
The latest migrant was on 20 December 1998 at Staines Reservoir.
There have been several winter records, including one at Waltham Abbey on 9 January 1983.
The largest count was 25 birds at Rainham Marshes on 1 September 1990.

Stonechat *Saxicola rubicola*

An uncommon partial migrant and localised breeder

Range: Europe, Asia and Africa.

Historical

In the 19th century Stonechats were seen throughout the year but were less common in winter. In 1831 Doubleday stated that in Epping Stonechat 'used to abound on the furze bushes by the sides of the forest, but, what is very singular, it has within the last three years totally disappeared'.

Harting noted that they were scarce breeders, mainly due to the lack of suitable habitat in north London. Few breeding locations were specified but they were known to breed at Greenford, Hampstead Heath and Perivale Wood. Elsewhere, they bred at Banstead Common, Epsom Common, Headley Heath, Mitcham Common, Orsett Heath and Wimbledon Common.

20th century

In the first quarter of the century the bulk of the breeding population was to the south of the Thames. There were up to 11 pairs on Dartford Heath and six pairs on Walton Heath, and they also bred at Epsom Common, Hayes Common, Mitcham Common and Richmond Park. Elsewhere, they bred on the banks of Queen Mary and Staines Reservoirs (up to 1907) and irregularly in the open parts of Epping Forest. Occasional pairs bred elsewhere in the London Area.

In the 1930s Glegg knew this as a local and uncommon resident in the rural areas, which was rarely seen in the suburbs. Across the London Area it was a partial migrant with many of the breeding pairs vacating their territories in autumn and returning in spring while birds from elsewhere wintered in the London Area. The spring passage was mainly during March. In 1933 it was specifically surveyed throughout the London Area; breeding was reported south of the river from seven locations on the North Downs, on Mitcham Common, in Richmond Park and on Wimbledon Common. A pair bred near Black Park on the western border; at three sites near Mill Hill; at Ruislip Lido; and at Barn Hill near Wembley. No nests were located anywhere in the east although they may have been missed on the Thames-side marshes as these were underwatched at the time. The overall results showed a small decline since the beginning of the century. The following year a pair probably bred on Swanscombe Marshes.

Between 1933 and 1950 there was a substantial decline in the population, not just in the London Area but also in many areas of the country. This was mainly due to the exceptionally cold winters of 1938/39, 1939/40 and 1946/47. The last of these winters was so severe that only one pair was found breeding in the London Area in the summer of 1947, on Walton Heath. There was a partial recovery in numbers during the late 1940s and early 50s and up to four pairs bred, but this small revival was not sustained and they went into another decline. After 1955, when a pair bred at King George V Reservoir, no breeding was reported between 1956 and 1960. Three pairs bred in 1961–62 but after the exceptionally cold winter of 1962/63 there were no further breeding records until 1966. There was a very slow recovery over the next few years, mainly on the Lower Thames Marshes.

At the beginning of the 1970s the breeding population had increased to five pairs and by 1974 it had increased to 25 pairs, a return to the level of the early 1930s. There were now two main populations: on the Lower Thames Marshes and on the North Downs, with another four pairs in Richmond Park. The breeding population expanded in 1979 with pairs penetrating the suburbs, where they bred at Brent Reservoir and Wembley.

In the early 1980s they had ceased breeding in north-west London but elsewhere the population was still increasing and it peaked at 38 pairs in 1984, including 16 at Rainham Marshes. Three successive cold winters reduced the breeding population to just six pairs in 1987, five at Rainham and one on Hounslow Heath. At the start of the 1990s the breeding population was expanding again and had reached a minimum of 30 pairs; the largest concentration was at Rainham Marshes with 25 territories, the highest number ever recorded at a London Area site. However, cold weather in 1991 reduced the population to about 15 breeding pairs, the level the species remained at before increasing from the middle of the decade.

21st century

At the start of the 21st century there were 27–28 breeding pairs at six sites with more than half being at Rainham Marshes. Despite relatively mild winters in subsequent years the population decreased and by 2005 was down to just 11 pairs. It remained around this level for the next few years but after two successive cold winters in 2009/10 and 2010/11, the number of Stonechats has significantly decreased. The breeding population has fallen to just a few pairs and the wintering population has also diminished. Traditional sites for autumn passage and overwintering birds, such as Rainham Marshes, Staines Moor, Tottenham Marshes and Wormwood Scrubs, have recorded reductions of up to 90 per cent since the first of these cold winters.

Facts and Figures

The highest count was 40 birds at Richmond Park on 6 October 2008.

Northern Wheatear *Oenanthe oenanthe*

A common passage migrant and rare breeder

Range: breeds Europe, Greenland and Asia, winters Africa.

Historical

Wheatears used to breed in south London in the 19th century, along the North Downs, in Croydon in 1874–75 and occasionally on Wimbledon Common. They were also captured in immense numbers and sold in the markets for food. North of the river they were widely seen on passage, especially on rabbit warrens as well as in old gravel and chalk pits.

20th century

At the beginning of the century a few pairs continued to breed across the North Downs and surrounding heaths. They also bred in the Darenth Valley and within the metropolis on Wimbledon Common and in Richmond Park. At the last site they last bred in 1908; their decline was a direct result of the elimination of rabbits which occurred in 1904 when they were all killed and their burrows filled in, leaving nowhere for the wheatears to nest.

Elsewhere, they declined rapidly and breeding was only known at Epsom and Purley in 1912; on Walton Heath in 1914; and at Caterham in 1917. There were no further breeding occurrences south of the Thames until 1941 when a pair bred on Walton Heath. Just after the Second World War there was a small series of breeding records: in 1946 two pairs nested on bomb-damaged rubble at Blackheath; a pair bred in the vicinity of Woolwich Common in the following year; and in 1949 a nest was found at Upper Warlingham in the council dump.

North of the Thames the only confirmed breeding records were in 1920 at Edgware, and in 1930 at Bushy Park and Harefield Place. Breeding possibly took place in 1944 in the vicinity of Brent Reservoir; breeding was also claimed at the Wembley Exhibition in 1924 but no further details were known.

In the second half of the century Northern Wheatears bred only sporadically. A pair bred in Richmond Park in 1955. There were also several instances of attempted breeding, such as a pair on a bombed site in Cripplegate in 1959, but their likely nest site – an old wall – was demolished. There were no further successful breeding attempts until the late 1970s when a pair bred in east London for three consecutive years from 1978 to 1980.

21st century

In 2000 a pair attempted to breed on Walton Heath but later deserted the site, possibly due to the eggs being predated; since then there has been no further breeding attempt.

Migrants are usually seen from the middle of March through to the middle of May, and again from late July to September with a few in October; the largest flocks are typically seen in spring. Although they are common on passage throughout the London Area, most occur on the Lower Thames Marshes, on the banks of the large reservoirs, at Beddington, the Wetland Centre and Wanstead Flats.

Facts and Figures

The earliest migrant was on 14 February 1834 in Middlesex.
The latest migrant was on 28 December 1990 at Walton Reservoirs.
The largest flock was of 200 birds in Hyde Park, on 8 April 1902.

'Greenland Wheatear' *O. o. leucorrhoa*

Glegg listed three definite specimens taken at Hampstead in 1869, a female on 5 May and a pair on the 7th. He knew of other sight records but did not accept them, although in all probability some of these were genuine.

From the late 1920s onwards they were reported in most years and are now annual in small numbers with the vast majority recorded in May; any apparent increase in frequency is likely to reflect improved optical equipment and increased expertise amongst birders. Two were trapped at Romford SF on 1–2 May 1954 and their measurements conformed to this subspecies. The largest count was 16 birds at Aldenham Reservoir on 1 May 1934.

Desert Wheatear *Oenanthe deserti*

A very rare vagrant (1)

Range: North Africa, Middle East and Asia; vagrant to UK.

There has been a single record of this rare species in the London Area, of a first-summer male in 1989 that spent two days at Barn Elms Reservoirs in the middle of April and was even heard singing occasionally. This was the thirtieth individual in Britain and Ireland and followed one in Kent a week earlier. Out of the 139 accepted records in the British Isles up to 2011, only nine were in spring whereas the vast majority arrived in late autumn with some remaining through the winter.

1989: Barn Elms Reservoirs, male on 13–14 April

Dunnock *Prunella modularis*

A very common resident and widespread breeder

Range: Europe and Middle East.

Historical

The earliest reference was by Gilbert White who heard one singing at South Lambeth on 19 February 1782. In the 19th century Harting noted that this species was common everywhere, despite its nests being 'plundered without remorse'.

20th century

In the 1930s Glegg recorded that this was one of the most common residents and could be found in central London where it was a fairly common breeder, as well as throughout the suburbs and rural areas.

In the 1950s this was considered to be one of the most widespread birds in the London Area, breeding in the central parks, suburban gardens and rural areas. After the extremely cold winter of 1962/63 the breeding population in Dollis Hill decreased from about 48 pairs to 25–30 pairs. In 1966 there were about 50 breeding pairs in Regent's Park, an increase on the 39–45 pairs in 1959.

In 1977 large increases were noted at some sites compared to the long hot summer the previous year; for example, the population increased from 20 territories on Bromley Common to more than 40, and from 46 in Regent's Park to 64. After a spell of severe weather in February 1991 there was a decline in reported breeding numbers of up to one-third. The second breeding Atlas showed an increase in the breeding distribution of 12 per cent compared to the first Atlas. Some of this was due to under-recording in the first survey but Dunnocks appeared to have colonised much of central London in the intervening period.

21st century

Dunnock is widely distributed throughout the London Area, although within built up areas it is confined to parks and green spaces with sufficient cover in the form of shrubberies, hedgerows or scrub. There is some evidence of a decline in the central area in recent years, as Dunnocks were found in 90 per cent of 20 small open spaces in a survey of Inner London in 1987–88 but only 9 per cent in a study by the London Biodiversity Partnership in 2004. The largest monitored breeding population is at Rye Meads where the number of territories has slightly increased from 35 in 2000 to 39 in 2008–09; elsewhere, the population has been fairly stable. Although numbers appear to fluctuate quite sharply, the Breeding Bird Survey in London has shown a 15 per cent increase from 1995–2010.

Movements

Movements of Dunnocks have often been recorded. For example, on 18 April 1868 many were seen in London Zoo in the morning after apparently arriving during the night; and on 5 October 1904 there was such an influx in Kennington Park that Power stated that it 'seemed to be alive with them'. He also added that in October they could be seen in parties of up to six moving westwards from garden to garden.

There was a widespread irruption in 1961 which was particularly noticeable at Rye Meads where numbers increased from 10 on 4 November to about 100 the next day, and increases were also noted at a number of other sites, including in central London. At the end of the year an influx of about 60 was noted at Brent Reservoir during heavy snowfall. Ringing at Kempton Park established that there was a spring and autumn passage; for example, 66 birds were ringed in September 1975. The following year 70 were ringed at Queen Mary Reservoir in September.

Alpine Accentor *Prunella collaris*

A very rare vagrant (2)

Range: breeds montane Europe, Asia and North Africa, winters lowlands; vagrant to UK.

The first London Area record, of a bird shot in Epping Forest in 1817, was also the first occurrence of this species in Britain. It was shot by James Pamplin in a garden among a flock of Chaffinches; as he did not know what species it was, he gave the specimen to someone who took it to London where it was identified by Mr Gould, a naturalist.

In about 1883 one was caught in a trap in the Penge garden of collector Arthur Butler. He later described that he first thought 'it might be only an unusually large, brownish and somewhat aberrant variety of the Hedge-Sparrow: it was evidently a young bird, as the white throat-patch was barely indicated'. He kept it in a cage with two Dunnocks and remarked that 'by the side of their rare relative, [they] looked insignificant, much as a Song thrush by the side of a Missel-Thrush'. It died in such a poor condition that Butler did not preserve the skin.

1817: Epping Forest, one shot at Forest House in August
c. 1883: Penge, captured alive in about September

House Sparrow *Passer domesticus*

A very common resident and widespread breeder but sharply diminished in number in the past 25 years

Range: Europe, Asia and North Africa.

Historical

The earliest documented sighting in the London Area was in 1779 when Gilbert White saw House Sparrows nest-building in South Lambeth on 27 February.

In the 19th century this was 'the commonest of all common birds' and acquired the name 'Cockney Sparrow'. It would have been a familiar sight to all London residents despite suffering persecution and even being despised. Thousands were captured every year by bird-catchers, mainly in winter, and were used either for shooting from traps or for decorating hats after being dyed. In 1884 they were extremely common in London Zoo and 'at dusk, the evergreen trees swarmed with them, the keepers seizing the opportunity to net them, to supply some animals with food'.

Their staple diet was the scraps dropped by horses from their nosebags onto the roads, and Hudson stated that they would follow people in the parks in the hope of obtaining food. Large numbers were resident in all of the central parks; for example, several thousand resided in Battersea Park. Hudson estimated the population in 1898 to be about two or three million in London.

Large roosts were known in central London, such as on the Bank of England and on Southwark Bridge. At one particular roost in Bethnal Green, a very severe rainstorm on 25 July 1867 killed nearly all the House Sparrows roosting in the trees and about 200 were found dead the following morning.

20th century

For much of the first half of the 20th century there was virtually no change in the status of House Sparrow. In the 1930s Glegg described it as 'one of the most abundant and certainly the most generally distributed bird in Middlesex'. Several thousand roosted in Richmond Park each winter, in the rhododendrons. Coordinated counts in the winter of 1949/50 found 551 in St James's Park and 432 in Green Park.

House Sparrows had to time their breeding carefully; if they were late with their second brood they would not be in time for the harvest and the fields of stubble. In the late 19th century it was believed that they left central London at the end of summer and went to feed on the stubble in nearby cornfields. However, a definitive study by Southern in 1945 concluded that they foraged within three kilometres and that the birds in the cornfields were most likely to be from the suburban areas and not from the central London population which was sedentary.

On the Thames, House Sparrows flocked around the grain barges as they were being unloaded to feed on any spill, and they sometimes fed along the tide-line. They were often observed entering buildings to exploit new food sources. At one time, for example, they were regularly seen in the wards of Central Middlesex Hospital, perching on beds and collecting food from the floor. In one London brewery there was so much available food in the form of spilt grain that they rarely left the building. They were also seen at a dairy in St Pancras feeding on the scraps of fat left in used milk bottles.

By the start of the 1950s, although they were still extremely common residents, there were fewer in central London. They were nevertheless at their densest in central London with the highest numbers being found in parks, gardens, bombed sites, docks and other open spaces rather than completely built-up areas where there were few trees. Two studies in 1950 showed that there was a density of 4.0 to 4.3 Sparrows per acre in Lambeth and Bloomsbury, respectively, compared to a rural area like Harlington where a 1948 study returned 0.2 per acre. In some of the central London parks the densities were even higher: surveys in Battersea Park in 1945 and 1950 showed a density of five per acre, and winter counts in Kensington Gardens showed nine per acre in 1925–26 (a total of 2,603 birds), although this had dropped to three per acre in 1948–49.

The results of this last survey, along with the views of several observers, showed that a real decline of House Sparrows in central London had started although it was to take another 50 years before they would disappear altogether from many localities. The decline was blamed on the reduction of available food as horse traffic had been completely replaced by motor vehicles. However, this may not have been the only reason as the birds diet in Kensington Gardens largely consisted of bread and it was possible that the over-reliance on this food meant that they were not getting enough nutrients to sustain them.

In the mid-1950s there was a roost of some 5,000 on Mitcham Common outside the breeding season; about 5,000–6,000 were counted at Kenwood, Hampstead; and about 10,000 roosted near Croydon in the autumn, with some travelling up to 5 kilometres to the roost. In 1960 some 18,000 roosted on Banstead Downs on 22 November.

The population in the outlying areas which had become suburbia would have increased as House Sparrows were very quick to colonise newly built-up areas. Due to the spread of London during the first half of the 20th century its new House Sparrow colonies probably more than compensated for the reduction in numbers in central London. In 1966 the population in central London was estimated to be 10,000–20,000 pairs. In the first breeding Atlas in 1968–72 they were found breeding in 93 per cent of the London Area; most of the gaps were in east London where there was some under-recording. In 1975 the count in Kensington Gardens was 544 on 2 November, more than 2,000 fewer than the count 50 years earlier. In Dollis Hill the population declined from 900 pairs from 1951 to 1965 to 550 by 1980; Eric Simms believed that some of this decline was due to a loss of nest sites as houses were renovated.

At Dartford Marsh there were some exceptionally large flocks during the 1970s, including one of about 7,000 birds in November 1971 and another of 11,000 at three roosts in December 1974, but by 1984 the roost had declined to 2,000. Elsewhere, the highest counts were 3,000 at a roost in Hornchurch in 1980; 4,000 at Harold

Hill in 1981; 4,000 at Dagenham in 1982; and 2,000 at both Sewardstone and Little Parndon in 1984. The largest concentration during the 1990s was 2,500 at a roost in Romford in 1991.

By the beginning of the 1990s House Sparrows were rapidly disappearing from the central London parks. In 1991 a spring bird census in Battersea Park found a total of 92 compared to a previous census in 1950 when 1,058 were present. In a survey in August 1996 only one sparrow was seen in St James's Park compared to 124 in 1988, and none were found in Regent's Park compared to 167 in 1988. The 1988-94 Atlas showed no decrease in range, indeed House Sparrows were recorded in slightly more tetrads than 1968-72, though the difference may partly reflect differences in recording.

By 1994 it had become apparent that some local suburban populations had declined. For example, in the Dagnam Park area many breeding colonies had disappeared over the previous 10 years, and at Essex Filter Beds flocks had decreased in size from up to 100 a few years earlier to 20-odd.

The biggest losses were in central London; by 1996 the population in Hyde Park and St James Park had fallen to about 20 pairs and three pairs, respectively (there had been 157 birds counted at the latter site in 1976). The following year only four or five pairs bred in Holland Park compared to about 30 two years earlier, and the breeding population had declined in all the Royal Parks. House Sparrows have not been seen in St James's Park since May 1999.

21st century

By the start of the 21st century the population of House Sparrows had drastically declined in Inner London; they had disappeared from Holland Park and St James's Park and none were recorded in a survey in Hyde Park. Six pairs bred in Kensington Gardens in 2000 (though four of the young were taken by Sparrowhawk) and one or two pairs bred in Regent's Park; and they were only recorded at a further dozen sites in Inner London. There has also been a marked decline in many suburban areas. A survey of 61 private gardens from 1995–2003 has tracked the species' decline, showing a fall from an average peak monthly count of 13 in 1995 to about three by 2002 in the 12 most studied gardens, with several gardens losing all their sparrows during this period (Baker 2004).

An example of the level of decline, particularly in Inner London, can be seen by comparing the Autumn Bird Counts in Kensington Gardens. On the first count in 1925 there were 2,603, but only eight were present in 2000. The Breeding Bird Survey in the London Area recorded a drop of almost 60 per cent between 1994 and 2000. Even the Prime Minister remarked in 2000 that 'The House Sparrow, once more cockney than the cockneys, is now a rarity in London'. In 2005 the Autumn Bird Count in Kensington Gardens failed to record a single House Sparrow for the first time ever. However, they had not entirely vacated Inner London as a flock of up to 170 was discovered at Surrey Docks in 2006 and there are small colonies around the Tower of London and at London Zoo; apart from occasional birds from the latter colony which are sometimes seen in Regent's Park they are still absent in the central London parks.

Reasons for the decline in the House Sparrow population are not entirely clear but may include the following: changes to house maintenance which block birds entering the eaves and reduce the availability of other crevices for nesting and replacement of old industrial buildings by office blocks with plate glass walls; the huge loss of brownfield sites to development, whose seeding vegetation was a major food source in late summer; loss of front gardens to paved drives further reduces foraging habitat; increased predation as there have been several anecdotal reports of Sparrowhawk or Kestrel nesting near some of the last remaining colonies in some parts of the capital; and research at Leicester suggests a shortage of invertebrate food for nestlings is a key factor. The RSPB is currently carrying out a research study to provide additional food for breeding colonies in London.

Nesting Places

House Sparrows have nested in the London Area for centuries, but there has been very little documentation about historic nesting sites. In 1831 Rennie wrote that they bred in unglazed pots under the eaves of houses in the vicinity of London. They have used many odd sites for nesting; a pair once built a nest in the hollow of the lock attached to the entrance gates of the Hanwell Lunatic Asylum. The fact that these gates were locked and unlocked rarely less than 300 times in the course of a day renders the circumstance especially remarkable; the eggs were laid, and the young successfully reared.

When the original Crystal Palace was erected in Hyde Park for the Great Exhibition in 1851 it was said that so many House Sparrows got inside the building that Queen Victoria sought the advice of the Duke of

Wellington on how to get rid of them; he apparently advised her to try Sparrowhawks.

Fitter listed many strange places in which they nested, including various London statues: in the mouth of a lion that used to stand in Northumberland House at Charing Cross; in the folds of George Canning's toga in Parliament Square; in the figure of St George and the Dragon on a Pimlico vicarage; and inside the Duke of Wellington's arm, the birds entering through a hole in his finger. Other unusual sites were in the wire netting surrounding the cages of birds of prey in London Zoo and on a large gun at Woolwich Arsenal which was fired twice a day.

House Sparrows have nested in every month of the year except January; a completed nest was found on 22 February 1877 and nesting was in progress at Kilburn on 4 December 1913. Hudson recorded that House Sparrows in central London usually remained paired all year round and during winter would roost in their nest-hole, often with the young from their last brood. An unfledged juvenile was found dead on 8 December 1948, having fallen out of a nest at Mill Hill. The following year, young were heard calling from a nest in Harrow Weald on 22 December.

Facts and Figures

House Sparrows have been predated by Mallard, Sparrowhawk, Kestrel, Moorhen, Black-headed Gull, Little Owl, Tawny Owl, Jay, Jackdaw and a pinioned Grey Heron, which makes them the most widely predated bird in the London Area.

Tree Sparrow *Passer montanus*

An uncommon partial migrant and localised breeder

Range: Europe, Asia and North Africa.

Historical

In the 19th century this was a rare breeding bird with the first record coming from Sudbury Grove where one or two pairs bred from 1828 to 1832. The next confirmed breeding record was at Hampstead in 1871 and they bred annually in Harrow from 1888 to at least 1893. Large numbers summered at Brent Reservoir from 1881 to 1883 and were assumed to be breeding. From 1899 to 1903 they bred along the upper stretches of the Yeading Brook. The only breeding colony known from south of the river was near Epsom where about 20 pairs bred until 1896; there was also a single breeding record near Wimbledon prior to 1900.

20th century

At the start of the 20th century the species began to increase; breeding occurred along the River Brent between Greenford and Hanwell in 1901; in Richmond Park and near Watford from 1904 onwards; at Horsenden Hill in 1905; and at Staines in 1906.

In 1909 Ticehurst stated that the colonies in the Beckenham, Bromley, Hayes, Keston and West Wickham areas were the largest in the entire county of Kent. At the Brent Valley Bird Sanctuary, they was scarce in 1896 but by 1910 had become a numerous breeder, helped by the provision of nest-boxes. A dozen of these were occupied in 1910, which appears to be the earliest evidence of the use of nest-boxes by any species in the London Area. Breeding was first recorded at Beddington in 1912. The first recorded breeding in the Lea Valley was in 1914 when a pair bred in a haystack. At Romford they bred annually to at least 1919 and in 1929 breeding occurred at Danson Park, Dartford and Greenhithe.

By the 1930s Tree Sparrows were common breeding residents in the rural areas and outer suburbs but rarely occurred in the more built-up areas. The closest colonies to central London were at Barn Hill, Edmonton, Edgware and Perivale Wood. In 1936 there were about six pairs at Beddington. In the Lea Valley small colonies bred in drainpipes at King George V and Walthamstow Reservoirs in 1938.

In the 1950s they were fairly widespread, albeit spread thinly, with some colonies being resident around a small group of trees. They were not common in any area but they were most frequent in the Colne Valley, in Epping Forest and at Beddington. They had found gravel pits and sewage farms to their liking and had formed thriving colonies at many of them. They were mainly absent from the most heavily built-up areas but they could be found in the suburbs at Elmers End, Beddington, Mitcham, Richmond Park, Twickenham, Osterley Park, Horsenden

Hill, Barn Hill, Hendon, Finchley, Enfield, Chingford, Walthamstow, Woodford Bridge and Upminster. They were thought to be resident in many rural area, but were known to be scarce along the North Downs.

The largest colony was at Beddington where there were 50 pairs on the sewage farm and another 15 in a nearby park. Between Barnet and St Albans about 50 pairs bred in eight colonies in 1950. In 1952–53 about 100 pairs were estimated to be breeding in an area bordered by Romford, Rainham, Stifford and Bulphan, including about 20 pairs at Romford SF; there were also 19 pairs at Aveley, 15–20 at Hutton, near Brentwood, and smaller colonies elsewhere in rural east London.

In the Lea Valley 15–20 pairs bred between Cheshunt and Broxbourne, all of which were in pollarded willows. In the Staines area between Perry Oaks SF and Stanwell Moor, 87 nests were found in 1955. At Elmers End SF breeding commenced in 1949 and up to two pairs bred until 1953 before they started to increase; by 1959 the colony had reached 20 pairs. The breeding colony at Epsom SF expanded to 30 pairs by 1958.

By the beginning of the 1960s the only colony in the east London suburbs was at Walthamstow Reservoirs where the population was eight pairs in 1960, although this colony soon died out. A breeding colony survey conducted in 1961–62 found a minimum of 745 breeding pairs. The largest populations were at Beddington (70 pairs), Old Parkbury (45), Tyttenhanger Park (40), Nazeing Park (30), Addlestone (20–30), Elmers End (22–25), Addlestone to Weybridge (24), Panshanger (23), Amwell (20), Dyrham Park (20), Fishers Green to Nazeing GP (20) and Monks Green (20). The closest breeding site to the centre of London was 11 kilometres away at Brent Reservoir where three pairs bred. Additional breeding records in 1964 included about 88 pairs in the Northaw area, 58 pairs in Lullingstone Park, 20 pairs at Hainault Lodge and 20 pairs at Hulberry.

During the first breeding Atlas in 1968–72 they were found breeding in 411 tetrads, almost half of the area. Their distribution covered the rural areas of north and east London, and to a lesser extent of south and west London, with a reasonable scattering in the suburbs.

A decline had set in by the beginning of the 1980s. In central London, at the recently decommissioned Surrey Docks, a pair bred in 1984. Further reductions in the total population were noted in 1985, especially at Rye Meads where a once-thriving colony of 50 pairs had died out, and by 1986 they had disappeared from Dagenham and Romford. By the time of the second breeding Atlas in 1988–94 their range had contracted by two-thirds with the largest declines being in the south and west of London.

By the beginning of the 1990s there were just 85 breeding pairs at 13 localities, about half of which were at Beddington. The following year the breeding population had declined further to 71 pairs. The colony at Beddington was stable, however, with at least 40 pairs, and a programme was started to erect nest-boxes to compensate for the lack of sufficient nesting places. In 1992 at least 50 pairs bred, more than half using the new nest-boxes, and they reared 350 young. By 1994 it had increased to 89 pairs, but it slumped to 27 pairs by 1998. In 1998/99 supplementary winter feeding was provided in an attempt to maintain the population at Beddington since the numbers decreased when natural food ran out during autumn. In the Lea Valley at least six colonies were known during the 1980s but the last pair to be recorded breeding there was at Middlesex Filter Beds in 1999.

21st century

The decline in London has mirrored the national pattern. A sharp decline set in from the mid 1970s, and although numbers have tended to increase since the late 1990s, the population remains at a tiny fraction of what it was before the crash. At the start of the 21st century there were just two colonies left in the London Area. At Beddington the population had started to increase again and there were 51 pairs in 2000. By 2003 it had doubled to 104 pairs and continued to increase, reaching 135 pairs in 2007. Although the number of pairs has decreased since then, breeding success has been good with at least 624 young hatched in 2010. In the London Colney area at least 20 broods were reared in 2000 and by 2010 the population at the two small colonies at Coursers Farm and Tyttenhanger Farm had increased to a total of 31 pairs. Elsewhere Tree Sparrows are extremely scarce winter visitors and passage migrants.

Movements and Winter Flocks

In the 19th century Harting stated that they were 'an uncertain visitant, occasionally appearing in small flocks in autumn, and remaining till the spring'. Several records from Brent Reservoir illustrate this point: in autumn 1857 they were so plentiful 'that a dozen might have been killed at a shot'; one was caught in October 1862 and three were killed in November 1864. Elsewhere, small flocks were observed in winter at Elstree and Harrow. In

October 1893 a small party of migrants dropped in on Tooting Common. Migrants were regularly seen flying over Brixton in autumn in the years 1874–98 but they were not recorded after this.

Wintering flocks had increased in size by the 1930s with peaks of 200–250 at Elmers End on 21 October 1934; 130 at Colnbrook in 1935; more than 100 at Beddington in 1936; and about 150 on Staines Moor in 1939. In 1945 about 100 were at Brent Reservoir.

A regular movement of Tree Sparrows in late autumn was recorded in the 1950s. They were noted entering the Thames Estuary, presumably from the Continent. As well as this regular passage there were also wandering birds such as the one ringed at Guilford, Surrey, on 1 March 1954 which was recovered at Gidea Park on 4 April. Mixed flocks of Tree and House Sparrows were sometimes seen on the marshes between Erith and Northfleet. In 1959 there was a large passage through Regent's Park with up to 47 recorded between 23 September and 2 November.

Outside the breeding season they were more widespread and visited sewage farms and fields to feed. They were sometimes seen in large groups and flocks of 100 or more were seen during the 1950s at Beddington SF, Perry Oaks SF, Romford SF and Stone Marshes. In 1953 about 200 were at Romford SF on 7 February.

Exceptional counts were made in the late 1960s and early 70s. At Romford SF there were up to 2,500 in 1966/67 and up to 4,000 in 1967/68. At Chorleywood up to 2,750 were seen feeding in a field with several thousand finches in 1970/71. There was also a roost of 1,500 at Roydon on 13 February 1971. There have been no further counts of more than 1,000 and the highest counts during the rest of the 1970s were 500–600 at Bromley Common, Cheshunt GP, Mill End and Swanscombe.

Since 1980 the highest counts have been at Beddington where the peak count was 250 in 1985. Migrants or wanderers continue to be seen annually in very small numbers, most often in late autumn.

Facts and Figures

The largest ever count was about 4,000, at Romford SF in January 1968.

Yellow Wagtail *Motacilla flava*

A common summer visitor and scarce breeder

Range: breeds Europe, Asia and North Africa, Western Palearctic population winters Africa. The subspecies breeding in Britain is *M. f. flavissima*.

Historical

In the 19th century it was known as 'Ray's Wagtail' and was a regular breeding summer visitor. In 1842 Meyer stated that it 'abounded upon the islands of the Thames'. Harting wrote that 'the eggs are seldom obtained, for the nest is always very carefully concealed. It is very compact in form, and is usually placed in a hole in the ground (oftentimes the depression made by a horse's hoof), surrounded with tall herbage'.

20th century

At the beginning of the century it was a reasonably common breeding bird in suitable habitat, particularly along river valleys. It bred in large numbers on the water meadows in the Colne Valley; in the Lea Valley at Edmonton SF; and on the eastern stretch of the Thames where it was said to be plentiful near Woolwich, and it was a common breeding bird on the marshes east of Erith. Six pairs bred on Wimbledon Common in 1902 and they bred in bracken in Richmond Park up to 1912. Yellow Wagtails first started breeding in good numbers at Walthamstow Reservoirs in 1911 and from there spread along the Lea Valley.

During the 1920s numbers increased in the Lea Valley as new habitats became available, such as the newly constructed King George V Reservoir and a series of gravel pits and sewage farms north from this reservoir up to Stanstead Abbots. In the 1930s Glegg described it as 'a not uncommon but local summer resident in the rural parts of the county', adding that it was an occasional visitor to central London.

In the Second World War years they bred in Regent's Park for four consecutive years. During the 1940s they were a common breeding bird at Barn Elms and Brent Reservoirs. Up to five pairs bred annually on Blackheath until 1947, after which it was partly converted into football pitches. In 1947 they bred at Greenwich Park and

Woolwich Common and a pair summered on a bombed site in Silvertown. From 1943 to 1953 they bred on rough ground by the Thames between Greenwich and Plumstead with at least 30 pairs present in 1952.

During the 1950s it bred in cornfields, allotments and grassy areas around Mill Hill, and in 1951 up to 10 pairs bred in disused cornfields near the canal at Greenford. Also in 1951 the peak counts of migrants were about 400 at King George VI Reservoir on 29 April and 300–350 at King George V Reservoir on 16 August. By 1958 the population at Beddington had reached about 35 breeding pairs. Two pairs nested at Old Ford, Bethnal Green, in 1960, these being the nearest breeding ones to the centre of London for many years. By the end of the 1960s the only breeding site in the inner suburbs was at Brent Reservoir.

The breeding population was fairly stable by the beginning of the 1970s but Yellow Wagtails had just started to breed in Inner London; two pairs nested on waste ground near Vauxhall Bridge and a pair probably bred in Surrey Docks. They continued to breed in Surrey Docks throughout the decade and the population grew to 12–15 pairs. The population began to decline on the Lower Thames Marshes, mainly due to development, for example the construction of a power station at Littlebrook.

In 1980 the breeding population was a minimum of 123 pairs, including more than 50 in the Lea Valley and over 30 on the Thames marshes between Aveley and Beckton. The second breeding Atlas in 1988–94 found evidence of breeding in 147 tetrads, an increase of a quarter on the first Atlas. The main areas of increase were in rural east London, which may have been under-recorded during the first survey, and in the upper Colne Valley. There were some losses in west London. At the start of the 1990s the breeding population was about 113 pairs; of these, 85 pairs were in the eastern area with the largest concentration being 30 pairs at Bulphan Fen. At Surrey Docks the population had decreased to two pairs and they finally ceased breeding there in 1992. During the decade the breeding population declined in the UK but in the London Area it crashed to just 11 pairs by 1999.

21st century

By the beginning of the 21st century there were only nine pairs breeding at four sites. One of these was Perry Oaks SF, where there were six pairs, and this was soon buried under Terminal 5 at Heathrow Airport. The breeding population was down to about six pairs in 2009, all in the far east of the area.

At most sites across the London Area Yellow Wagtail is a scarce and declining passage migrant. The main arrival in spring is from mid-April with a few exceptionally early birds recorded in the last few days of March. The first autumn migrants are seen from July onwards with the bulk passing through in August and the first half of September; one or two are seen in most years in October.

Facts and Figures

The earliest migrant was on 15 March 1994 at Queen Mother Reservoir.
The latest migrant was on 27 December 1960 at Petts Wood.
Overwintering has occurred at Beddington on four occasions. The first of these, in 1956/57, was also the first time one had overwintered anywhere in the UK. There were two in 1959/60 and singles in 1960/61 and 1984/85.
The largest concentration was of about 400 birds at King George VI Reservoir on 29 April 1951.

'Blue-headed Wagtail' *M. f. flava*

There are four 19th-century records of this subspecies in the London Area, when it was known as 'Grey-headed Wagtail': a summer-plumage male taken in Finsbury in April 1837 which appeared in Yarrell's *British Birds*; an immature shot at Brent Reservoir in May 1864; a pair on Stanmore Common in May 1868; and three on Wimbledon Common on 2 June 1890.

In the first three decades of the 20th century it was still a rare visitor and there were only three records. Since then it has been recorded annually, usually in small numbers. Although it is typically seen between April and June there have been several breeding records. In 1933 a pair nested between New Eltham and Sidcup. In 1947–48 several intergrades bred on Staines Moor. During the 1950s there were several breeding records at Beddington SF. Mixed pairings with *flavissima* have occurred as recently as 1999.

'Grey-headed Wagtail' — *M. f. thunbergi*

The first one was seen and sufficiently described at Brent Reservoir on 7 April 1945. Since then another 17 have been identified, almost half of which were at Beddington. There have also been a few intergrades resembling this subspecies.

'Sykes's Wagtail' — *M. f. beema*

Birds showing characteristics of this subspecies have been recorded on several occasions, including at Brent Reservoir on 30 April and 6 May 1950, and on 14 April 1963. A juvenile first ringed at Beddington in 1956 was re-trapped there the following year and found to be an adult male Sykes's Wagtail. In 1957 and 1958 eight of the breeding males at Beddington resembled this subspecies and there were also a number of Blue-headed types and some variants. Males also summered or bred with other races at Beddington in 1963; Walthamstow Reservoirs in 1973; and Helicon GP in 1977. There have been another 14 since then; the most recent being on Staines Moor from 12 to 15 April 2003.

Some progeny from mixed pairs (*flava* x *flavissima*) that breed in France and are known as 'Channel Wagtail' closely resemble this race and it is possible that some of these records relate to these intergrades.

'Ashy-headed Wagtail' — *M. f. cinereocapilla*

A male showing characteristics of this subspecies was found among the breeding Yellow Wagtails at Perry Oaks SF in May and June 1959; it was not clear if it was paired with a female but it was seen feeding juveniles. In 1963 four or five were at Perry Oaks during the summer. There have been another eight records as follows: Beddington from 19–22 May 1976; Girling Reservoir on 3–4 May 1980; Rainham from 19–21 April 1984; King George V Reservoir on 7 May 1991; Tyttenhanger GP on 28–29 May 1991; Amwell on 28 May 1994; King George V Reservoir on 23 April 1996; and Hersham GP on 20 May 1999.

'White-headed Wagtail' — *M. f. leucocephala*

A Yellow Wagtail with a completely white head was seen at Beddington on 15 May 1958. It resembled the 'White-headed' subspecies, but as a full description was not taken it cannot be accepted as a definite record. Another similar male was present at Hersham SF in 1961 and was paired with a female. Similar-looking birds have been seen breeding in France and are thought to be intergrades of *flava* x *flavissima*.

Citrine Wagtail — *Motacilla citreola*

A very rare vagrant (2)

Range: breeds eastern Europe and Asia, winters southern Asia; vagrant to UK.

The first record occurred in 1993 when a juvenile was present for five days at Beddington in August. Unlike most of the clean monochrome first-winters that occur in Britain in autumn this individual still retained much of its brown-toned juvenile plumage. The following year one in first-winter plumage was seen on the evening of 22 August at King George V Reservoir. The vast majority of national records are in autumn, typically on the coast with few inland records.

1993: Beddington, on 24–28 August
1994: King George V Reservoir, on 22 August

Grey Wagtail *Motacilla cinerea*

A common partial migrant and localised breeder

Range: Europe, Asia and North Africa.

Historical

During the 19th century this was colloquially known in London as 'Winter Wagtail' which was an apt description of its status. In 1836 Blyth stated that it was a rare winter visitor to Tooting; it was also said to be fairly common along the Thames in winter; and Harting described it as 'an uncertain winter visitant; in some years tolerably numerous, in others scarce'.

Harting knew of two instances when Grey Wagtail had been recorded in summer but it was not known to breed. Breeding was first reported at Rickmansworth in 1879 and was followed in 1893 on Wimbledon Common and in 1898 on Barnes Common.

20th century

At the beginning of the century the only place it bred was in the Cray Valley where five nests were discovered. It gradually began to spread into the London Area and breeding occurred on the River Chess, at Bushy Park and Twickenham. It also colonised the Godstone area from 1912 onwards, breeding by Ivy Mill and Leigh Mill. During the 1920s it bred on the outskirts of London on the River Colne at Hamper Mill, in the Darenth Valley and on the River Gade near Cassiobury Park. The only breeding in the suburbs was on the Silk Stream in an industrialised part of Hendon in 1929.

It became established along many of the rivers of south London during the 1930s; it also bred on the water meadows at Mitcham, at Beddington and Epsom SF, and among the waterfalls and lakes in parks around Beckenham and at Elmers End SF. The only place where breeding occurred north of the Thames, where it was still known mainly as a passage migrant and winter visitor, was on the Chess near Chorleywood. In 1933 about 20 were seen at Edmonton SF in October, the largest concentration known at the time.

In the 1940s Fitter stated that there was an annual passage through urban London from mid-September to mid-November. The numbers fluctuated during the first half of the century but by 1949 it had become a regular autumn passage migrant. By the 1950s it was breeding annually in small numbers, albeit in widely separated localities; one of its strongholds was on the streams on the North Downs around Tandridge, Oxted and Limpsfield. In 1952 a pair bred in a bombed site in Cripplegate, the first breeding record in Inner London. In 1959 the breeding population had increased to 14 pairs throughout the London Area.

The breeding population suffered during the severe winter of 1962/63 and in the following summer only five or six pairs could be found, all south of the Thames. There was a very slow recovery and by 1969 only ten pairs were reported in the LBR. However, clearly many more pairs were present as the first breeding Atlas found in 1968–72 which found evidence of breeding in 70 tetrads. The majority of these were in south-west London and along the Colne Valley; it was particularly scarce in the eastern half of London. It had also started breeding in Inner London and in 1977 three or four pairs bred.

A survey in 1979–81 put the total population at a maximum of 137 pairs. There had also been an increase in the population in Inner London with breeding proven at 17 locations in the innermost 10 kilometre squares compared to just five during the Atlas years. During the second breeding Atlas in 1988–94 Grey Wagtails were found breeding in 183 tetrads, an increase of 160 per cent since the first Atlas. They had spread eastwards but were still scarce between the Lea Valley and the Thames.

21st century

Since the beginning of the 21st century the population appears to have remained fairly stable with at least 80 breeding pairs at over 70 localities, including up to four pairs in central London; this compares with a decrease of 15 per cent nationally from 1995-2010 as measured by the Breeding Bird Survey. The species is also seen regularly on passage across the London Area in autumn, and to a much lesser extent, in spring.

Facts and Figures

The highest count was 25 birds at Rye Meads on 12 August 2001.

Pied Wagtail *Motacilla alba*

A very common resident and widespread breeder

Range: Europe, Asia and North Africa.

Historical

Harting described this as a common resident but partially migratory 'for although common in summer, and most numerous in autumn, when it may be observed in small parties of seven or eight, yet very few are to be seen here throughout the winter, and I have no doubt that a great number of Pied Wagtails move southwards as that season approaches'.

20th century

By the 1930s it was still a common resident but it had become more numerous outside the breeding season. Flocks of around 50 were regular on some sewage farms in autumn and winter, and there was an exceptionally large flock of 200–250 at Edmonton SF on 30 July 1933. In November 1943, 100–150 were seen on ploughed land near Claygate.

The overall population remained fairly stable during the second half of the century with a slight increase noted between the two breeding Atlases. During the second survey in 1988–94 they were found breeding in 57 per cent of the tetrads in the London Area; the breeding population in central London was about 10 pairs.

Wintering numbers continued to increase and the highest counts occurred in the 1990s. In 1990 there was an exceptionally large feeding flock of up to 355 at Rammey Marsh in November and December, and 470 were seen at Sevenoaks Wildfowl Reserve on 18 February; there was also a feeding flock of 348 at Queen Mary GP on 22 November 1991.

21st century

Pied Wagtail is a common breeding bird throughout much of the London Area apart from in central London where only a few pairs are recorded breeding. It is regularly seen on passage migrating over the London Area especially during autumn and is widespread in winter.

Roosts

Pied Wagtails were first discovered roosting in large numbers in the 1930s. While most of these roosts were active during winter there were also a few in late summer after the breeding season. In 1937 about 150 roosted in holly trees under a powerful street lamp in Balham High Road, 150 roosted in Richmond Park and 129 roosted on Esher Common.

In later years additional roosts were located and the peak counts continued to increase. In 1949 a roost of about 300 was found in plane trees in Thornton Heath; most of the birds were seen to fly across from Beddington SF. In 1950 about 460 roosted in a reed bed at Crayford Creek on 1 August, and in 1952 about 500 roosted in a reed bed in the Roding Valley in November and December. In 1967 about 1,500 were discovered roosting at Thorndon Park during winter. Although most roosts are located in the outer suburbs and rural areas there have been several large roosts in central London, including one of 400–500 birds at Walbrook in the City in 1980; there were 500 on Hammersmith Broadway in 1976.

The largest roost was in use during the 1970s, in bushes by the Civic Hill in Orpington; it peaked at 4,300 birds in 1977. The only other roosts to contain more than 1,000 birds were Maple Cross SF with 1,500 birds in 1978 and at Rye Power Station with 1,109 in 1980. By the start of the 21st century the numbers of roosting birds was well down compared to previous decades and the maximum count was about 250 at Croydon shopping centre; however, a roost of at least 2,000 was discovered in trees by Terminal 5 at Heathrow Airport in the winter of 2011/12.

Nest Sites

Although Pied Wagtails quite often choose artificial nest sites, one of the most unusual was under a rail at the end of Colliers Wood station. The nest was later deserted, not due to the succession of trains passing over it but because of disturbance by people after the news became public. Also, in 1994 a pair nested at Walton-on-Thames on top of the engine in a tractor that was in daily use and successfully reared four young.

Facts and Figures

The largest count was 4,300 birds at a roost in Orpington on 22 January 1977.

'White Wagtail' *M. a. alba*

The earliest record was of one shot at Dartford in the winter of 1829/30.

Harting considered that 'owing to its close resemblance to the Pied Wagtail, this [sub] species has been much overlooked, and that it visits this country regularly every summer'. Bond found two pairs frequenting the banks of Brent Reservoir in late May 1841 and shot three of them; this was the first British record. He also saw another there in the spring of 1843. This was clearly a good site as two more were shot there in 1859 and another pair was seen there in 1862. There was subsequently an increase in the number of sightings, possibly as this subspecies became more widely known.

From the 1920s onwards it was recognised as a regular double-passage migrant in small numbers, most often in spring. In 1937 a male was seen at Staines Reservoir on the unseasonable date of 31 January. In the mid-1950s around a dozen were seen every year at reservoirs and sewage farms, either singly or in groups of up to four. In the spring of 1989 there was a particularly good passage with a peak of 43 on 25 April, including 21 at Barn Elms Reservoirs. There was another large passage in 2000 when about 150 were seen. A female nested by Wraysbury Reservoir in 2008; it was unclear whether her mate was a White or Pied Wagtail.

Richard's Pipit *Anthus richardi*

A scarce passage migrant (29)

Range: Asia; scarce passage migrant in Europe and UK.

Richard's Pipit was first identified in the UK in 1812 by Vigors who illustrated a specimen that had been taken alive at Copenhagen Fields where there was a large cattle market; this is modern-day Kings Cross. The second record for the London Area was of a bird also captured at the same site, shortly prior to 1831.

In the spring of 1836 there were records of two taken 'near London' and one taken at Tooting Common; it is likely that these are the same. A specimen received at the British Museum the following year was said to have been taken at Bermondsey but could possibly be one of the two 'near London'.

There were no records in the first half of the 20th century and the next one was seen on the causeway at Staines Reservoir twice in April 1956. Two years later there were another two spring records. The only record in the 1960s was from Brent Reservoir in October 1963 and was belatedly accepted. In the early 1970s four were seen, including two together at Beddington.

From the mid-1980s onwards they have been seen more frequently and have become almost annual visitors with most records in autumn.

1812: Kings Cross, one captured in October
Pre-1831: Kings Cross, one captured
1836: 'near London', two taken in spring
1837: Bermondsey, killed
1956: Staines Reservoir, on 10 and 21 April

1958: Beddington SF, on 17–18 April
1958: King George VI Reservoir, on 30 April
1963: Brent Reservoir, on 19 October
1970: Beddington SF, two on 23–24 October with one until the 26th
1972: Swanscombe SF on 15–17 April
1973: Latton Common, Harlow, on 25 October
1986: Staines Reservoir, on 26 September
1987: Fairlop Waters, on 20 September
1988: Dartford Marsh, on 29 September
1988: Wraysbury GP, on 13–20 October
1989: Dagnam Park, on 2–3 October
1992: Fairlop Waters, on 12 May
1994: Barn Elms Reservoirs, on 26 September
1995: Queen Mother Reservoir, on 30 April–2 May
1996: Barking, on 4–5 October
1996: Hampstead Heath, on 5 October
1998: Wormwood Scrubs, on 29 September
1998: Chingford Plain, on 23 October–1 November
2001: Lake Farm, Hayes, on 10 November
2004: Staines Moor, on 11 December
2007: Wormwood Scrubs, on 23 September
2008: Rainham Marshes, on 27–28 April
2008: Crossness, on 5 May

Unacceptable Records

In the autumn of 1866, there was a remarkable series of records from Highgate: one was caught on 4 October and this was followed by a further four, which were taken between then and the end of November. There are no doubts about the identification of these birds as two of them ended up in Bond's collection. A further three were apparently caught at nearby Hampstead in October 1869. One can only speculate as to the validity of these records, bearing in mind the rarity of this species at the time and the fact that no more were seen for almost 90 years. The possibility of fraud cannot be discounted and these records should not be considered acceptable.

Tawny Pipit *Anthus campestris*

A rare vagrant (9)

Range: breeds Europe, Asia and North Africa, Western Palearctic population winters Middle East and Africa; scarce passage migrant in UK.

There have been nine in the London Area, all but one in the west and south-west. The first three occurred at Perry Oaks SF, including two in 1963 when there was a considerable influx into the UK. There have also been three at Beddington, including two together in 1992. The only one in spring spent two days on the grassy banks of King George VI and Staines Reservoirs in May 1990. The only one on the east side of London was at Barking Bay in August 2012, 20 years after the previous record. Six of the nine birds arrived in mid to late September.

1954: Perry Oaks SF, immature on 17 October
1963: Perry Oaks SF, immature on 15–18 September
1963: Perry Oaks SF, adult on 22 September
1969: Hayes, on 29 September
1989: Beddington, on 30 September–3 October
1990: King George VI and Staines Reservoirs, on 3–4 May

1992: Beddington, two on 6 September
2012: Barking Bay, on 19 August

Olive-backed Pipit *Anthus hodgsoni*

A very rare vagrant (1)

Range: Asia; vagrant to UK.

The only record occurred in October 1992 when one was present at Chingford for four days. It arrived at the end of a notable influx into the country and still remains one of the few inland records.

1992: Chingford on 23–26 October

Tree Pipit *Anthus trivialis*

An uncommon visitor and scarce breeder

Range: breeds Europe and Asia, winters Africa.

Historical

In the 19th century this was a common and widespread species, even in central London where Hamilton described it as a regular summer visitor in Hyde Park in the 1850s. Harting stated that few nature lovers could fail to notice this bird while walking through country lanes and fields, giving us an insight as to how widespread it must then have been.

20th century

By the beginning of the century it had ceased breeding in central London.

In the 1930s Glegg described it as a fairly common but localised breeder in the rural areas, suggesting that urbanisation had begun to push this species further away from central London. In Epping Forest, where it was previously common, a decline was noted by the end of the decade and this was thought to be due to overgrazing in some areas and the loss of open heath in other places. In the 1940s it increased in the Addington, Croydon, Sevenoaks and Staines areas. At Kenwood, on Hampstead Heath, a pair bred in 1946 – the only breeding record in the inner suburbs since 1910; breeding was suspected there in subsequent years up until 1951 and a pair nested again in 1954.

During the 1950s and 60s it was common across parts of the London Area, particularly in open woodlands, grasslands with scattered trees, along the North Downs, on the commons and heaths, and along many railway cuttings. In 1962, for example, there were 18 pairs around Addington and Warlingham, and in Epping Forest there were at least 20 pairs in 1968. The closest breeding to central London was on Wimbledon Common and in Richmond Park; a pair also bred at Barn Hill, Wembley, in 1969. By the beginning of the 1970s it was still fairly common and widespread although some declines were noted, e.g. none bred in Richmond Park.

In 1985 a total of 86 pairs were located at 30 sites. This was probably a fairly accurate assessment of the breeding population as most were found during a special woodland survey; the highest concentration was 20 pairs in Epping Forest. There then followed a decline and by 1989 the population had declined to 62 pairs at 28 sites. By the start of the 1990s the decline had accelerated; the number of territories had fallen to 28, a reduction of two-thirds in just five years. The majority of birds were in the eastern part of London with just three south of the Thames. The second breeding Atlas in 1988–94 recorded a reduction in the distribution of two-thirds since the first Atlas; this decline was significantly above the national decrease of 15 per cent over the same period. Although losses were recorded across the London Area between the two atlases, the largest decrease was in the southern half of London. By 1999 there were only 16–18 pairs. The national population suffered a loss of 75 per cent between 1966 and 1999.

21st century

Tree Pipits have continued to decline as a breeding species in the London Area since 2000 and are likely to join the list of former breeding species. In 2004 none were seen for the first time ever at Limpsfield Chart, a former breeding stronghold. There was a maximum of six pairs in 2009, all but two north of the Thames. Elsewhere, they are a scarce but decreasing passage migrant, with the vast majority occurring in late August and September.

Facts and Figures

The earliest migrant was on 15 March 1981 at Stocker's Lake.
The latest migrant was on 26 October 2003 at Wimbledon Common.
The largest group of migrants was of 13 birds at Hilfield Park Reservoir on 29 April 1956.

Meadow Pipit *Anthus pratensis*

A very common partial migrant and localised breeder

Range: Europe, Asia and North Africa.

Historical

In the 19th century this was a common species, resident throughout the year; Harting described it as preferring 'moist situations, such as water-meadows, brook-sides, and turnip-fields, after rain' and added that 'it may be observed in company with wagtails, following cattle and seizing the flies which they disturb'. It was known to be breeding as close to central London as Hampstead Heath in 1878.

20th century

In 1909 Dixon stated that it was more common than Tree Pipit within 25 kilometres of St Paul's Cathedral and that 'it would be impossible to mention all the places in the more rural suburbs where [it] nested'. However, it did not breed in all the districts on the edges of the London Area as it was only a winter visitor around Staines, and the only documented nesting record from the northern extremities was at Watford in 1902. In the east it was only recorded breeding around Epping Forest at Chigwell in 1906 and Loughton in 1907. In 1909 it was comparatively rare around Bromley but it did breed at Keston, Hayes and on Mitcham Common. Its stronghold was the rough grasslands on the North Downs. In Richmond Park it was predominantly a winter visitor and was seldom seen in summer. In the 1920s and 30s there was very little change in status. Meadow Pipits colonised Richmond Park in the 1920s and continued to breed at Dulwich. The only decline was at Hayes and Keston Commons where they ceased to nest after about 1939 because of the growth of silver birches. In 1948 a small area of Mimmshall Woods was felled and replanted with young conifers; by the following year 12 pairs of Meadow Pipits had colonised it and they bred annually until 1951, by which time the trees and undergrowth had grown up and rendered the habitat unsuitable. During the first half of the 20th century Meadow Pipits were a numerous and widespread winter visitor, particularly around reservoirs and sewage farms. In some years large counts were made, such as in 1933 when there were 700–750 at Edmonton SF on 24 December.

By the 1950s the breeding population had actually increased since the beginning of the century. This was due to more breeding habitat being available, such as the grassy banks of newly constructed reservoirs and sewage farms, and the reversion of some arable land to pasture. In 1950 the species was suspected of breeding on Hampstead Heath. Breeding also occurred in marshy fields alongside the Colne and Lea; in open areas such as at Chingford and Stanmore; on the Thames marshes from Barking and Plumstead eastwards; and on the Downs and commons in south London. A survey of breeding Meadow Pipits in 1957 resulted in almost 70 pairs being found; more than half of these were in south London.

At the start of the 1970s the majority of the breeding population was in south London; most of the rest were on the Thames marshes with some in the Staines area and a few scattered ones elsewhere. In the 1970s up to 100 wintered in Inner London at Surrey Docks; in 1975 three remained throughout most of May with one until 16 June. Two pairs bred there in 1977, this being the first confirmed breeding record in Inner London. A male remained on territory there the following year but there was no evidence of breeding; however, the species continued to breed there in subsequent years until at least 1985.

The breeding population in 1983 was a minimum of 110 pairs, including at least 60 on the marshes between Aveley and Beckton. There was virtually no change in the breeding distribution between the two Atlases. At the time of the second Atlas in 1988–94 Meadow Pipits bred in 123 tetrads; the majority were in the Thames corridor with a smaller population in south-west London. At the beginning of the 1990s there were at least 215 pairs or singing males, almost half of which (109) were found at Rainham Marshes during a complete census of the area. In 1992 more than 500 migrant birds were feeding on the lagoons at Rainham Marshes on 19–20 September.

21st century

Meadow Pipit remains sparsely distributed during the summer months with breeding reported at about 30 sites; the main population is on the Lower Thames Marshes. The species has decreased in the suburbs and no longer breeds at Dagenham Chase, Hounslow Heath or Wimbledon Common. This decrease is also apparent nationally with the Breeding Bird Survey showing a 23 per cent decline from 1995–2010. Within the built up area, breeding is now restricted to a handful of sites such as Wanstead Flats and Wormwood Scrubs. It is more widespread in winter and especially on passage.

Movements

The passage of Meadow Pipits over the London Area was noted by several commentators in the 19th century, including Gould who in 1873 stated that there was a partial migration in spring and autumn over Primrose Hill. The peak autumn passage is from mid-September to the end of October; the movement is on a broad front and can be seen over central London, mainly from tall buildings and in the parks.

It has not been established where all these passage Meadow Pipits originate. During westerly winds many are seen flying west; for example, 346 were counted flying west along the Thames in one hour towards the end of October and these may therefore be continental birds. However, some are seen flying east when the wind is blowing from that direction. At Brent Reservoir in north-west London, the highest numbers are recorded when the wind is from the north-west and birds are passing over southwards; additionally, more are seen when the previous night's temperature in the north of England has been low suggesting that these birds originate in this country.

In 1959 about 900 flew over King George V Reservoir on 1 November. In 2000 there was a particularly strong passage between the end of September and early October with a peak of 505 over Regent's Park on 5 October. In 2001 the peak count was 574 over the London Wetland Centre on 3 October.

The return passage takes place in March and April and much less is known about the routes taken or the intended destination, though most of the birds seen over Brent Reservoir during this time are flying north. Although most migrating birds usually just pass straight through the London Area, large flocks sometimes stop off. In Hyde Park about 50 seen on 29 March 1926 were so tired that they allowed an approach of less than 3 metres.

In 1960 there was an extraordinary passage in late March with several sites reporting groups in the hundreds while at Beddington about 2,000 were seen on 27 March. In 1995 there was an exceptional passage of 1,400 over Beddington on 17–18 March. In 2001 some 700 flew over Hampstead Heath on 24 March and in 2006 about 600 flew over Queen Mother Reservoir on 14 April.

Cold weather often brings an influx into the London Area and this was observed in February 1917 when some were seen feeding along the Thames at Chiswick at low tide, then retreating to the pavements at high tide. In 1938 they were seen along the tide-line at Hammersmith and in snow-covered streets at Epsom, Hampstead and Upper Norwood.

Facts and Figures

The largest-ever flock was of 500 birds at Rainham Marshes on 19–20 September 1992.
The largest count of migrants was about 2,000 birds over Beddington on 27 March 1960.

Red-throated Pipit　　　　　　　　　　　　　　　　　　　　　　　　　　　　*Anthus cervinus*

A rare vagrant (11)

Range: breeds Arctic Circle, winters Africa and Middle East.

The first record was at Staines Reservoir in April 1964. No more were seen until the 1980s when another three spring records occurred. During the 1990s a further six were seen, five of which were in successive years. Unlike the earlier sightings, all of these were in autumn. In 1996 one spent a week at Beddington but was very elusive during its stay. In 2000 yet another was seen at Beddington in autumn, the fourth sighting there in six years.

1964: Staines Reservoir, on 17–18 April
1980: Sevenoaks WR, on 3 May
1982: Rainham Marshes, on 1 May
1988: Barn Elms Reservoirs, on 13 May
1992: Barn Elms Reservoirs, on 28 September
1993: Rainham Marshes, on 19 September
1994: Barn Elms Reservoirs, on 14 October
1995: Beddington SF, on 2 November
1996: Beddington SF, on 6–13 October
1998: Beddington SF, on 3 October
2000: Beddington SF, on 15 October

Rock Pipit　　　　　　　　　　　　　　　　　　　　　　　　　　　　　　　　*Anthus petrosus*

An uncommon winter visitor

Range: coastal western and northern Europe.

Historical

In the 19th century Harting knew this to be a coastal species rarely found inland. He listed several occurrences of birds being shot at Brent Reservoir, with the first being in October 1843 when Bond killed several; and there were two on 20 March 1866, one of which was shot. Elsewhere, one was taken by a bird-catcher in Harrow in November 1862 and there was one in 1871.

20th century

Nothing was known about the status of Rock Pipit on the Thames-side marshes until 1921 when Horn stated that it was regular in winter at Thurrock. Away from the river, there were no records until 1923 when one was identified at Aldenham Reservoir on 28 October. Rock Pipits continued to be scarce away from the Thames marshes in the 1930s with a handful of migrants seen in most years. In 1931 there were six at Barn Elms Reservoirs on 16 October. On the marshes at Swanscombe 30–40 were counted on 11 December 1949.

In the 1950s the species was regular in winter on marshes along the southern side of the Thames from Crayford eastwards, and about 50 birds were counted between Dartford and Northfleet from October to December 1950. Up to 80 were seen on Swanscombe Marshes on 26 October 1953. There was little information available from the north side of the Thames but it was likely to be wintering there. Elsewhere, it was just an occasional visitor on passage, mainly in October, and in winter. Most of these visitors were from the large reservoirs at Barn Elms and Staines, and in the Lea Valley. Typically, migrants were seen singly but at Barn Elms there was a party of six on 5 October 1954.

In the autumn of 1960 two flew right over the centre of London at St Paul's Cathedral on 7 November. In 1968 about 12 were seen by the Thames at Barn Elms on 9 January during cold weather. During the 1970s the peak count was 40 at both Dartford Marsh and Rainham Marshes; elsewhere, there were 10 on the banks of Staines Reservoir on 11 October 1975.

Far fewer were seen in the 1980s when there was a series of cold winters. However, there was a good autumn passage at Barn Elms Reservoirs in 1981, which peaked at 13 on 18 October.

In the early 1990s there were up to 26 wintering on the Lower Thames Marshes; the largest count away from the Thames was nine at King George VI Reservoir on 11 October 1990. In 1994/95 the wintering birds at Swanscombe Marsh were carefully studied as they began to moult, and before they left in spring it became apparent that they were of the Scandinavian race *littoralis*; a colour-ringed bird from Norway was also seen in October 1995 at King George V Reservoir. This helped to back up the thought at the time that most, if not all, of the Rock Pipits seen in the London Area were of the migratory Scandinavian race rather than the resident British race.

21st century

Since the beginning of the 21st century the only regular wintering sites have been along the Lower Thames with up to 29 at Swanscombe Marsh and 14 at Rainham. Elsewhere, small numbers can be seen on passage in spring and autumn, particularly at the large reservoirs but also occasionally in central London.

Facts and Figures

The earliest migrant was on 10 September 2008 at the London Wetland Centre.
The latest migrant was on 19 April 2003 at Beddington SF.
The largest-ever count was 80 birds on Swanscombe Marshes on 26 October 1953.

Water Pipit *Anthus spinoletta*

An uncommon winter visitor

Range: breeds montane Europe and Asia, winters lowlands.

Historical

The first record for the London Area was of a specimen illustrated by Edwards in 1760 found 'in the neighbourhood of London'; although he referred to it as 'Pennsylvanian Lark', the name at the time for the North American Buff-bellied Pipit, the description matches Water Pipit. Montagu also described this species from a specimen killed in Middlesex which he named 'Red Lark' and stated was sometimes found in London.

20th century

For much of the 20th century this was regarded as a race of Rock Pipit and there were no further records until 1938 when one was seen at Walthamstow Reservoirs on 30 March. Up to 1954 it was seen on a further five occasions but it then became more frequent with 34 seen between 1955 and 1963. Some of this increase may have been due to observers being able to identify this bird rather than a general increase in numbers. The birds were usually seen on passage at the reservoirs or in winter on the Lower Thames Marshes.

In early 1965 up to 13 were present in the Colne Valley area between Watford and West Hyde, and one in Regent's Park was observed feeding on a butter wrapper on a rubbish dump – the first to be identified in Inner London. Water Pipits were first recorded wintering at Beddington SF in 1968 when up to 13 were present between January and mid-April. By the beginning of the 1970s small numbers were wintering at favoured sites, for example at Beddington, Hersham SF and Rainham Marshes, and in the Colne and Gade Valleys. There were occasional high peaks, including 20 at Beddington on 28 November 1971.

By 1986, when this species was split from Rock Pipit by the BOU, there were single wintering birds at Cassiobury Park and Rainham Marshes, with a few other sightings in winter and on passage. Wintering numbers remained low until the 1990s. There was a large increase in sightings in 1991 with a total of at least 65, notably at Rainham Marshes where there was a marked passage during March with a peak of 20 on the 29–30th; elsewhere, there were up to eight at Beddington in February.

21st century

At the beginning of the 21st century Water Pipits wintered regularly at Beddington and Rainham Marshes, with up to eight seen at both sites; a few also wintered at Rye Meads and at other sites along the Lower Thames They began wintering on Staines Moor in 2000/01 and within two years up to 16 were present in the area, also being seen on the adjacent King George VI Reservoir. They have also started to overwinter at Crossness and the London Wetland Centre. Elsewhere, they are scarce passage migrants, typically in October and April.

Facts and Figures

The earliest migrant was on 26 September 2004 at Tyttenhanger GP.
The latest migrant was on 2 May 2000 at Staines Moor.
The largest concentration was of 30 birds at Rainham Marshes on 7 March 1992.

Buff-bellied Pipit *Anthus rubescens*

A very rare vagrant (2).

Range: Americas and Asia; vagrant to UK.

The first record was of a bird found at Queen Mother Reservoir on 12 December 2012 where it was joined by a second bird on 26 December; both remained into 2013. This species was previously extremely rare in the UK with only four occurrences prior to 2000. Since then there has been a large increase in sightings with a total of 27 accepted records up to 2011; another eight birds were reported in the autumn of 2012. Most records are from the Scottish islands or Scilly and these are just the second and third inland birds following one in Oxfordshire in 2007.

2012: Queen Mother Reservoir, on 12 December with two on 26 December to 22 January 2013.

Unacceptable records

This species was known in the 19th century under various other names, including Red Lark and Pennsylvanian Pipit. A bird illustrated and described as being this species by Edwards in his 1760 book refers to a Water Pipit, which was unknown in the London Area at the time. Jenyns (1835) quotes a record from Col Montagu of one taken near Woolwich in the winter of 1812 in a net with other larks. No further details are known about this record.

Chaffinch *Fringilla coelebs*

A very common resident and widespread breeder

Range: Europe, Asia and North Africa, northern population migratory.

Historical

The earliest record was of a leucistic bird at Havering in 1732. Fifty years later Gilbert White heard one singing in South Lambeth. In the 19th century Harting described this as a common breeding resident but by the end of the century it was in retreat from central London; it had ceased breeding in the central parks and in the squares around Bloomsbury. This decline may have been attributable to the bird-catchers for whom Chaffinch was one of their main targets; one bird-catcher was found with 40 Chaffinches in Richmond Park.

20th century

By the 1930s it had recovered to be described as 'an abundant resident'. Glegg also stated that it became more common 'as we proceed outwards through the suburbs and in the rural parts of the county it is a very numerous bird'. He also noted that it often gathered in large flocks after the breeding season, for example particularly large numbers wintered at Hendon SF.

In the 1950s this was considered to be the most common and widespread bird in the London Area, as although House Sparrow was certainly the most numerous bird in the built-up areas, it was much less common elsewhere. Chaffinches could be found breeding in small numbers in central London in all of the main parks and in some large gardens in Campden Hill and St John's Wood, although they had not recolonised any of the squares. The only habitat away from central London where they were not found was on some of the most easterly marshes alongside the Thames. The breeding population was virtually sedentary, with the only ringing recovery away from the nesting area being found just 20 kilometres away. At Bookham Common declines were reported of 60 per cent in oak wood and 30 per cent in open scrub and grassland over a 10-year period during the 1950s.

They remained relatively common breeding birds in the Royal Parks in central London during the 1960s and 1970s, e.g. in Regent's Park 18 pairs were located in 1963 and 19 pairs in 1978. Elsewhere the population showed signs of a decrease and there was a 10 per cent decline on 15 Common Bird Census plots from 1968 to 1969. In February 1984 about 1,000 were seen on the marshes at Thamesmead.

During the second breeding Atlas in 1988–94 they were found breeding in 82 per cent of the area, virtually unchanged since the first survey in 1968–72. Most of the areas where they were absent were in the more heavily built-up parts of London, particularly the inner suburbs. The Breeding Bird Survey showed a substantial increase of more than 50 per cent in the London population between 1994 and 2000 compared to a national increase of just 6 per cent. In 1991 there was a feeding flock of 1,500–2,000 at High Beech, Epping Forest, on 3 February.

21st century

By the start of the 21st century, this species appeared to be less common in central London, with no more than 5–10 pairs reported in the LBR annually. However, the Breeding Bird Survey shows a steady increase from 1994–2009 in the London Area, with an overall 164 per cent increase from 1995–2009 compared with an 11 per cent increase for the UK as a whole.

Movements

The autumn passage of Chaffinches has been studied and documented since the late 19th century. They were first described by Power who watched these movements from his Brixton home between 1874 and 1909. He wrote 'On a favourable occasion flock after flock, containing half a dozen to fifty or more birds, will pass in an almost continuous stream for hours i.e. from 8 a.m. to noon or thereabouts. One of the most remarkable of these passages occurred in 1902, when the first flock passed over on the 7th of October, and others almost daily until the 5th of November; ... during this time many thousands must have passed before my eyes.' In 1903 Eagle Clark studied these movements in the Thames Estuary and recorded vast numbers flying west along the Thames which had come in from the Continent. No further studies were made until the 1920s when several observers recorded movements of Chaffinches flying over central London during October.

In the late 1940s and early 50s a coordinated watch took place from rooftops around Trafalgar Square and other vantage points in and around the London Area. This series of observations showed that there was a passage of Chaffinches over the London Area from late September through to early November. Birds travelled on a broad front in a westerly direction but not necessarily all following the Thames. An example of one day's passage, on 14 October 1950, showed that 75 flocks totalling 715 Chaffinches passed west over Trafalgar Square between 07:30 and 10:00 hrs into a light breeze from the west-south-west. Some flocks were seen to land although it was not known if they later continued their journey.

In the autumn of 1958 there was a large passage of several species on 19 October and about 2,000 Chaffinches were seen over Staines; a similar number was counted from a train between Dartford and London Bridge. On 1 November 1959 there was a massive passage of a number of species, including an estimated 60,000 Chaffinches. In the autumn of 1963 the peak count was about 1,300 over Dollis Hill on 18 October. In 1968 the peak passage occurred on 15 October when about 1,100 flew over Worcester Park, and 1,000 flew over Danson Park and Hyde Park. The highest count in the 21st century was in 2001 when 963 flew over the London Wetland Centre on 11 November.

The population in winter outnumbers the breeding population and it is likely that many birds from the Continent spend the winter in the London Area. There have been several recoveries of continental birds, including one ringed in Cricklewood on 31 December 1949 and recovered in north Sweden the following August which appeared to be of the Scandinavian race *coelebs*. Further ringing recoveries during the 1950s showed that wintering birds in the London Area had come from Finland, Norway and Sweden. The wintering flocks are predominantly single sex; for example, a flock of 150 at High Beech on 14 April 1941 was virtually all-female in composition, while another at the same site on 10 January 1942 was 70 per cent male.

The return passage in spring is much less notable than in autumn. There have been large numbers noted occasionally in February, such as in Bushy Park on 27 February 1934 and again on 18 February 1935, when there was an influx of many hundreds, but it is not clear if these movements relate to northbound migrants or not.

Brambling *Fringilla montifringilla*

An uncommon winter visitor

Range: breeds northern Europe and Asia, winters Europe and southern Asia.

Historical

Bramblings were first recorded in October 1840 at Epping. Harting knew them as late-autumn visitors in small flocks. They were regularly seen at Kenwood, Hampstead, where they fed on beechmast, and small flocks 'may generally be found in the autumn along the banks of Kingsbury [Brent] Reservoir'. At the latter site half a dozen bird-catchers would spend all day with their nets in October trying to catch Bramblings among the large mixed flocks of finches and other birds that fed on seeds. Power recorded Brambling flying over his Brixton home in October fairly regularly between 1874 and 1909.

There were occasional records at other times of year, for example one was seen in February and another was caught at Edgware on 18 April; and at Hertingfordbury about 55 were caught on 8 January 1872. It is interesting that they were only seen in small numbers as just beyond the London boundary a correspondent of *The Times* stated that an immense flock visited Stoke Park, near Slough, in November 1865.

20th century

During the early years of the century very little information was recorded about the status of Brambling. There was an influx in the winter of 1905/06 and large flocks were seen in the Watford district and south London. Further large influxes were noted in the winters of 1920/21 and 1922/23.

Glegg considered that it was an irregular winter visitor, not one that was recorded annually. The largest numbers were about 300 at Tadworth on 9 March 1930; about 360 at Beddington SF on 22 March 1931; 500–600 feeding on ploughed fields that had recently been covered by hop-manure at Osterley from December 1931 to January 1932; and 200–300 at Norwood Green on 3 March 1932. Large numbers were seen towards the end of the 1936/37 winter, including 200–300 at Beddington.

By the beginning of the 1950s Bramblings were recorded annually in winter but in varying numbers. In some years the species was common in Epping Forest, with flocks of up to 100, and on the North Downs. In other years it was virtually absent from these favoured areas, where there are good stands of beech and hornbeam. There were several influxes during the middle of winter with occasional flocks of 200–300 seen at various sites, typically on the Lower Thames Marshes and at sewage farms. Much bigger flocks were sometimes seen: in 1951 about 3,000 roosted in rhododendron bushes at Westerham during February and March; and in 1956 at least 1,000 were seen at Dartford on 18 March. In early 1959 there were several large flocks, including 600–800 at Chipstead on 14 March, about 350 at Croydon from 23 to 25 March and about 300 at Carshalton on 7 March.

During the very cold winter of 1962/63 they were more numerous, albeit in small numbers across the London Area. In central London about 40 were seen in total with some even feeding on scraps of bread with House Sparrows in Archbishop's Park in Lambeth. Large numbers were seen in early 1967 with the peak count being 750 at Romford SF on 18 February, while in 1969 there were up to 800 at Rainham during March.

There were flocks of 500 at Beddington in 1970, 1974 and 1979; elsewhere in the 1970s the highest count was about 1,000 at Crossness SF on 27 February and 1 March 1976. During the 1980s and 90s there were fewer large flocks and the maximum count was 400 in the Dartford area in early 1986 and in Norbury Park in early 1994.

21st century

Since the beginning of the 21st century Bramblings have been seen in fairly small numbers during winter apart from in 2003 when several hundred were in Epping Forest between January and April, and in early 2007 when there was a massive flock of up to 2,500 at Moorhouse. They are most commonly recorded flying over in autumn; the peak count was 53 over the Wetland Centre on 27 October 2003. In autumn the first arrivals are usually in early October although in some years none have been seen until November or even December. In spring there is often a small passage in March and very few are seen after the middle of April.

Summer Records

There have been occasional records in summer: a male shot at Enfield on 7 July 1923; on Hampstead Heath between 28 June and 9 July 1926; a male in Epping Forest near Loughton from 12 to 25 June 1932; and one on Wimbledon Common on 12 June 1938. In 1967 a female with a blue ring on one leg, suggesting a captive origin, was paired with a male Chaffinch in Battersea Park; the female was seen on a nest from 10 to 12 June but the nest was subsequently disturbed. A singing male at Grovelands Park on 11 July 1987 was possibly also an escape.

Facts and Figures

The earliest migrant was on 14 September 1997 at Regent's Park.
The latest migrant was on 15 May 1995 at Dulwich Park.
The largest ever group was of 3,000 birds at Westerham during February and March 1951.

Greenfinch *Carduelis chloris*

A very common resident and widespread breeder

Range: Europe, Asia and North Africa.

Historical

In the middle of the 19th century Harting stated that it was 'a common resident, flocking with Sparrows, Chaffinches and Yellowhammers in winter'. It bred in Battersea Park prior to 1879 but by the 1890s it had become rare in the central parks, probably due to the activities of the bird-catchers as Hudson remarked that 'scores, nay hundreds may be bought any Sunday morning in the autumn at the bird-dealers' shops in the slums of London, at about two pence per bird, or even less'.

20th century

At the beginning of the 20th century it was a relatively common and widespread breeding bird except in central London where it was just a winter visitor. There was a gradual increase in the population, starting in some areas in 1905 then more generally from 1910 to 1915.

In the 1930s Glegg described it as a common resident in the rural areas and as an increasing resident in central London where it bred annually in small numbers. He also noted that large flocks, numbering in the hundreds, were seen in winter, often mixed in with Chaffinches, Tree Sparrows and Yellowhammers, etc. In the Harrow district 24 nests were discovered in 1930 and large numbers were also recorded at sewage farms, including flocks of over 150 in spring. In February 1947 a flock frequented a cornfield at Walton Heath and peaked at about 500; the same year a roost of about 150 was found in Richmond Park on 22 November.

It had become much more common by the 1950s and was mainly known as a breeding bird in the suburbs and rural areas, being absent from many of the most built-up areas although it did breed in the large parks. In autumn and winter most Greenfinches moved away from their breeding sites to more open areas to feed in mixed flocks with other finches, sparrows and buntings on fields and weedy ground.

In 1960 about 75 were seen in Kensington Gardens on 18 January, attracted by hop-manure. During the cold winter of 1962/63 larger numbers than usual were reported from some gardens, including in Esher where 295 were trapped and ringed between December and March. In 1967 there were several unprecedented counts, notably 1,000 at Wraysbury Reservoir on 21 January. There was a series of records from Potters Bar in early March, thought to have been birds leaving a roost shortly after dawn; the highest of these were 900 on 4 March and at least 500 at Maple Cross on 28 January. In addition, 481 were trapped and ringed in a garden in Weybridge in the first three months of the year. The following year, 1968, there was an even larger count, 1,200 seen at Walton Reservoirs heading to roost at St George's Hill, Weybridge, on 4 February. By 1968 the breeding population in central London had increased to at least 55 pairs, virtually double the number after the severe winter in 1963.

During the 1970s the largest counts were some 1,500–2,000 feeding in a field at Chorleywood in December 1970; more than 1,000 at Beddington on 2 October 1977; a roost of up to 557 by Kensington Palace in 1977; and at least 1,000 throughout September at Radlett in 1978. At Potters Bar 2,860 were trapped and ringed during 1981. There were peak counts of 1,000 at Batlers Green in September 1980 and at London Colney in December 1985.

During the second breeding Atlas Greenfinch were found breeding in 90 per cent of the area, an increase of 5 per cent since the first survey in 1968–72. However, the population in central London was in decline. In Hyde Park and Kensington Gardens there were eight pairs in 1994 compared to 16–30 pairs in 1967–80; in Buckingham Palace Gardens there was just one pair in 1996, down from 10–18 pairs in 1961–63; in Regent's Park it declined from 37 pairs in 1976 to four in 1997; and in Battersea Park the population fell from 25–30 pairs in the 1970s to 5–10 by 1995. By contrast the Breeding Bird Survey in 1994–2000 showed an increase in the overall London population of 29 per cent.

21st century

Since the beginning of the 21st century Greenfinch has remained a common and widespread bird throughout the London Area, and there is a small breeding population in central London. The Breeding Bird Survey in London indicated an increase of 139 per cent from 1994 to 2007, but numbers have since fallen sharply perhaps because the Trichomonosis disease has significantly affected this species. The highest count was 300 on Hampstead Heath in September and October 2004.

Movements

The earliest ringing recovery relates to a bird that was first ringed in Streatham on 20 July 1923 and recovered at Lower Edmonton on 19 December the same year. Some suburban breeding birds move south-west after the breeding season as evidenced by ringing recoveries from Ewhurst, Reading and Weymouth. In late autumn, during October and early November, a diurnal passage takes place over the London Area. Birds are regularly seen flying between south-west and north-west in flocks of up to 30, usually during the early morning.

Facts and Figures

The largest count was 1,500–2,000 birds at Chorleywood on 23 December 1970.

Serin *Serinus serinus*

A rare visitor (25)

Range: Europe and North Africa; scarce visitor and rare breeder in UK.

The first record was of a male captured at Hampstead in October 1859 after a severe storm; the specimen eventually ended up in Bond's collection.

It was almost a century before the next record when a male was seen in Dollis Hill in November 1951. The identity was accepted by the LNHS but the record was placed in square brackets as the possibility of it being an escape had not been ruled out, although the observer had stated that it did not allow a close approach. Given the number of subsequent records this is now considered acceptable.

From the late 1960s onwards there was a gradual increase in sightings in the UK and this was also reflected in the London Area. In 1969 a pair was seen at Sevenoaks in July and two more were seen in the mid-1970s. There was a further increase during the 1980s when five males were seen, one of which spent three days in a garden in Wilmington and was heard singing.

There were two in 1991, one in winter for a week at various sites in the Lea Valley, and a singing male at Brent Reservoir in April. Another two were seen in 1996, one of which was present for almost two weeks in Charlton in early winter. The only one seen in Inner London occurred in 1998 when a female was present for just over two hours on the morning of 23 April in Regent's Park. The following year a male was present on Hounslow Heath for 11 days in February.

In 2002 a singing male was present on wasteland at Sidcup between 28 February and 2 March. There was an unprecedented influx at Rainham Marshes late in 2008; up to three were present from 22 October with seven seen on 13 November. The three remained until the end of the year and presumably two of these returned the following winter.

1859: Hampstead, captured in October
1951: Dollis Hill, male on 13 November
1969: Sevenoaks, pair on 13 and 20 July
1973: Hilfield Park Reservoir, on 21 March
1974: Twickenham, male on 15 March
1982: Stanstead Abbots, male on 26 May
1985: Streatham, male on 10 February
1986: Wilmington, male on 17–19 May
1986: Barn Elms Reservoirs, male on 2 June
1988: High Elms, male on 7 February
1991: Rye Meads, on 6–12 January; also seen at Roydon GP and Stanstead Abbots GP
1991: Brent Reservoir, singing male on 26–27 April
1996: Gillespie Park, on 26 March
1996: Charlton, male on 27 November–9 December
1998: Regents Park, female on 23 April
1999: Hounslow Heath, male on 13–23 February
2002: Sidcup, male on 28 February–2 March
2008: Rainham Marshes, up to seven on 22 October–31 December; also two on 21 November–5 December 2009

Unacceptable Records

A record of one at Paddington Green on 17 September 1998 was originally accepted but later withdrawn by the observer after discovering that a local resident had released a number of finches in the area.

Goldfinch *Carduelis carduelis*

A very common resident and widespread breeder

Range: Europe, Asia and North Africa.

Historical

In the early part of the 19th century Goldfinches were common and widespread breeding birds but many were captured for the cage-bird trade. An old bird-catcher related how, as a youth, he once caught almost 150 in a single morning on the site where Paddington station now is.

Harting remarked that they were 'at one time common, but now seldom seen; owing, no doubt, to the increased cultivation of waste lands, and the disappearance of gorse and thistles'. He added that although a few were seen in autumn and early spring, they had ceased to breed. They had become extinct as a breeding bird in the inner suburbs by around the 1880s, probably due to the activities of bird-catchers, and the only documented breeding record in the last two decades of the 18th century was of a pair in Pinner in 1882.

20th century

They were reported to be absent from Harrow between about 1899 and 1903, and elsewhere in the London Area by 1900. The decline that had been going on for 30 or 40 years had made Goldfinch one of the rarest breeding birds in suburban London. A few continued to be seen in autumn and winter, and these may have come in from outside the area.

After legislation curtailed the activities of the bird-catchers in the early years of the 20th century, Goldfinches began to return. In Enfield immense numbers were seen in 1905, including a large percentage of young birds thought to have bred in nearby orchards; breeding also took place at Winchmore Hill. A more general recovery was noted from about 1907–08 onwards. In 1909 Ticehurst stated that it was particularly common around Sevenoaks and Westerham, and in the Upper Darenth Valley. It started to become re-established in the Harrow district in the late 1920s.

Increases were noted in some areas up until the 1930s when Glegg described it as an uncommon resident in the more rural areas, becoming more noticeable in winter when small flocks could be found. Occasionally very large flocks were reported, including one of more than 500 birds in Bushy Park on 16 October 1937 and 200 mobbing a Little Owl near Warlingham prior to 1932. By the end of the decade it had become a widespread breeding bird in the outer areas, particularly along the North Downs where it was common, and in west London where it bred abundantly in orchards. Closer into London, breeding occurred at Beddington, Kew Gardens, Mitcham Common and Richmond Park, and birds were present during the breeding season at Hampstead Heath, Horsenden Hill, Putney and Wembley. It did not breed in central London at the time, being only an uncommon visitor in autumn and winter.

By the 1950s it had become a reasonably widespread breeding bird, mainly in the suburbs but also in some central London locations. Breeding took place in Battersea Park from 1949 to 1950 and pairs probably also bred in Brompton Common, Chelsea Hospital, Regent's Park and St John's Wood. At least two pairs bred in Regent's Park in 1953. During the 1960s-70s this species was particularly associated with seeding herbs e.g. thistles on wasteland plots awaiting development and alongside railways.

Goldfinches continued to increase in central London and in 1965 a roost was discovered in plane trees in South Kensington; on 4 January it was used by 10 birds and by the 23rd there were 105 roosting there. The roost was occupied until at least early April. The following winter only a few birds used the roost but another containing 50–60 birds was found in plane trees in Warwick Avenue; this was abandoned early in 1966 and may have moved to Ecclestone Square since up to 52 were found roosting there between February and April. Later, peak counts became considerably higher with at least 600 at Waltham Abbey in early 1977 and about 1,000 at Rainham Marshes in 1990. In the autumn of 1977 a total of 1,265 flew over Dulwich during the period 3–20 October.

During the second breeding Atlas in 1988–94 they were found breeding in just over 70 per cent of the area, virtually unchanged from the first survey in 1968–72.

21st century

Since 2000 this has remained a widespread breeding bird apart from in central London where only a few pairs breed. Wasteland had become far less available by this time, with faster turnover of vacant plots, but it has become a regular visitor to garden bird feeders, especially where niger seed is provided. It was the 10th most common species in RSPB's Big Garden Birdwatch in the London Area in 2013. The Breeding Bird Survey for the London Area overall showed a massive increase of 242 per cent from 1995–2010.

The peak counts are made in autumn when family parties often join together to feed on teasel and other seeds; these have been much smaller than in previous decades. In late autumn small numbers are regularly recorded passing overhead. They have become much commoner in winter, even in built up areas, where large numbers can be seen feeding on the seed heads of the London Plane; presumably this is partly an effect of generally milder winters. Although most large flocks break up in winter, there are occasional large concentrations such as one of about 400 birds at Holmethorpe in January 2008.

Facts and Figures

The largest count was about 1,000 birds at Rainham Marshes on 6 October 1990.
Juveniles ringed in the London Area during the autumn have later been recovered abroad in winter in Belgium, France and Spain.

Siskin *Carduelis spinus*

A common winter visitor and rare breeder

Range: Europe and Asia.

Historical

The earliest record was prior to 1812 when a pair was shot during the summer months in a garden in Dartford. In 1831 Montagu stated that the London bird-catchers referred to Siskin as 'Aberdevine'; they used to ask a rather high price for Siskins, on account of both their rarity and their value as a pairing bird with Canaries, the hybrid offspring apparently being remarkable for their soft and sweet song.

Apart from a few isolated breeding records it was described as 'an uncertain visitant, appearing occasionally towards the end of autumn, in company with Linnets and Lesser Redpolls'. It was 'formerly plentiful during most winters among the alders in the brook at Hendon' and was also found rarely at Kingsbury, Elstree, Stanmore and Harrow Weald, but in west London it was rather more common during winter. There was a particularly large influx during the autumn of 1865.

20th century

It was a very rare visitor to central London and one feeding on an alder in St James's Park in December 1908 was the first known record. Little had changed by 1935 when Glegg described it as 'a local winter resident' whose appearance was subject to weather conditions, cold weather producing more birds which tended to stay longer. The largest flocks were of 70–100 birds in Richmond Park in March 1937; and about 150 in Bushy Park on 19 February 1941. Prior to 1935 it was typically observed between December and March although there were a few records in October, November and April. However, by the 1950s the arrival time was usually noted between the third week of September and the first week of November with a few seen in late August. At the beginning of spring there was often a passage in March and April when the last birds were seen.

The numbers seen each autumn and winter varied considerably. In 1959/60 the winter population was about 550–600. During the autumn of 1960 there was a considerable passage but by December only around 100 were present. In the exceptionally cold winter in early 1963 larger than usual numbers were present, and in early April well over 150 were seen at Oxshott Woods. In early 1965 many more than usual were present with the highest count being up to 350 in Gunpowder Park in the Lea Valley.

During the 1970s there was an increase in the wintering population and in early 1976 there were peaks of 450–500 at Pett's Wood, 400 on Hampstead Heath and a mixed flock of 1,000 Siskins and Redpolls in Ruislip. In 1970, 451 were ringed in a garden in Weybridge during the year. In 1991 the autumn migration began particularly early with an influx in July that was also recorded at many East Coast sites.

21st century

Although Siskins are regular winter visitors, their numbers vary from year to year. They are most commonly seen along the river valleys where there are large numbers of alders, for example along the Colne and Lea. They are also regularly recorded on passage over the London Area, particularly in autumn.

Breeding

The earliest reference to possible breeding was prior to 1812 when a pair was shot during the summer months in a garden at Dartford. Confirmed breeding has only occurred on a few occasions. In 1836 a pair bred in a gorse bush at Anne's Hill, Chertsey. In 1838 a juvenile taken near Tooting was thought to have been locally bred. Siskins bred on Wimbledon Common on two occasions in the early 19th century and also in Kenwood, on Hampstead Heath, in 1853, although there was some speculation that the latter could have been the result of escaped cage birds and Glegg considered the Kenwood breeding record unconvincing.

In the first half of the 20th century the only confirmed breeding record was in 1901 when a pair built a nest in Sundridge Park near Bromley; it was not known if it bred successfully as the nest was found to have been destroyed when next visited. In 1924 a male held territory at Weybridge in May and June. In 1969 two adults were accompanied by a juvenile in Warren Wood, Epping Forest, possibly representing the first successful breeding record for over 130 years.

There were several instances of possible breeding during the 1990s. During the summer of 1990 a pair was present at Havering and may have bred, and a single bird was seen in Hainault Forest in June. The following year a pair was seen feeding juveniles on Hampstead Heath in mid-July and probably bred nearby. In 1994 a pair summered at Hainault Forest and another pair possibly bred in Havering CP. In 1997 birds were seen at five localities in the breeding season although there was no evidence of local breeding.

Juveniles at Kenley in May 2000, Sanderstead in June 2002, and Garston and St Albans in June 2004 all suggest local breeding. At least one pair bred in Black Park in 2009–10 and a pair bred in the outer suburbs in Barnet in 2013 where a family party regularly came to garden feeders.

Linnet *Carduelis cannabina*

A common partial migrant and localised breeder

Range: Europe, Asia and North Africa.

Historical

In the 19th century this was described by Harting as a resident throughout the year, but not as common as it once was due to a loss of uncultivated lands and waste ground. In autumn Harting observed Linnets 'in small flocks upon the stubble and cleared bean-fields, and in much greater numbers among the weeds on the banks of the Reservoir [Brent]'. In north-west London they bred at Hampstead Heath, Harrow Weald Common and Stanmore Common. Elsewhere, they were described as 'pretty numerous' on Wimbledon Common in 1881 and were fairly abundant on Streatham Common; large numbers were taken by bird-catchers on Wanstead Flats.

20th century

In the first 20 years of the century there was an increase in the population, probably attributable to legislation which put an end to the activities of the infamous bird-catchers. There was clear evidence of an increase in the Staines district where it could only be found in one spot in 1896, but 10 years later 'every bush and hedge contained a nest'.

By the 1930s it had become 'a common resident throughout the rural portions of the county', Glegg adding that it had increased although it was still a localised breeder. In Harrow, it was rare in 1903 but had become widespread, though nowhere common, by 1930. There was a sizeable breeding population on the banks of Staines Reservoir in 1935. In central London it was an occasional autumn and winter visitor between 1900 and 1935, and a pair bred in Hyde Park in 1918.

By 1949 Linnets were breeding in the suburbs as close to central London as Finchley, Brent Reservoir, Ham, Wandsworth Common, Elmers End and Barnehurst. During the 1950s they were considered to be a widespread and fairly common bird across the London Area. The population was generally stable although there were some local losses due to the ploughing up of commons and rough ground. As they preferred more open areas than some other finches, Linnets remained absent from the inner suburbs but a pair was present during the breeding season in Chelsea in 1951. They were fairly common on the marshes beside the Thames from Greenwich eastwards.

Between the late 1960s and early 90s they bred in Inner London. A pair bred successfully in Regent's Park in 1968 but there was no further breeding until 1971 when at least three pairs bred in the recently decommissioned Surrey Docks. By 1982 up to 10 pairs were breeding at Surrey Docks but redevelopment work later reduced the population although they continued to breed until 1991. They also bred by Bankside Power Station and at Wapping.

In the second breeding Atlas in 1988–94 they were found breeding in 63 per cent of the area, almost unchanged from the first survey in 1968–72. Although they were found most commonly in the rural areas, there was a big decline in south-west London from Wimbledon Common out to the North Downs.

21st century

By the start of the 21st century the breeding population had become much more localised and they had ceased breeding at a number of former breeding sites in the suburbs, e.g. Brent Reservoir and Woolwich Common. The national population had declined by over half since the 1970s and Linnet was placed on the UK 'Red List'. At Rye Meads the numbers fell to three pairs in 2000 from 10 the previous year. They have continued to decline and now only breed at a handful of suburban sites, for example at Wormwood Scrubs where ten pairs nested in 2010.

Autumn and Winter Counts

After the breeding season most Linnets move away from their nesting territory and are joined by immigrants from northern Britain and the Continent. In late autumn, a notable westerly movement of Linnets has been recorded in the Thames Estuary and these birds from Europe are likely to be those seen passing through the London Area. Locally raised Linnets have also been recovered in France in late autumn.

In severe weather in February 1956 large numbers were recorded in central London, including at least 82 at Battersea Park and more than 40 in Regent's Park. In 1960 there was a large influx into central London at the beginning of the year and numbers started to build up by the end of the month. Peak counts were made on 11 February when there were about 220 in Battersea Park and 213 in Regent's Park; a good proportion of these flocks lingered well into March.

In 1965 they increased on Clapham Common from about 100 to about 400 ahead of snowy weather on 3 March, and in the autumn 300 were counted flying over Danson Park in one hour on 27 September. The following year about 400 were at Romford SF on 8 January. In 1967 about 500 were counted at Swanscombe on 29 January. During the 1970s several exceptionally large flocks were seen, including some 2,500–3,000 birds feeding in a field at Chorleywood in 1970; 1,500 at Navestock in September 1972; 1,000 roosting at both Queen Mary Reservoir in January 1971 and Rye Meads in November 1972.

Large autumn movements over the London Area are occasionally noted, such as the 200 birds that flew south over Hyde Park/Kensington Gardens on 5 October 1976; and in 2004 a total of 461 flew over the London Wetland Centre on 7 October.

Since the beginning of the 1980s the largest flocks have been mostly on Rainham Marshes where they peak in October or November. In the winter of 2000/01 several large flocks were present with the peak count being 500 birds at Rainham.

Facts and Figures

The largest ever flock was of 2,500–3,000 birds at Chorleywood in 1970.

Twite *Carduelis flavirostris*

A scarce winter visitor

Range: breeds northern Europe and Asia, European population winters central and western Europe; breeds northern UK.

Historical

The earliest reference was in 1788 when Latham stated that they could be taken in the migration season 'near London', a definition now generally accepted to be within the recording area. In the 1830s they were regular winter visitors to Tooting.

Harting described them as rare autumn visitors and he knew of specimens obtained at Hampstead, Harrow, Kilburn and Kingsbury, as well several which were taken by bird-catchers around Brent Reservoir. In 1891 one was seen in St James's Park. They were also known as occasional visitors to Epping Forest. South of the river, Power recorded a small flock flying over Brixton in October 1893, and they regularly wintered on Banstead Downs during the 1890s where some were captured each year.

20th century

In the first half of the century there were just four records, in each case of two birds being seen: in Kensington Gardens on 21 November 1925; in the Roding Valley near Barking in about 1930; at Barn Elms Reservoirs on 28 October 1946; and at Staines Moor on 29 September 1948. An earlier record of one on Clapham Common on 1 January 1918 was later discounted due to the lack of identification details. It is surprising that there were no records in this period from the Thames-side marshes; however, they were particularly under-recorded during this time.

From the mid-1950s they were seen almost annually, mostly in small numbers apart from an exceptional flock of 22 at Sevenoaks Wildfowl Reserve on 2 April 1964. During the 1970s they started to winter at Rainham Marshes; in 1971/72 up to 50 wintered, feeding on sea aster. In 1974/75 unprecedented numbers wintered at Rainham with up to 200 in February and there were still 60 on 11 April. Occasional groups were seen elsewhere on the Lower Thames, including one of 30 at Barking in December 1975; 20–30 at Swanscombe on 4 February 1979; and 18 at Northfleet on 24 January 1979.

The peak counts at Rainham varied from year to year during the 1980s but the trend was downwards; there were 150 birds on 30 January 1982 but by 1989/90 only five overwintered. The only large flock seen away from Rainham was of 25 birds at Dartford Marsh on 12 March 1987. At the start of the 1990s only two or three were seen at Rainham but in the following two winters 20–25 were present. Occasional flocks were seen elsewhere, with eight birds at both King George V Reservoir on 16 October 1994 and Bury Farm, Edgware, on 27 October 1999; and 15 flew over Brent Reservoir on 7 December 1996.

21st century

Since the beginning of the 21st century Twite have become much scarcer and are not recorded every year. Most of the records are of short-staying birds on the Lower Thames Marshes. In early 2000 one overwintered at Hersham SF. The largest group was of four birds at Rainham Marshes on 1 December 2007.

Facts and Figures

The earliest migrant was on 4 October 1869, at Hampstead.
The largest flock was 200 at Rainham Marshes on 27 February 1975.

Lesser Redpoll *Carduelis cabaret*

A common winter visitor; former breeder

Range: western and central Europe.

Historical

The earliest record was in 1831 when Montagu stated that they were 'frequently taken about London by bird-catchers during the winter months'. Harting stated that they generally appeared 'towards the end of autumn in small flocks, and may then be found among the weeds along the banks of Kingsbury Reservoir, on Hampstead Heath, Wormwood Scrubs, Harrow Weald Common and Stanmore Common'. In the last quarter of the 19th century they were frequently recorded from central London, mainly in the Royal Parks.

The first documented breeding was in Cricklewood in 1857; there were no further breeding records until 1884 in Highgate Cemetery where they bred annually until 1887. There were also breeding records from Headstone Lane in 1888; near Cobham in 1890; near Epsom in 1894; Farnborough in 1896; Coombe Wood, Wimbledon, in 1897; near Headley in 1897–98; Winchmore Hill in 1898; and on Wimbledon Common in 1899. Breeding probably also occurred in Orsett in 1887 and at Dulwich from 1894 to 1896.

20th century

Early in the 20th century there was a general increase in the breeding population in the country and this was mirrored in the London Area. The first birds bred in the Staines area in 1903 and in the Darenth Valley they were considered to be a common breeding species in 1905 with more than 20 nesting pairs. Around Bromley, Walpole-Bond stated that they 'had increased wonderfully during the last few years' and by 1907 there were 20–30 pairs breeding in the area. On Hampstead Heath, several pairs bred from 1907 onwards, increasing to seven pairs by 1910.

Breeding also occurred at many other locations in the outer suburbs and rural areas but by the 1920s a decline had set in and breeding had ceased at many sites.

By the 1930s the breeding population had reduced further; breeding was confirmed from Epping Forest, Harrow Weald, Hayes and Heston Commons and Stanmore, and probably also occurred on Hampstead Heath, at Northwood and Staines, and at a dozen or more sites in south London. Elsewhere it was mainly a common winter visitor. Flocks were regularly seen at Epping Forest during winter, sometimes numbering in excess of 100, especially at High Beach; in 1936 about 250 were counted there on 25 November. In the 1940s they bred regularly in the inner suburbs at Hampstead and Wimbledon.

By the 1950s they had become a sparse local breeding species, particularly on heaths and commons, and in the outer suburbs in orchards and large gardens. Most bred in south London; the only breeding areas north of the river were at Epping Forest and Hampstead Heath. During the 1960s the breeding population began to increase; there were 19 pairs on Banstead and Walton Heaths, eight pairs at Headley Heath, seven pairs at Pett's Wood and up to eight pairs in the north-west section of Epping Forest. In 1969 three pairs summered at Brent Reservoir and one or two were seen in Regent's Park during the breeding season but there was no evidence of any breeding. On Wimbledon Common there was a flock of 800–1,000 on 30 January 1966.

During the first breeding Atlas in 1968–72 they were found to be breeding in 20 per cent of the area with a fairly even spread between north and south of the Thames. There were about 6,000 in the Weybridge area in early March 1970. Wintering numbers were exceptionally high in early 1976 with maximum counts of 1,100 at Pett's Wood in January, 750 at Eastcote Common, 600 at Thorndon Park and at least 500 in Epping Forest.

In 1980 the breeding population was about 34 pairs with the highest concentration being 15 pairs in the northern section of Epping Forest. In Inner London a pair probably bred in Surrey Docks in 1989; the following year there were two pairs there and another pair in St John's Wood. By the time of the second breeding Atlas in 1988–94 there had been a decrease in the breeding range of 13 per cent since the first survey. There had also been a shift in the distribution with a move northwards; less than a quarter of the breeding tetrads were located south of the Thames and they had vacated much of the North Downs.

The decline continued and by 1995 the breeding population had almost totally collapsed with just one pair reported breeding, at Hainault; the only other potential breeding was of a pair at Orpington in June. There was a slight recovery over the next few years and in 1999 four pairs bred at Ongar Park Wood. Meanwhile there were reasonable numbers wintering with occasional influxes as in 1995/96.

21st century

By the start of the 21st century Lesser Redpoll had become close to extinction as a breeding species in the London Area. The only possible breeding records are of single males holding territory in Hainault Forest in 2000 and in Ongar Park Wood in 2003.

There is a notable passage during spring and autumn, and flocks are commonly seen during winter, often in the company of Siskins and Goldfinches on alders and birches. There were far fewer overwintering birds after the 1990s. However, in the winter of 2008/09, there was a notable influx with about 400 redpolls in Thorndon CP, many of which were Lesser Redpolls. In November 2011 there were 150 Lesser Redpolls in Wanstead Park.

Facts and Figures

The largest flock was of 1,100 birds at Pett's Wood in January 1976.

Common (Mealy) Redpoll *Carduelis flammea*

A scarce winter visitor

Range: Europe, Asia and North America; uncommon winter visitor in UK.

Historical

During the 19th century it was quite scarce and records most commonly referred to birds taken by bird-catchers. It was first recorded in the winter of 1833 near Croydon where about six were captured. In December 1855 about 30 were taken by two bird-catchers at Hammersmith. There were also records from Brent Reservoir, Cobham, Hampstead, Harrow, Kilburn, Kingsbury, Tooting and Wimbledon.

Harting saw a pair that was obtained near Elstree and the owner informed him that the species was there every year; he explained that Lesser Redpoll was very scarce there, which he considered curious since he knew the reverse to be the case in Kingsbury. Harting thought this may have been due to there being comparatively more alders in Elstree. Around Epping Forest many were trapped prior to 1885 where it was stated that it 'only appears at long intervals, and, like the Crossbills, in considerable numbers'.

20th century

During the first half of the century it was rarely recorded and none were seen until 1909 when a flock of 30 or more remained on Hampstead Heath from the middle of January to early April. In November 1913 more than a dozen were caught in a garden at Chingford. There was also one at Hampstead Heath on 29 March 1925; a pair at Tadworth on 6 April 1941; and two at Ruislip on 24 January 1942.

In the 1950s it was considered probably to be an annual visitor but one that often went unrecorded. During the 1960s there was a slow increase in sightings with a peak of 10 in Epping Forest on 15 October 1967. At Sevenoaks Wildfowl Reserve two were trapped and ringed in November 1973, and there was a flock of at least 140 on 30 April 1978. They were particularly scarce in the 1980s and the only records were of seven birds in 1986 and four in 1989.

There was an influx in December 1995, with small groups seen at Broadwater Lake, Hainault Forest and Thorndon CP where at least 40 were seen; by the beginning of 1996 up to 60 were present in Hainault Forest and up to 100 were at Thorndon. Only small numbers were seen in subsequent years apart from a flock of 22 at Paynes Lane GP in January 1998.

21st century

Since 2000 they have been seen almost annually and sightings have increased; this may be partly due to more interest in this bird after it was elevated to full species level. The first in Inner London was seen in Regent's Park on 6 October 2000. Up to five were present at a garden feeder in New Malden between 2 February and 21 April 2002. In 2006 about 32 were seen at 12 sites between January and early April. There was an influx during the winter of 2008/09 with small numbers seen widely across the London Area.

Facts and Figures

Two of the race *rostrata* were in Orpington on 6 February 1996.

Arctic Redpoll *Carduelis hornemanni*

A rare winter visitor (14)

Range: Arctic Redpoll; rare winter visitor to UK.

The first record was in 1976 at Thorndon CP in a flock of 600 Lesser Redpolls. Two years later one was seen in a flock of at least 140 Mealy Redpolls at Sevenoaks Wildfowl Reserve. During a large national influx in 1991 there was one at Godstone. Over half the London Area records occurred in the winter of 1995/96 when there was an even larger influx in the country along with exceptional numbers of Lesser and Common Redpolls. The only ones since then were in 2008 and 2011.

1976: Thorndon CP, on 4 April
1978: Sevenoaks WR, on 29 April

1989: Stanstead Abbots GP, on 25 February
1991: Godstone, on 10–11 March
1995: Thorndon CP, up to four on 17 December–28 January 1996
1995: Hainault Forest, on 26 December
1996: Chingford, on 1 February
1996: Broxbourne Woods, on 18 February and 15 March
1996: Hainault Forest, on 30 March
2008: Thorndon CP, on 1 December
2011: Weybridge, on 15 March

Two-barred Crossbill *Loxia leucoptera*

A very rare vagrant (4)

Range: northern Europe, Asia and North America; vagrant to UK.

There are just three records in the London Area, all of which were in the 19th century. There was an influx into the country in 1846 with at least 11 in Cumbria and three in Suffolk, and Doubleday shot one in his garden in Epping. There was one at East Molesey at the beginning of February 1876 and two were shot near Croydon in late November 1889.

1846: Epping, immature shot
1876: East Molesey, in February
1889: near Croydon, two shot in November

Common Crossbill *Loxia curvirostra*

An uncommon visitor and rare breeder

Range: northern hemisphere; in some years irruptions occur when the spruce cone crop fails in northern Europe.

Historical

Described by Harting as a 'rare and uncertain visitant', he recorded that 'great flights appeared in the neighbourhood of London in 1756–57'. In 1791 there were six in a garden at Dartford in July and one was shot at Erith in August.

They were described as being plentiful in England from 1835 to 1839; during this period several were shot at Bentley Priory and Harrow; they were also seen at Tooting, and hundreds were seen around Epping Forest, many of which were nestlings. In 1847 Ruegg stated that they were occasionally seen in the Woolwich area. The next influx appears to have been in 1855 when a flock of 30 was seen at Beckenham and seven were taken by a bird-catcher near Hampstead. There were also later records from Bushey, Cobham, Dulwich, Gatton Park and Reigate.

There were seven instances of breeding during the 19th century: a pair bred near Dartford prior to 1812; a male that was accompanied by a brood of five was shot at Harrow on 19 March 1839; also in 1839 a pair nested at Theydon Garnon but the nest and eggs were taken; a young bird was picked up after apparently falling from a nest near Muswell Hill in the middle of the century; a pair bred near Epping railway station prior to 1885; a pair bred at Orpington prior to 1893; and a nest was found in a pine tree near Dartford but no eggs were laid, probably late in the century.

20th century

Apart from an isolated breeding record at Keston prior to 1909 it remained rare until 1909/10 when a major irruption occurred. This produced a flock of 150 near Woodmansterne along with about 50 at both Northwood and Watford, and the first records in Inner London when a flock flew over Kensington Gardens on 21 November.

Following this irruption six pairs bred at Bostall Woods in Woolwich in 1910, and birds also bred at Croydon, Gerrards Cross, Keston, Walton Heath and Weybridge.

In 1914 a pair bred again on Walton Heath but after this very few were recorded until 1924 when there were a few scattered records. A pair bred at Addington in 1926 and there was a small influx in 1927–28 when large numbers started arriving in July at Northwood and up to 60 were seen in August. Another small influx occurred in 1930–31; this culminated with a pair breeding on Iver Heath. This was followed by a large irruption in 1935–36 and many birds were present throughout the breeding season; the only confirmed breeding was at Kew Gardens but breeding almost certainly occurred in the Ashtead and Oxshott area.

During the 1940s it was particularly scarce although in 1943 a pair was seen nest-building on Reigate Heath. There were no further influxes until 1953–54. Following another small influx in the autumn of 1956 up to 25 remained in the Esher Common and Oxshott Heath area and at Limpsfield during spring but there was no evidence of breeding. There were several further small irruptions in subsequent years and in 1960 a pair may have bred in Black Park.

In 1962 there was a particularly large irruption from the end of June onwards and up to 200 were recorded; the highest count was of more than 100 at Kingswood. Most of these birds had departed by the end of the year but during the severe winter of 1962/63 small groups of up to six were seen at six sites. During the spring of 1963 more crossbills were recorded and a flock of up to 32 remained in the Panshanger area throughout April and May; some birds probably bred. Further influxes occurred during the next four autumns. A pair bred successfully in Kew Gardens in 1967, and at least five pairs were present during the breeding season in 1968.

There were fewer irruptions during the 1970s and none were seen in 1976, the first year with no sightings since 1955. Breeding may have occurred at Thorndon CP in 1977–78 and also at Badgers Mount in 1977. In 1980 up to 15 wintered in Black Park and in 1984 a group of 12 was seen there in March, some of which had nesting material; it is possible that they bred there as the area is underwatched. An irruption in 1985 was the largest since 1972; all sightings were between 8 June and 26 October. Up to 14 were present in Black Park in 1987–88.

In 1990 there was an exceptionally large influx into the country from late June onwards. By the end of the year at least 100 had been recorded in the London Area, and while most birds moved on quickly small flocks of up to 16 remained at Broxbourne Woods and Black Park. The following year breeding probably took place at four sites: Black Park, Ongar Park Wood, Potters Crouch Plantation and Warley. Despite there being hardly any crossbills in 1993, six pairs bred the following year; there was an influx in March, an atypical time, and many stayed on during the summer, again fairly unusual as they typically breed early in the year. The largest flock was of at least 60 birds in Black Park on 7 May where at least two pairs bred. Probably the largest invasion in the London Area occurred in 1997; by the end of June some 554 had been recorded, including 225 in Black Park, 100 at Thorndon Park and 80 in Broxbourne Woods. In July between 841 and 1,100 were recorded, with a peak of 75 in Titsey Woods; most had passed through by the middle of the month and relatively few were seen during the remainder of the year apart from a few isolated flocks. Oliver (1998) calculated a minimum total of 1,600 birds in the 1997 irruption between 30 June and 31 December. Two pairs bred in 1998 following this influx, at Havering CP and Walton Heath, and a pair may also have bred at Ongar Park Wood.

21st century

Small irruptions, beginning in the summer and going through to autumn, have become almost an annual feature since the start of the 21st century. Most birds are seen flying over and very few stop to feed. There have been a few larger influxes, which have led to a flock overwintering in Black Park. In early 2000 a few were present at Limpsfield Chart until spring and at least one pair bred. In 2011, after a small influx in the summer months, there was a second irruption in October.

Facts and Figures

The largest ever flock was of 225 birds in Black Park on 30 June 1997.

Parrot Crossbill *Loxia pytyopsittacus*

A very rare vagrant (4)

Range: northern Europe and Asia; breeds Scotland, also rare irruptive visitor to other parts of UK.

The only records are all from the 19th century. The first of these was of a bird killed in the winter of 1831 at Tooting; a description was published in *The Field Naturalist* but a cat ate the specimen. The other records were of an immature male and female shot in Southgate in 1864, and a female shot at Plumstead in 1868.

1831: Tooting, killed
1864: Southgate, two shot in November
1868: Bostal Common, Plumstead, shot in January

Unacceptable records

Three records have no supporting details: one of a bird listed by Blyth that was taken in Epping Forest in the autumn of 1835; a male shot on 21 January 1850 at Harrow-on-the-Hill that was recorded by Yarrell in his *History of British Birds* and accepted by Harting; and one referred to by Ticehurst that was shot at Eltham prior to 1859. None of these were deemed acceptable by Catley and Hursthouse in a paper in *British Birds* due to lack of details; they also did not include the Tooting record although full details were published.

Common Rosefinch *Carpodacus erythrinus*

A rare vagrant (11)

Range: breeds Europe and Asia, winters Middle East and southern Asia; scarce visitor to UK with occasional influxes which have led to successful breeding. It was removed as a BBRC description species after 1982.

The only historical record was an of immature female taken at Kenwood, Hampstead, in 1870 and this was just the second British record. The specimen was illustrated by Gould in his *Birds of Great Britain* and was purchased by Bond for his collection. No more were recorded until just over a century later when a female was trapped and ringed at Kempton Park on 6 February; it remained there until the end of the month. Another 15 years elapsed before the next record, of a female or immature which spent two weeks in a garden feeding on a bird table at Abbots Langley in the middle of winter, disappearing only when very cold weather arrived.

The 1990s was a good time for this nationally scarce finch and six were seen in the London Area. In 1992 there was an unprecedented influx into the country and a singing adult male was present at Lullingstone Park on 21 June. Another singing male spent a morning at Beddington in 1995; either side of this record were females or immatures at Hainault Forest and Perry Oaks. Two more males were seen in the subsequent two years, then there was a gap of 13 years until the next one when a singing male spent three days at Tottenham Marsh. Following on fairly quickly, an immature bird spent two days at Wormwood Scrubs in autumn 2013. Late summer and the autumn is the peak period in the London Area for this species with 36 per cent of all records in June and July and 45 per cent in September and October; this mirrors the national trends.

1870: Hampstead Heath, taken at Kenwood on 5 October
1971: Kempton Park, on 6–28 February
1986: Abbots Langley, on 29 December–11 January 1987
1992: Lullingstone Park, on 21 June
1994: Hainault Forest, on 2 September
1995: Beddington SF, on 13 July
1995: Perry Oaks SF, on 8 October
1996: Walton Heath, on 26 July
1997: Gillespie Park, on 26 September

2010: Tottenham Marsh, on 13–15 June
2013: Wormwood Scrubs, on 8–9 September

Pine Grosbeak *Pinicola enucleator*

A very rare vagrant (1)

Range: northern Europe, Asia and North America; vagrant to UK.

The only record is of a female shot at Harrow-on-the-Hill some time prior to 1843. This was the second British record, following one shot in Durham before 1831. The specimen was owned by Yarrell before being passed to Bond.

Pre-1843: Harrow-on-the-Hill, shot

Bullfinch *Pyrrhula pyrrhula*

An uncommon resident and localised breeder

Range: Europe and Asia.

Historical

In the early part of the 19th century this was a very common species but by the 1860s Harting had noticed a decline 'since it became the custom to lay the hedges', adding that 'it used to breed here regularly at one time, but now it is more frequently observed in winter and early spring, leaving us for some favourable locality when the breeding season arrives'. However, a pair of Bullfinches did breed in a summer-house at Hampstead despite it being in constant use. The species was also known from Battersea Park and it bred at Beaulieu Heights and Streatham in the 1880s. It was common in the Epsom area and Bucknill once found 16 nests in the area in one afternoon.

20th century

There was some evidence of an increase in the mid-1920s; breeding was first recorded in Richmond Park after 1925, while in Harrow although just one pair bred between 1925 and 1929 eight pairs bred in 1930. Glegg described it as 'a not uncommon resident in the rural parts of the county' and as an occasional visitor to central London. He noted that there was an especially good population along the Colne Valley between Uxbridge and Harefield.

In the 1930s and 40s the only sites where it bred in the inner suburbs were at Hampstead and Wimbledon. In the 1940s a pair was seen perched on a pram in Victoria Gardens adjacent to the Houses of Parliament. The only large group during this period was of about 30 birds, mostly males, at Strawberry Hill, Epping Forest on 23 January 1943.

In the 1950s it was thought that there had been little change in its status since 1900, apart from a small loss in the inner suburbs where large gardens had been built upon. It bred as close to the centre of London as Shooters Hill, Dulwich, Ranelagh and Hampstead Heath. It was fairly common throughout the rural areas and in suitable parts of the outer suburbs such as in woods, parks, heaths, commons and large gardens. One of its favoured habitats was the overgrown hedgerows along green lanes to the north of Epping.

It remained a scarce visitor in central London throughout the 20th century and had apparently never nested there until 1959 when a pair bred in Regent's Park. From the 1960s onwards breeding occurred in most years in the Royal Parks and the population peaked at nine pairs in 1969.

By the beginning of the 1990s there were about five pairs in central London; elsewhere the largest breeding concentration was nine pairs at Alexandra Park. During the second breeding Atlas it was found breeding in 63 per cent of the area, a slight increase on the first survey in 1968–72. It was fairly well distributed across the London Area apart from in the inner suburbs where it was mostly absent. By the end of the decade there had been a noticeable decrease in the population. On Wimbledon Common it declined from 11 pairs in 1986 to just one in 1997, and it decreased in Hainault Forest from 15 pairs in 1992 to six in 1998. By 1999 the national population had decreased by more than 50 per cent in 25 years and Bullfinch was added to the UK 'Red List'.

21st century

At the start of the 21st century there was still a reasonable breeding population of more than 100 pairs but this has subsequently decreased. Bullfinches are fairly scarce breeders in the suburbs and had completely disappeared from central London by 2000. They are slightly more widespread during winter.

Movements

Although they are largely sedentary in the London Area, some family parties wander in winter and irruptions from the Continent occasionally occur. An irruption took place in early 1962 and several sites reported much larger numbers than usual, including Rye Meads where a peak of 55 was seen on 14 January and Brent Reservoir where they peaked at 40 on 3 January; small numbers were also seen in central London. Two large flocks were seen in 1965: of about 70 birds at Ridge, near Potters Bar, on 13 February and of about 50 by Queen Mary Reservoir on 4 December. In 1982 a flock of more than 30 flew over Regent's Park on 1 October and in 1989–90 up to 50 were seen at Hainault Forest.

Facts and Figures

The largest-ever flock was of about 70 birds at Ridge on 13 February 1965.

'Northern Bullfinch' *P. p. pyrrhula*

A pair of the northern race from Siberia was released in Kensington Gardens and later bred there. No wild birds were recorded in the 20th century. In 2004 there was a national influx and presumed birds of this race were seen at 12 sites between 20 November and the end of the year, including up to 23 in Epping Forest, up to 12 at Thorndon CP, 10 at Warley CP and five on Hampstead Heath; however none of these were trapped to verify their identification.

Hawfinch *Coccothraustes coccothraustes*

An uncommon winter visitor and rare breeder

Range: Europe, Asia and North Africa.

Historical

The earliest reference dates back to the 17th century when it was suggested that Hawfinches in Epping Forest were 'as common as are pigeons in Guildhall Yard'.

In the early part of the 19th century it was an occasional visitor, in some years numerous, in other years scarce. In 1807 one was captured in Southall; following this it was said to be an annual visitor to Hackney Marshes up to 1813. Yarrell recorded that one was shot in Notting Hill Gate in January 1825. In 1832 it was first proven to have bred in Britain when a nest was found in Epping Forest and this area was to become a stronghold for over 150 years. Within a few years of this discovery it was found that large numbers congregated in winter and groups of 100 or more were regularly seen feeding on the hornbeam seeds in Epping Forest. The warm spring of 1842 gave rise to a large seed crop the following winter and hundreds were seen, with many staying to breed. Flocks of 200–300 were seen prior to 1881. In south London a pair bred on Tooting Common in 1833, at Roehampton in around 1835 and in Richmond Park in 1836.

By the middle of the 19th century it had started to increase, becoming a regular migrant in spring and autumn in Tottenham, and Harting noticed that it was most numerous during February in Kingsbury, 'when there appears to be a partial migration'. It was considered to be common in Chiswick in 1854, and in the winter of 1859 it was abundant in Hampstead; in 1864 it was particularly numerous in parts of north-west London.

Hawfinches started to breed in Chiswick and Muswell Hill before 1860, and were also recorded breeding in Ealing, Enfield, Finchley and Harrow. They bred annually in the woods around Hampstead, Harting revealing that he obtained a nest containing five eggs in Bishop's Wood in 1863 and that in the previous year three nests were taken in Kenwood. Hampstead appears to have been a breeding stronghold; they persisted there when they had become scarce in other areas as they had towards the end of the 19th century. Read described it as scarce in the Lower Brent Valley in 1896, whereas it had been common 30 or 40 years previously. In south London it was quite widespread, breeding at Ashtead, Claygate, Dulwich, Epsom, Ewell, Godstone, Leatherhead, Mortlake, Reigate, Streatham Common and Sutton.

20th century

During the first half of the century there appeared to be a general increase in the overall population, although there were some local declines. In the early 1900s they were abundant in the fruit-growing districts around Orpington and in the Darenth Valley, and in 1906 more than 50 were shot by a fruit grower at Swanley. There were large influxes in some years, notably in 1909 when hundreds were seen on migration on 19 April. The largest population was in Epping Forest where one observer knew of more than 100 nests up to 1917.

During the 1920s and 30s they bred regularly in the outer districts. Flocks of up to 20 were not uncommon but larger numbers were occasionally noted, such as several hundred at Leatherhead on 20 February 1927. In the middle of the century Hawfinch was a scarce, thinly distributed resident in wooded areas around the London Area. It continued to breed, however, in the inner suburbs at Dulwich Woods, Hampstead, Highgate, Putney Heath, Tooting Bec Common and Wimbledon Common.

It was rarely recorded in central London between 1900 and 1954 with just the occasional migrant being seen, along with a pair in Kensington Gardens which frequented the sheepfold between 23 April and 2 May 1947. Breeding was suspected in St John's Wood in 1950 when two juveniles were seen on 24 August. In 1951 a family party was seen in Regent's Park in July and a single bird was seen in St James's Park on 22 August. In June of the following year, young birds were watched being fed in Regent's Park and five birds, possibly a family party, were seen in St John's Wood on 30 May.

The largest numbers for many years were recorded in 1963 at Epping Forest where up to 62 were present between 29 August and 9 October, but there was no repeat of these numbers in subsequent years. The largest breeding populations were eight pairs Mimms Wood and at least seven pairs in Northaw Great Wood. During the first breeding Atlas in 1968–72 they were found breeding in 25 tetrads north of the Thames and 10 to the south. In 1971 they bred in the Chess Valley and were considered to be abundant. The population continued to fluctuate throughout the 1970s and 80s.

During the 1980s they began wintering in large numbers at several sites. In 1983 up to 12 wintered at Amwell and 1985 saw the start of a regular flock in Dagnam Park; by the following year up to 40 were wintering there. Also in 1986 up to 14 wintered at Bookham Common.

During the second breeding Atlas in 1988–94 an increase in the breeding distribution of just over 30 per cent was noted since the first survey, but they had almost become extinct south of the Thames with breeding recorded in just six tetrads compared to 40 to the north. By the start of the 1990s the breeding population was at least 28 pairs, the highest number to be recorded for several years. The largest concentrations were 13 pairs at Coopersale, near Epping, and six in Hainault Forest.

However, they went into decline after this time and in 1997 no breeding pairs were reported, probably for the first time in well over a century. The wintering flock in Dagnam Park had also decreased due to disturbance and habitat changes; by 1993 they had ceased breeding there and the maximum winter count was just 15.

21st century

At the start of the 21st century the only known breeding pair was in the Bayfordbury area. There was also a small wintering population of up to 11 at Broxbourne Woods with one or two elsewhere. Although the occasional bird has been seen in the breeding season since 2000, the species is now considered extinct as a breeding bird in the London Area; there has also been a large decline in the national breeding population.

It is still recorded every year, however. Since 2003/04 small numbers have overwintered on Bookham Common in most years. In autumn 2005 there was a good passage, including four over central London in October; some of these remained during the winter and there was a maximum count of 24 at Bookham Common.

Snow Bunting *Plectrophenax nivalis*

A scarce winter visitor

Range: breeds northern Europe, Asia and North America, European population winters Europe.

Historical

Early in the 19th century this was a rare winter visitor to the London Area; it was first recorded in Harrow prior to 1838. From 1840 onwards it started to be seen more often and large flocks were occasionally recorded, e.g. 50–60 birds on Hampstead Heath in November 1871.

20th century

During the first quarter of the century they were quite rare and there were only two records: some on Hampstead Heath in the winter of 1901/02 and one at Brockley in 1905. In the mid- to late 1920s it slowly started to increase and was seen at Epping Forest, Mill Hill and Staines Reservoir. Between the 1930s and mid-50s it had become a scarce passage migrant, being recorded most years in November; almost all records were of single birds apart from two small flocks. All the records in this period came from north of the Thames apart from those of two single birds seen on Dulwich Common in 1951 and 1952. There was one in Inner London, in Kensington Gardens, in 1935.

During the late 1950s there were about five or six records annually, the largest group recorded being of at least six birds at North Mimms on 13 December 1959. There were no documented records on the Lower Thames Marshes until 1960 when nine birds were seen at Swanscombe and they may have gone unrecorded there previously. In the autumn of 1962 there was a relatively large passage with 17 seen; 40 were at Swanscombe Marsh 2 December and 16 remained throughout the severe winter until March. In subsequent years a few migrants were seen annually apart from in 1969 when up to two were present at Rainham from 9 January to 3 February. There was little change during the 1970s with most birds being seen on the large reservoirs; occasional birds remained for several weeks.

In the early 1980s up to four overwintered at Rainham Marshes in 1980/81, 1981/82 and 1982/83, and at Stone Marsh in 1984/85. In 1985/86 two were present at Queen Mother Reservoir, sometimes also being seen at Wraysbury Reservoir; these were the first overwintering birds away from the Lower Thames Marshes. Additionally, two more overwintered at Rainham, this being the last year that they wintered. During the 1990s a few migrants were seen in most years, the peak being 10 in 1996.

21st century

Since the beginning of the 21st century it has remained a scarce annual visitor with most sightings being in November. It is typically seen on the Lower Thames Marshes or on the large concrete-banked reservoirs.

Facts and Figures

The earliest migrant was on 12 September 1929 at Mill Hill.
The largest flock was of 50–60 birds on Hampstead Heath in November 1871.

Lapland Bunting *Calcarius lapponicus*

A scarce winter visitor

Range: breeds northern Europe, Asia and North America; European population winters western and eastern Europe.

Historical

This was a rare visitor in the 19th century with just seven records. The first was of a bird caught in Copenhagen Fields (now Kings Cross) in September 1828 and was the third-ever British record. This was followed by a female taken in Battersea Fields in the winter of 1830; one caught at Highgate in the autumn of 1866; a male captured at Lewisham in 1867; one captured alive 'a few miles south of London' in October 1869; one shot at Brent Reservoir in 1892; and two on Wimbledon Common during severe weather in February 1895. An adult male was caught at Shoreham on 22 June 1889, but given the unusual date it may have been an escape.

20th century

There were no further records until the 1950s when four were seen. One was flushed at Perry Oaks SF on 10 October 1953; the record was placed in square brackets at the time because no plumage details were noted, although the two observers heard it call and were familiar with this species. This was followed by one on Wimbledon Common on 3 November 1957; and two in April 1958, at Beddington SF and at Staines Reservoir.

No more were seen until October 1973 when singles were seen on Latton Common in Harlow and at Beddington SF. It started to become more frequent and there was one at Queen Mary Reservoir in September 1976 and three there in 1979. From the beginning of the 1980s it increased dramatically and about 35 were seen during that decade, mostly between October and December. They were mainly seen in east London, particularly along the Lower Thames Marshes, including three at Rainham on 7 February 1987.

The increase continued at the start of the 1990s when 11 were seen in 1990, coinciding with an influx along the East Coast of the UK; all were between 6 October, when five flew over Beddington, and 23 December. During the 1990s they were annual visitors and one or two overwintered at Rainham Marshes in some years.

21st century

Since the start of the 21st century it has become scarcer again and only nine more were seen up until 2009, all of which were in October and November apart from one in January. Most were recorded flying over and none remained for more than a few days. In the autumn of 2010 there was an unprecedented influx into the UK; 34 were seen in the London Area between 15 September and 21 November, most of which were flying over.

Facts and Figures

The largest flock was of nine birds over Stoke Newington Reservoirs on 16 September 2010.

Pine Bunting *Emberiza leucocephalos*

A very rare vagrant (1)

Range: Asia; vagrant to UK.

The only record occurred in 1992 when a male spent almost five weeks at Dagenham Chase. It was a very popular bird and was seen by several thousand people during its stay.

1992: Dagenham Chase, on 12 February–17 March

Yellowhammer *Emberiza citrinella*

An uncommon partial migrant and localised breeder

Range: Europe and Asia.

Historical

Harting described this as 'our commonest species of Bunting, and is resident throughout the year'. He noted that although there had been a considerable decrease in the numbers of many birds due to 'an interference with their nesting-places' and 'increased cultivation of waste lands', this did not apply to the Yellowhammer, adding that 'while other and less sociable birds seek their living in the open fields, he confidingly visits our rick yards and poultry yards to pick up the scattered grain'.

In 1898 Hudson stated that they could still be found breeding on Hampstead Heath and Wandsworth Common although he clearly had something against the inhabitants of the latter area as he remarked that they did not deserve to have such an attractive bird breeding in their midst.

20th century

During the early part of the 20th century they began to disappear from some of the innermost breeding areas; they ceased breeding on Wandsworth Common in the early 1900s but continued to breed on Hampstead Heath until about 1930. By the mid-1930s they were still a common and widespread breeding resident in the rural areas but had been pushed out of the inner suburbs. During the winter they were often seen in small flocks and in December 1947 a flock of more than 40 was seen on Barnes Common.

In the 1950s it was still a fairly common resident in the outer suburbs and rural areas. There had been some local increases, such as at Beckenham and Mill Hill, which were attributed to an increase in agriculture during and after the Second World War. In the suburbs, its range encircled the more heavily built-up areas and it could be found at Shooter's Hill, Eltham, Beckenham, Wimbledon Common, Richmond Park, Horsenden Hill, Harrow, Hendon, Mill Hill, Totteridge, Enfield, Chingford and Upminster. Inside this area it was a scarce visitor in winter or on passage. On Hampstead Heath there were five territorial males in 1954 and breeding may have occurred; Yellowhammers bred there successfully in 1959–60.

In central London it was very rare with only seven records between 1900 and 1954, mostly in winter or on passage, but between 1954 and 1970 it was recorded more frequently.

The population began to decrease from the 1970s onwards; in the Tadworth, Walton and Headley areas of south London there was a 50 per cent decline in the population between 1974 and 1978; and on Wimbledon Common they declined from 10 pairs in 1976 to just one pair by 1984 when breeding was last recorded. The closest breeding birds to central London were at Barn Hill, near Wembley, where two pairs bred in the mid-1970s. There was a roost of up to 200 at Rye Meads in the 1970s and early 80s.

During the second breeding Atlas in 1988–94 Yellowhammers were found breeding in almost half of the London Area tetrads, mostly beyond the suburbs. There was only a small decrease from the first survey in 1968–72 with fewer breeding birds found in the south-west of the area.

The species continued to decrease during the 1990s with some populations going into a terminal decline; at Hainault there had been a 70 per cent reduction over the previous 20 years. At least 148 pairs/singing males were located across the London Area in 1998.

21st century

The breeding population has continued to decline since the start of the 21st century. Despite good coverage in 2002, less than 100 pairs were located, a fall of one-third in four years; by 2010 the population had decreased only slightly with 87 territories reported. Most of the population is located in the northern, eastern and southern rural areas. Yellowhammers are more numerous and widespread in winter but are rarely seen in the suburbs.

Facts and Figures

The largest count was 250 birds at Rainham on 25 January 1982.

Cirl Bunting *Emberiza cirlus*

A scarce visitor; former breeder

Range: Europe and North Africa; in UK now breeds only in south-west.

Historical

Cirl Bunting was first identified and described as a British bird by Colonel Montagu in 1830 from specimens observed in the neighbourhood of Knightsbridge. In 1836 Blyth stated that they were 'occasionally taken in the nets during the winter months' near Tooting.

Harting knew it as a species from southern counties and considered it to be a rare visitor north of the Thames where specimens had been taken at Brent Reservoir, Hampstead Heath and Harrow. In the second half of the 19th century there were two breeding records in north-west London: near Wembley Park in May 1861 and in Hendon in 1871. South of the river breeding occurred at Reigate Hill in 1871 and 1877; Gatton in 1873; Croydon in 1878; Lewisham in 1884; Reigate Heath in 1887; Wimbledon Common in 1890; Epsom in 1892; and Coombe Wood, Wimbledon in 1898.

20th century

In the first half of the century it continued to breed irregularly, mainly on the North Downs and surrounding countryside. Breeding took place at Banstead in 1900; Shoreham in about 1902; Epsom in 1905; two pairs at Tadworth in 1922; Biggin Hill in 1935; Chipstead from 1942 to 1943; Betchworth in 1947; and Chaldon in 1949. The only other breeding record south of the Thames, away from this area, was in Weybridge in 1905. North of the river, the only confirmed breeding was in Harrow between 1919 and 1928, and at Croxley in 1935.

It was widely seen across south London during the first half of the century, including at Tadworth from 1923 to 1925 and Sutton from 1930 to 1935; breeding may have occurred at these sites. It was occasionally present during the breeding season in Richmond Park and on the adjacent Ham Common. The only other sightings north of the Thames were on Hampstead Heath in the winter of 1901–02; at Elstree in 1923; at New Edgware on 6 October 1930; at Colney Street in 1947; and at Staines Reservoir on 15 October 1949. Breeding was suspected at Rickmansworth in 1948–49 and also, in 1949, a male held territory at Old Parkbury between April and June with a female seen there on 23 April. There were a few records on the Thames-side marshes, including of three at Dartford on 4 December 1949.

After 1950 it became scarce and there were no more confirmed breeding records, but singing males were occasionally still detected, such as around Betchworth and Chipstead, and breeding may have gone unrecorded. Pairs were also seen at Moor Mill and Ruislip in 1954. In 1960 a pair was present at Chipstead in June and July but breeding was not thought to have taken place. There were potential signs of breeding in the mid-1960s when a male was seen carrying food at Brands Hatch on 7 June 1965; a male was on territory on Wimbledon Common in 1966–67; a male was on territory at Orpington in 1967; and two males were singing at Pebblecombe on 2 May 1969.

By the start of the 1970s it had become rare; the only records since then were of a singing male at Juniper Top on 30 July 1970; one at Moor Park Golf Course on 15 June 1977; a male on territory at North Ockendon between 7 May and 10 July 1978; a male on Walton Heath on 6 July 1980; and an immature at Bricket Wood in September 1981.

Ortolan Bunting *Emberiza hortulana*

A scarce passage migrant (35)

Range: breeds Europe, Asia and North Africa, winters Africa; scarce passage migrant in UK.

This was first described as a species from a specimen captured alive in Marylebone Fields prior to 1776 and pictured by Brown in his *Illustrations of Zoology* in which he referred to it as 'Green-headed Bunting'. There were four instances of it being captured in the 19th century: a male caught 'near London' in the winter of 1836/37 was placed in London Zoo; a bird-catcher caught three in Kilburn in 1838; a female was caught at Hampstead in 1867; and one was taken near Brent Reservoir in the following year.

There were only two further records in the first half of the 20th century but during the 1950s five were seen. After this burst of records they resumed their rarity status with just one more sighting in the next 27 years. Two were seen in the 1980s, both in spring, including a male that spent all day feeding in a disused filter bed at Wood Green Reservoirs; this is the only April record.

Since the start of the 1990s they have been seen much more frequently despite the breeding population decreasing in north-west Europe. In 1993 a male was present for three days in Richmond Park in spring and was even heard to sing on occasions; there was some suspicion that this may have been an escaped cage bird as it was said to have had abraded tail feathers.

Ortolan Buntings are mostly seen in autumn with just a handful in May and one late migrant on 11 June. Four were seen in the autumn of 2003, including two together at Sewardstone between 28 September and 1 October, the first sighting of more than one bird for almost 50 years. In 2007 a juvenile was trapped and ringed at Wraysbury GP.

Pre-1776: Marylebone, caught
1836-37: near London, caught in winter
1838: Kilburn, three caught in October
1867: Hampstead, caught
1868: Brent Reservoir, caught nearby in October
1908: Plaistow, on 6 May
1947: Nore Hill, on 23 August
1955: Epsom SF, on 1 October
1956: Brent Reservoir, on 2 September
1958: Epsom SF, two on 3–4 May with one remaining until the 10th
1958: Weald Country Park, on 17 May
1972: Beddington, on 22 August
1985: Barn Elms Reservoirs, on 23 May
1987: Wood Green Reservoirs, on 24 April
1993: Richmond Park, on 7–9 May
1994: Beddington, on 9 October
1995: Beddington, on 13 September
1996: King George V Reservoir, on 3 September
1996: Hounslow Heath, on 20 September
1999: London Wetland Centre, on 6 May
1999: Beddington, on 11 June
2000: Queen Elizabeth II Reservoir, on 3–4 May
2002: Pinner Park Farm, on 2 September
2003: Wormwood Scrubs, on 2 September
2003: Sewardstone, on 3 September
2003: Sewardstone, two on 28 September–1 October
2004: Beddington, on 27 August
2007: Wraysbury GP, on 5 September
2008: Rainham Marshes, on 30 August
2010: Rainham Marshes, on 31 August
2010: Brent Reservoir, on 12 October

Rustic Bunting *Emberiza rustica*

A very rare vagrant (2)

Range: breeds northern Europe and Asia, winters east Asia; vagrant to UK.

There are two London Area records, separated by over a century. The first was of a bird caught in 1882 by a bird-catcher near Aldenham Reservoir. The only modern record was of a bird present at Beddington for over a month in

1993. It was the first time one had been seen in Britain in winter and it arrived with an influx of Reed Buntings. Their nearest wintering grounds are in Turkestan, as are those of Little Bunting, one of which was present at the same time.

1882:   near Aldenham Reservoir, caught in November 1882
1993:   Beddington SF, on 9 February–13 March

Little Bunting *Emberiza pusilla*

A rare vagrant (11)

Range: northern Europe and Asia, winters Asia; scarce visitor to UK, occasionally overwinters.

The first records were in 1956 at Beddington and Staines Reservoir. Like these first two records, all of the subsequent ones were in spring until a bird overwintered at Beddington in 1993; this individual arrived three days after a Rustic Bunting at the same site. The following year one was present briefly at Coppetts Wood in October, the first time one had been seen in autumn. In 2007 one was regularly seen at the feeding station at Amwell between January and mid-April, the longest-staying bird in the London Area. The birds at Iver and Amwell were heard singing during their stay. 81 per cent of all birds were between February and May.

1956:   Beddington SF, on 31 March–3 April
1956:   Staines Reservoir, two on 7 April
1960:   Hilfield Park Reservoir, on 20 April
1965:   Perry Oaks SF, on 2–19 May
1978:   Poyle, on 8–11 April
1985:   Rye Meads, trapped and ringed on 3 May
1987:   Iver, on 17 March–26 April
1993:   Beddington SF, on 12 February–17 April
1994:   Coppetts Wood, on 2 October
2007:   Amwell GP, on 31 January–13 April

Unacceptable Records

Originally, the LNHS accepted a record of three birds at Beddington in March and April 1956, but two of these were deemed unproven by the BBRC when they reviewed all records from 1950 to 1957.

Reed Bunting *Emberiza schoeniclus*

An uncommon partial migrant and widespread breeder

Range: Europe and Asia.

Historical

The earliest record was in 1666 of a bird in reeds near Kingston. In the 19th century it was a common resident about which Harting wrote 'usually to be found by reedy ponds, and along the brook sides'.

20th century

For much of the 20th century Reed Bunting was a widespread breeding bird beyond central London, and the old-style sewage farms were a favoured habitat. Hampstead Heath was formerly a breeding stronghold but the construction of a rifle range near one of the ponds in 1921 destroyed its breeding habitat. After the breeding season many birds left their breeding grounds and joined together in large flocks often well away from breeding areas, such as on the North Downs.

By the 1950s there had been no change of status in the rural areas where it was a widespread but local breeding bird. The main change to the population was in the suburbs where many had disappeared as their habitats were built on or became surrounded by development. However, in areas such as east London they had actually increased and at Bookham Common the population doubled in the 1950s. They also started to spread to other habitats and could be found breeding on arable land north of Epping, as well on the drier parts of Ashtead Common.

Large numbers were also seen on passage and in severe weather; for example, in 1958 about 250 were at Rye Meads on 19 March, of which about 90 per cent were males. They were also seen in the grounds of the Natural History Museum in South Kensington and at Cripplegate, the first ones seen in Inner London. During the very cold winter of 1962/63 they were more often seen in gardens. Large roosts were also formed and at least 400 roosted in a reed bed in Thorndon Park in October 1967 and 300 roosted at Rye Meads in October 1971. A flock of 400 was present at Beddington SF in February and March 1974.

In the first breeding Atlas in 1968–72 they were found breeding in about 38 per cent of the area; the largest breeding population was about 100 pairs at Rye Meads. In 1972 a pair held territory in Surrey Docks and the following year a pair bred there, the first breeding record in Inner London; they continued to breed there in subsequent years and there were up to 10 pairs in 1982. At the start of the 1990s the breeding population was thriving, especially at Rainham Marshes where there were 148 pairs.

By the time of the second breeding Atlas in 1988–94 there had been a small decline in the breeding distribution of 6 per cent since the first survey; much of the decrease was on the North Downs and in farmland in the far north-east of the London Area.

21st century

The national population declined sharply from the mid 1970s to the 1980s, according to CBC and BBS monitoring, leading to 'Red' listing as a species of conservation concern. However, there has been a significant increase since the beginning of the 21st century, partly reflecting the species' ability to take advantage of vast areas of oil seed rape, and it has now been moved to the 'Amber' list. This recovery has yet to be seen in the London Area; since 2000 the breeding population appears to be slowly declining. In 2000 there were about 265 pairs reported at 52 localities, the largest population being 116 pairs at Rainham. The numbers reported have continued to decline since then and in 2010 only about 125 pairs were reported (although under-recording could be a contributing factor). Although mostly absent from the inner suburbs and central London, a pair probably bred in Regent's Park from 2004 to 2006. It is most widespread in winter and small numbers are regularly recorded flying over on migration.

Facts and Figures

The largest concentration was of 500 birds at Ockendon on 7 March 1981.

Black-headed Bunting *Emberiza melanocephala*

A very rare vagrant (1)

Range: breeds eastern Europe and Middle East, winters southern Asia; vagrant to UK.

The only occurrence was in August 1986 when an adult male was seen at Bromley-by-Bow. The urban location coupled with the unusual time of year suggests that this may have been an escape as large numbers were imported into Britain for the cage-bird trade, but there are other August records in the UK and this was accepted by the BBRC.

1986: Bromley-by-Bow, one on 24 August

Corn Bunting *Emberiza calandra*

An uncommon partial migrant and scarce breeder

Range: Europe, Asia and North Africa.

Historical

In the 19th century this was known as 'Common Bunting'; Harting described it as 'resident throughout the year; but not being so generally distributed in the county as the Yellow Bunting [Yellowhammer], it is less entitled to be called "common". He added that it is 'to a certain extent, migratory, for there is a perceptible increase to the numbers in summer, and as perceptible a decrease in winter'.

There were only two records in Inner London during the century, at Regent's Park prior to 1879 and at South Kensington on 6 January 1897. The only population change was around Harrow where it was common in 1838 but had become scarce by the end of the century.

20th century

In the first decade of the century Corn Buntings bred in west and north-west London, in the Brent and Colne Valleys. In the south and south-west of the area they were not uncommon at the start of the century, but they rapidly disappeared from many areas as farmland was built over. There were also scattered pairs along the Lower Thames Marshes. Once the breeding season finished they often moved away and gathered in small groups. They were regularly seen in stubble fields near the Thames in late summer and by September they had completely vacated the marshes and did not return until March.

There were further losses south of the Thames during the 1930s; they were rarely seen in the inner suburbs and there were no records in the first half of the 20th century from Inner London.

In the 1950s it was a localised breeding species, mainly to the north of the Thames where the population was stable or even increasing, although it had disappeared from the Brent Valley. However, there had been a serious decline in south London and it had vanished from most of the North Downs; the last colony was at Ewell. The main stronghold was on the Lower Thames Marshes on the east side of London. There, they were present from Abbey Wood Marshes eastwards and there were 20 pairs at Rainham, but by 1959 there had been a big decline in the numbers breeding on the marshes between Crayford and Swanscombe due to development. They had increased their range in east London and were common around Hainault and between East Horndon and Upminster where about 25 pairs bred.

The total population continued to decline during the 1960s except in east and south-east London; in the mid-1960s there were 23 pairs around North Ockenden and 20–30 between Darenth and Lullingstone. Flocks of 100 or more were occasionally seen on the Lower Thames Marshes in winter during the 1970s. Elsewhere, there were also a few large roosts, including 450 birds at Maple Cross SF; 150 at Ham Moor Marsh, Weybridge; and 100 at Rye Meads. In 1979 there were estimated to be 60 territories during the breeding season in the West Hyde area.

The second breeding Atlas in 1988–94 showed breeding in 13 per cent of all the tetrads in the London Area, virtually unchanged since the first survey in 1968–72. However, they had virtually disappeared from the North Downs and most of the population was situated in the Thames corridor and Darenth Valley, with a few birds in the far west and north-west of the area. The largest breeding populations were 20 pairs at Berwick Ponds, 15 at Bulphan Fen and 12 at Fairlop.

Nationally, the population decreased by more than half between 1971 and 1995, and there was a further decline of 35 per cent between 1994 and 2000. This decline was also evident in the London Area in the 1990s; for example, the winter roost at Springwell reed bed had declined from 250 in 1985 to just 19 by 1994, and there was a reduction in the nearby breeding population of 90 per cent.

21st century

At the beginning of the 21st century the breeding population was still in free-fall and was probably no more than 20 pairs. Corn Buntings had totally disappeared in the west and south-west areas of London. By 2009 there were about 10 territories, mostly in rural east London apart from two in south-east London, at Crockenhill. Outside the breeding season they are seen slightly more widely but rarely closer to central London than Rainham Marshes.

Facts and Figures

The largest count was 450 birds at a roost at Maple Cross SF on 27 January 1979.

APPENDIX 1 – ESCAPES

The following species, which have occurred in London, are all deemed to be escaped birds. Captive wildfowl in the collections in the Royal Parks are not generally listed here unless otherwise stated. Taxonomy follows *Clements Checklist of Birds of the World*.

White-faced Whistling Duck *Dendrocygna viduata*
There were several records between 1996 and 2002 with a maximum of five birds at Beddington in 2000.

White-backed Duck *Thalassornis leuconotus*
Recorded in 2002 and 2006.

Black Swan *Cygnus atratus*
These have been recorded breeding in London since the late 19th century in St James's Park. A few were present on the Thames from the mid-1920s until the 1950s; the population declined during the Second World War and virtually died out after the severe winter in 1947.

There were few further records until the late 1980s when they became more widespread. In 1989 a pair nested at Brent Reservoir, and up to five were seen in November. A pair bred at Upminster in 1996 and 1998. In 2000 up to eight were at Thorpe Water Park and a pair bred at Claremont Lake. In 2003 a pair bred in St James's Park. Since 2004 two or three pairs have bred annually.

Black-necked Swan *Cygnus melanocoryphus*
There were a few isolated records from 1963 to 2002.

Trumpeter Swan *Cygnus buccinator*
This species was recorded in 1993 and 1996 at Weald CP.

Coscoroba Swan *Coscoroba coscoroba*
One in Kensington Gardens on 12 November 2009.

Swan Goose *Anser cygnoides*
This was first recorded in 1967 when it bred at Godstone and Norwood. There have been isolated sightings since, with occasional small groups such as five at Godstone in 1996 and four at Walthamstow Reservoirs in 2000.

Lesser White-fronted Goose *Anser erythropus*
First recorded in the Stocker's Lake area from 6 to 12 April 1975, and odd ones were seen at various sites from 1980 onwards; some of these were present for several months. In 1986/87 one wintered in the Lea Valley and there were four at Walthamstow Reservoirs in February.

Bar-headed Goose *Anser indicus*
First recorded at Barn Elms Reservoirs in 1965, small numbers are recorded annually. A pair bred at Stockley Park in 1995. In 1998 there were six pairs and three broods at Kew Gardens. Single pairs nested at Fishers Green in 1999 and in the Teddington area in 2001.

Emperor Goose *Chen canagica*
First recorded in 1989 when there were up to four at Walthamstow Reservoirs; occasional sightings since.

Cackling Goose *Branta hutchinsii*
Various birds have been seen since 1986 when there were two at Beddington SF from August to September; then at Broadwater Lake between April and December 1987; at Amwell on 19 September 1987; on Wood Green Reservoirs on 16 August 1988; and at Walthamstow Reservoirs on 26 April 2001.

Spur-winged Goose — *Plectropterus gambensis*
In 2006 one was at the Wetland Centre.

Cape Barren Goose — *Cereopsis novaehollandiae*
Recorded in 1985.

Blue-winged Goose — *Cyanochen cyanoptera*
Recorded at three sites in 2002.

Upland Goose — *Chloephaga picta*
One of the race *leucoptera* recorded in 1986.

Ruddy-headed Goose — *Chloephaga rubidiceps*
Recorded in the Lea Valley in 2004.

Ruddy Shelduck — *Tadorna ferruginea*
There have been many sightings since the late 19th century, all of which are presumed to have been of escapes. However, a fuller account of this species is given as some may relate to immigrants.

It was first recorded in 1891 when one was caught in Wanstead Park. A pair bred in St James's Park in 1932 and raised two young. In about 1944 one flew through a window into the ruined House of Commons but despite a thorough search there was no sign of a nest. In 1950 two flew over Brent Reservoir on 9 May.

In the 1960s singles were seen on Staines Reservoir on 10 October 1964; at Swanscombe on 28 April, then later on the North Kent Marshes until 22 May 1967; and on the lower Thames at Woolwich and Swanscombe between February and 5 April 1969. While the last was likely to have been an escape it did coincide with unprecedented numbers of Common Shelduck on the Thames marshes. In the mid-1970s singles were seen at four sites, the only contenders for wild birds being on the Thames at Barking on 21 January 1975 and at Swanscombe on 16 May 1976.

From the late 1970s onwards long-staying birds were seen regularly, including pairs at Stoke Newington Reservoirs and Walthamstow Reservoirs. By 1982 there were up to four resident at Walthamstow and eight in Clissold Park. Breeding occurred in the Stoke Newington area in 1983 and 1988–89. There were also full-winged birds in the collection in St James's Park. They remained in the Stoke Newington area until at least 1992.

In 1994 there appeared to be a national influx in the autumn, possibly from the Continent, and some of the birds seen in London may have been connected to this, for example three on Queen Mary Reservoir from 20 July into August and one at Rainham and Barking Bay between October and December. Up to four were present in the Lea Valley during 1996. In 1999 four flew south-west over Queen Mother Reservoir on 8 August. Since 2000 only occasional birds have been seen apart from in 2007 when three were seen in Regent's Park on 19 August.

South African Shelduck — *Tadorna cana*
Several were released from Kew Gardens in 1939 and a pair bred on Barnes Common in 1946. Odd ones have been seen since 1981, and in 1995 up to nine were on Queen Elizabeth II Reservoir for most of the year, with a flock of six at three other sites. Up to seven were seen in 1996 although some of these may have been hybrids.

Australian Shelduck — *Tadorna tadornoides*
Occasional records from 1984 into the 1990s.

Paradise Shelduck — *Tadorna variegata*
First recorded in 1965 at Barn Elms Reservoirs, there were also occasional sightings in the 1990s. A number of hybrids involving this and the previous two species have been noted since the mid-1990s.

Muscovy Duck *Cairina moschata*
Single birds or small groups have been seen almost annually since 1972 when a pair bred at Godstone. An unsuccessful attempt was made to introduce them at Shepperton GP in 1985. There were up to 10 on Wimbledon Common in 2000 and on Verulamium Lake, St Albans, in 2002. The peak count was 16 at Ridge on 26 January 2008.

Ringed Teal *Callonetta leucophrys*
First recorded in 1985, then almost annually since 1990.

Wood Duck *Aix sponsa*
The species was first seen in London at Moor Mill in December 1891. A pair nested at Sunbury Cross in 1987. Nine at Bushy Park on 13 February 1994 were believed to have been introduced there. It has been seen fairly regularly since the late 20th century.

Falcated Duck *Anas falcate*
In 2000 a female was on Connaught Water on 26 August; various wildfowl species have been seen at this lake and are thought to have escaped from a collection at nearby Copped Hall. There was one at Swanscombe Marsh on 21 February 2007 and perhaps the same throughout 2008 at Northfleet.

Chiloe Wigeon *Anas sibilatrix*
One or two seen in most years since first recorded in 1962.

Philippines Duck *Anas luzonica*
Up to two were seen in 2000–01.

Cinnamon Teal *Anas cyanoptera*
A pair was seen at Hilfield Park Reservoir on 3 September 1955; also recorded in eight years between 1983 and 2004.

Red Shoveler *Anas platalea*
Up to three were seen in various years since 1990.

Australian Shoveler *Anas rhynchotis*
Recorded in most years since 1987; some of these were the New Zealand race *variegata*.

White-cheeked Pintail *Anas bahamensis*
Frequently seen since first noted in 1973.

Red-billed Duck *Anas erythrorhyncha*
Recorded between 1996 and 2004.

Yellow-billed Pintail *Anas georgica*
A handful seen since first recorded in 1990.

Silver Teal *Anas versicolor*
First recorded in 1988; up to three on Connaught Water.

Puna Teal *Anas puna*
Seen at several sites between 2001 and 2005.

Hottentot Teal *Anas hottentota*
Recorded in 1993 and 2001.

Baikal Teal — *Anas formosa*
A ringed drake at Tyttenhanger GP on 9 December 2001; also mentioned in *Checklist of the Birds of the London Area, 1901–77*.

Yellow-billed Teal — *Anas flavirostris*
Recorded almost annually since 1989.

Cape Teal — *Anas capensis*
First seen in 1997 with a few records since.

Bernier's Teal — *Anas bernieri*
One in 2008 at Beddington SF and Queen Elizabeth II/Walton Reservoirs on various dates.

Grey Teal — *Anas gracilis*
Recorded only in 1991.

Chestnut Teal — *Anas castanea*
Only seen twice since the first record in 1983.

Pink-eared Duck — *Malacorhynchus membranaceus*
Recorded at various sites between 1994 and 1999.

Marbled Teal — *Marmornetta angustirostris*
Only a few have been seen since the first at Staines Reservoir in September 1984. The others were at Battersea Park from 26 September 1989 into 1990; at Barn Elms Reservoirs from 3 to 30 August 1991 with probably the same bird at Beddington on 11 August; at Hilfield Park Reservoir on 21–22 April 1988; at Perry Oaks on 15 and 28 May and 7 June 1998 (accepted on to the BOU Category D list); and at Bushy Park on 29 November 2007.

Rosy-billed Pochard — *Netta peposaca*
A pair of full-winged Rosy-bills, which were thought to have originated in St James's Park, bred in Regent's Park in 1946. It has been recorded in most years since 1987.

Baer's Pochard — *Aythya baeri*
Recorded in 2002–03.

New Zealand Scaup — *Aythya novaeseelandiae*
Recorded annually since 1997.

Barrow's Goldeneye — *Bucephala islandica*
An unringed bird was on Queen Mary Reservoir from 10 September into October 1972.

Hooded Merganser — *Lophodytes cucullatus*
A pair resided on the lake at Copped Hall from 2001 to 2008.

White-headed Duck — *Oxyura leucocephala*
In 1990 there were three at Lonsdale Road Reservoir on 9 April with one there on 1 July. One was at Walthamstow Reservoirs on 11 September 1999 and again from 29 December to 20 February 2000. A first-winter was present in 2002 at Broadwater Lake from 4 to 18 December; it was seen again in 2003 at Staines Reservoir on 7 July, then wintered at Broadwater Lake and Hilfield Park Reservoir until 29 January 2006.

Lake Duck — *Oxyura vittata*
Recorded in various years since 1993.

Rock Partridge *Alectoris graeca*
One with Red-legged Partridges at South Ockenden in May 1975.

Chukar *Alectoris chukar*
The species was first recorded in London in 1985. Singles were then seen at various sites and there were up to nine at Gorhambury in 1992, when further releases were made illegal, although occasional birds are still seen.

Kalij Pheasant *Lophura leucomelanos*
Recorded in 1991.

Silver Pheasant *Lophura nycthemera*
Five were released into Richmond Park in 1929; also recorded in 1993.

Reeve's Pheasant *Syrmaticus reevesii*
Two males were at Banstead on 21 February 1996.

Golden Pheasant *Chrysolophus pictus*
There have been various escapes seen since 1978, none of which are thought to be from a self-sustaining population; additionally up to four have been seen in Kew Gardens. The records are as follows: one at Amwell in May 1978; the feathers of one were found in a garden in Fetcham in April 1980; another at Amwell on 1 November 1981; a very tame female at Stocker's Lake between 11 August and 20 September 1982; one in Hatfield Park throughout February 1991; in 1993 one in Denham Woods in January and at two sites in the Colne Valley in April; one in the Wildlife Garden, Kensal Town, in 1994; and a male at Copped Hall on 3 May 2003.

Lady Amherst's Pheasant *Chrysolophus amherstiae*
Some were released into Richmond Park in 1928–29 and a further 24 were released there in 1931–32. Although they were still present in 1937 they were not known to have bred.

During survey work for the Hertfordshire Breeding Atlas (1982) one was seen in the Whippendell Wood area and four males were seen at Chorleywood; however, most of Chorleywood is outside London. Although the species was possibly breeding about 20 kilometres away from Whippendell Wood there were no further sightings from this area and it is not clear if this originated from the naturalised population or was an escape. A male was seen in Kew Gardens in 1993.

Common Peafowl *Pavo cristatus*
A pair bred successfully near Barn Elms in 1999; the birds were removed from the area in April 2000. There were up to eight in Hatfield Park from 2005 to 2009, five at Bayford in 2008 and occasional singles seen elsewhere.

Wild Turkey *Meleagris gallopavo*
During the 18th century thousands were introduced into Richmond Park by order of King George I. They were eventually killed during the reign of King George III.

Helmeted Guineafowl *Numida meleagris*
First recorded in 1993; large flocks have been seen in recent years, including 19 birds at Bayford in 2008, 13 at Redwell Wood Farm, South Mimms, in 2008 and 14 near Brookmans Park in 2009.

California Quail *Callipepla californica*
In 1889 or 1890 some were released in Aldenham Park but were not reported again. One shot near Watford in late December 1897 was thought to be from a release at Chalfont, Buckinghamshire. One at Kempton NR on 4 February 2007.

Northern Bobwhite *Colinus virginianus*
They were introduced on various estates in the country during the 19th century; the only one in London was shot near Chelsham Court in October 1845. More recent records were in 1972, 1996 and 1998.

Greater Flamingo *Phoenicopterus roseus*
There was one at King George V Reservoir on 14–15 July 1934 and two at Leatherhead on 8 October 2008.

Chilean Flamingo *Phoenicopterus chilensis*
First recorded in 1970 and others seen in 1989–90.

An escaped flamingo species from Windsor was present in Richmond Park in the winter of 1932/33 and another flew over East Molesey in 1965.

Great White Pelican *Pelecanus onocrotalus*
Pelicans have been in the collection at St James's Park at least as far back as 1665. In 1940 one of these pelicans had regrown its flight feathers and was sometimes seen flying around the park. In 1996 a full-winged adult was seen at Southend, Essex, in February, then picked up in March and released in St James's Park where it has remained since; it often flies around on warm, sunny days and has been seen as far as Hampstead Heath and the Wetland Centre.

There have also been a few sightings of birds that did not come from St James's Park. Three adults flew in and landed on Staines Reservoir on 18 April 1971, two of which flew off after 20 minutes; and in 1972 two adults were seen on Wraysbury Reservoir on 10 November and at Kempton Park Reservoirs the following day. Both records were accepted to Category D of the British List. Singles were also seen at Barn Elms Reservoirs from 22 to 24 May 1973; and at Fairlop Lake on 13 November 2000.

Puna Ibis *Plegadis ridgwayi*
One at Coopers Green Lane on 20 August 1995; this was the bird that had been at large in the Home Counties for 10 years.

Sacred Ibis *Threskiornis aethiopicus*
First recorded in 1983; others seen since in 1990, 1994–97 and 2005–07.

White-rumped Vulture *Gyps bengalensis*
One was seen at three sites in south London on 10–11 November 2006.

Harris's Hawk *Parabuteo unicinctus*
Various birds have been seen in the 21st century, including one resident at Radlett for about three years.

Red-tailed Hawk *Buteo jamaicensis*
Various escaped falconers' birds have been seen since 1991.

Eagle sp. *Aquila* sp.
One, believed to be a Tawny Eagle *A. rapax*, flew north over Brent Reservoir on 26 August 1963.

Lanner Falcon *Falco biarmicus*
Escaped falconers' birds were seen at Staines Moor and West Thurrock in 1991; also recorded in 2006 and 2009.

Saker Falcon *Falco cherrug*
One at Holyfield Marsh on 9 September 2001; another with jesses in 2002; and one at Queen Mother Reservoir on three dates between November 2007 and February 2008

Crowned Crane sp. *Balearica* sp.
Singles were at Brent Reservoir on 28 September 1974 and at Cobham in 1979. In 1985 two escaped from Chessington Zoo and were seen at various locations in west London. In 2006 another was seen at Cobham before being recaptured.

Demoiselle Crane — *Anthropoides virgo*
One in 1960 at Carshalton on 31 March and at Beddington on 10 April.

Peruvian Thick-knee — *Burhinus superciliaris*
One at Barn Elms Reservoirs on 28 April and 1 May 1994 was later seen in Norfolk; it returned to London in 1995 to Wood Green Reservoir from 10 to 12 January. In 1997 a juvenile at Hampton Court Park on 27 November was considered to be a hybrid.

Spur-winged Lapwing — *Vanellus spinosus*
Two escaped from the collection at Crystal Palace in 1988 and were then seen at various locations between June and September, and again at Barking Creek/Thamesmead from 12 to 22 July 1989.

Southern Lapwing — *Vanellus chilensis*
One remained in the Tyttenhanger area between September and March 2002/03.

Silver Gull — *Larus novaehollandiae*
An escape from London Zoo was later seen at Ruislip Lido and Barn Elms Reservoirs in May 1988. In 2001–02 there was one in Regent's Park and possibly the same one in St James's Park in 2003 and 2006.

Grey-hooded Gull — *Larus cirrocephalus*
One or two were seen at several sites in 2001–02 and in 2008.

Hartlaub's Gull — *Larus hartlaubii*
One was at East India Dock Basin in 2000.

African Collared Dove — *Streptopelia roseogrisea*
Recorded in 1985, 1987–88 and 1993.

Red-eyed Dove — *Streptopelia semitorquata*
Some were introduced into St James's Park in 1907 and remained for several months before disappearing.

Crested Pigeon — *Ocyphaps lophotes*
These were released into Regent's Park early in the 20th century and were said to be 'doing well' by 1911 but there were no further records.

Diamond Dove — *Geopelia cuneata*
Recorded in 2005 and 2009.

Galah — *Eolophus roseicapilla*
Recorded in 2006.

Sulphur-crested Cockatoo — *Cacatua galerita*
Recorded in 1997 and 2009.

Cockatiel — *Nymphicus hollandicus*
One of the most frequently reported escapes; first seen in 1985.

Red Lory — *Eos bornea*
Recorded in the Hersham parakeet roost in 2003.

Papuan Lorikeet — *Charmosyna papou*
Recorded in 2004.

Eastern Rosella — *Platycercus eximius*
One of the 'Pale-headed' race seen in 1987; also recorded almost annually since 2000.

Red-rumped Parrot — *Psephotus haematonotus*
Recorded in 1985, 1994 and 1999.

Budgerigar — *Melopsittacus undulatus*
Three frequented Lincoln's Inn in 1863–64; a pair spent the winter of 1864/65 around the Tower of London; and one flew over the Oval in 1876. A pair lived in a London square for a number of years early in the 20th century. There are also many other recent records of single birds.

Superb Parrot — *Polytelis swainsonii*
Recorded in 1998.

Princess Parrot — *Polytelis alexandrae*
Recorded in 2004.

Alexandrine Parakeet — *Psittacula eupatria*
First recorded in 1997 and seen almost annually since then. A pair bred at Foots Cray Meadows in 2001.

Plum-headed Parakeet — *Psittacula cyanocephala*
Recorded in 1996.

Blossom-headed Parakeet — *Psittacula roseata*
Recorded several times since 1983.

Rosy-faced Lovebird — *Agapornis roseicollis*
Recorded in 2003 and 2007.

Fischer's Lovebird — *Agapornis fischeri*
Recorded several times since 1985.

African Grey Parrot — *Psittacus erithacus*
Various escapes frequently recorded since 1983.

Senegal Parrot — *Poicephalus senegalus*
Several recorded since 1997; three were resident at Shirley during 2007–08.

Nanday Parakeet — *Nandayus nenday*
Two were at Fairlop Waters in 2000.

Burrowing Parakeet — *Cyanoliseus patagonus*
Recorded in 1999.

Monk Parakeet — *Myiopsitta monachus*
In the early years of the 21st century there were three distinct populations. The longest established of these is in Borehamwood where they were first seen in 1993 and there were up to 51 in 2006–08; however, in 2009 the peak count was only 33. On the Isle of Dogs a pair built a nest in 2003 and there were up to 25 in 2006–08; this colony was probably started by escapes from the aviaries in Mudchute Farm. There were also about eight at a colony in Southall in 2008 but these were culled. There have been other isolated pairs, for example a pair built a nest at Lonsdale Road Reservoir in 1996–97 and these birds were still in the Barnes area in 2000.

Blue-crowned Parakeet — *Aratinga acuticaudata*
This was first recorded in Bromley in 1997 with up to eight in 1999. A pair bred at Beckenham Place Park in 2001, the first breeding in the wild in the UK. Singles were seen in Shirley in 2003 and in Beckenham in 2004 with three there in 2005.

Scarlet-fronted Parakeet — *Aratinga wagleri*
One was at Dagenham Chase in 1989/90.

Mitred Parakeet — *Aratinga mitrata*
Recorded at Holyfield Lake in 2004.

Sun Parakeet — *Aratinga solstitialis*
One in Mar Dyke Valley on 22 July 2009.

Military Macaw — *Ara militaris*
Recorded in 2000.

Blue-and-yellow Macaw — *Ara ararauna*
Recorded in 1983.

Blue-fronted Amazon — *Amazona aestiva*
Up to two in the Hersham parakeet roost between 2000 and 2006.

Orange-winged Amazon — *Amazona amazonica*
One in the Hersham parakeet roost in 1998–2006 and one in Stoke Newington and surrounding areas from 2009 to 2010.

Violet Turaco — *Musophaga violacea*
Recorded in 2006–07.

Eurasian Eagle-owl — *Bubo bubo*
There have been many reports of Eurasian Eagle-owls seen at large in London; according to the BOU there is no proof that this species was ever a native bird in Britain and all records are treated as escapes.

One was taken in the spring of 1841 at Hornsey and Harting reported that a female was caught in Hampstead on 3 November 1845. A male shot at Bayfordbury on 23 December 1881 was traced to an escape from Welwyn.

Since 1993 there have been another eight records, including one of a bird in central London from 6 December 1996 into 1997 when it was found dead by St Paul's Cathedral on 8 July; a pair nested at Bedwell Park Golf Course in 2002; and a pair at Essendon in February 2005 when one was captured, the other disappearing a few weeks later.

Rock Eagle-owl — *Bubo bengalensis*
One in Hertford on 30 September 2008.

San Blas Jay — *Cyanocorax sanblasianus*
Recorded in 1991.

Blue Magpie — *Urocissa erythroryncha*
Recorded in 1986.

Red-billed Chough — *Pyrrhocorax pyrrhocorax*
There are a number of records from the 19th and early 20th centuries, but they are all considered to be escapes as the species was held in captivity at the time. The earliest was shot on Mitcham Common in or before 1836.

Although choughs were more widely distributed in Britain around this time, the closest breeding ones to London were on the Isle of Wight. Further records include one of a bird near Beechbottom Wood, St Albans, in 27 May 1884; one killed in Balham in 1893; and one shot near Hendon in January 1900.

Later presumed escapes were seen at Sutton on 3 May 1961; flying with Jackdaws at Wraysbury GP on 18 May 1980; and flying with a small flock of Black-headed Gulls over Maple Cross SF on 26 August 1980.

Red-vented Bulbul *Pycnonotus cafer*
A pair remained in a Rickmansworth garden in the winter of 1956/57.

Red-whiskered Bulbul *Pycnonotus jocosus*
Recorded in 1995 and 2001.

White-eared Bulbul *Pycnonotus leucotis*
Recorded in 2006.

White-throated Laughing Thrush *Garrulax albogularis*
One at Inner Temple Gardens in 1998.

Red-tailed Laughing Thrush *Garrulax milnei*
Recorded at Chingford in 1990.

Red-faced Liocichla *Liocichla phoenicea*
One at Hoddesdon in January and February 2009.

Red-billed Leiothrix *Leiothrix lutea*
An attempt to introduce these was made in April 1905 when three or four dozen were released into Regent's and St James's Parks but none were seen after a few months. Also recorded several times between 1988 and 1997.

Red-backed Thrush *Zoothera erythronota*
One at Wraysbury on 29 October 2009.

Cape White-eye *Zosterops pallidus*
Recorded in 2003.

Common Hill Myna *Gracula religiosa*
Recorded in 1984.

Common Myna *Acridotheres tristis*
Recorded several times between 1991 and 1998; also up to five free-flying birds at Chessington World of Adventure in 2003.

Greater Blue-eared Glossy Starling *Lamprotornis chalybaeus*
Recorded in 1991.

Red-crested Cardinal *Paroaria coronata*
One present for four months in Kensington Gardens in 1926; also recorded several times since 1984.

Chestnut Bunting *Emberiza rutila*
A male in Bushey on 8–9 August 2002.

Red-headed Bunting *Emberiza bruniceps*
Males at Brent Reservoir on 13 August and 8 September 1967 in the Garston area in the last three weeks of May 1991.

Northern Cardinal — *Cardinalis cardinalis*
Recorded in 2006.

Rusty Blackbird — *Euphagus carolinus*
One in St James's Park in July 1938 was found dead the following month.

Long-tailed Rosefinch — *Uragus sibiricus*
A male was at Nags Head SF, Harold Wood on 14 December 1991.

Canary — *Serinus canaria*
First seen in Brixton in 1877; recorded regularly since 1983.

Yellow-fronted Canary — *Serinus mozambicus*
Recorded most years between 1990 and 2005.

Sudan Golden Sparrow — *Passer luteus*
Recorded in 2006.

African Golden Weaver — *Ploceus subaureus*
Recorded in 2002.

Northern Brown-throated Weaver — *Ploceus castanops*
Recorded in Theobalds Park, Cheshunt, in 1956.

Village Weaver — *Ploceus cucullatus*
Recorded in 1983 and 2000; a pair built a nest in Romford in 1987.

Red-billed Quelea — *Quelea quelea*
Recorded in 1996 and 2003.

Zanzibar Bishop — *Euplectes nigroventris*
Recorded in 1999.

Yellow-crowned Bishop — *Euplectes afer*
Recorded in 2004 and 2008.

Fan-tailed Widowbird — *Euplectes axillaris*
Recorded in 2005.

Orange-cheeked Waxbill — *Estrilda melpoda*
Recorded at various sites between 1990 and 2003, including in the Hersham parakeet roost in 2000.

Common Waxbill — *Estrilda astrild*
Recorded several times between 1988 and 2006.

Cut-throat Finch — *Amadina fasciata*
Recorded in 1997.

Red Avadavat — *Amandava amandava*
In the late 1960s and early 70s these were regularly seen in small numbers during autumn at Brent Reservoir.

Zebra Finch *Taeniopygia guttata*
Recorded several times between 1993 and 2002.

Indian Silverbill *Euodice malabarica*
Recorded in 2004.

Nutmeg Mannikin *Lonchura punctulata*
Recorded in 2002.

Java Sparrow *Lonchura oryzivora*
Singles in St Albans in April 1889 and at Brent Reservoir on 1 August 1967.

Pin-tailed Whydah *Vidua macroura*
Recorded in 2004–05.

Shaft-tailed Whydah *Vidua regia*
Recorded in 1997.

APPENDIX 2 – SPECIES NOT ACCEPTED

American Wigeon *Anas americana*
There were two records of this North American duck in the mid-1970s, both at Surrey Docks: a male on 29 August 1973 and a pair on 5 September 1975, with the male present again from 9 to 15 September. They were originally accepted by the BBRC, although acceptance of the 1975 pair came with the caveat that they may have been escapes from captivity. However, a recent examination of the photographs from 1975 showed that the male had been pinioned. The circumstances also suggested a possible captive origin. This also casts a doubt over the earlier record; neither is now deemed acceptable.

Wilson's Storm-Petrel *Oceanites oceanicus*
A specimen in the collection of Mr Doubleday was said to be of this species and it may refer to a bird found near Epping in November 1840 which was identified as a Leach's Storm-petrel. The specimen was later housed in the Chelmsford and Essex Museum but no longer exists so its identity and provenance cannot be confirmed.

American Bittern *Botaurus lentiginosus*
Ticehurst lists a record of one that was shot in Lullingstone Park in about 1861; the specimen had been kept in the castle. However, this record is no longer deemed acceptable.

Semipalmated Sandpiper *Calidris pusilla*
A juvenile bird at Sevenoaks Wildfowl Reserve on 10 September 1967 was originally accepted as this species but later reviewed and rejected as it was thought to be a Little Stint. One at Barking on 4 May 1974 was accepted by the BBRC but later reviewed and rejected as no call was heard.

Short-billed Dowitcher *Limnodromus griseus*
An immature was reportedly shot on the River Brent at Stonebridge in October 1862. The specimen was in Harting's collection and is now in the Natural History Museum. At the time only one species of American dowitcher was known and it was not split until 1932. All records were reviewed in *British Birds* in 1961 to try and assign them to species. This record was then accepted as the first Short-billed Dowitcher in the UK by the BOU in 1971, but in 1991 the BOU reviewed all the accepted records again and decided that this record was insufficiently documented and therefore deemed unacceptable.

Passenger Pigeon *Ectopistes migratorius*
In Meyer's *British Birds* there is a short reference to this species having been seen in London. In April 1843 two males and a female were seen flying in and landing in the woods of Littleton Common. The observers, who were familiar with the species, noted their long tails, the colour of the birds and that their call resembled that of a Pheasant. There were also two records from Hertfordshire, one in July 1844 and another around the same time. At this time, Passenger Pigeons were the most common migratory bird in North America so it seems rather surprising that there were no acceptable records of them being seen in the UK. However, many were imported into Britain from the 1830s onwards.

Northern Hawk Owl *Surnia ulula*
In 1926 an owl was seen in a tree at Walton Reservoirs on 27 December by two observers who were in no doubt that it was a Hawk Owl. The details were passed to *British Birds* but it was considered to be only 'probable'.

Middle Spotted Woodpecker *Dendrocopos medius*
Harting gave a full account of a pair shot by Mr Spencer at Kenwood, Hampstead, in June 1846; he had spent more than 30 years studying birds and was very familiar with all plumages of the woodpeckers in Britain. The specimens were described as being somewhat smaller than Great Spotted Woodpecker and considerably larger than Lesser Spotted and 'the red on the head extends to the top of the crown…there is rather more white on the

scapulars, and the red of the under-tail coverts is not so brilliant'. Yarrell examined the specimens and concluded that they were 'old' birds [adults rather than juveniles].

Although this description is reasonably accurate for Middle Spotted Woodpecker, this species is not on the British List so the record must be considered unproven.

Black Woodpecker *Dryocopus martius*

Harting knew of two occurrences of this species in London. One bird was shot in 1805 'on the trunk of an old willow tree in Battersea Fields' and was written up by Montagu in a supplement to his *Ornithological Dictionary*; however, no further details or the whereabouts of the specimen are known. The second record was related directly by the observer, Mr Spencer, to Harting who had no reason to doubt his word. In May 1845 Mr Spencer was walking through Kenwood, Hampstead, early one morning when he 'was suddenly startled to see a Black Woodpecker dart between the trees and alight upon an oak at some distance. It was extremely shy, and he was scarcely ever able to approach within a hundred yards of it. On the following morning he again visited the spot, on the chance of getting a shot at the bird, and again saw it; but it was too wary to allow of a sufficiently near approach...on the morning of the fourth day he saw the bird again for the third and last time.'

Unfortunately no description of this bird was ever written and as this species is not on the British List this record remains unproven. Fitter stated that if this species had already been admitted to the British List then there would be no doubts as to the authenticity of these records. Another was claimed from Hatfield Golf Course in July 1937 but this was considered to be doubtful.

White-winged Lark *Melanocorypha leucoptera*

In 1955 there was a record of a female at Hilfield Park Reservoir from 12 to 17 August; it was seen by a number of observers and was accepted at the time. Following a review of all British records in 1995, it was considered that this was probably an aberrant Skylark and could no longer be accepted.

Black Lark *Melanocorypha yeltoniensis*

A specimen, probably of this species, was caught near Highgate in about 1737 and illustrated by Albin in *Natural History of Birds*. The illustration clearly shows white spots on the head and pale fringes to the breast feathers. At the time of capture this species was unknown to European ornithologists. As the official British List only includes species seen from the beginning of the 19th century this record cannot be considered by the BOURC.

Purple Martin *Progne subis*

Yarrell documented two of these North American birds that were said to have been shot in September 1842 at Brent Reservoir by John Calvert of Paddington. They were an adult male and a juvenile, which were said to have been shot during the same week but several days apart. However, Harting made further inquiries and received such unsatisfactory information regarding these birds that he would not even have noted the occurrence in his book were it not for Yarrell's mention of them. As we know the specimens were critically examined by Bond and Harting so were obviously this species, it must be assumed that Harting believed the record to be fraudulent. There has since been one accepted record of this species in Britain.

Great Reed-warbler *Acrocephalus arundinaceus*

In 1852 a male was alleged to have been shot near Dartford on 8 May and was documented by Newman in *The Zoologist* at the time. However, it later transpired that the dealer who shot it had not identified it correctly and Ticehurst stated that 'it is evident that this bird was not a Great Reed Warbler'. Christy stated that a known dealer of rare bird specimens had claimed to have caught one at Dagenham on 16 June 1853 but he rejected the record. Another one, allegedly shot at Erith between 1853 and 1856, was likely to have been imported. An accepted record of one at Amwell in May 2008 was just beyond the London boundary.

Black-eared Wheatear *Oenanthe hispanica*

A male of the black-throated form was reported in Regent's Park on 23 April 1951. It was seen by two observers, including Eric Simms, and a full description was taken along with drawings which appeared to support the identification. Although the record was originally accepted by the LNHS, it was much later deemed unproven by the BBRC when all records between 1950 and 1957 were reviewed.

Hermit Thrush *Catharus guttatus*
A record of one in a garden in Chipping Ongar between 28 October and 3 November 1994 was initially accepted by the BBRC; it was later withdrawn by the observer who admitted that it was a hoax.

Greenish Warbler *Phylloscopus trochiloides*
There were two accepted records of this species: at Perry Oaks SF from the end of 1960 to 26 February 1961; and at Dollis Hill on 1 October 1964. However, in a major review of all Greenish Warbler records by the British Birds Rarities Committee in the mid-1980s, after the identification features were clarified, both of these birds (along with many other records) were rejected. They were considered most likely to have been Chiffchaffs with a wing-bar of one of the northern/eastern races.

Western Orphean Warbler *Sylvia hortensis*
In June 1866 a juvenile warbler which was unable to fly was chased and caught by a boy in Holloway and sold to H. Hanley. Blyth saw the bird in December and identified it as a female Orphean Warbler. It must be presumed that this bird was captured as a nestling abroad and brought to this country where it either escaped or was released.

Red-winged Blackbird *Agelaius phoeniceus*
There are three records, the first of which was of an adult male shot near London and illustrated by Albin in 1738. In the 19th century this species was on the British List as it had been recorded in the UK a number of times. One of these records was of a male shot in the autumn of 1844 in a reed bed at Shepherd's Bush, described at the time as 'a swampy situation, about three miles west of London, on the Uxbridge Road'. The specimen was described by Yarrell in his *History of British Birds* and was retained by Bond in his collection. An escaped bird frequented Hampstead Heath in the early 20th century.

These records, along with all the other British records, are no longer accepted as being wild birds and this species was removed from the British List.

BIBLIOGRAPHY

The following books and significant papers relating to the birds of London, as well as other ornithological publications, have been used as reference material. A fuller list of notes and short papers can be found in Glegg's books (1929a, 1929b, 1930, 1935).

Adams, H.G. and H.B. 1894. *The Smaller British Birds*. Gibbings, London.

Albin, E. 1731–38. *A Natural History of Birds*. London.

Baker, H. 1984. The Status of the Canada Goose in the London Area. *London Bird Report* 49: 111–136.

Baker, H. 1991. The Mute Swan Survey. *London Bird Report* 55: 160–163

Baker, H. 2004. LNHS House Sparrow monitoring 1995–2003. *London Bird Report* 69: 158–165.

Banks, A.N., Burton, N.H.K., Calladine, J.R. and Austin, G.E. 2007. *Winter Gulls in the UK: Population Estimates from the 2003/04 – 2005/06 Winter Gull Roost Survey*. BTO Research Report No. 456

Barrett-Hamilton, G.E.H. 1892. *Harrow Birds*. Harrow.

Batten, L.A. 1972. The Past and Present Bird Life of Brent Reservoir and its Vicinity. *The London Naturalist* 50: 8–62.

Batten, L., Beddard, R., Colmans, J. and Self, A., eds 2002. *Birds of Brent Reservoir*. Welsh Harp Conservation Group, Barnet.

Betton, K.F. 1994. The Birds of Bushy Park and Hampton Court Park. *London Bird Report* 58: 164–182.

Blyth, E. 1836. On the Species of Birds Observed During the Last Four Years at Tooting, Surrey. *Loudon's Magazine of Natural History*: 622–638.

Bond, F. 1843. (A) Note on Water-birds Occurring at Kingsbury Reservoir. *The Zoologist* 1843: 102.

Bond, F. 1843. (B) Note on the Occurrence of Rare British Birds. *The Zoologist* 1843: 148.

Bond, F. 1844. Note on Rare Waders Occurring at Kingsbury Reservoir. *The Zoologist* 1844: 767.

Bond, F. 1860. Occurrence of the Serin Finch in England. *The Zoologist* 1860: 7105.

Bowlt, C. 2008. London's Missing Starlings. *London's Changing Natural History*: 27–28. LNHS, London.

Bridgeman, F.C and G.O. 1864. Birds of Harrow. *The Flora of Harrow*. J.C. Melvill, London.

Brown, P. 1776. *New Illustrations of Zoology*. B. White, London.

Bucknill, J.A. 1900. *The Birds of Surrey*. R.H. Porter, London.

Bunyard, P.F. 1931. Great Skua in Middlesex. *BOC Bulletin* 51: 104.

Catley, G.P. and Hursthouse, D. 1985. Parrot Crossbills in Britain. *British Birds* 78: 482–505.

Cecil, E. 1907. *London Parks and Gardens*. London.

Chipperfield, H. 1900. Birds of North-East London. *Nature Notes* 11: 70–74.

Clark Kennedy, A.W.M. 1868. *The Birds of Berkshire and Buckinghamshire*. Simpkin, Marshall and Co., London.

Cocker, M. and Mabey, R. 2005. *Birds Britannica*. Chatto & Windus, London.

Cornish, C.J. 1902. *The Naturalist on the Thames*. London.

Cramp, S. and Gooders, J. 1967. The Return of the House Martin. *London Bird Report* 31: 93–98.

Cundall, J.W. 1898. London Birds. London. *A Guide for the Visitor*: 88–90.

Davis, W.J. 1907. *The Birds of Kent*. Dartford.

Dennis, M.K. 1995. Yellow-legged Gulls Along the River Thames in Essex. *British Birds* 88: 8.

Dixon, C. 1909. *The Bird-Life of London*. London.

Donovan, E. 1794–1820. *The Natural History of British Birds*. London.

Bibliography

Doubleday, H. 1836. *A Nomenclature of British Birds*. Wesley and Davis, London.

Dresser, H.E. 1871–1881. *A History of the Birds of Europe*. London.

Dudley, S.P. 2010. *Non-native Bird Species and the British List*. BOU Proceedings – The Impacts of Non-native Species.

Dutton, J. 1871. Great Bustard in Middlesex. *The Zoologist* 1871: 2473.

Edwards, G. 1743–51. *A Natural History of Birds*. London.

Edwards, G. 1758–64. *Gleanings of Natural History*. London.

Fitter, R.S.R. 1949. *London's Birds*. Collins. London.

Fuller, R.J., Noble, D.G., Smith, K.W. and Vanhinsbergh, D. 2005. Recent Declines in Populations of Woodland Birds in Britain. *British Birds* 98: 116–143.

Glegg, W.E. 1929 (A). The Thames as a Bird-Migration Route. *The London Naturalist* 8: 3–15.

Glegg, W.E. 1929 (B). *A History of the Birds of Essex*. London.

Glegg, W.E. 1930. The Birds of Middlesex Since 1866. *The London Naturalist* 9: 3–32.

Glegg, W.E. 1935. *A History of the Birds of Middlesex*. H. F. & G. Witherby, London.

Gould, J. 1862–73. *The Birds of Great Britain*. London.

Graves, G. 1811–21. *British Ornithology*. London.

Gunn, D. 1924. Aquatic Warbler at Staines. *The Field* 144: 640.

Hamilton, E. 1879. The Birds of London: Past and Present, Residents and Casuals. *The Zoologist* 1879: 273–291.

Hamilton, E. 1889. Wild Bird-Life in London, Past and Present. *Murray's Magazine* 5: 661–671.

Hardwick, M.A. and Self, A.S.M. 1991. Rare Birds in the London Area 1900 to 1991. *London Bird Report* 56: 183–211.

Harrison, B.R. 1893. The Birds of Highgate Woods. *The Naturalists' Journal* 1: 30.

Harrison, J.M. 1953. *The Birds of Kent*. Witherby, London.

Harrison, C.J.O. 1959. Woodlark Population and Habitat. *London Bird Report* 24: 71–80.

Harrisson, T.H. 1931. Birds of the Harrow District. *The London Naturalist* 10: 82–120.

Hartert, E. 1898. A Hitherto Overlooked British Bird. *The Zoologist* 1898: 116.

Harting, J.E. 1866. *The Birds of Middlesex*. John van Voorst, London.

Harting, J.E. 1872. *A Handbook of British Birds*. John van Voorst, London.

Harting, J.E. 1875. *Our Summer Migrants*. Bickers and Son, London.

Harting, J.E. 1886. On the Former Nesting of the Spoonbill in Middlesex. *The Zoologist* 1886: 81.

Harting, J.E. 1888. Bird-Life in Kensington Gardens. *The Field* 71: 52.

Harting, J.E. 1889. The Birds of Hampstead. *Hampstead Hill* 87–96.

Harting, J.E. 1901. *A Handbook of British Birds*. Nimmo, London.

Hayes, W. and family. 1794–99. *Portraits of Rare and Curious Birds, with their Descriptions, from the Menagery of Osterly Park, in the County of Middlesex*. London.

Hewlett, J., ed. 2002. *The Breeding Birds of the London Area*. LNHS, London.

Hibberd, S. 1865. London Birds. *The Intellectual Observer* 7: 167–175.

Holling, M. and the Rare Breeding Birds Panel. 2011. Rare Breeding Birds in the United Kingdom in 2009. *British Birds* 104: 476–537.

Hudson, W.H. 1898. *Birds in London*. Longmans, Green and Co., London.

Hussey, H. 1864. Wild-fowl in the London waters. *The Zoologist* 1864: 9049–9053.

Jennings, J. 1829. *Ornithologia*, 2nd edn. London.

Jenyns, Rev. L. 1835. *A Manual of British Vertebrate Animals*. Longman and Co., London.

Jesse, E. 1832–35. *Gleanings in Natural History*. London.

Kerr, G.W. 1908. The Birds of the District of Staines. *The Zoologist* 1908: 137.

Kidd, B. 1891. The Birds of London. *The English Illustrated Magazine* 9: 38–45.

Latham, J. 1781–1802. *A General Synopsis of Birds*. London.

Latham, J. 1790. *Index Ornithologicus*. London.

Latham, J. 1821. *A General History of Birds*. Jacob and Johnson, Winchester.

Lilford, L. 1883. Rustic Bunting Near London. *The Zoologist* 1883: 194–195.

London Natural History Society. 1964. *The Birds of the London Area*. Rev. Ed. Ruprt Hart-Davis, London.

Macpherson, A.H. 1929. A List of the Birds of Inner London. *British Birds* 22: 222.

Macpherson, A.H. 1891–1927. Some London Birds. *Nature Notes 2: 110 et seq; London Naturalist* vols 4–7.

Macpherson, A.H. 1928. London Reservoirs and their Influence on Bird Life. *London Naturalist* 7: 5–11

Meinertzhagen, R. 1893. Black Stork in Middlesex. *The Zoologist* 1893: 396.

Meyer, H.L. 1842-50. *Coloured Illustrations of British Birds and their Eggs*. G.W. Nickisson, London.

Miller, W.J.C. 1894. Bird-Life in London. *Nature Notes*.

Mlikovsky, J. 2006. Black Lark or black Lark? An Historical Record from England. *British Birds* 99: 262–263.

Montagu, G. 1802 and 1813. *Ornithological Dictionary of British Birds*. White, London.

Montagu, G. and Rennie, J. 1831. *Ornithological Dictionary of British Birds*. Hurst, London.

Montier, D.J., ed. 1977. *Atlas of Breeding Birds of the London Area*. Batsford, London.

Moon, A.V. 1983. Pelagic Seabirds in the London Area. *London Bird Report* 48: 106–119

Moon, A.V. 1985. The Last Fifty Years. *London Bird Report* 50: 124–139.

Moon, A.V. 1987. The Influx of Sabine's Gulls and Other Seabirds in October 1987. *London Bird Report* 52: 121–132.

Morant, P. 1768. *The History and Antiquities of the County of Essex*. T. Osborne, Essex.

Morgan, D.H.W. 1993. Feral Rose-ringed Parakeets in Britain. *British Birds* 86: 561–564.

Morris, F.O. Rev. 1850–57. *A History of British Birds*. London.

Mountford, G. 1957. *The Hawfinch*. Collins, London.

Nichols, J.C.M. 1941. *Shooting Ways and Shooting Days*. Herbert Jenkins, London.

Nicholson, E.M. 1995. *Bird-Watching in London: a Historical Perspective*. LNHS, London.

Oliver, P.J. 1982. The Decline of the Mute Swan in the London Area. *London Bird Report* 46: 87–91.

Oliver, P.J. 1985. Breeding Tufted Ducks in the London Area. *London Bird Report* 49: 104–110.

Oliver, P.J. 2006. People, Crows and Squirrels – Some Recent Changes at St James's Park. *London Naturalist* 85: 105–113.

Oliver, P.J. 2011. The Establishment of Breeding Common Buzzards. *London Bird Report* 72: 199–205.

Osborne, K.C. 1980. Checklist of the Birds of the London Area 1901–77. *London Bird Report* 43: 71–84

Parker, E. 1952. *Surrey Naturalist*. Robert Hale, London.

Parkin, D.T. and Knox, A.G. 2010. *The Status of birds in Britain and Ireland*. Helm, London.

Pennant, T. 1812. *British Zoology*. London.

Pigott, T.D. 1902. *London Birds and Other Sketches*. Edward Arnold, London.

Pithon, J. A. and Dytham, C. 2002. Distribution and Population Development of Introduced Ring-necked Parakeets in Britain between 1983 and 1998. *Bird Study* 49: 110–117.

Power, F.D. 1910. *Ornithological Notes from a South London Suburb 1874–1909*. Henry J. Glaisher, London.

Read, R.H. 1896. The Birds of the Lower Brent Valley, *Annual Report of the Ealing Microscopical and Natural History Society*.

Ruegg, R. 1847. *Summer's Evening Rambles Round Woolwich*. Woolwich.

Sage, B.L. 1959. *A History of the Birds of Hertfordshire*. Barrie and Rockliff, London.

Sage, B.L. 1963. The Breeding Distribution of the Tree Sparrow. *London Bird Report* 27: 56–65.

Sage, B.L. and Cornelius, L.W. 1977. Rook Population of the London Area. *London Bird Report* 40: 66–73.

Sanderson, R.F. 1997. Annual Bird Counts in Kensington Gardens 1925–1995. *London Bird Report* 60: 170–176.

Self, A.S.M. 2004. Honey Buzzard Influx Autumn 2000. *London Bird Report* 65: 191–198.

Sharrock, J.T.R. (Editor) 1976 *The Atlas of Breeding Birds in Britain and Ireland*. Poyser, London.

Sibley, P., Hewlett, J.F., Vickers, D., Gannaway, C., Morgan. K., and Reeve, N. 2005, *London's small parks and squares - a place for nature?* London Biodiversity Partnership.

Snow, D.W. and Perrins, C.M. 1998. *The Birds of the Western Palearctic*, Concise Edition. Oxford University Press, Oxford.

Swann, H.K. 1893. *The Birds of London*. Sonnenschein, London.

Sweet, R. 1828–32. *The British Warblers*. London.

Taylor, D.W., Davenport, D.L., Flegg, J.J.M. eds. 1981. *Birds of Kent*, 2nd edn. Meresborough Books, Rainham.

Ticehurst, N.F. 1909. *A History of the Birds of Kent*. Witherby, London.

Torre, H.J. 1838. List of Birds Found in Middlesex. *The Naturalist* 3: 420–422.

Tristram-Valentine, J.T. 1895. *London Birds and Beasts*. Horace Cox, London.

Tuck, W.H and Smith, H. 1886. Birds in London. *The Field* 47: 528, 769.

Warren, S.F. 1884. Little Auk in the Thames. *The Field* 43: 100.

Watt, H.B. 1908. Changes in the London Avifauna. *Nature Notes* 19: 135.

Webb, W.M. 1911. *The Brent Valley Bird Sanctuary*, 3rd edn. London.

Webster, A.D. 1911. *The Regent's Park and Primrose Hill*. Greening, London.

Wheatley, J.J. 2007. *Birds of Surrey*. Surrey Bird Club, Hersham.

White, G. 1789. *The Natural History and Antiquities of Selborne*. Various.

Whiting, J.E. 1900. Some Notes on the Birds of Hampstead. *The Hampstead Annual* 1900: 37–45.

Whiting, J.E. 1912. The Birds of Hampstead. *The Annals of Hampstead* 3: 177–188.

Wood, S. 2007. *The Birds of Essex*. Christopher Helm, London.

Yarrell, W. 1826. Notice of the Occurrence of Some Rare British Birds. *The Zoological Journal* 2: 24–27.

Yarrell, W. 1843. *A History of British Birds*, 1st edn. London.

Yarrell, W. 1871–85. *A History of British Birds*, 4th edn. London.

GAZETTEER

The following gazetteer includes the main sites mentioned in this book but omits obvious place names of towns and villages. Each site name is followed by the vice-county (abbreviated as: BU – Buckinghamshire, EX – Essex; HE – Hertfordshire; KE – Kent; MX – Middlesex; and SY – Surrey as well as IL, signifying the LNHS Inner London recording zone) and a six-figure Ordnance Survey Grid Reference enabling sites to be pinpointed on a map. The Grid Reference number refers either to the centre of the site or a point of particular ornithological interest. Note that the inclusion of a site in this gazetteer does not imply public access; indeed some sites are strictly private or access is by permit only.

It should be noted that this list of sites is current at the time of publishing and that the names may differ from those of historical sites mentioned in the text; additionally, sites that no longer exist will not be found here. A fuller list is published each year in the annual London Bird Report.

Site	VC	Grid Ref	Site	VC	Grid Ref
Addington Hills	(SY)	TQ352644	Bessels Green	(KT)	TQ505555
Aldenham Res	(HR)	TQ169955	Betchworth	(SY)	TQ218505
Alexandra Park	(MX)	TQ302900	Beverley Brook	(SY)	TQ213726
Amwell GP	(HR)	TL380125	Bexley Wood	(KT)	TQ483737
Apps Court Farm GP	(SY)	TQ110673	Biggin Hill	(KT)	TQ457572
Arbrook Common	(SY)	TQ145630	Bishop's Park	(SY)	TQ241761
Ashtead Common	(SY)	TQ175595	Bishop's Wood	(HR)	TQ066920
Balls Wood	(HR)	TL344106	Black Park	(BU)	TQ010836
Banbury Res	(EX)	TQ362915	Blackheath	(KT)	TQ390766
Banstead Down	(SY)	TQ252610	Blackwall Basin	(MX)	TQ381802
Banstead Heath	(SY)	TQ235545	Blue Lake	(KT)	TQ622737
Banstead Wood	(SY)	TQ260560	Bookham Common	(SY)	TQ130565
Barking Bay	(EX)	TQ451816	Bowman's Heath	(KT)	TQ518738
Barnes Common	(SY)	TQ222758	Bowmansgreen Farm	(HR)	TL189041
Barrack Wood	(EX)	TQ597915	Bowyers Water	(HR)	TL367015
Barwell Court Farm	(SY)	TQ170630	Box Wood	(HR)	TL353096
Batchworth Lake	(HR)	TQ058940	Boxer's Lake, Enfield	(MX)	TQ305962
Batler's Green	(HR)	TQ158985	Brent Park	(MX)	TQ240889
Battersea Park	(IL)	TQ282772	Brent Res	(MX)	TQ215870
Bayhurst Wood	(MX)	TQ065892	Bricket Wood Common	(HR)	TL130010
Beaulieu Heights	(SY)	TQ334696	Broad Colney Fields	(HR)	TL180030
Beckenham Place Park	(KT)	TQ383708	Broad Colney Lakes	(HR)	TL178034
Beddington Park	(SY)	TQ292654	Broadwater Lake	(MX)	TQ045892
Beddington SF	(SY)	TQ290662	Brockwell Park	(SY)	TQ316740
Bedfont Lakes	(MX)	TQ078726	Bromley Common	(KT)	TQ415655
Bedford Park	(MX)	TQ210791	Brooklands	(SY)	TQ068620
Bedfords Park	(EX)	TQ518925	Brookmans Park	(HR)	TL253038
Beech Farm GP	(HR)	TL190086	Broomfield Park	(MX)	TQ304927
Belair Park	(SY)	TQ328733	Broxbourne GP	(HR)	TL379078
Bell Lance Fields	(HR)	TL198040	Broxbourne Woods	(HR)	TL340080
Bencroft Wood	(HR)	TL330064	Bruce Castle Park	(MX)	TQ336908
Bentley Priory	(MX)	TQ155927	Brunswick Park	(IL)	TQ331769
Berrybushes Wood	(HR)	TL069007	Buckhurst Hill GP	(EX)	TQ425934
Berwick Ponds	(EX)	TQ543835	Buckland SP	(SY)	TQ227510

418

Bugsby's Reach	(KT)	TQ398798	Crane Park	(MX)	TQ126730
Bulphan Fen	(EX)	TQ633864	Crayford Marsh	(KT)	TQ532775
Burgess Park	(IL)	TQ335778	Crayford Ness	(KT)	TQ532781
Burwood Park	(SY)	TQ100644	Creekmouth, Barking	(EX)	TQ457814
Bury Lake	(HR)	TQ053938	Crews Hill	(MX)	TL312000
Bushy Park	(MX)	TQ160690	Crossness	(KT)	TQ478815
Camley St NR	(IL)	TQ298836	Crossness East	(KT)	TQ492809
Cannon Hill Common	(SY)	TQ238683	Croxley Green	(HR)	TQ070955
Canons Park	(MX)	TQ182915	Croxley Hall GPs	(HR)	TQ068943
Carshalton Beeches	(SY)	TQ272638	Croxley Moor	(HR)	TQ065956
Cassiobury Park	(HR)	TQ090970	Crystal Palace Park	(KT)	TQ347707
Chalfont Lodge	(BU)	TQ013899	Cuddington GC	(SY)	TQ239615
Chalfont Park	(BU)	TQ013896	Curtismill Green	(EX)	TQ519965
Chandlers Hill, Iver	(BU)	TQ035834	Dagenham Chase	(EX)	TQ514858
Charlton GP	(MX)	TQ087692	Dagenham Corridor	(EX)	TQ495875
Chelsfield	(KT)	TQ482642	Dagnam Park	(EX)	TQ550933
Chelsham	(SY)	TQ373589	Danson Park	(KT)	TQ473748
Chertsey Meads	(SY)	TQ060662	Darenth Lake	(KT)	TQ559710
Cheshunt GP	(HR)	TL370030	Darenth Valley Walk	(KT)	TQ560704
Cheshunt Marsh	(HR)	TL370008	Darenth Wood	(KT)	TQ580727
Chigwell Res	(EX)	TQ460937	Darlands Lake	(MX)	TQ243943
Chipstead Lake	(KT)	TQ505565	Dartford Heath	(KT)	TQ516733
Chiswell Green	(HR)	TL125047	Dartford Marsh	(KT)	TQ544774
Chiswick Eyot	(SY)	TQ219779	Delaford Park	(BU)	TQ043818
City Road Basin	(IL)	TQ320830	Denham Aerodrome	(BU)	TQ033887
Clapham Common	(SY)	TQ286748	Denham CP	(BU)	TQ048865
Clapton Common	(MX)	TQ342878	Denham GC	(BU)	TQ027883
Clay Tye Wood	(EX)	TQ595587	Denham Mount	(BU)	TQ025865
Claygate Common	(SY)	TQ161632	Denham Place	(BU)	TQ040873
Claygate Green	(SY)	TQ157637	Denham Quarry Lake	(MX)	TQ045886
Clissold Park	(MX)	TQ326864	Desborough Island	(SY)	TQ084664
Cobbins Brook	(EX)	TL409018	Dews Farm Lake	(MX)	TQ052880
Coldfall Wood	(MX)	TQ276903	Ditton Field	(SY)	TQ157679
Cole Green	(HR)	TL282114	Dobb's Weir	(HR)	TL384043
Colne Brook, Iver	(BU)	TQ042813	Dog Kennel Covert	(MX)	TQ103783
Colney Heath	(HR)	TL205062	Dulwich GC	(SY)	TQ339728
Connaught Water	(EX)	TQ404933	Dulwich Mill Pond	(SY)	TQ333731
Coopers Green Lane	(HR)	TL193098	Dulwich Park	(SY)	TQ335735
Coopersale	(EX)	TL480025	Dulwich Woods	(SY)	TQ340725
Copped Hall	(EX)	TL430010	Durant's Park	(MX)	TQ357968
Coppetts Wood	(MX)	TQ277916	Eagle Pond	(EX)	TQ390884
Copse Wood	(MX)	TQ085897	East India Dock Basin	(MX)	TQ391808
Cornmill Meads	(EX)	TL380011	East Iver Lakes	(BU)	TQ046803
Coursers Lane Fields	(HR)	TL200040	Edgwarebury Park	(MX)	TQ190934
Cowley Lake	(MX)	TQ052814	Enfield GC	(MX)	TQ313962
Cowley Peachy	(MX)	TQ054814	Enfield Lock	(MX)	TQ368982

Epping Forest	(EX)	TQ420985	Harmondsworth CP	(MX)	TQ050779
Epsom Common	(SY)	TQ190605	Harold Wood SF	(EX)	TQ565915
Epsom Downs	(SY)	TQ211580	Hartnips Wood	(KT)	TQ547637
Erith Marsh	(KT)	TQ488805	Hatfield Aerodrome	(HR)	TL205085
Esher Common	(SY)	TQ135625	Hatfield Park	(HR)	TL240080
Eynsford	(KT)	TQ540655	Hatherop Park	(MX)	TQ124704
Fairlop Waters CP	(EX)	TQ459905	Havering CP	(EX)	TQ505930
Fairmile Common	(SY)	TQ125617	Hawk's Wood	(BU)	TQ014862
Farleigh	(SY)	TQ367602	Hawkshead Wood	(HR)	TL220030
Farlows Lake	(BU)	TQ047810	Hawkwood	(KT)	TQ442695
Field Common GP	(SY)	TQ128671	Headley Heath	(SY)	TQ200535
Fishers Green	(EX)	TL378026	Helicon Lake	(HR)	TQ039909
Foots Cray Meadows	(KT)	TQ480715	Hendon Park	(MX)	TQ233884
Foots Cray Woods	(KT)	TQ479721	Henley Wood	(SY)	TQ374587
Forster Park	(KT)	TQ387722	Hersham GP	(SY)	TQ128663
Forty Hall, Enfield	(MX)	TQ337987	Hersham SF	(SY)	TQ127657
Frays Carp Lake	(MX)	TQ054865	Hertford Heath	(HR)	TL349107
Fray's Farm Meadow	(MX)	TQ055862	Highgate Wood	(MX)	TQ283887
Frays Middle Lake	(MX)	TQ056866	Hilfield Park Res	(HR)	TQ158959
Frays Northern Lake	(MX)	TQ056868	Hill End	(MX)	TQ050918
Frogmore GP	(HR)	TL150033	Hinchley Wood	(SY)	TQ157650
Fryent CP	(MX)	TQ194874	Hither Green NR	(KT)	TQ390740
Furzefield Wood	(HR)	TL103055	Hoblingwell Wood	(KT)	TQ450690
Gatton Park/Lake	(SY)	TQ271525	Hoddesdonpark Wood	(HR)	TL352084
Gernon Bushes	(EX)	TL478030	Hogsmill SF	(SY)	TQ197682
Gladwin's Wood	(BU)	TQ019864	Hogwood	(KT)	TQ555631
Godstone SP	(SY)	TQ345518	Holland Park	(IL)	TQ248796
Grays Chalk Pits	(EX)	TQ609788	Hollow Pond	(EX)	TQ414873
Great Halings Wood	(BU)	TQ032895	Holmethorpe SP	(SY)	TQ295515
Great Soloms Wood	(SY)	TQ272588	Holyfield Hall Farm	(EX)	TL384038
Green Park	(IL)	TQ290800	Hooks Marsh	(EX)	TL373023
Greenhill Wood	(KT)	TQ538602	Hornchurch CP	(EX)	TQ535825
Greenwich Park	(KT)	TQ390775	Horton CP	(SY)	TQ190627
Grovelands Park	(MX)	TQ305944	Horton Fields	(BU)	TQ015764
Gunnersbury Triangle NR	(MX)	TQ201787	Horton GP	(BU)	TQ005753
Gutteridge Wood	(MX)	TQ091843	Hounslow Heath	(MX)	TQ123745
Hadley Common	(HR)	TQ265972	Howell Hill, Ewell	(SY)	TQ238621
Hainault Forest	(EX)	TQ476932	Hunston Mead	(EX)	TL422114
Hall Marsh	(EX)	TL373017	Hunton Bridge Hill	HR)	TL076006
Hall Place Gardens	(KT)	TQ505743	Huntsmoor Park	(BU)	TQ047815
Ham Lands	(SY)	TQ165725	Hurst Green, Oxted	(SY)	TQ396512
Hampermill Lake	(HR)	TQ095942	Hyde Park	(IL)	TQ270803
Hampstead Heath	(MX)	TQ273866	Hythe End	(SY)	TQ020718
Hampton Court Park	(MX)	TQ166676	Ingrebourne Valley	(EX)	TQ538843
Hampton Hill	(MX)	TQ145708	Island Barn Res	(SY)	TQ140670
Hampton Waterworks	(MX)	TQ127686	Isleworth Ait	(SY)	TQ167757

Itchingwood Common	(SY)	TQ416506	Moorhall Lake	(MX)	TQ048889
Iver Heath	(BU)	TQ033835	Moorhouse	(SY)	TQ433532
Jersey Farm	(HR)	TL174097	Morden Hall Park	(SY)	TQ264687
Joyden's Wood	(KT)	TQ500715	Navestock Lake/Park	(EX)	TQ538985
Jubilee Gardens	(IL)	TQ318803	Nazeing GP	(EX)	TL385072
Juniper Hill	(HR)	TQ060931	Netherhall GP	(EX)	TL394083
Juniper Wood	(BU)	TQ025895	Nightingale Wood	(BU)	TQ038888
Kelsey Park	(KT)	TQ376688	Noke Farm	(HR)	TL125037
Kempton Park NR	(MX)	TQ116706	Nonsuch Park	(SY)	TQ232638
Kenley Aerodrome	(SY)	TQ328587	Norbury Park	(SY)	TQ310699
Kensington Gardens	(IL)	TQ270803	North Cray	(KT)	TQ490723
Keston Ponds	(KT)	TQ426645	Northaw Great Wood	(HR)	TL285044
Kew Gardens	(SY)	TQ182769	Norwood Grove	(SY)	TQ333704
King George V Res	(EX)	TQ374964	Nower Wood	(SY)	TQ195547
King George VI Res	(MX)	TQ041732	Nunhead Cemetery	(SY)	TQ355756
Korda Lake	(MX)	TQ045886	Nutfield Ridge	(SY)	TQ293540
Kynaston Wood	(KT)	TQ482672	Oak Hill Park	(HR)	TQ278949
Ladywell Park	(KT)	TQ372740	Oak Hill Wood	(HR)	TQ280952
Laleham Park	(MX)	TQ053680	Oakend Wood	(BU)	TQ014888
Langley Park	(BU)	TQ013814	Oakmere Park	(MX)	TL263013
Limpsfield Chart	(SY)	TQ445523	Old Parkbury	(HR)	TL163023
Little Britain Lake	(MX)	TQ049813	Old Slade Lake	(BU)	TQ040780
Littlebrook Lake	(KT)	TQ553756	Old Wood, Richings Park	(BU)	TQ031781
Littlebrook PS	(KT)	TQ563764	One Tree Hill, Honor	(SY)	TQ354743
London Wetland Centre	(SY)	TQ228770	Ongar Park Wood	(EX)	TL495025
Long Coppice	(BU)	TQ031849	Orlitts North Lake	(BU)	TQ041780
Long Reach SF	(KT)	TQ553768	Orlitts South Lake	(BU)	TQ039775
Lonsdale Road Res	(SY)	TQ218775	Orsett Fen	(EX)	TQ628833
Lower Feltham	(MX)	TQ094720	Osterley Park	(MX)	TQ145780
Lullingstone Park	(KT)	TQ523644	Oxhey Wood	(HR)	TQ105925
Lynsters Lake	(HR)	TQ038916	Oxleas Wood	(KT)	TQ450686
Mad Bess Wood	(MX)	TQ077893	Oylers Farm	(HR)	TL350005
Maple Lodge NR	(HR)	TQ036924	Panshanger Park	(HR)	TL283130
Martens Grove	(KT)	TQ505752	Park Downs	(SY)	TQ267585
Maryon Wilson Park	(KT)	TQ419785	Park Street GP	(HR)	TL149024
Mayfield Lakes	(MX)	TQ052789	Park Wood	(MX)	TQ092891
Mayow Park	(KT)	TQ358718	Parkland Walk	(MX)	TQ300878
Merstham SW	(SY)	TQ303523	Parndon Mead	(HR)	TL433113
Middlesex FB	(MX)	TQ359865	Paynes Lane GP	(EX)	TL380052
Mill Green SF	(HR)	TL245100	Peckham Rye Park	(SY)	TQ348750
Millwall Docks	(MX)	TQ377795	Pen Ponds	(SY)	TQ200730
Mitcham Common	(SY)	TQ290675	Petersham Meadows	(SY)	TQ180736
Moat Mount OS	(MX)	TQ215944	Petts Wood	(KT)	TQ445670
Molesey GP	(SY)	TQ129671	Pickett's Lock	(MX)	TQ362937
Molesey Heath	(SY)	TQ132672	Pilvage Wood	(MX)	TL235028
Moor Mill	(HR)	TL145030	Pinner Park	(MX)	TQ132905

Polhill	(KT)	TQ500612	Scratchwood OS	(MX)	TQ205945
Pond Wood	(HR)	TL279006	Sevenoaks WR	(KT)	TQ522570
Ponders End Lake	(MX)	TQ362946	Seventy Acres Lake	(HR)	TL374030
Poplar Docks	(MX)	TQ382803	Sewardstone Marsh	(EX)	TQ379985
Potters Crouch Plantation	(HR)	TL103050	Shadwell Basin	(MX)	TQ352807
Prae Wood	(HR)	TL121068	Sheen Common	(SY)	TQ196746
Primrose Hill	(IL)	TQ276839	Sheepwash Pond	(MX)	TQ224928
Princes Coverts	(SY)	TQ160610	Shepperton GP	(MX)	TQ065674
Purfleet Chalk Pits	(EX)	TQ566785	Slipe Lane GP	(HR)	TL370049
Putney Heath	(SY)	TQ230739	Smallford GP	(HR)	TL198071
Pygro Park	(EX)	TQ524935	Sopwell Mill	(HR)	TL154054
Pynesfield North Lake	(HR)	TQ038912	South Norwood CP	(SY)	TQ353684
Pynesfield South Lake	(HR)	TQ035910	South Norwood Lake	(SY)	TQ341693
Queen Elizabeth II Res	(SY)	TQ120670	Sparrows Wood	(KT)	TQ436667
Queen Mary GP	(MX)	TQ059700	Spring Ponds	(MX)	TQ165932
Queen Mary Res	(MX)	TQ070695	Springfield Marina	(MX)	TQ347878
Queen Mother Res	(BU)	TQ017773	Springwell Lake	(MX)	TQ041925
Queens Wood	(MX)	TQ288886	St James's Park	(IL)	TQ294798
Rainham GP	(EX)	TQ549829	St Saviour's Creek	(IL)	TQ340799
Rainham Marshes	(EX)	TQ525800	Stain Hill Res	(MX)	TQ124693
Rammey Marsh	(MX)	TQ374996	Staines Moor	(MX)	TQ033734
Ranston Covert	(BU)	TQ041890	Staines Res	(MX)	TQ051731
Ravensbury Park	(SY)	TQ268681	Stanborough Lakes	(HR)	TL230108
Ravenscourt Park	(MX)	TQ223790	Stanmore Common	(MX)	TQ156940
Raynes Park	(SY)	TQ231684	Stanmore CP	(MX)	TQ173928
Regent's Park	(IL)	TQ281828	Stanstead Abbotts GP	(HR)	TL391109
Richings Park	(BU)	TQ030794	Stanwell Moor	(MX)	TQ040743
Richmond Park	(SY)	TQ200730	Stanwell Moor GP	(MX)	TQ034744
Ridgehill	(HR)	TL202025	Stocker's Lake	(HR)	TQ046935
Riverside Open Space	(SY)	TQ143693	Stocker's West Lake	(MX)	TQ042934
Roding Valley Meadows	(EX)	TQ435956	Stockley Park	(MX)	TQ080790
Rowdow Wood	(KT)	TQ547595	Stoke Newington Res	(MX)	TQ326876
Royal Albert Dock	(EX)	TQ425807	Stone Chalk Pit	(KT)	TQ566753
Royal Oak Lake	(HR)	TQ034915	Stone Lake	(KT)	TQ565752
Royal Victoria Dock	(EX)	TQ410807	Stone Marsh	(KT)	TQ568755
Ruislip Common	(MX)	TQ087887	Streatham Common	(SY)	TQ305709
Ruislip Gardens	(MX)	TQ095863	Sundridge Park	(KT)	TQ416706
Ruislip Lido	(MX)	TQ089891	Surrey Docks	(IL)	TQ360797
Rush Green	(HR)	TL350127	Sutton at Hone Lakes	(KT)	TQ560700
Ruxley GP	(KT)	TQ473700	Swanscombe Marsh	(KT)	TQ605760
Rye Meads RSPB	(HR)	TL383103	Symondshyde Great Wood	(HR)	TL195110
Savay Lake	(MX)	TQ049880	Syon Park	(MX)	TQ175765
Saxten's Wood	(KT)	TQ585648	Telegraph Hill	(SY)	TQ159647
Scadbury Park	(KT)	TQ455700	Ten Acre Wood	(MX)	TQ099838
Scotsbridge Meadows	(HR)	TQ064955	The Basin, Edgware	(MX)	TQ188919
Scratch Wood, Woodmansterne	(SY)	TQ271593	The Causeway NR	(MX)	TQ105754

Gazetteer

The Clump	(BU)	TQ022845	Watermeads	(SY)	TQ274677
Theobalds Park	(HR)	TL345005	Waterworks NR	(EX)	TQ363868
Thistledene	(SY)	TQ155673	Watts Wood	(EX)	TQ565788
Thorndon Park	(EX)	TQ620916	Weald Park	(EX)	TQ570945
Thorney CP	(BU)	TQ048790	Welham Green	(HR)	TL235055
Thorney Weir Lake	(BU)	TQ051799	Wells Park	(KT)	TQ345717
Thorpe Water Park	(SY)	TQ030681	West End Common	(SY)	TQ125632
Thrift's Pit	(HR)	TL366132	West Hyde Fields	(HR)	TQ030910
Tilehouse North Lake	(BU)	TQ038898	West India South Dock	(MX)	TQ376800
Tilehouse South Lake	(BU)	TQ038896	West Ruislip GC	(MX)	TQ081872
Tolpits Lake	(HR)	TQ085943	West Thurrock Marshes	(EX)	TQ583767
Tooting Bec Common	(SY)	TQ293720	Westerham Heights	(KT)	TQ430560
Tottenham Lock	(MX)	TQ348895	Wey Meadows	(SY)	TQ067645
Tottenham Marsh	(MX)	TQ354910	Whippendell Wood	(HR)	TQ075980
Totteridge Long Pond	(MX)	TQ234941	Whitewebbs Park	(MX)	TQ325997
Tower Wood	(KT)	TQ448528	Whyteleafe	(SY)	TQ337582
Trent Park	(MX)	TQ290970	Willett Wood	(KT)	TQ452685
Troy Mill Lake	(HR)	TQ039905	William Girling Res	(EX)	TQ367945
Turnford Marsh GP	(HR)	TL370044	Wimbledon Common	(SY)	TQ247723
Tyler's Common	(EX)	TQ568907	Wimbledon Park	(SY)	TQ246723
Tyttenhanger GP	(HR)	TQ191865	Wimbledon Park Lake	(SY)	TQ247724
Verulamium Lake	(HR)	TL140070	Windsor OS	(MX)	TQ241905
Victoria Park	(IL)	TQ363840	Winterdown Woods	(SY)	TQ123622
Waddon Ponds	(SY)	TQ309650	Wintry Wood	(EX)	TL475035
Walsingham Wood	(HR)	TL215035	Wood Green Res	(MX)	TQ304900
Waltham Cross	(HR)	TL368006	Woodford GC	(EX)	TQ395928
Waltham Marsh	(EX)	TL373013	Woodlands Park	(BU)	TQ038830
Walthamstow Marsh	(EX)	TQ350878	Woodoaks Farm	(HR)	TQ033933
Walthamstow Res	(EX)	TQ353890	Worcester Park	(SY)	TQ222655
Walton Heath	(SY)	TQ232540	Wormley Wood	(HR)	TL322058
Walton Res	(SY)	TQ122685	Wormwood Scrubs	(MX)	TQ221818
Wandsworth Common	(SY)	TQ275740	Wraysbury GP	(BU)	TQ015735
Wanstead Flats	(EX)	TQ410864	Wraysbury Res	(MX)	TQ025745
Wanstead Park	(EX)	TQ415875	Yeading Brook Meadows	(MX)	TQ105835
Waterlow Park	(MX)	TQ286872	Yiewsley Lake	(MX)	TQ050804

INDEX OF SCIENTIFIC NAMES

Accipiter gentilis 130
   nisus 131
Acrocephalus agricola 320
    dumetorum 320
    paludicola 319
    palustris 321
    schoenobaenus 319
    scirpaceus 322
Actitis hypoleucos 187
   macularius 188
Aegithalos caudatus 291
   c. caudatus 292
Aegolius funereus 254
Aix galericulata 52
Alauda arvensis 295
Alca torda 234
Alcedo atthis 258
Alectoris rufa 85
Alle alle 235
Alopochen aegyptiaca 49
Anas acuta 59
   carolinsis 57
   clypeata 62
   crecca 56
   discors 62
   penelope 53
   platyrhynchos 58
   querquedula 61
   strepera 54
Anser sp. 44
   albifrons 41
   anser 43
   brachyrhynchus 40
   caerulescens 45
   fabalis 39
   (serrirostris) rossicus 40
Anthus campestris 366
   cervinus 370
   hodgsoni 367
   petrosus 370
   pratensis 368
   richardi 365
   rubescens 372
   spinoletta 371
   trivialis 367
Apus apus 256
   melba 257

Aquila chrysaetos 135
Ardea alba 106
   cinerea 107
   purpurea 110
Ardeola ralloides 104
Arenaria interpres 195
Asio flammeus 253
   otus 252
Athene noctua 249
Aythya affinis 74
   collaris 67
   ferina 65
   fuligula 70
   marila 72
   nyroca 68

Bombycilla garrulus 323
Botaurus stellaris 101
Branta bernicla 47
   b. hrota 48
   b. nigricans 49
   canadensis 45
   leucopsis 46
   ruficollis 49
Bubo scandiacus 249
Bucephala clangula 78
Bulbulcus ibis 104
Burhinus oedicnemus 153
Buteo buteo 133
   lagopus 134

Calandrella brachydactyla 293
Calcarius lapponicus 392
Calidris acuminata 171
   alba 166
   alpina 173
   bairdii 170
   canutus 165
   ferruginea 171
   fuscicollis 169
   maritima 172
   mauri 167
   melanotos 170
   minuta 168
   temminckii 169
Caprimulgus europaeus 254
Carduelis cabaret 382

Index of scientific names

cannabina 380
carduelis 377
chloris 375
flammea 383
flavirostris 381
hornemanni 384
spinus 379
Carpodacus erythrinus 387
Catharus minimus 332
Cecropis daurica 303
Cepphus grylle 235
Certhia brachydactyla 326
  familiaris 325
Cettia cetti 303
Charadrius alexandrinus 158
  dubius 155
  hiaticula 156
  morinellus 159
  vociferus 158
Chlidonias hybrida 225
  leucopterus 227
  niger 225
Chordeiles minor 255
Chroicocephalus philadelphia 207
  ridibundus 207
Ciconia ciconia 111
  nigra 111
Cinclus cinclus 331
Circus aeruginosus 127
  cyaneus 128
  pygargus 129
Clangula hyemalis 75
Coccothraustes coccothraustes 389
Coccyzus americanus 247
Columba livia 238
  oenas 239
  palumbus 241
Coracias garrulus 260
Corvus corax 281
  cornix 280
  corone 278
  frugilegus 276
  monedula 274
Coturnix coturnix 87
Crex crex 145
Cuculus canorus 246
Cursorius cursor 154
Cyanistes caeruleus 285
Cygnus columbianus 37
  cygnus 38
  olor 35

Delichon urbicum 301
Dendrocopos major 264
  minor 265

Egretta garzetta 105
Emberiza calandra 397
  cirlus 394
  citrinella 393
  hortulana 394
  leucocephalos 392
  melanocephala 397
  pusilla 396
  rustica 395
  schoeniclus 396
Eremophila alpestris 297
Erithacus rubecula 342

Falco columbarius 139
  naumanni 136
  peregrinus 141
  rusticolis 141
  subbuteo 140
  tinnunculus 137
  vespertinus 138
Ficedula hypoleuca 346
  parva 345
Fratercula arctica 236
Fringilla coelebs 372
  montifringilla 374
Fulica atra 148
Fulmarus glacialis 92

Galerida cristata 294
Gallinago gallinago 178
  media 179
Gallinula chloropus 147
Garrulus glandarius 272
Gavia arctica 90
  immer 91
  stellata 89
Gelochelidon nilotica 224
Glareola pratincola 155
Grus grus 149

Haematopus ostralegus 150
Haliaeetus albicilla 126
Himantopus himantopus 151
Hippolais icterina 318, 319
  polyglotta 318, 319
Hirundo rustica 300
Hydrobates pelagicus 95
Hydrocoloeus minutus 209

Hydroprogne caspia 224

Ixobrychus minutus 102

Jynx torquilla 262

Lanius collurio 267
   cristatus 267
   excubitor 269
   isabellinus 267
   minor 269
   senator 270
Larus argentatus 216
   atricilla 211
   cachinnans 218
   canus 212
   delawarensis 213
   fuscus 214
   f. intermedius 216
   glaucescens 220
   glaucoides 218
   g. kumlieni 219
   hyperboreus 220
   marinus 221
   melanocephalus 211
   michahellis 217
   pipixcan 211
   schistisagus 220
Limicola falcinellus 175
Limnodromus sp. 180
   scolopaceus 180
Limosa lapponica 183
   limosa 182
Locustella luscinioides 317
   naevia 316
Lophophanes cristatus 287
Loxia curvirostra 385
   leucoptera 385
   pytyopsittacus 387
Lullula arborea 294
Luscinia megarhynchos 343
   svecica 344
Lymnocryptes minimus 177

Melanitta fusca 77
   nigra 76
Mergellus albellus 79
Mergus merganser 82
   serrator 81
Merops apiaster 259
Milvus migrans 124
   milvus 124
Morus bassanus 96

Motacilla alba 364
   a. alba 365
   cinerea 362
   citreola 362
   flava 360
   f. beema 362
   f. cinereocapilla 362
   f. flava 361
   f. leucocephala 362
   f. thunbergi 361
Muscicapa striata 341
Netta rufina 64
Nucifraga caryocatactes 273
Numenius arquata 185
   phaeopus 184
Nycticorax nycticorax 103

Oceanodroma leucorhoa 95
Oenanthe deserti 354
   oenanthe 353
   o. leucorrhoa 354
Onychoprion anaethetus 223
   fuscatus 222
Oriolus oriolus 266
Otis tarda 149
Otus scops 249
Oxyura jamaicensis 83

Pandion haliaetus 135
Panurus biarmicus 292
Parus major 286
Passer domesticus 355
   montanus 358
Perdix perdix 86
Periparus ater 288
Pernis apivorus 122
Phalacrocorax aristotelis 99
   carbo 97
Phalaropus fulicarius 198
   lobatus 197
   tricolor 197
Phasianus colchicus 88
Philomachus pugnax 175
Phoenicurus ochruros 347
   phoenicurus 348
Phylloscopus collybita 308
   coronatus 304
   fuscatus 306
   humei 306
   ibericus 309
   inornatus 305
   proregulus 304

schwarzi 306
sibilatrix 306
trochilus 310
Pica pica 271
Picus viridis 263
Pinicola enucleator 388
Platalea leucorodia 113
Plectrophenax nivalis 391
Plegadis falcinellus 113
Pluvialis apricaria 160
    dominica 160
    fulva 160
    squatarola 161
Podiceps auritus 120
    cristatus 117
    grisegena 119
    nigricollis 121
Podilymbus podiceps 115
Poecile montana 289
    palustris 290
Porzana parva 144, 145
    porzana 143
    pusilla 144, 145
Pratincole sp. 155
    Glareola 155
Prunella collaris 355
    modularis 354
Psittacula krameri 245
Puffinus baroli 94
    mauretanicus 94
    puffinus 93
Pyrrhula pyrrhula 388
    p. pyrrhula 389

Rallus aquaticus 142
Recurvirostra avosetta 152
Regulus ignicapilla 283
    regulus 282
Remiz pendulinus 285
Riparia riparia 298
Rissa tridactyla 205

Saxicola rubetra 350
    rubicola 351
Scolopax rusticola 181
Serinus serinus 376
Sitta europaea 324
Somateria mollissima 74
    spectabilis 75
Stercorarius longicaudus 202
    parasiticus 200
    pomarinus 199

skua 203
Sterna dougallii 231
    hirundo 229
    paradisaea 232
    sandvicensis 228
Sternula albifrons 223
Streptopelia decaocto 243
    turtur 244
Strix aluco 251
Sturnus roseus 330
    vulgaris 327
Sylvia atricapilla 311
    borin 312
    cantillans 316
    communis 314
    curruca 313
    melanocephala 316
    nisoria 312
    undata 315
Syrrhaptes paradoxus 238

Tachybaptus ruficollis 115
Tadorna tadorna 50
Tetrao tetrix 84
Tringa erythropus 190
    flavipes 192
    glareola 193
    nebularia 191
    ochropus 188
    solitaria 189
    stagnatilis 192
    totanus 194
Troglodytes troglodytes 326
Tryngites subruficollis 175
Turdus iliacus 338
    merula 334
    migratorius 340
    naumanni 336
    philomelos 337
    pilaris 336
    torquatus 333
    viscivorus 339
Tyto alba 247
    a. guttata 248

Upupa epops 261
Uria aalge 233

Vanellus gregarius 163
    leucurus 163
    vanellus 163

Xema sabini 204

INDEX OF ENGLISH COMMON NAMES

Accentor, Alpine 355
Auk, Little 236
Avocet 152

Bee-eater, European 259
Bittern 101
Bittern, Little 102
Blackbird 334
Blackcap 311
Bluethroat 344
Brambling 374
Brant, Black 49
Bullfinch 388
   Northern 389
Bunting, Black-headed 397
   Cirl 394
   Corn 397
   Lapland 392
   Little 396
   Ortolan 394
   Pine 392
   Reed 396
   Rustic 395
   Snow 391
Bustard, Great 150
Buzzard, Common 133
   Rough-legged 134

Chaffinch 372
Chiffchaff 308
   Iberian 309
Coot 148
Cormorant 97
Corncrake 145
Courser, Cream-coloured 154
Crake, Baillon's 144, 145
   Little 144, 145
   Spotted 143
Crane 149
Crossbill, Common 385
   Parrot 387
   Two-barred 385
Crow, Carrion 278
   Hooded 280
Cuckoo 246

   Yellow-billed 247
Curlew 185

Dipper, 331
Diver, Black-throated 90
   Great Northern 91
   Red-throated 89
Dotterel 159
Dove, Collared 243
   Rock 238
   Stock 239
   Turtle 244
Dowitcher sp. 180
   Long-billed 180
Duck, Ferruginous 68
   Long-tailed 75
   Mandarin 52
   Ring-necked 67
   Ruddy 83
   Tufted 70
Dunlin 173
Dunnock 354

Eagle, Golden 135
   White-tailed 126
Egret, Cattle 104
   Great White 106
   Little 105
Eider 74
   King 75

Falcon, Red-footed 138
Fieldfare 336
Firecrest 283
Flycatcher, Pied 346
   Red-breasted 345
   Spotted 341
Fulmar 92

Gadwall 54
Gannet 96
Garganey 61
Godwit, Bar-tailed 183
   Black-tailed 182
Goldcrest 282

Index of English common names

Goldeneye 78
Goldfinch 377
Goosander 82
Goose, Barnacle 46
  Bean 39
  Brent 47
  Brent (Pale-bellied) 48
  Canada 45
  Egyptian 49
  White-fronted 41
  Grey goose sp. 44
  Greylag 43
  Pink-footed 40
  Red-breasted 49
  Snow 45
  Tundra Bean 40
Goshawk 130
Grebe, Black-necked 121
  Great Crested 117
  Little 115
  Pied-billed 115
  Red-necked 119
  Slavonian 120
Greenfinch 375
Greenshank 191
Grosbeak, Pine 388
Grouse, Black 85
Guillemot 233
  Black 235
Gull, Black-headed 207
  Bonaparte's 207
  Caspian 218
  Common 212
  Franklin's 211
  Glaucous 220
  Glaucous-winged 220
  Great Black-backed 221
  Herring 216
  Iceland 218
  Kumlien's 219
  Laughing 211
  Lesser Black-backed 214
  Little 209
  Mediterranean 211
  Ring-billed 213
  Sabine's 204
  Scandinavian Lesser Black-backed 216
  Slaty-backed 220
  Yellow-legged 217
Gyrfalcon 141

Harrier, Hen 128
  Marsh 127
  Montagu's 129
Hawfinch 389
Heron, Grey 107
  Purple 110
  Squacco 104
Hobby 140
Honey Buzzard 122
Hoopoe 261

Ibis, Glossy 113

Jackdaw 274
Jay 272

Kestrel 137
  Lesser 136
Killdeer 158
Kingfisher 258
Kite, Black 124
  Red 124
Kittiwake 205
Knot 165

Lapwing 163
Lark, Crested 294
  Shore 297
  Short-toed 293
Linnet 380

Magpie 271
Mallard 58
Martin, House 301
  Sand 298
Merganser, Red-breasted 81
Merlin 139
Moorhen 147

Night Heron 103
Nighthawk, Common 255
Nightingale 343
Nightjar 254
Nutcracker 273
Nuthatch 324

Oriole, Golden 266
Osprey 135
Ouzel, Ring 333
Owl, Barn 247

Barn (Dark-breasted) 248
Little 249
Long-eared 252
Scops 249
Short-eared 253
Snowy 249
Tawny 251
Tengmalm's 254
Oystercatcher 150

Parakeet, Ring-necked 245
Partridge, Grey 86
Red-legged 85
Peregrine 141
Petrel, Leach's 95
Phalarope, Grey 198
Red-necked 197
Wilson's 197
Pheasant 88
pigeon, feral 238
Wood 241
Pintail 59
Pipit, Buff-bellied 372
Meadow 368
Olive-backed 367
Red-throated 370
Richard's 365
Rock 370
Tawny 366
Tree 367
Water 371
Plover, American Golden 160
Golden 160
Grey 161
Kentish 158
Little Ringed 155
Pacific Golden 160
Ringed 156
Sociable 163
White-tailed 163
Pochard 65
Red-crested 64
Pratincole sp. 155
Collared 155
Puffin 236

Quail 87

Rail, Water 142
Raven 281

Razorbill 234
Redpoll, Arctic 384
Common (Mealy) 383
Lesser 382
Redshank 194
Spotted 190
Redstart, Black 347
Common 348
Redwing 338
Robin 342
American 340
Roller, European 260
Rook 276
Rosefinch, Common 387
Ruff 175

Sanderling 166
Sandgrouse, Pallas's 237
Sandpiper, Baird's 170
Broad-billed 175
Buff-breasted 175
Common 187
Curlew 171
Green 188
Marsh 192
Pectoral 170
Purple 172
Sharp-tailed 171
Solitary 189
Spotted 188
Western 167
White-rumped 169
Wood 193
Scaup 72
Lesser 74
Scoter, Common 76
Velvet 77
Serin 376
Shag 99
Shearwater, Balearic 94
Macaronesian 95
Shearwater, Manx 93
Shelduck 50
Shoveler 62
Shrike, Brown 267
Great Grey 269
Isabelline 267
Lesser Grey 269
Red-backed 267
Woodchat 270

Siskin 379
Skua, Arctic 200
   Great 203
   Long-tailed 202
   Pomarine 199
Skylark 295
Smew 79
Snipe 178
   Great 179
   Jack 177
Sparrow, House 355
   Tree 358
Sparrowhawk 131
Spoonbill, (Eurasian) 113
Starling 327
   Rose-coloured 330
Stilt, Black-winged 151
Stint, Little 168
   Temminck's 169
Stone-curlew 153
Stonechat 351
Stork, Black 111
   White 111
Storm Petrel 95
   Storm-petrel sp. 96
Swallow 300
   Red-rumped 303
Swan, Bewick's 37
   Mute 35
   Whooper 38
Swift, Common 256
   Alpine 257

Teal 56
   Blue-winged 62
   Green-winged 57
Tern, Arctic 232
   Black 225
   Bridled 223
   Caspian 224
   Common 229
   Gull-billed 224
   Little 223
   Roseate 231
   Sandwich 228
   Sooty 222
   Whiskered 225
   White-winged Black 227
Thrush, Grey-cheeked 332
   Mistle 339

   Naumann's 336
   Song 337
Tit, Bearded 292
   Blue 285
   Coal 288
   Crested 287
   Great 286
   Long-tailed 291
   Marsh 290
   Northern Long-tailed 292
   Penduline 285
   Willow 289
Treecreeper 325
   Short-toed 326
Turnstone 195
Twite 381

Wagtail, Ashy-headed 362
   Blue-headed 361
   Citrine 362
   Grey 363
   Grey-headed 362
   Pied 364
   Sykes's 362
   White 365
   White-headed 362
   Yellow 360
Warbler, Aquatic 319
   Barred 312
   Blyth's Reed 320
   Cetti's 303
   Dartford 315
   Dusky 306
   Eastern Crowned 304
   Garden 312
   Grasshopper 316
   Hume's 306
   Icterine 318, 319
   Marsh 321
   Melodious 318, 319
   Paddyfield 320
   Pallas's 304
   Radde's 306
   Reed 322
   Sardinian 316
   Savi's 317
   Sedge 319
   Subalpine 316
   Willow 310
   Wood 306

Yellow-browed 305
Waxwing 323
Wheatear, Desert 354
　　Greenland 354
　　Northern 353
Whimbrel 184
Whinchat 350
Whitethroat, Common 314
　　Lesser 313
Wigeon 53

Woodcock 181
Woodlark 294
Woodpecker, Great Spotted 264
　　Green 263
　　Lesser Spotted 265
Wren 326
Wryneck 262

Yellowhammer 393
Yellowlegs, Lesser 192